Springer Texts in Business and Economics

More information about this series at http://www.springer.com/series/10099

Bilash Kanti Bala • Fatimah Mohamed Arshad • Kusairi Mohd Noh

System Dynamics

Modelling and Simulation

Springer

Bilash Kanti Bala
University Putra Malaysia
Serdang, Selangor, Darul Ehsan
Malaysia

Fatimah Mohamed Arshad
University Putra Malaysia
Serdang, Selangor, Darul Ehsan
Malaysia

Kusairi Mohd Noh
University Putra Malaysia
Serdang, Selangor, Darul Ehsan
Malaysia

ISSN 2192-4333 ISSN 2192-4341 (electronic)
Springer Texts in Business and Economics
ISBN 978-981-10-2043-8 ISBN 978-981-10-2045-2 (eBook)
DOI 10.1007/978-981-10-2045-2

Library of Congress Control Number: 2016951454

Printed on acid-free paper

This Springer imprint is published by Springer Nature
The registered company is Springer Nature Singapore Pte Ltd.
The registered company address is: 152 Beach Road, #22-06/08 Gateway East, Singapore 189721, Singapore

Foreword

The world, today, is passing through a period of great turmoil, socially, politically and environmentally, in spite of the numerous technological wonders that are taking place almost everyday. One needs to take a systems view of the influencing factors and their interactions and impacts in order to find the root causes of these problems and to arrive at viable policy options. System dynamics provides such an approach. The book authored by Professor Bala, Professor Fatimah and Professor Noh presents the principles of system dynamics in very simple language and illustrates its use with the help of five real-life case studies.

This book is divided into two parts. The first part of the book presents, in a very simple way and starting with the fundamental principles, how complex interactions among the interacting forces can be modelled by capturing their cause–effect interrelations through dynamic models, how the models can be simulated and evaluated to depict reality and how policy interventions can be tested for testing their viability. Although the material covered in this part of the book is not new, the examples supporting the theoretical nuances of the subject covering population growth, grain storage, food security, commodity production, food relief, crop livestock, shrimp farming, crop irrigation and pollution are very interesting and appealing.

In the second part of the book, the authors discuss case studies related to the areas of agriculture, aquaculture and environment in Bangladesh and Malaysia. Both hilsa fish and rice are important for the economy of Bangladesh, just as food security and cocoa production for Malaysia. The case study for solid waste management is well chosen as it is a perennial problem in third-world countries. This part of the book is illustrative of the power of system dynamics methodology as to how it can address many complex issues of today very easily.

I believe that a newcomer to the field of system dynamics will find the book extremely useful and will be highly motivated to use system dynamics and systems thinking in understanding and addressing the issues that arise out of the behaviour of systems that are integral part of their lives.

Professor of Industrial Engineering and Management
Indian Institute of Technology
Kharagpur, West Bengal, India
September 2014

P.K.J. Mohapatra

Preface

This book *System Dynamics: Modelling and Simulation* is a totally new book with numerous examples and case studies for better understanding the complex systems and their changes through modelling and simulation to aid in policy formulation and developing management strategies for sustainable development. This book provides a comprehensive introduction to systems thinking and modelling of complex systems with application to agricultural, aquacultural, environmental and socio-economic systems. Also this book essentially provides the principles of system dynamics with numerous examples and a good number of case studies in agricultural, aquacultural, environmental and socio-economic systems. It covers all aspects of system dynamics starting from systems thinking to participatory model building to provide a tool for policy planning, and the main focus is to aid in policy design.

This book has been written primarily for undergraduate and postgraduate courses on system dynamics, systems engineering, system simulation, agricultural systems and multidisciplinary courses on agricultural, aquacultural, environmental and socio-economic systems. This book can be adopted for courses in electrical engineering and computer science. It will also serve as an excellent reference for practicing system dynamists, system dynamics researchers and policy planners. It is the outcome of several years of teaching and research in system dynamics modelling and simulation with applications in agricultural, aquacultural, environmental and socio-economic systems and also is an updated and a new book on principles of system dynamics.

This book covers the wide spectrum of system dynamics methodology of modelling and simulation of complex systems: systems thinking, causal diagrams, system structure of stock–flow diagrams, parameter estimation and tests for confidence building in system dynamics models with a good number of worked-out examples in diverse fields using STELLA and VENSIM. In case studies, problem statement with dynamic hypothesis is followed by causal loop diagrams, stock–flow diagrams, parameter estimation, model validation and policy design. Exercises have also been included at the end of each chapter for further practices.

The authors have a great pleasure in expressing the acknowledgements which they owe to many persons in writing this book. Professor B K Bala warmly recognises the continuing debt to his teacher, Dr. Donald R. Drew, W. Thomas

Rice professor of systems engineering, Virginia Polytechnic Institute and State University, USA, who introduced him to system dynamics at the Asian Institute of Technology, Bangkok, Thailand. Thu authors also express sincere acknowledgements to Professor P K J Mohapatra, Indian Institute of Technology, Kharagpur, India who is the Father of System Dynamics in India has written the foreword of this book. The authors have a great pleasure in expressing the acknowledgements to Dr. Serm Janjai, Department of Physics, Silpakorn University, Nakhon Pathom, Thailand, for his encouragement and support in the preparation of this book and my colleague Professor Ashraful Haque, Department of Farm Power and Machinery, who read the manuscript and made many helpful suggestions. I owe my thanks to Mrs. Emmy Farhana Alias, Institute of Agricultural and Food Policy Studies, Universiti Putra Malaysia, Malaysia, for her help in the preparation of the manuscript and Dr. Itsara Masiri of the Department of Physics, Silpakorn University, Nakhon Pathom, Thailand, for the assistance in graphics and in drawing the beautiful figures.

Selangor, Malaysia

Bilash Kanti Bala
Fatimah Mohamed Arshad
Kusairi Mohd Noh

Contents

Part II Cases and Applications

Part I

Concepts, Methodology and Techniques

Introduction 1

This chapter introduces the complexity and change of the dynamic systems, systems thinking and system dynamics as a methodology of systems thinking. Feedback system concepts and mode of behaviour of dynamic systems are highlighted. Model and simulation are defined, systems thinking methodology outlined and usefulness of models are discussed.

1.1 Introduction to Complexity and Change of the Dynamic Systems

We live in a complex world, and it is always at a change. Also we are confronted with complex agricultural, biological, aquacultural, environmental, technological and socio-economic problems which we need to understand and manage for sustainable development. In global context we debate seriously on global warming and its impacts on agriculture, energy and environment and also find reports on economic cycles causing financial panics, and in regional and local levels, we often find price fluctuations and food insecurity in the developing countries to mention some of the problems of complex and dynamic systems. Figure 1.1 shows the price fluctuations of wheat, maize and rice in the international markets as an example of the dynamic and complex problem which caused devastating consequences for hundreds of millions of people and negatively impacted many more. How can we understand and manage these complex systems? If yes, how can we do it? We need improved knowledge and analytical capabilities to understand and manage food price volatility and reduce hunger and malnutrition (Hajkowicz et al. 2012). System dynamics methodology based on feedback concepts to handle non-linearity, multi-loop and time-lag characteristics of complex dynamic systems can be applied to model and simulate such complex dynamic systems to understand the dynamics of systems and design management policy for sustainable development.

Indeed, we can understand and design management strategies, but we need some structures or guiding principles to understand and manage the complexity and

B.K. Bala et al., *System Dynamics*, Springer Texts in Business and Economics,
DOI 10.1007/978-981-10-2045-2_1

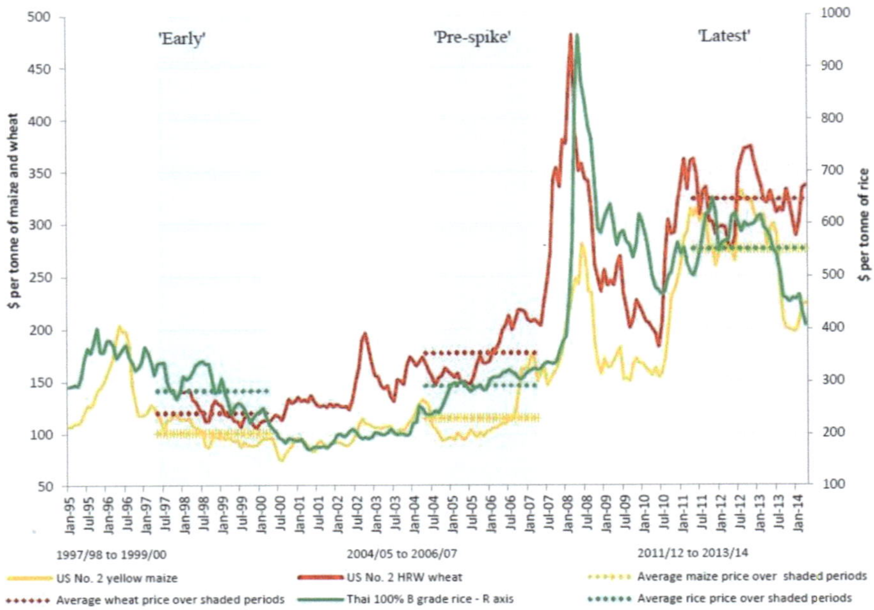

Fig. 1.1 Price fluctuations of wheat, maize and rice in the international market (Steve and Sharada 2014)

changes of complex dynamic systems based on systems approach which considers the whole systems rather than in isolation. Systems approach is rather a rational rather intuitive approach. It depends on some formalised methodology consisting of methods of problem definition, dynamic hypothesis, modelling, policy analysis, etc., and theoretical techniques which are useful for solving models and sub-models of the problem. In essence the systems must be modelled and simulated to understand the systems and design management strategies. Also these must be done before implementation of the management strategies. Forrester's system dynamics methodology provides the methodology—the guiding principles to develop computer models to simulate such complex and dynamic systems to understand the complex systems and design the management strategies. In essence, systems thinking is a formalised methodology consisting of methods of problem definition, dynamic hypothesis, modelling and policy analysis to understand and manage complex and dynamic systems.

As the complexity of our world increases, systems thinking is emerging as a critical factor for success, and even survival. How then can people become skilled systems thinkers? But in the world of complex dynamic systems, everyday experience fails because the time horizon and scope of the systems are so vast—we never experience the majority of the effects of our decisions. When experiments in the real world are impossible, simulation becomes the main way we can learn effectively about the dynamics of complex systems. System dynamics is the most appropriate technique to simulate complex and dynamic systems based on systems thinking to develop policy scenarios and learn to effectively manage the systems.

1.2 Concepts of Systems and System Dynamics

System dynamics is a methodology based on feedback systems borrowed from control theory, and it can handle easily the non-linearity and time-delay and the multi-loop structures of the complex and dynamic systems. Forrester's methodology provides a foundation for constructing computer models to do what the human mind cannot do—rationally analyse the structure, interactions and modes of behaviour of complex social systems, thus providing a framework whereby strategies can be tested and trade-offs can be performed, while options are still open. Many software such as STELLA, VENSIM, and POWERSIM are nowadays available which have revolutionised the system dynamics modelling. Furthermore, these software are icon operated and allow us to model virtually any process or system.

1.3 Open and Feedback Systems

We have discussed about some complex and dynamic systems in Sect. 1.1. What is meant by a system? Here system means a grouping of parts that operate together for a common purpose. For a simple example, a tractor is a system of components that work together for providing cultivation. A system may include people as well as physical parts. A family is a system for living and raising children. Systems may include biological as well as economic components, and such systems are known as bioeconomic systems. Agricultural and aquacultural systems are examples of bioeconomic systems. Crop irrigation system is an example of an agricultural system, while prawn production system is an example of aquacultural system. Also price forecasting is an economic system. In aquacultural system aquatic animals are raised to maturity in an economic manner. A system may include physical, economic, social, biological, technological and political components, and such a system is highly complex. For example, agricultural production systems with climate change in an agricultural system consists of physical, biological, social, technological, environmental, economic and political components and their interactions.

Systems may be classified as (a) open systems and (b) feedback systems. In open systems the output responds to input, but output has no influence on the input. Also the input is not aware of its own performance. In an open system, past action does not affect the future action. For example, a watch is not aware of its inaccuracy and does not correct the time itself. In an open system, the problem is perceived and action is taken, but the result does not influence action (Fig. 1.2). Filling up a water tank without controlling the valve is an example of open loop system (Fig. 1.3). When the hand valve is opened, water starts flowing in the tank and the water level

Fig. 1.2 Open system concept

Fig. 1.3 Filling up a water tank without control of the valve

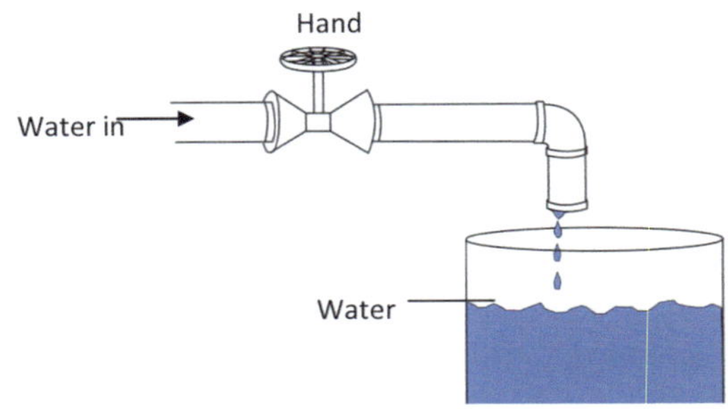

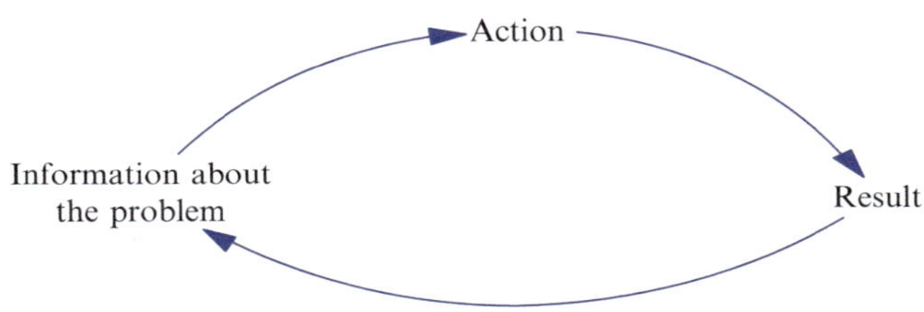

Fig. 1.4 Closed loop system concept

in the tank continues to increase until the hand valve is closed by the user who is not part of the system.

Feedback systems are closed loop systems, and the inputs are changed on the basis of output. A feedback system has a closed loop structure that brings back the results of the past action to control the future action. In a closed system, the problem is perceived, action is taken and the result influences the further action (Fig. 1.4). Thus, the distinguishing feature of a closed loop system is a feedback path of information, decision and action connecting the output to input.

Feedback system may be classified as either positive feedback systems or negative feedback systems. Positive feedback systems generate growth, and negative feedback systems are goal seeking. Population growth system is an example of positive feedback systems. Population multiply to produce more population which increases the growth rate at which the population is increased (Fig. 1.5). Thus, positive feedback system generates growth as shown in Fig. 1.5.

When the room temperature is controlled by a thermostat or temperature controller, the system tries to maintain the set temperature. When the heater is turned on, the temperature increases to set point and then heater is turned off. The temperature reaches ultimately equilibrium and thus, the system is goal seeking. The temperature control system in a room is an example of a negative feedback system (Fig. 1.6).

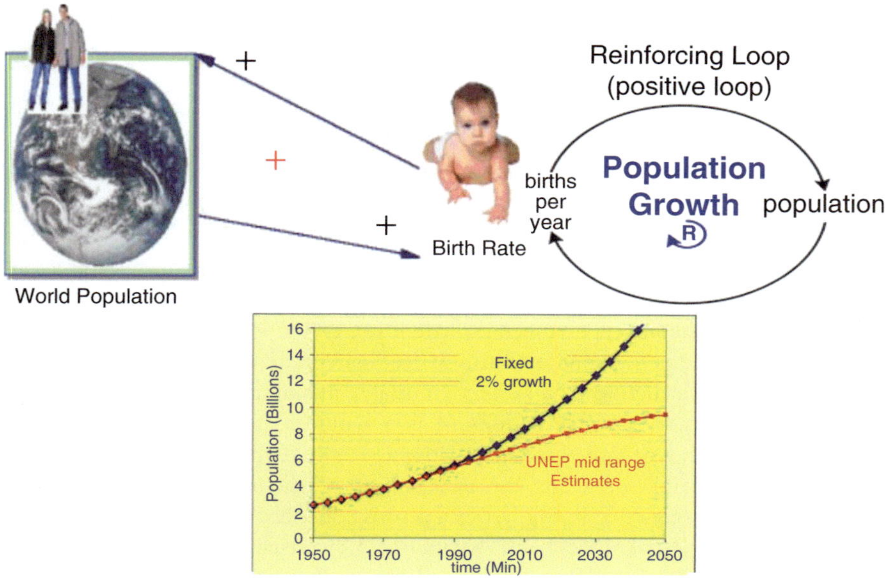

Fig. 1.5 Population growth system

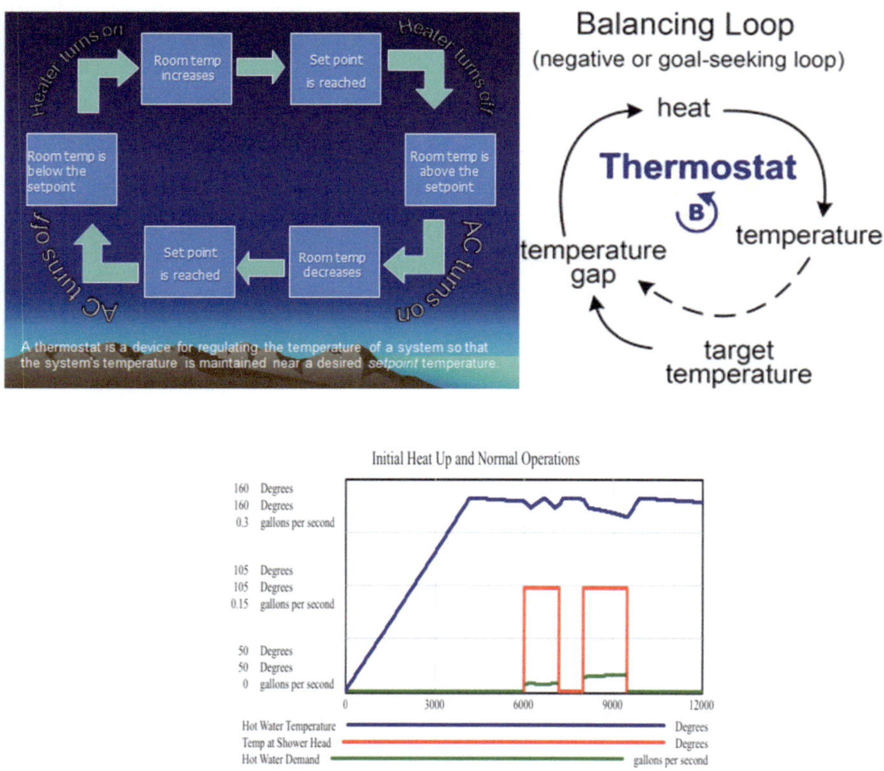

Fig. 1.6 Temperature control system in a room

1.4 Modes of Behaviour of Dynamic Systems

The basic structure of a feedback loop is shown in Fig. 1.7, and it is a closed path in a sequence consisting of a decision that controls action (based on the present state of the system and the desired goal) which results in flow, the stock or level of the system and information about the stock of the system, the latter returning to the decision-making point for further action. The current available information about the level or stock of the system and the goal is basis for current decision that controls action. The action changes the condition of the system. The single feedback loop structure is the simplest form of feedback system.

Feedback loop structure of a system simulates dynamic behaviour, and all the dynamics arises from the interactions of two types of feedback loops: positive feedback loops and negative feedback loops. Positive feedback loop generates growth, i.e. self-reinforcing and the causal loop consisting of population, birth and population in sequence in Fig. 1.8 is a positive feedback loop, and it reinforces population level. Negative feedback loop in Fig. 1.8 consisting of population, death and population is goal seeking. Figure 1.9 shows the dynamic behaviour of the positive and negative feedback systems. Positive feedback system generates exponential growth as shown in Fig. 1.9a, while the negative feedback system is goal seeking as shown in Fig. 1.9b. Figure 1.10 shows the control theory block diagram of a negative feedback system which is analogous to the causal loop diagram in system dynamics, and essentially system dynamics is a control theory for social systems.

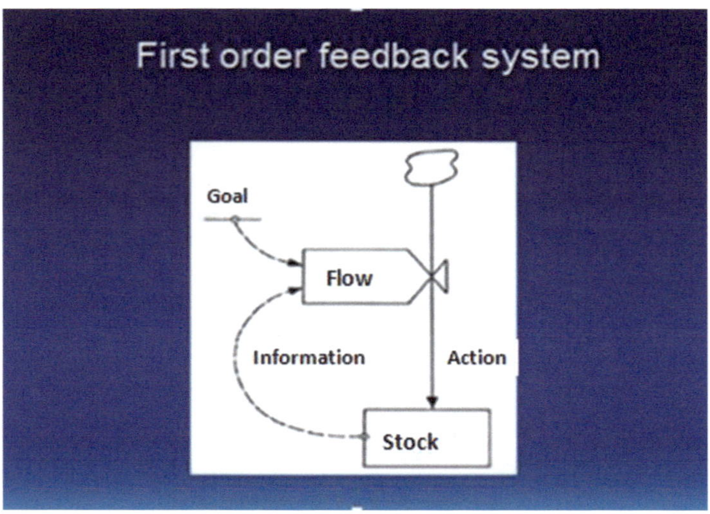

Fig. 1.7 Feedback loop

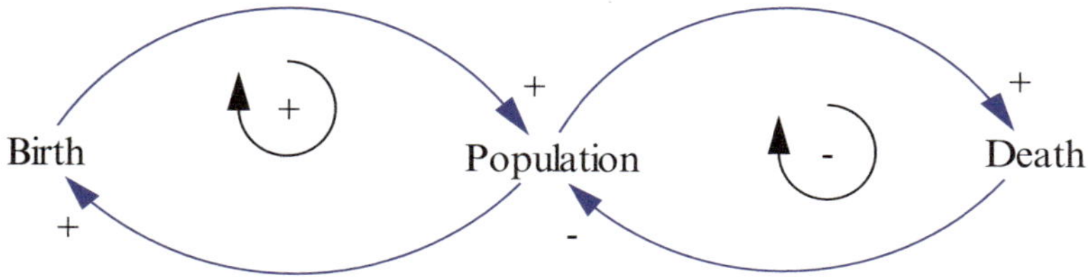

Fig. 1.8 Causal loop

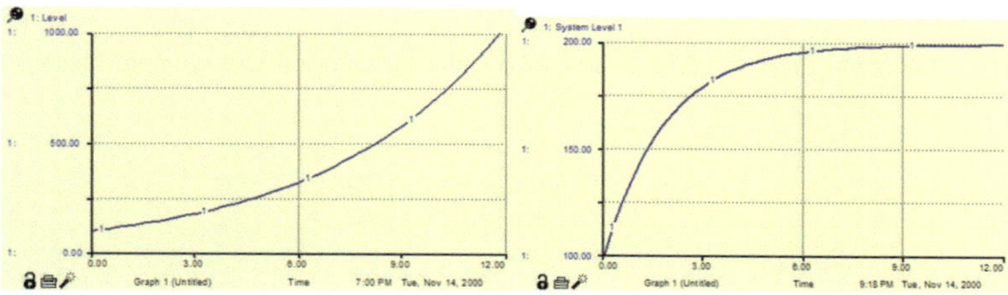

(a) (b)

Fig. 1.9 Dynamic behaviour of (**a**) positive feedback and (**b**) negative feedback systems

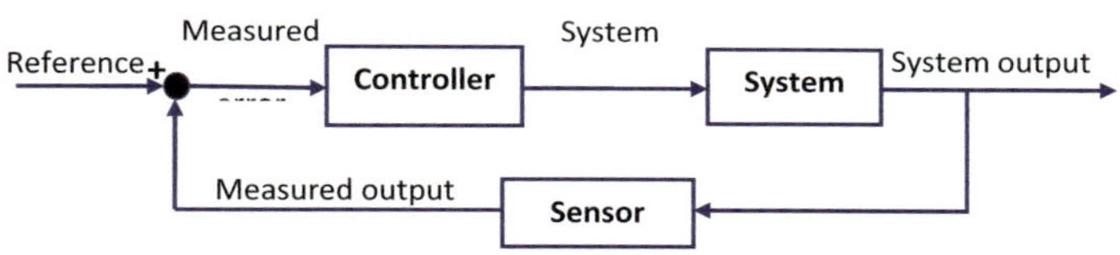

Fig. 1.10 Control theory block diagram of a negative feedback system

Figure 1.11a shows a second order feedback closed loop system characterised by two stocks: employment and inventory, and Fig. 1.11b represents the response of a second order system. The second order feedback loop system generates oscillation.

Figure 1.12 represents the responses of some complex systems. Figure 1.12a shows s-shaped growth, Fig. 1.12b shows s-shaped growth with overshoot and Fig. 1.10c shows overshoot and collapse of complex systems. First order and higher order positive feedback systems generate growth. First order negative feedback system is goal seeking, but the higher order negative system oscillates with a search for goal. Models in practice may contain thousands of interconnected loops with time delay and non-linearities.

1.5 Models and Simulation

We can study the dynamic behaviour of a physical system by experimentation with the system itself. Sometimes it may be expensive and time consuming. An alternative to this method is to construct a number of prototypes of physical models to

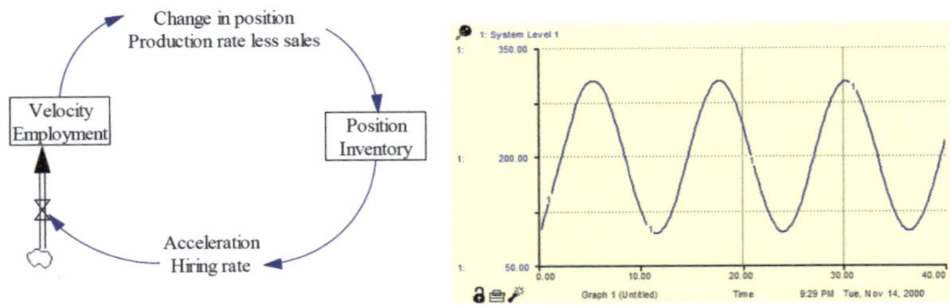

(a) (b)

Fig. 1.11 (**a**) Causal loop diagram of a second order system and (**b**) dynamic behaviour of a second order feedback system

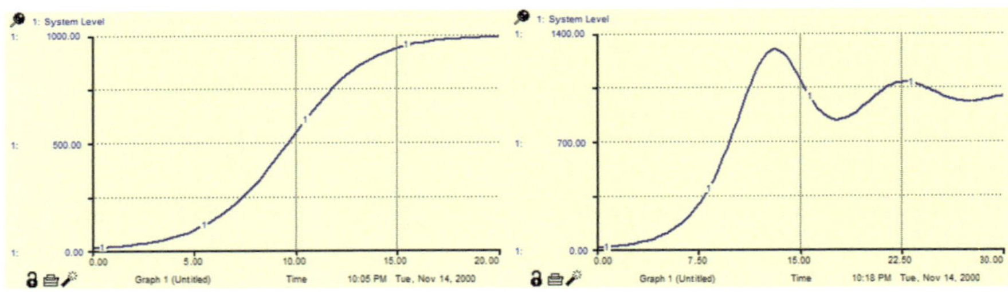

(a) (b)

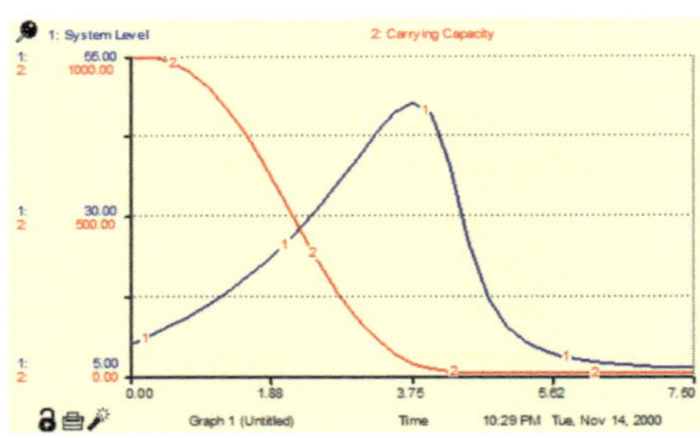

(c)

Fig. 1.12 Dynamic behaviour of complex systems: (**a**) s-shaped growth, (**b**) s-shaped growth with overshoot and (**c**) overshoot and collapse of complex systems

experiment with it. Sometimes even it may not be possible or practical to experiment with the existing system or construct a physical model for experimentation. Consequently the most inexpensive and less time consuming method is to use a mathematical or computer model.

A model may be defined as a substitute of any object or system. Everyone uses models in their daily life. A mental image used in thinking is a model, and it is not the real system. A written description of a system is a model that presents one aspect of reality. The simulation model is logically complete and describes the dynamic behaviour of the system. Models can be broadly classified as (a) physical models and (b) abstract models as shown in Fig. 1.13. Children's model of car and aeroplane are examples of physical models. Mental models and mathematical models are examples of abstract models.

Mathematical models can be classified as shown in Fig. 1.13 (Bala 1999). A model is a dynamic if it portrays time-varying characteristics; otherwise, the model is static. Steady state and transient characteristics are the properties of the dynamic systems, whereas in static systems, the characteristics do not undergo substantial change in time. In dynamic systems the system states change substantially in time.

It is sometimes possible to solve the mathematical models by analytical methods. But for complex systems, the solution of the mathematical model of the systems by analytical methods is extremely difficult, or it may be beyond the reach of today's mathematics. For such complex systems, only the step-by-step numerical solution is possible. This process of step-by-step numerical solution is called simulation. Simulation models are used in place of real systems. The computer simulation is an inexpensive and rapid method of experimenting with the system to give useful information about the dynamics of the real system. Scenarios based on simulated results can provide guidelines for policy planning and management of complex and dynamic systems.

Forrester's system dynamics methodology provides a foundation for constructing computer models to do what the human mind cannot do—'rationally' analyse the structure, the interactions and mode of behaviour of complex

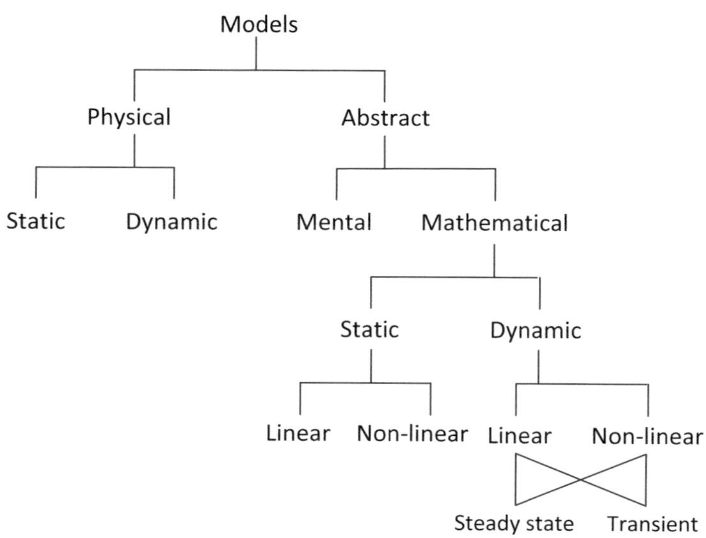

Fig. 1.13 Classification of mathematical models

socio-economic, technological, biological and political systems. The advantages of a computer model over a mental model are (Forrester 1968):

1. It is precise and rigorous instead of ambiguous and unquantified.
2. It is explicit and can be examined by critics for consistency and error.
3. It can contain much more information than any single model.
4. It can proceed from assumption to conclusions in logical error-free manner.
5. It can easily be altered to represent different assumptions or alternate policies.

1.6 Systems Thinking and Modelling

Indeed, we need some structures or guiding principles to model and simulate complex dynamic systems and the systems approach which considers the whole systems rather than in isolation refers to a set of conceptual and analytical methods used for systems thinking and modelling (Cavana and Maani 2000). The general methodological approach towards systems thinking and modelling is discussed here. Many contributions have been reported on systems thinking and system dynamics. The systems thinking and modelling essentially consist of problem statement, causal loop diagram, stock–flow diagram, scenario planning and modelling and implementation and organisation learning. The character of systems thinking makes it extremely effective on the most difficult types of problems to solve: those involving complex issues, those that depend a great deal dependence on the past or on the actions of others and those stemming from ineffective coordination among those involved. The steps for simulating a system dynamics model are summarised below:

- Identify the problem and formulate the mental model in the form of a verbal description (problem identification/conceptualisation) and develop a dynamic hypothesis to account for problematic behaviour in terms of causal loop diagrams and stock and flow structure of the system.
- Create basic structure of the causal diagram from the verbal model.
- Augment causal loop diagrams into system dynamics flow diagrams.
- Translate the system dynamics flow diagrams into STELLA or VENSIM or a set of simultaneous difference equations.
- Estimate the parameters.
- Validate the model, analyse the sensitivity and analyse the policy.
- Application of the model.

1.7 Usefulness of Models

The validity and usefulness of dynamic models should be judged not against imaginary perfection, but in comparison with other mental and descriptive models which are available. The usefulness of a mathematical simulation model should be judged in comparison with the mental image or other models which would be used instead (Forrester 1968). There is nothing in either physical or social science about which we have perfect knowledge and information. We can never say that a model is a perfect representation of the reality. On the other hand, we can say that there is nothing of which we know absolutely nothing. So, models should not be judged on an absolute scale but on relative scale if the models clarify our knowledge and provide insights into systems.

1.8 Structure of the Book

This book is organised into two parts—Part I: Concepts, methodology and techniques and Part II: Case studies. Part I consists of seven chapters, and Part II has five chapters. Chapter 2 provides the over review of systems thinking methodology with an example of systems thinking in action. In Chap. 3 we introduce causal loop diagram, a qualitative methodology with examples, and in Chap. 4, we present stock–flow diagrams, a computer simulation methodology with examples. We discuss parameter estimation and model validation in Chaps. 5 and 6, respectively. Chapter 7 provides scenario development and policy planning. In Part II we present five case studies of systems thinking and modelling, and these are modelling of boom and bust of cocoa production systems in Malaysia in Chap. 8, modelling of hilsa fish in Bangladesh in Chap. 9, modelling of food security in Malaysia in Chap. 10, modelling of supply chain of rice milling systems in Bangladesh in Chap. 11 and modelling of solid waste management in Chap. 12.

Exercises

Exercise 1.1 What is meant by complexity and always at a change of a dynamic system? How such a system can be understood and policy options to manage can be designed?

Exercise 1.2 What is system dynamics? Discuss the potentials of system dynamics methodology to handle complex dynamic socio-economic systems.

Exercise 1.3 Define feedback systems and describe the basic structure of a feedback loop. Discuss the different modes of behaviour of dynamic systems.

Exercise 1.4 What is meant by model? Describe the classification of models. Mention the advantages of a computer model over a mental model. Also discuss the usefulness of models.

Exercise 1.5 What is meant by simulation? What is simulation based on systems thinking and systems approach? Discuss the steps to be followed for simulating a system dynamics model.

References

Bala BK (1999) Principles of system dynamics, 1st edn. Agrotech Publishing Academy, Udaipur

Cavana RY, Maani KE (2000) A methodological framework for systems thinking and modelling (ST&M) interventions. Ist international conference on systems thinking in management. Internet

Forrester JW (1968) Principles of systems. MIT Press, Cambridge, MA

Hajkowicz S, Negra C, Barnett P, Clark M, Harch B, Keating B (2012) Food price volatility and hunger alleviation: can Cannes work? Agric Food Secur 1(8):1–12

Steve W, Sharada K (2014) Food prices 2013/2014. ODI Annual Report, May, 2014, 27 pages

Bibliography

Maani KE, Cavana RY (2000) Systems thinking and modelling: understanding change and complexity. Prentice Hall, Auckland

Manetsch TJ, Park GL (1982) Systems analysis and simulation with applications to economic and social systems. Department of Electrical Engineering and System Science, Michigan State University, East Lansing

Mohapatra PKJ, Mandal P, Bora MC (1994) Introduction to system dynamics modelling. Universities Press, Hyderabad

Sterman JD (2000) Business dynamics: systems thinking and modeling for a complex world. McGraw-Hill Higher Education, Boston

Systems Thinking: System Dynamics 2

The concept of systems thinking was introduced in Chap. 1 and this chapter presents systems thinking based on systems approach. Systems thinking methodology, participatory systems thinking and systems thinking in action are presented to demonstrate the potentiality of systems thinking to study complex and dynamic systems. Participatory systems thinking is highlighted.

2.1 Introduction

Systems thinking is a method of studying the dynamic behaviour of a complex system considering the systems approach, i.e. considering the entire system rather than in isolation, and system dynamics is a tool or a field of knowledge for understanding the change and complexity over time of a dynamic system. In isolation a complex system may give a false impression of the dynamic behaviour which is far from the real behaviour of the actual system. Thus, systems thinking should consider all the interacting components influencing the dynamics of the complex system, and system dynamics methodology based on the feedback concepts of control theory developed by Forrester (1968) is the most appropriate technique to handle such complex systems to enhance systems thinking and systems learning.

2.2 Systems Thinking Methodology

To enhance systems thinking and systems learning, the system must be modelled and simulated. Basically, there are six important steps in building system dynamics model. It starts with the problem identification and definition, followed by system conceptualisation, model formulation, model testing and evaluation, model use, implementation and dissemination and design of learning/strategy/infrastructure. There is a feedback in this step and it is illustrated in Fig. 2.1. Therefore, the steps

B.K. Bala et al., *System Dynamics*, Springer Texts in Business and Economics,
DOI 10.1007/978-981-10-2045-2_2

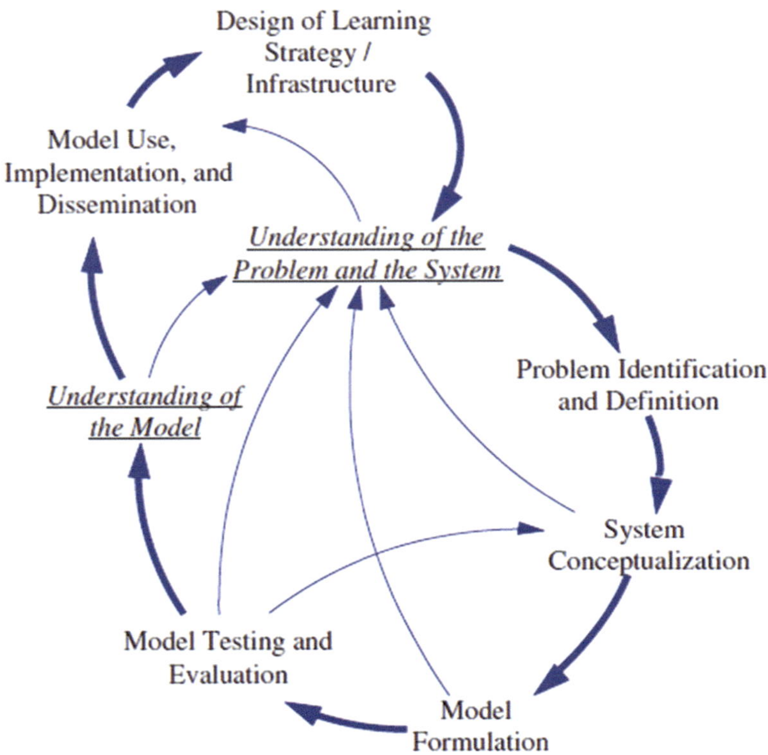

Fig. 2.1 Overview of the system dynamics modelling approach (Source: Martinez-Moyano and Richardson 2013)

needed for modelling and simulating of complex systems based on systems thinking are:

1. Identify the problem.
2. Develop a dynamic hypothesis explaining the cause of the problem.
3. Create a basic structure of a causal graph.
4. Augment the causal graph with more information.
5. Convert the augmented causal graph to a system dynamics flow graph.
6. Translate a system dynamics flow graph into STELLA or VENSIM programs or equations.

These steps of systems thinking are discussed below in detail.

2.2.1 Problem Identification

The first step in the model building is to identify the problem, set its boundary and state the specific objectives. The problem should be clearly identified and it is important for a successful modelling to solve the real problem. Systems thinking should be used for addressing the problem. Neither the whole system nor the part of it should be considered to draw the boundary of the model rather systems approach of considering the entire system that endogenously responsible to cause the problem from the feedback structure of the stated entire system. Therefore, the system boundary should encompass that portion of the whole system which includes all the important and relevant variables to address the problem and the purpose of policy analysis and design. The scope of the study should be clearly stated in order to identify the causes of the problem for clear understanding of the problem and policies for solving the problem in the short run and long run.

To recognise the problem, prepare a detailed description of the system based on available reports and studies, expert opinions and past behaviour of the system and identify the important variables generating the observed dynamic behaviour of the system. The problem of system identification is the problem of system operation. Thus, the problem identification should include clear statement of the problem based on different reports, historical and statistical records and previous studies. The problem statement should clearly describe the major factors influencing the dynamics of the system behaviour with facts and figures. Next, it should include the purpose and clearly defined objectives. Discussion with all the stakeholders such as focus group discussions should be conducted to justify their opinions on the existing problems, their views on the data collected and also their views on the solution of the problems.

The verbal description is the simplest way to communicate with others about the system. The more detailed is the description, the more it becomes easier to model the system. Major subsystems and their relationships within and between the subsystems of the system as a whole should be clearly described. The model should include only the relevant aspects of the study objectives. The verbal description is in practice a qualitative model of the system.

In selecting the variables to be included in the model, all the variables or factors relevant to the study objectives should be included, and unnecessary restrictions must be avoided. The accuracy of the information gathered should be considered. A further factor to be considered is the extent to which the number of individual entities can be grouped together into large entities. The boundary should be such that nothing flows across the boundary except perhaps a disturbance for exciting the system, and the factors needed to address the problem must be included inside the system boundary for the proper comprehensiveness of the model with adequacy. Figure 2.2 illustrates the closed boundary concept. Formulating a model of a system should start with a boundary that encompasses the smallest number of components within which the dynamic behaviour under study is generated.

Often, it may be difficult to comprehend the whole system, especially when it is very large and complex. It is convenient to break up such system into sectors or

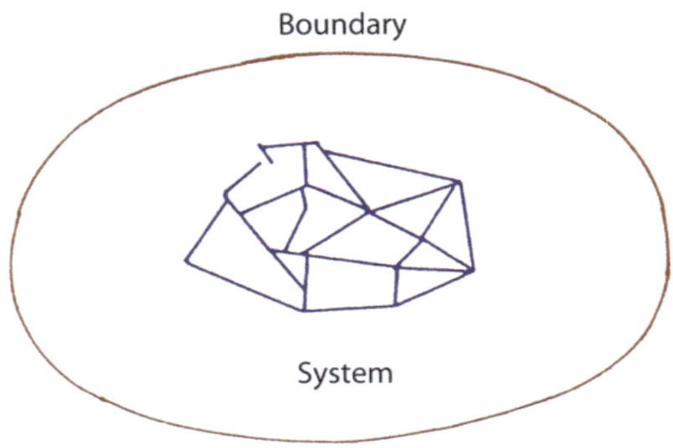

Fig. 2.2 Closed boundary concept

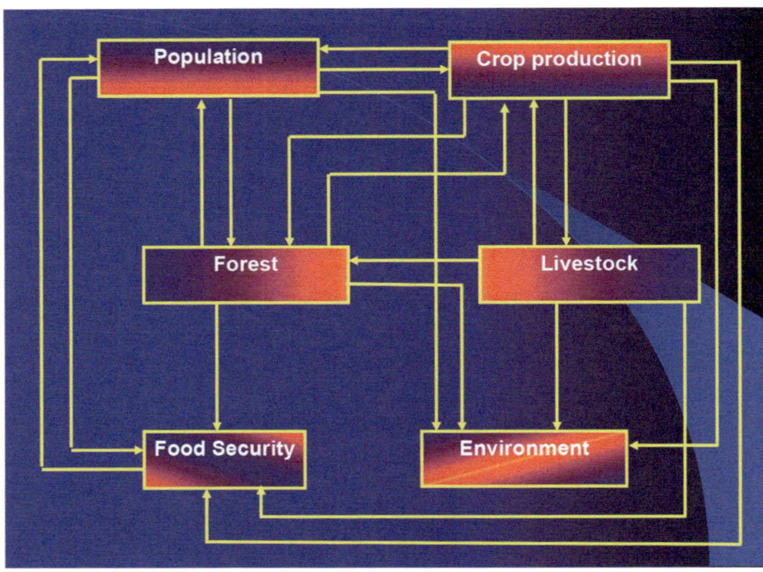

Fig. 2.3 Structure of food, energy and environment model

blocks. The description of the system should be organised in a series of blocks. The aim in constructing the blocks is to simplify the specification of the interactions within the system boundary. Each block describes a part of a system that depends upon a few, preferably one, input variable and results in a few output variables. The system as a whole can be described in terms of interconnections between the blocks. Correspondingly, the system can be represented graphically as a simple block diagram. Figure 2.3 shows the overall structure of food, energy and environment model and it is a typical example of a block or sectorial diagram. The model is about the study of food, energy and CO_2 production in Bangladesh. The six sectors of the model are food, forest, population, cattle population, energy and CO_2. The major influences to a sector from other sectors and its influences on other sectors are shown in the diagram.

System dynamics model of endogenous structure of feedback loops is simulated to generate the problem dynamically, i.e. the observed dynamic behaviour. This pattern of change of the behaviour with time is termed as reference mode behaviour or historical behaviour. We need the observed reference mode behaviour to understand the problem and hence variables are selected accordingly. Figure 2.4a shows the observed reference mode behaviour of crude palm oil (CPO) price and the observed and simulated reference mode behaviours of boom and bust of cocoa production systems in Malaysia. The time horizon of the reference mode and policy are also important and must be sufficient to cover the problem symptoms and policy issues addressed (Fig. 2.4b).

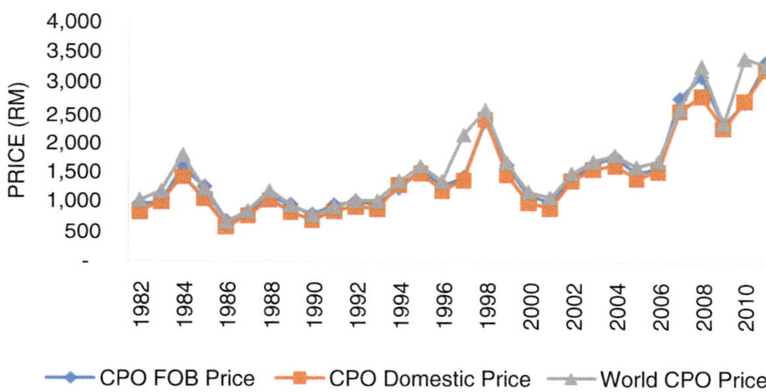

Fig. 2.4a CPO FOB Price, CPO Domestic Price and World CPO Price (1982–2012) (Source: MPOB 2012)

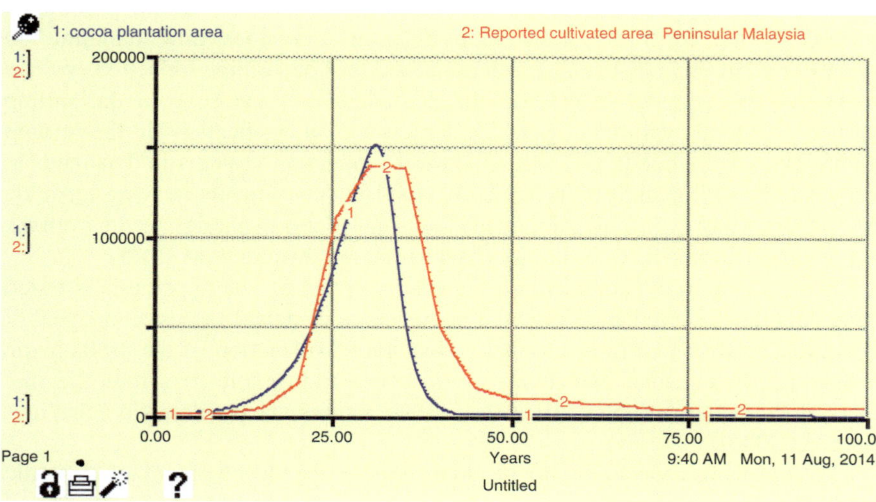

Fig. 2.4b Observed and simulated reference mode behaviour of cocoa production systems in Malaysia (Source: Fatimah et al. 2015)

The following aspects are to be addressed in the development of problem identification:

1. Definition of the problem
2. Purpose of the model
3. Systems approach
4. Reference mode
5. Time horizon

2.2.2 Dynamic Hypothesis

Once the problem is identified, the next step is to develop a theory called dynamic hypothesis based on the reference mode behaviour over a time horizon. The dynamic hypothesis in terms of causal loop diagram and stock–flow diagram of the system can explain the dynamics of the problem. The hypothesis is provisional and is subject to revision and rejection which solely depends on the observed and simulated reference mode of behaviour over a time horizon (Sterman 2000).

The dynamic hypothesis is a conceptual model typically consisting of a causal loop diagram, stock–flow diagram or their combination. The dynamic hypothesis seeks to define the critical feedback loops that drive the system's behaviour. When the model based on feedback concept is simulated, the endogenous structure of the model should generate the reference mode behaviour of the system, and thus, the endogenous structure causes the changes in the dynamic behaviour of the system (Sterman 2000). For example, the boom and bust of shrimp production systems can be represented by causal loop diagram and stock–flow diagram, and the simulation model based on the causal loop diagram and stock–flow diagram can generate dynamic behaviour of the shrimp production systems. The shrimp production systems in the form of causal loop diagram and stock–flow diagram are hypothesised to generate the observed boom and bust of shrimp production systems in the reference mode. In essence the degradation of the soils in the shrimp aquaculture ponds resulting from the large-scale intensification of the shrimp culture caused the boom and bust of shrimp production systems in Thailand. In fact when the shrimp industry is prone to exceed and consume its carrying capacity, the boom and bust type of development results in and this dynamics results from the endogenous consequences of the feedback structure (Arquitt et al. 2005).

The next step in dynamic hypothesis is how to test it. The hypothesis is tested using both the observed and simulated reference mode data. In essence, the goal of dynamic hypothesis is to develop an endogenous explanation of the problematic behaviour. Endogenous explanation is that the endogenous structure, i.e. the interactions of the variables inside the system, causes the problematic behaviour (Sterman 2000).

The following aspects are to be addressed in the development of dynamic hypothesis:

1. Endogenous feedback structure
2. Observed and simulated reference mode behaviour
3. Theory to explain the reference mode behaviour

2.2.3 Causal Loop Diagram

The system boundary covers the key variables inside the boundary and variables crossing the boundary. The variables inside the boundary are endogenous variables and the variables outside the boundary are exogenous variables. The next step in the systems thinking is to search the relationships between the variables and the developments of feedback loops. These feedback structures are represented in the form of causal loop diagrams in system dynamics (Sterman 2000) and in the form of control theory block diagram in systems analysis (Manetsch and Park 1982). Figure 2.5 shows the causal loop diagram of a simple irrigation model. In this simple irrigation model, the major variables are irrigated area, irrigated area increase rate and also irrigation area discard or abandon rate. Irrigated area increase rate decreases with the increase in irrigated area and increase in irrigated area increase increases the irrigated. This forms the negative feedback loop B1. The irrigation area discard rate increases with the increase in irrigated area and in turn this causes to decrease the irrigated area. This forms the negative feedback loop B2. Thus, the irrigated area forms two negative feedback loops. The causal loop diagram represents feedback loop structure of the system and causes the dynamic behaviour of the system. Causal loop diagram represents the feedback structure of systems to capture the hypotheses about the causes of dynamics and the important feedbacks. The causal loop structure generating the reference behaviour of the system is hypothesised to be the dynamic hypothesis. The following steps are followed in the development of causal loop diagram:

1. Define the problem and the objectives.
2. Identify the most important elements of the systems.
3. Identify the secondary important elements of the systems.
4. Identify the tertiary important elements of the systems.
5. Define the cause–effect relationships.

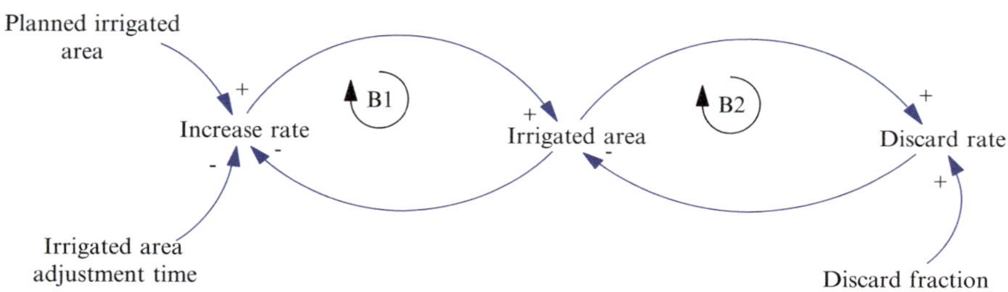

Fig. 2.5 Causal loop diagram of a simple irrigation model

6. Identify the closed loops.
7. Identify the balancing and reinforcing loops.

The details of construction of causal loop diagrams are described in Chap. 3.

2.2.4 Stock–Flow Diagram

The stock–flow diagram is the underlying physical structure of the system in terms of stock and flow. Stock–flow diagram is usually followed after the causal loop diagram. However, the causal loop diagram can follow the stock–flow diagram. The stock represents the state or condition of the system, and the flow is changed by decisions based on the condition of the system. It is essentially the physical structure of the system and can be simulated to generate the dynamic behaviour of the system. The stock–flow diagram represents integral finite difference equations involving the variables of the feedback loop structure of the system and simulates the dynamic behaviour of the system. But differential equations are formulated in systems analysis based on the control theory block diagram (Manetsch and Park 1982). The stock–flow diagram or the system of the differential equations representing the feedback structure of systems captures the hypotheses about the causes of dynamics and the important feedbacks. The stock–flow diagram or the system of the differential equations representing the feedback structure of the system generating the reference behaviour of the system is hypothesised to be the dynamic hypothesis. Figure 2.6 shows the stock–flow diagram of a simple irrigation model. The three main variables are irrigated area, irrigated area increase rate and irrigation area discard rate. Here, we have one stock variable irrigated area stating the condition of irrigation, and it is increased by one inflow-increasing rate and decreased by one outflow discard rate. Also, the irrigated area has the unit of quantity, while the increase rate and discard rate have the unit of quantity per unit time. The following steps are followed in the development of stock–flow diagram:

1. Define the problem and the objectives.
2. Identify the most important variables of the systems.
3. Identify the secondary important variables of the systems.
4. Identify the tertiary important variable of the systems.
5. Identify the variables representing the stocks, i.e. accumulations.
6. Identify the variables representing the flows having a unit of per unit time of the stock.
7. Ensure the inflows entering the stock and outflow leaving the stock.

The details of construction of stock–flow loop diagrams are described in Chap. 4.

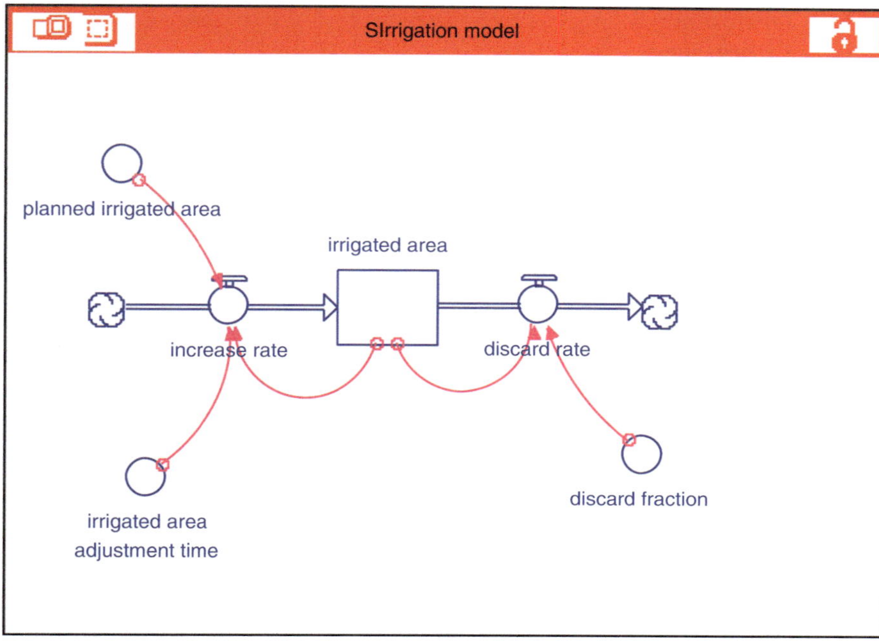

Fig. 2.6 Stock–flow diagram of the irrigation model

2.2.5 Parameter Estimation

Parameter estimation is one of the important steps in system dynamics modelling. Parameter estimation techniques can be classified into three categories: these are (i) estimation from unaggregate data, (ii) estimation from an equation and data at a level of aggregation of model variables and (iii) estimation from the knowledge of the entire model structure and data at a level of aggregation of model variables. The details are described in Chap. 5.

2.2.6 Model Validation, Sensitivity Analysis and Policy Analysis

The tests for building confidence in system dynamics model consist of validation, sensitivity analysis and policy analysis of system dynamics models. The two important notions of the building confidence in system dynamics models are testing and validation. Testing means the comparison of a model to empirical reality for accepting or rejecting the model, and validation means the process of establishing confidence in the soundness and usefulness of the model. The tests for building confidence in system dynamics models may be broadly classified as:

1. Tests for structure
2. Tests for behaviour
3. Tests for policy implication

The detailed tests for building confidence in models are described in Chap. 6.

2.2.7 Application of the Model

Because of the counterintuitive nature of the complex systems, the human mind is not capable of tracing the dynamic behaviour of the systems. System dynamics simulation model can provide better understanding and greater insights of such systems. Examining the alternative policies for the selection of best policy for the improved performance of the system is essential for policy planning. System dynamics simulation model can be used as a computer laboratory for policy analysis, and it can also be used to assist in the management and control policy design. Optimal management and control policy design of the system with regard to certain criterion and constraints is the ultimate goal of the optimisation of the system. A simulation model is essential for such optimisation of the system. Applications of system dynamics models in different areas of agricultural, biological, aquacultural, environmental and socio-economic systems are presented in Part II in Chaps. 8, 9, 10, 11 and 12.

2.3 Critical Aspects of Systems Thinking

The following aspects of systems thinking are very important for studying the dynamic behaviour of the complex system and need attention to develop the model based on systems thinking:

1. Thinking in terms of cause-and-effect relationships
2. Focusing on the feedback linkages among components of a system
3. Determining the appropriate boundaries for defining what is to be included within a system

We are interested to study and examine the dynamic behaviour of systems containing biological, agricultural, aquacultural, environmental, technological and socio-economic components. In formulating a model for this purpose in mind, formulating the model should start from the question 'where is the boundary of the dynamic system?'. In concept the feedback system is a closed system and the dynamic behaviour arises within the system.

2.4 Participatory Systems Thinking

The system dynamics uses simulation models for policy design and policy analysis and it is based on the feedback concepts of control theory. More specifically, the system dynamics uses feedback loops, stock and flow diagrams and non-linear differential equations. Stakeholders form an important part of the system dynamics methodology (Forrester 1961; Gardiner and Ford 1980; Vennix 1996, 1999; Hsiao 1998; Elias et al. 2000 and Maani and Cavana 2000). Group model building is defined as a model-building process which involves the client group deeply in the process of modelling (Vennix 1996, 1999; Andersen and Richardson 1997; Rouwette et al. 2011).

There are many reasons to take stakeholders into account in the model-building process (de Gooyert 2012). And the three distinct types of group model-building interventions are (1) modelling stakeholder behaviour, (2) modelling with stakeholders and (3) modelling stakeholder behaviour with stakeholders. Group model-building intervention needs several iterations. This intervention will support learning for the stakeholders, and learning will change the behaviour and hence the decision. The simulation model should be updated after each iteration to take the new stakeholder behaviour into account

Participatory modelling includes a broad group of stakeholders in the process of formal decision analysis. It is a process of incorporating stakeholders, often the public, and decision-makers into the modelling process (Voinov and Gaddis 2008). Non-scientists are engaged in the scientific process and the stakeholders are involved to a greater or lesser degree in the process. A fully participatory process is one in which participants help identify the problem, describe the system, create an operational computer model of the system, use the model to identify and test policy interventions and choose one or more solutions based on the model analysis. Involving the stakeholders in the model-building process can build trust among stakeholders (Tàbara and Pahl-Wostl 2007).

Participatory system dynamics modelling uses system dynamics perspective in which stakeholders or clients participate to some degree in different stages of the model-building process. Participatory system dynamics modelling is more than simply eliciting knowledge from clients about the problem and the system. It involves building shared ownership of the analysis, problem, system description and solutions or a shared understanding of the tradeoffs among different decisions. In other words, it may be termed as participatory systems thinking. The details of the participatory system dynamics modelling are discussed in Chap. 7.

2.5 Systems Thinking in Action

In the previous sections, the systems thinking methodology has been explained. Here in this section, the focus is on policy simulation and analysis to address how systems thinking-based modelling and simulation can assist in policy simulation and analysis. To achieve this goal, the focus is concentrated on the introduction of system, on differential equation model and stock–flow diagram and more importantly on policy simulation and analysis to demonstrate how systems thinking-based model can address the policy issues. To demonstrate how to apply systems thinking in action, we consider here the dynamics of the mangrove forest in the Sundarbans in Bangladesh as an example.

2.5.1 Introduction

The **Sundarbans** is the largest single block of tidal halophytic mangrove forest in the world located in the southern part of Bangladesh, and the Sundarbans was declared as the world heritage site by UNESCO in 1997. The Sundarbans is intersected by a complex network of tidal waterways, mudflats and small islands of salt-tolerant mangrove forests. The Sundarbans mangrove ecoregion is the world's largest mangrove ecosystem. The Sundarbans flora is characterised by the abundance of Sundari (*Heritiera fomes*), Gewa (*Excoecaria agallocha*) and Keora (*Sonneratia apetala*). Figure 2.7 shows a pictorial of Sundari, Gewa and Keora trees in the Sundarbans. The number of tree species in the mangrove forest of Sundarbans is very large (a total of 245 genera and 334 plant species). Based on their growth characteristics, tree species can be classified into three functional groups, and Keora, Gewa and Sundari are identified to represent these functional groups and the growth dynamics of the mangrove forest.

For proper management and understanding of the forest ecosystem, it must be modelled and simulated. Growth models of forests can assist in many ways. Some important uses of growth models are its ability to predict the future yields; it provides an efficient way to resource forecasts; and it can be used to prepare harvesting schedules for sustainable development. It can also provide a better understanding and greater insights into forest dynamics. The model presented here focuses on these three functional groups: pioneer (Keora), intermediate (Gewa) and climax (Sundari).

(a) (b) (c)

Fig. 2.7 (a) Sundari, (b) Gewa and (c) Keora trees of the Sundarbans

2.5.2 Differential Equation Model and Stock–Flow Diagram

In this example we demonstrate systems analysis approach of using differential equations and also the system dynamics approach of stock–flow diagram which essentially represents the integral finite difference equations. The model is based on a system of three coupled differential equations, which describe stem volume changes of the succession groups for every time steps. The equations are converted into stock–flow diagram of STELLA and solved by the Runge–Kutta fourth-order method.

The volume change for pioneer tree species turns out to be fundamentally different from that of the two other species. After a clear cut, pioneer tree species grows at the fastest pace at the beginning of the regrowth process, which can be described by the Michaelis–Menten kinetics (Haefner 1996). In contrast intermediate and climax tree species initially show a much slower growth and have their maximal volume increment when they have reached half of the volume they would have in a mature forest. This kind of dynamics can be described by logistic growth (Bossel 1994). Both approaches are consistent with the observation that on a scale of several hectares of the forest dynamics lead to a steady state, in which no changes occur any more over time, except from stochastic processes. Additionally, for each tree species group, competition exists within and between groups.

The following assumptions are made for the mangrove forest growth model:

1. Stand growth was considered in the model.
2. Aggregated growth of Sundari, Gewa and Keora is considered in the model.
3. Competition among these species exists.
4. The tree species are pioneer (Keora), intermediate (Gewa) and climax (Sundari).

Pioneer Tree Species

According to the Michaelis–Menten equation, the volume of pioneer tree species V_P at time t is calculated from the maximal volume $V_{mat,P}$, time t and constant K_M. This constant describes the time it takes to reach half of the maximal volume $V_{mat,P}$:

$$V_P(t) = \frac{V_{mat,P}t}{K_M + t} \tag{2.1}$$

The equation can be rewritten as a differential equation where only the first term is dependent on the volume. The first term can be interpreted as a growth term, and the second one as a constant input rate:

$$\frac{dV_P}{dt} = \frac{1}{K_M} V_P \left(\frac{V_P}{V_{mat,P}} - 2 \right) + \frac{V_{mat,P}}{K_M} \tag{2.2}$$

The growth of pioneer tree species is inhibited by trees of other successional groups. Tietjen and Huth (2006) reported that only the second term is affected by other species groups. Therefore, a competition factor C_P that reduces the constant input is introduced:

$$\frac{dV_P}{dt} = \frac{1}{K_M} V_P \left(\frac{V_P}{V_{mat,P}} - 2 \right) + \frac{V_{mat,P}}{K_M} C_P \tag{2.3}$$

with

$$C_P = 1 - \frac{\omega_{IP}}{\omega_{IP} + \omega_{CP}} \frac{V_I}{V_{mat,I}} - \frac{\omega_{CP}}{\omega_{IP} + \omega_{CP}} \frac{V_C}{V_{mat,C}}$$

where V_I and V_C describe the current volumes of intermediate and climax tree species, respectively, and $V_{mat,I}$ and $V_{mat,C}$ denote the corresponding volumes in a mature forest. The additional factors ω_{IP} and ω_{CP} weigh the influence of the particular competition and add up to 1. If the volume of intermediate and climax trees species is zero, then the competition factor equals one, and the growth of the pioneer trees is not inhibited. On the other hand, if both tree species groups reach the volume of a mature forest, the competition factor becomes zero and the constant input disappears.

Intermediate and Climax Tree Species

Both successional groups are described by logistic growth. The specific growth rates for intermediate and climax tree species are g_I and g_C, respectively. Competition within one group and in between the two groups is considered, and as before it is weighed by competition factors adding up to 1. An additional constant a_{input} is added to the equations of both groups to avert extinction. This constant is chosen small enough not to affect the main dynamics and can be interpreted as constant seedling input:

$$\frac{dV_I}{dt} = \left(a_{input} + g_I V_I \right) \times \left(1 - \frac{\mathrm{II}}{\omega_{II} + \omega_{CI}} \frac{V_I}{V_{mat,I}} - \frac{\omega_{CI}}{\omega_{II} + \omega_{CI}} \frac{V_C}{V_{mat,C}} \right) \tag{2.4}$$

$$\frac{dV_C}{dt} = \left(a_{input} + g_C V_C \right) \times \left(1 - \frac{\omega_{IC}}{\omega_{IC} + \omega_{CC}} \frac{V_I}{V_{mat,I}} - \frac{\omega_{CC}}{\omega_{IC} + \omega_{CC}} \frac{V_C}{V_{mat,C}} \right) \tag{2.5}$$

The total harvestable volume $V_{harv,total}$ is determined by adding the difference between current and remaining volume of both successional groups: climax and intermediate:

$$V_{harv,total} = V_{harv,I} + V_{harv,C} \tag{2.6}$$

$$V_{harv,X} = V_X - V_{rem,X}, \quad X = I, C \tag{2.7}$$

The stock–flow diagrams of the mangrove forest growth models are shown in Figs. 2.8 and 2.9 for undisturbed forest growth and forest growth with logging scenarios, respectively.

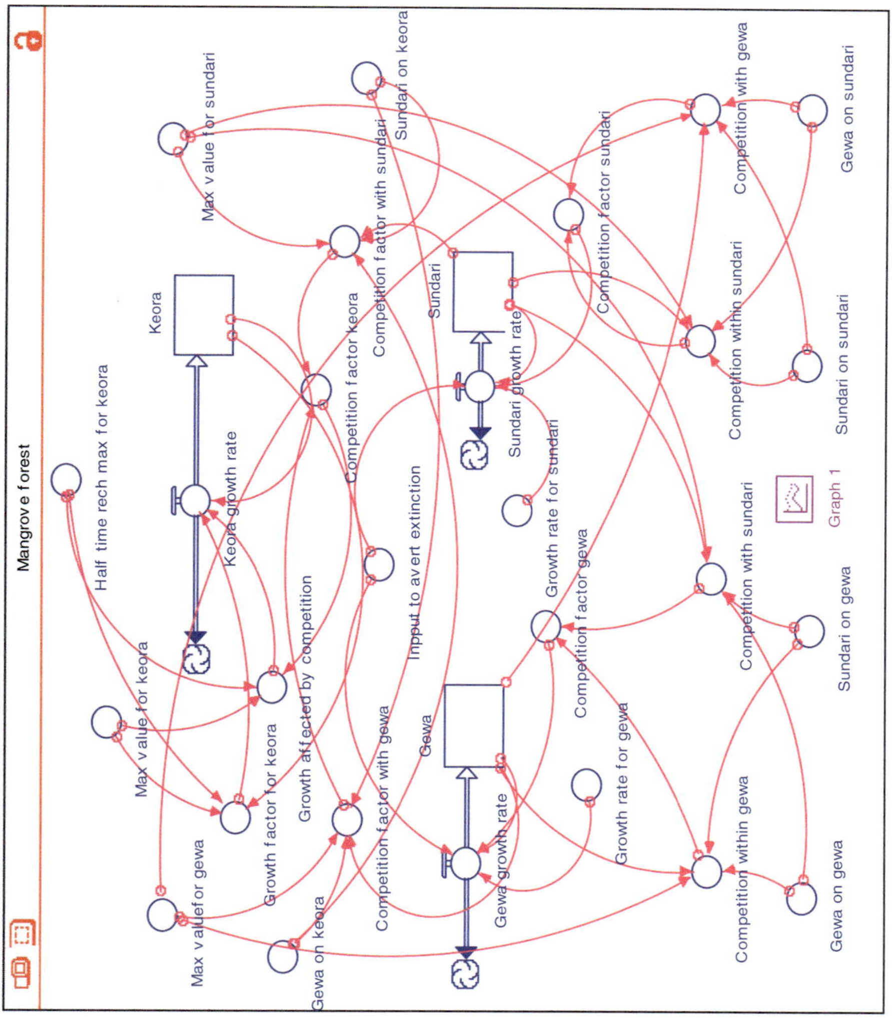

Fig. 2.8 Stock–flow diagram for undisturbed mangrove forest model

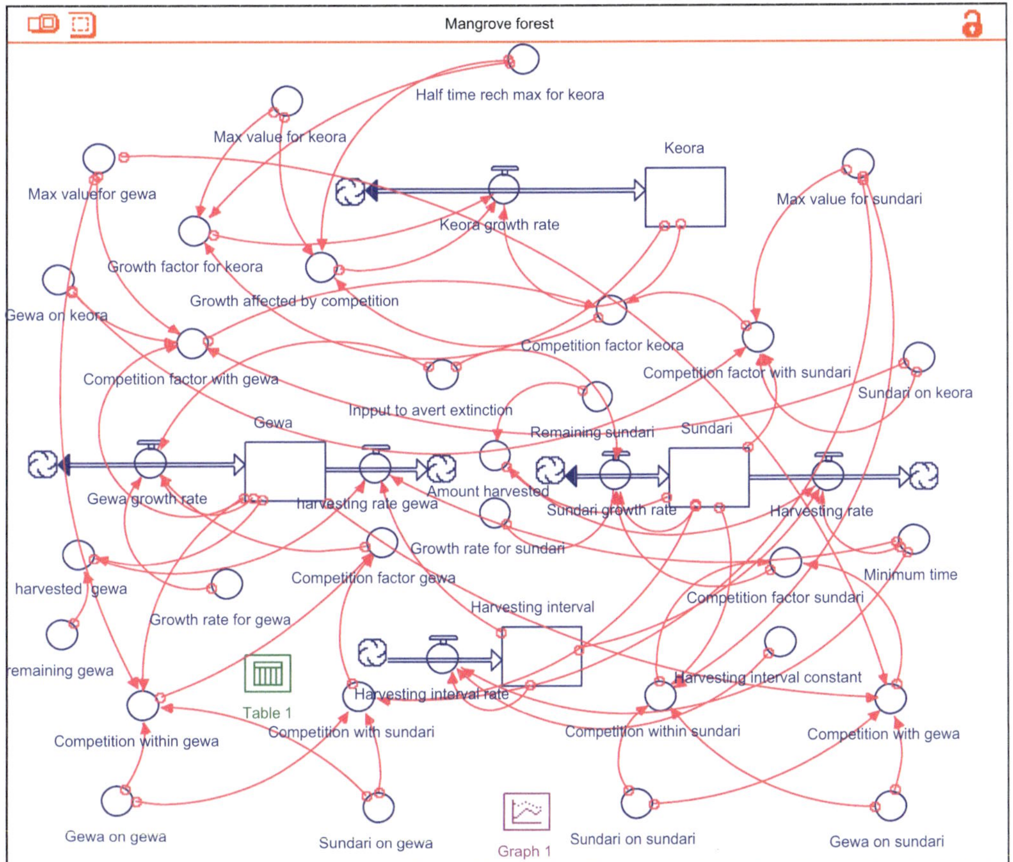

Fig. 2.9 Stock–flow diagram for mangrove forest model with logging operations

Table 2.1 Parameter values of the model

Parameter description	Values
Competition factor Gewa	Gewa on Gewa = 0.32
	Gewa on Keora = 0.32
	Gewa on Sundari = 0.01
Competition factor Sundari	Sundari on Gewa = 0.68
	Sundari on Keora = 0.68
	Sundari on Sundari = 0.99
Half-time reach maximum for Keora	3.65
Input to avert extinction	0.01
Maximum value for Gewa	80.0
Maximum value for Keora	196.11
Maximum value for Sundari	320.0

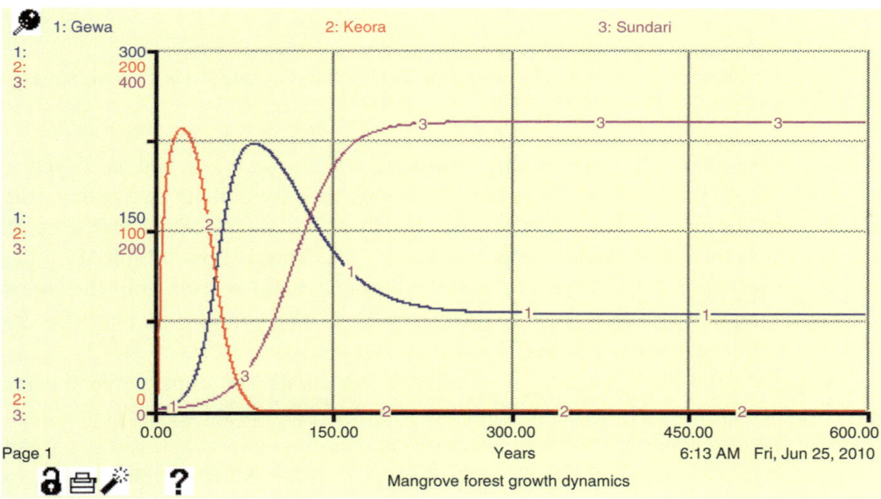

Fig. 2.10 Simulated volume for undisturbed mangrove forest (Keora, pioneer; Gewa, intermediate; Sundari, climax)

2.5.3 Simulation and Policy Analysis

The parameters of the model were collected from secondary sources such as reports, journal publications and personal communication with experts. The important parameter values are given in Table. 2.1. The models were assessed for structural consistency and both the model generate plausible behaviours.

The model was simulated to predict the forest stand of three major mangrove species Sundari, Gewa and Keora of the mangrove forest in the Sundarbans. Simulated total bole volumes of pioneer, intermediate and climax tree species for undisturbed forest growth with time are shown in Fig. 2.10. Pioneer tree species

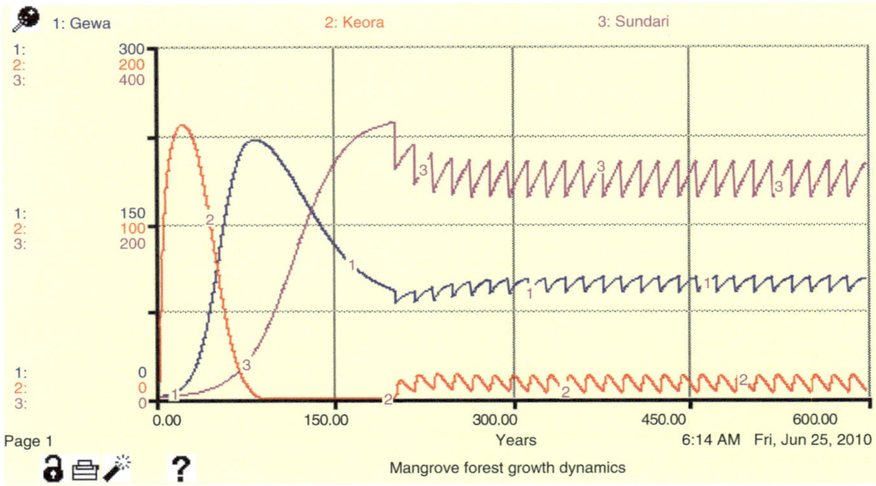

Fig. 2.11 Simulated volume of mangrove forest for 16 years logging cycle (Keora, pioneer; Gewa, intermediate; Sundari, climax)

(Keora) grows fast until a maximum volume of 160 m^3/ha is reached at 25 years. Because of light competition, these are suppressed by intermediate and climax tree species and disappear after 80 years; by this time, the intermediate tree species (Gewa) reach their maximum volume of 225 m^3/ha. This is now inhibited by the increasing appearance of climax tree species (Sundari) and decline until the forest reaches a steady state after another 190 years. Here, the volumes of intermediate and climax tree species are 80 and 320 m^3/ha, respectively.

Simulated total bole volumes of pioneer, intermediate and climax tree species for a logging cycle of 16 years are shown in Fig. 2.11. The first logging is practised after the forest reached a steady-state condition. For the undisturbed forest growth model, the pioneer tree species (Keora) disappears after a certain period due to competition with climax and intermediate tree species (Fig. 2.10), but after a logging operation, it reappears and if the logging operation continues at a certain interval, the pioneer species never disappear (Fig. 2.11). The maximum volume of the climax tree species for undisturbed model is 320 m^3/ha, but for 16 years logging cycle, it can't reaches this maximum volume, rather than volume changes in cyclical manner with a mean volume of 195 m^3/ha with a maximum volume is 275 m^3/ha.

When the logging cycle of 50 years is selected, the climax tree species reaches its maximum volume, that is, the forest reaches to its steady-state condition before each logging operation. Figure 2.12 shows simulated effects of 50 years logging cycle. The volume of the pioneer tree species changes in cyclic manner with a mean value of 10 m^3/ha.

Figure 2.13 simulates the effect of 100 years logging cycle. It shows similar effects of 50 years logging cycle. In both cases pioneer species develops a maximum volume of 15 m^3/ha and then it disappears and reappears in a cyclic manner. The changes in cyclic manner of intermediate tree species are small and remain almost same for both 50 and 100 years logging cycle.

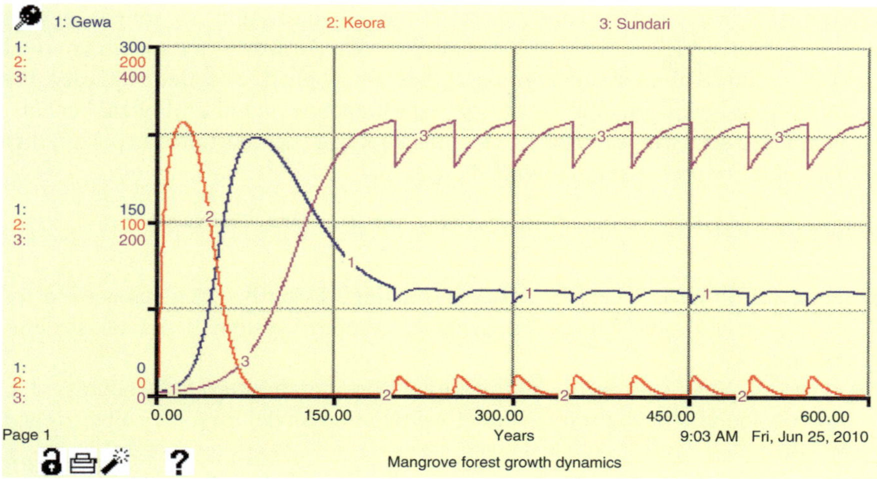

Fig. 2.12 Simulated volume of mangrove forest for 50 years logging cycle (Keora, pioneer; Gewa, intermediate; Sundari, climax)

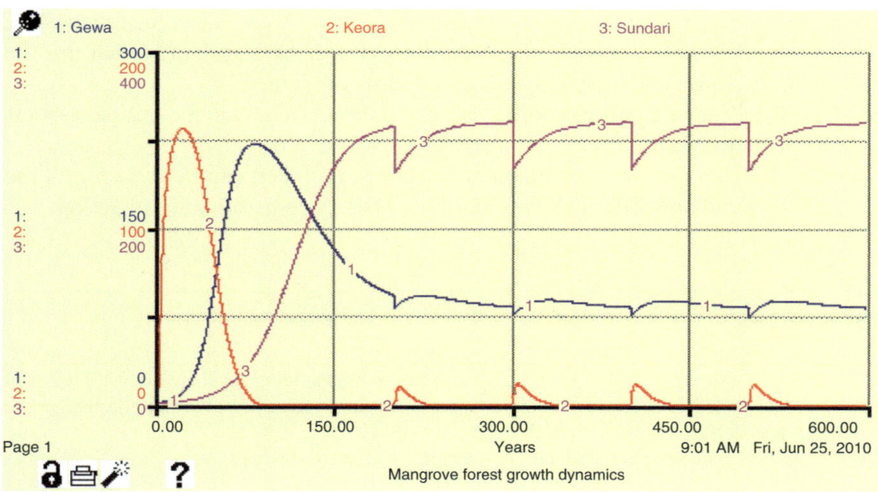

Fig. 2.13 Simulated volume of mangrove forest for 100 years logging cycle (Keora, pioneer; Gewa, intermediate; Sundari, climax)

Here, simulated harvesting strategies are presented using system dynamics model based on systems thinking. Simulated results show plausible behaviour, and this model is able to reproduce the volume dynamics of succession groups of the mangrove forest of the Sundarbans in Bangladesh. Simulated results for different logging strategies show that short logging cycles affect the forest more heavily in terms of damage and the ability to regrow to a mature stand. So, longer logging cycles are preferable to protect the mangrove forest. However, the maximum total yield is achieved if medium logging cycles are applied, and this suggests that medium cycle logging should be considered for economic and ecological benefits. The simulated scenarios for different logging strategies demonstrate the potentiality of the model for policy simulation and analysis.

Exercises

Exercise 2.1 What is meant by systems thinking? Describe the steps needed for modelling and simulation of a dynamic complex system based on systems thinking.

Exercise 2.2 What is problem identification? What are the steps to be addressed in problem identification for developing a simulation model for policy analysis and design?

Exercise 2.3 What is meant by dynamic hypothesis? What are the steps to be taken into account to develop a dynamic hypothesis?

Exercise 2.4 What is causal loop diagram? What are the purposes of drawing causal diagrams of a complex dynamics systems? Describe the steps to be followed to develop the causal loop diagrams.

Exercise 2.5 What is stock–flow diagram? What is the basic difference between a differential equation model and stock–flow model of a system? What are the steps to be followed to develop the stock–flow diagrams?

Exercise 2.6 What are participatory systems thinking? What are the critical aspects of systems thinking?

Exercise 2.7 Describe how participatory systems approach can be included in the modelling and simulation in the example of the systems thinking in action.

References

Andersen DF, Richardson GP (1997) Scripts for group model building. Syst Dyn Rev 13:107–129

Arquitt S, Hongang X, Johnstone R (2005) A system dynamics analysis of boom and bust in the shrimp aquaculture industry. Syst Dyn Rev 21:305–324

Bossel H (1994) Modeling and simulation. K.Peters, Wellesley, 484 pp

de Gooyert V, Rouwette FAJA, van Kranenburg HI (2012) A typology of stakeholder involvement in group model building. In: Proceeding of the 30th international conference of the System Dynamics Society

Elias AA, Cavana RY, Jackson LS (2000) Linking stakeholder literature and system dynamics: opportunities for research. In 1st international conference on systems thinking in management, pp 174–179

Fatimah MA, Bala BK, Alias EF, Abdulla I (2015) Modeling boom and bust of cocoa production systems in Malaysia. Ecol Model 310:22–32

Forrester JW (1961) Industrial dynamics. MIT Press, Cambridge, MA

Forrester JW (1968) Principles of system dynamics. MIT Press, Cambridge, MA

Gardiner PC, Ford A (1980) Which policy run is best, and who says so. TIMS Stud Manag Stud 14:241–257

Haefner JW (1996) Modeling biological systems: principles and applications. Kluwer Academic Publishers, New York, 473 pp
Hsiao N (1998) Conflict analysis of public policy stakeholders combining judgment analysis and system dynamics modeling. In: Proceeding of the 16th international conference of the System Dynamics Society
Maani KE, Cavana RY (2000) Systems thinking and modelling: understanding change and complexity. Prentice Hall, Auckland
Manetsch TJ, Park GL (1982) Systems analysis and simulation with applications to economic and social systems. Department of Electrical Engineering and System Science, Michigan State University, USA
Martinez-Moyano IJ, Richardson IJG (2013) Best practices in system dynamics modelling. Syst Dyn Rev 29(2):102–123
MPOB (Malaysian Palm Oil Board) (2012) Prices of palm oil products. Economics and Industry Development Division, Malaysian Palm Board, Kuala Lumpur
Rouwette EAJA, Korzilius H, Vennix JAM, Jacobs E (2011) Modeling as persuasion: the impact of group model building on attitudes and behavior. Syst Dyn Rev 27(1):1–21
Sterman JD (2000) Business dynamics: systems thinking and modeling for a complex world. McGraw-Hill Higher Education, Boston
Tàbara D, Pahl-Wostl C (2007) Sustainability learning in natural resource use and management. Ecological Society, 12, 3. Available online: http://www.ecologyandsociety.org/vol12/iss2/art12/. Accessed on 17 Apr 2010
Tietjen B, Huth A (2006) Modelling dynamics of managed tropical rainforests – an aggregated approach. Ecol Model 199:421–432
Vennix J (1996) Group model building. Irwin McGraw-Hill, New York
Vennix JAM (1999) Group model-building: tackling messy problems. Syst Dyn Rev 15 (4):379–402
Voinov A, Gaddis EJB (2008) Lessons for successful participatory watershed modeling: a perspective from modeling practitioners. Ecol Model 216:197–207

Bibliography

Bala BK (1999) Principles of system dynamics. Agrotech Publishing Academy, Udaipur
Cavana RY, Maani KE (2000) A methodological framework for systems thinking and modelling (ST&M) interventions. Ist international conference on systems thinking in management. Internet
Mohapatra PKJ, Mandal P, Bora MC (1994) Introduction to system dynamics modelling. Universities Press, Hyderabad
Yadama GN, Hovmand PS, Chalise, N (2010) Community driven modeling of social-ecological systems: lessons from Andhra Pradesh, India. In: Proceedings of the 28th international conference of the System Dynamic Society. Seoul, South Korea

Causal Loop Diagrams 3

The concept of causal loop diagram was introduced in Chap. 2, and this chapter presents concepts, methodology and techniques of causal loop diagrams. The cause–effect relationships and reinforcing and balancing loops are highlighted with examples. Steps to construct causal loop diagrams are provided. A good number of worked out examples are included to illustrate the techniques of constructing causal loop diagrams for dynamic systems.

3.1 Introduction

System dynamics methodology is based on feedback concepts of control theory, and causal loop is a convenient way to represent the feedback loop structure of systems. Causal loop diagram is used to represent the feedback loop systems diagrammatically, and it is a communication tool of feedback structure representing the principal feedback loops of the systems which generate the reference dynamic behaviour of the systems.

3.2 Causal Loop Diagrams

Causal loop diagrams identify the principal feedback loops of the systems. The causal loop diagrams are used to describe basic causal mechanisms hypothesised to generate the reference mode of behaviour of the system over time. A feedback loop contains two or more casualty related variables that close back on themselves. The relationship between one variable and next in the loop can be either positive or negative. A positive relationship means that if one variable increases, the other also increases. For example, in Fig. 3.1 the arrow from A to B means that an increase in A causes an increase in B. It can also mean if A decreases, B will also decrease. The arrow starting from A and terminating at B with a (+) sign at the end of the arrow means the cause–effect relationship is positive. In a negative relationship, the two

B.K. Bala et al., *System Dynamics*, Springer Texts in Business and Economics,
DOI 10.1007/978-981-10-2045-2_3

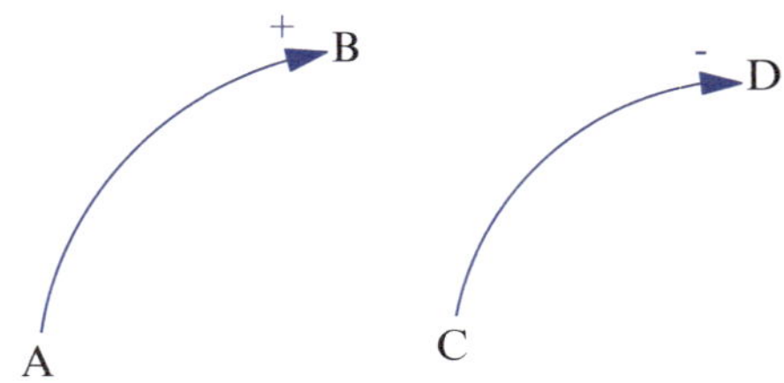

Fig. 3.1 Cause and effect relationships

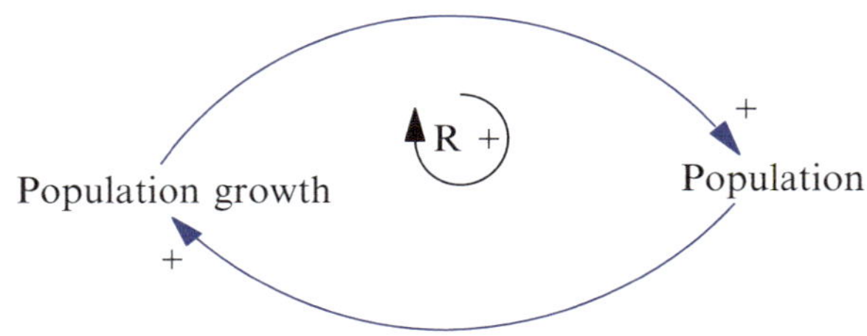

Fig. 3.2 Positive feedback loop

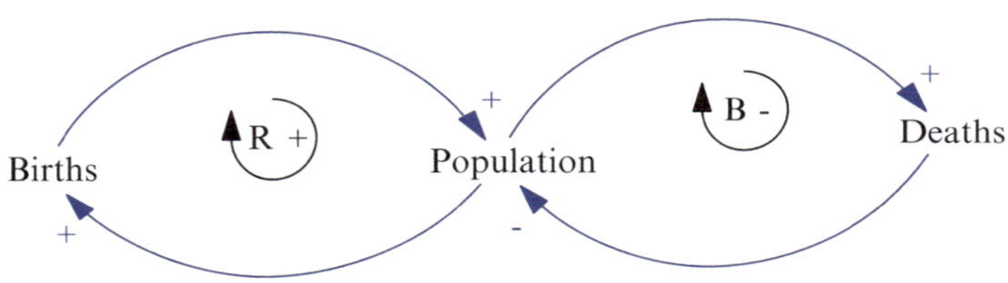

Fig. 3.3 Positive and negative feedback loops

variables change inversely. For example, in Fig. 3.1 the arrow in the direction of C to D means that if C increases, D will decrease. It can also mean if C decreases, D will also increase. The arrow starting from C and terminating at D with a (−) sign at the end of the arrow means the cause–effect relationship is negative.

Figure 3.2 shows an example of causal loop diagram with notation. This simple causal loop diagram of population consists of two variables, population growth and population. This figure illustrates a simple positive feedback loop consisting of two cause–effect relationships. In this example an increase in population will cause an increase in population growth. The cause–effect relationship is positive, and it is indicated by an arrow with a (+) sign starting at population and terminating at population growth. The cause–effect relationship between population growth and population is also positive. An increase in the population growth will cause an increase in population. This is indicated by an arrow with (+) sign staring at population growth and terminating at population. The loop formed from population to population growth and back to population is reinforcing loop and it is indicated by a (+) sign with an arrow inside the causal loop diagram. Figure 3.3 illustrates two feedback loops. In loop 1 the number of births increases with the population, and the births in turn increases population. These are positive relationships and

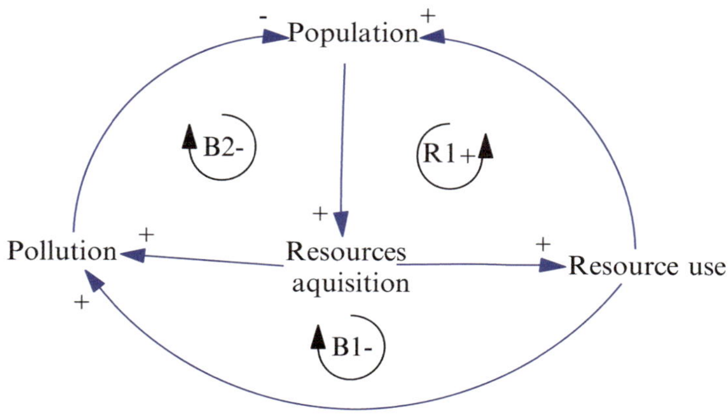

Fig. 3.4 Coupled feedback loops

represented by arrows with a (+) sign. In loop 1 there are two positive relationships. Hence, it is a positive and reinforcing loop. In loop 2 the number of deaths increases with population and the population decreases as the death increases. In loop 2 the first relationship is positive, and it is represented by an arrow (+) sign, while the second relationship is negative, and it is represented by an arrow with (−) sign. One can easily determine if a loop is positive or negative by counting the number of negative relationships in a loop. If there are an even numbers of negative relationships in total in a feedback loop, then the loop is positive; if there are odd numbers of negative relationships, the loop is negative. In fact, positive feedback loops generate growth, i.e. the loop is reinforcing and negative feedback loops are goal seeking. In loop 2 number of negative relationships is odd, i.e. 1. Hence, this loop is negative and goal seeking.

Let us now consider a coupled feedback loop of population, resource use and pollution as shown in Fig. 3.4. Loop at the right corner is positive, i.e. reinforcing, while the loops at the bottom and left corner are negative, i.e. balancing loop. In the positive loop, all the cause–effect relationships are positive, while in negative feedback loops, the number of (−) relationships is 1, i.e. an odd number.

3.3 Steps in Causal Loop Diagram

We must consider the description of the system and dynamic behaviour of the reference modes to construct the causal loop diagram, and these can aid in developing dynamic hypothesis. The following steps are to be followed for developing causal loop diagrams.

1. *Define the problem and the objectives.*

We must first of all study the system based on information collected through interview, focus group discussion, research report and case study. We must describe the system and define the problem with the reference mode of the behaviour of the system.

2. *Identify the most important elements of the systems.*

We should identify the key variables affecting the behaviour of the system, and it should be a good starting point to develop the causal loop diagram. Other variables can be added during later stages of causal loop development.

3. *Identify the secondary important elements of the systems.*

Secondary variables within the system boundary should be added after careful identification of the most important variables. This would provide an opportunity to consider the secondary variables of the system of importance in the causal loop diagram.

4. *Identify the tertiary important elements of the systems.*

Tertiary variables within the system boundary should be added after careful identification of the secondary variables. However, tertiary variables of little importance can be omitted in the later stages once it is established by simulated studies.

5. *Define the cause–effect relationships.*

Find the cause–effect relationships using arrows with polarity for the primary variables first, then for the secondary and tertiary variables.

6. *Identify the closed loops.*

Trace closed loops formed by cause–effect relationships for the variables describing the system.

7. *Identify the balancing and reinforcing loops.*

Identify the number of negative cause–effect relationships in each of the closed loops. The closed loops with odd number of negative relationships are negative, i.e. balancing loops, and the others are positive, i.e. reinforcing loops.

3.4 Examples

3.4.1 Population

Population has been growing exponentially throughout the history. Population increases by a fixed percentage per year, and also it decreases by a fixed percentage per year because of the fact that the human has a limited life. Draw the causal loop diagram of the simple population model.

Solution

Three key variables are population, birth rate and death rate. Birth rate increases with population, and also birth rate adds to population. These two are positive cause–effect relationships. Death rate increases with the increase in population, and it is a positive cause–effect relationship, but population decreases with the increase in death rate, and it is a negative cause–effect relationship. Hence, the model consists of two fundamental loops. The regenerating loop R generates new birth and adds it to population. The balancing loop B creates the death and depletes it from the population. The birth creates a positive loop since all the cause–effect relationships within this loop are positive, while there is an odd number of negative relationships within the balancing loop, and hence it a negative feedback loop. Figure 3.5 shows causal loop diagrams of a simple population model.

3.4.2 Carbon Metabolism in Green Plant

Carbon enters the plant by diffusion as carbon dioxide and is fixed. A labile pool of carbohydrate is produced. The entry and fixation rate is controlled by photosynthetic tissue represented chiefly by the leaves. The carbohydrate from the pool is allocated to the photosynthetic and non-photosynthetic parts of the plant body. Net loss of plant weight from the photosynthetic and non-photosynthetic parts is due to respiration. Draw the causal loop diagram for the description of carbon metabolism in green plant.

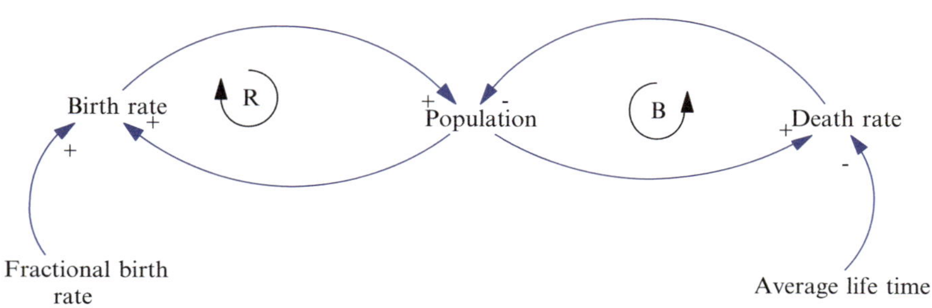

Fig. 3.5 Causal loop diagram of a simple population model

Solution

The carbon enters the plant in the form of carbon dioxide, and the rate of photosynthesis depends on solar radiation, temperature, weight of photosynthetic organs and photosynthetic area. The rate of photosynthesis increases with the increase of the weight of photosynthetic organs. The carbohydrate produced increases with the increase of photosynthesis rate and is translocated to the photosynthetic and non-photosynthetic organs. The carbohydrate pool, weight of photosynthetic organs and photosynthesis rate form a positive feedback loop (R1).

Again, respiration rate increases with carbohydrate in photosynthetic organs. The weight of the organs is decreased by respiration. The photosynthetic organs and respiration rate form a negative feedback loop (B1). Similarly the weight of non-photosynthetic organ and respiration form another negative feedback loop (B2). The resulting causal loop diagram of the carbon metabolism in green plant is shown in Fig. 3.6.

3.4.3 Food Security

Food security is a worldwide problem that has called the attention to governments and the scientific community. It particularly affects developing countries. The scientific community has had increasing concerns for strategic understanding and implementation of food security policies in developing countries, especially since the food crisis in the 1970s and 2009. The process of decision-making is becoming increasingly complex due to the interaction of multiple dimensions related to food

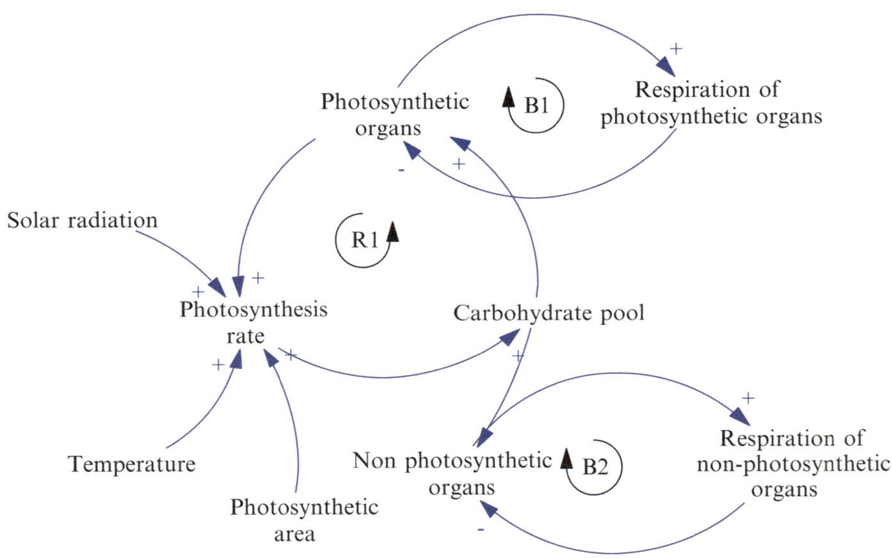

Fig. 3.6 Causal loop diagram of carbon metabolism in green plant

security (Giraldo et al. 2008). There is a need for models to examine the dynamics of this complex food security issues and design policies for sustainable development.

Food security is a situation in which people do not live in hunger or fear of starvation. FAO (1996a, b) defined the objective of food security as assuring to all human beings the physical and economic access to the basic food they need. This implies three different aspects: availability, affordability and access. Food security exists when all people at all times have access to sufficient, safe and nutritional food to meet their dietary needs and food preferences for an active and healthy life (FAO 2002). Draw the causal loop diagram of the food security model.

Solution

Food production depends on area cultivated and yield of the crop. Food production increases food availability. And also food production increases per capita income and hence food affordability. Food availability and food affordability would enhance the food access. Food security would depend on food availability, food affordability and food access.

As mentioned earlier the increased food production increases food availability, per capita food availability and hence food self-sufficiency. The increased food self-sufficiency would reduce the food price. The higher price would motivate the farmers to produce more food. This forms the negative feedback loop B1. Again, increased food self-sufficiency would reduce the deaths, and the increased deaths would reduce the population level. The increased population would reduce per capita food availability, and the increased per capita food availability would increase the food self-sufficiency. This forms the negative loop B2. Population forms one positive feedback loop R1 resulting from births and one negative feedback loop B3 resulting from deaths. The resulting causal loop diagram of the food security model is shown in Fig. 3.7.

3.4.4 Price Determination of a Commodity

Agricultural commodities are sold in highly competitive markets whose prices change yearly, daily, hourly and even by the minute. Farmers are affected by fluctuations of price—good times bring him an increase in income, and depression causes his cash income to drop away to very little. The price received by the farmers is important not only to him; the rate of production depends on it to a large extent. In addition, pigs, and to some extent cattle and chickens show fluctuations in price and supply of a persistent nature at a frequency consistent over long periods of time (Meadows 1970).

Most of the poorer nations are dependent upon the export of primary agricultural commodities as a source of foreign exchange, and the pronounced fluctuations in

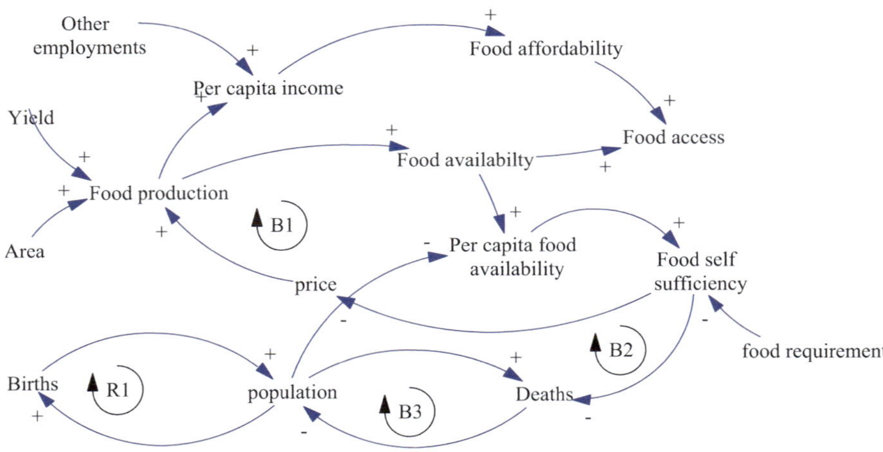

Fig. 3.7 Causal loop diagram of food security

price and supply have aroused much concern. Inflation, disruption of development programmes, loss of investor's confidence and political unrest are only a few of the results of the fluctuations in price and supply. Even developed countries find their economies affected.

Price determination is one of the most difficult challenges in economic modelling. Nerlove (1958) has given an excellent mathematical interpretation, and according to Nerlove farmers revise their previous expectations of normal price in each period in proportion to the difference between actual and what was previously considered normal. Sterman (2000) further modified this model, and according to Sterman the traders revise their expectations based on expectation adjustment and price adjustment. Draw the causal loop diagram of price determination based on the modifications suggested by Sterman.

Solution

The causal loop diagram of price determination of an agricultural commodity is shown in Fig. 3.8. The trader price is settled by price adjustment and expectation adjustment, and these form one positive feedback loop R1 and one negative feedback loop B1. The price depends on relative inventory coverage and trader price and the increase in price would reduce the consumption and thereby forms the negative feedback loop B2. On the production side, price, profitability and production also form a negative feedback loop B3. Also inputs, yield and production form a negative feedback loop B4. R & D, yield and production rate form a positive feedback loop (R2). The increase in population would increase the consumption rate and the population consists of one positive feedback loop R3 and one negative feedback loop B5.

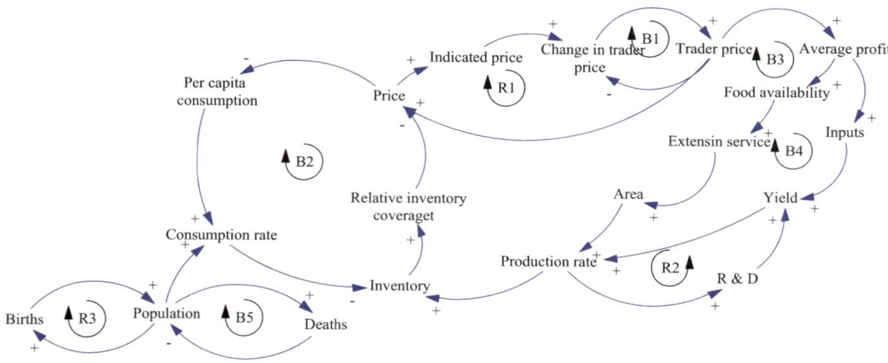

Fig. 3.8 Causal loop diagram of price determination of a commodity

3.4.5 Fishery Dynamics

Many fish stocks are in decline worldwide. Overfishing has led to the complete collapse of numerous fisheries including those for important and well-publicised species. In addition, over-exploitation of fisheries has led to reductions in biodiversity and modified ecosystem functioning. Despite management attempts to reduce overfishing, little progress has been made due to a general inability to endure the short-term economic and social costs of reduced fishing. Indeed, historical analysis reveals that resources are consistently and inevitably over-exploited. There is a need for models to examine the dynamics of this complex fishery issues and design policies for sustainable fishery development (Dudley 2008).

The model developed here is based on the biomass dynamic model (Graham 1935; Schaefer 1954, 1957), in which the population biomass increased due to both growth and the addition of new fish and decreases due to natural mortality and catch. Growth is equal to biomass fractional growth times existing biomass, and the natural decrease is growth multiplied by the ratio of biomass population to maximum biomass population possible. Catch depends on the current level of population biomass, number of gear units and fishing gear efficiency. The fleet entering or leaving depends on catch per unit effort and vessel replacement. In some fisheries, fishing activity decreases ecosystem carrying capacity. Draw the causal loop diagram of the fishery management system.

Solution

The causal loop diagram of fishery dynamics is shown in Fig. 3.9. Fish biomass increases by regeneration and decreases by catch and deaths. The regeneration forms a positive feedback loop (R1) with fish biomass, and the catch and deaths form negative feedback loops (B1 and B2) with fish biomass. Effect of relative fish biomass and deaths has a negative effect on fish biomass (B3). Fish catch, CPUE, fishing unit entry and units of fishing gear form a positive feedback loop (R2). Units

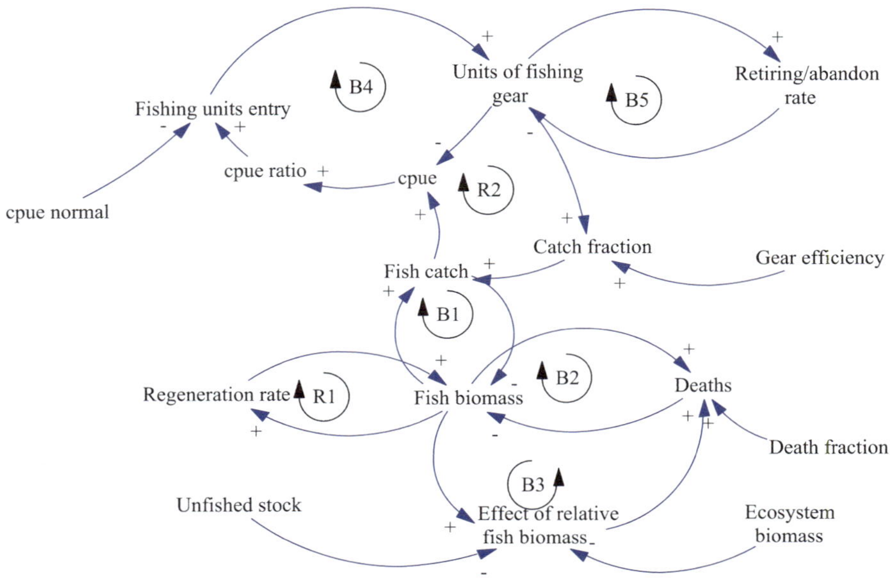

Fig. 3.9 Causal loop diagram of fishery dynamics

of fishing gear form two negative feedback loops with fishing unit entry and retiring/abandon rate (B4 and B5).

3.4.6 Forest Dynamics

Industrialisation, population growth and depletion of natural resources are threatening ecosystems all over the world. Climate change is one of the results of this development. But forests can play an important role in maintaining the global system in balance, and these forests are also the largest carbon sink above the soil. Furthermore, about 70–90 % of the global biodiversity are found in this ecosystem.

Over the last few decades, due to the increasing environmental awareness, the importance of the forest ecosystem has been acknowledged at the global level, which in turn enabled it to become a topic of relevance to the general public. Sustainable forest management, that is, the achievement of long-term economic, social and environmental goals has thus become the basic forestry principle. Forest management is a problem of increasing controversy and difficulty (Rosser Jr 2003).

Forest provides timber and fuel wood. Forest is also a renewable resource and is the largest carbon sink on the earth. Plantation of forest is motivated by profitability and demand for fuel wood and timber, and the harvesting of the mature forest for timbers is triggered by the demand. The saw mills demand logs from harvesting of

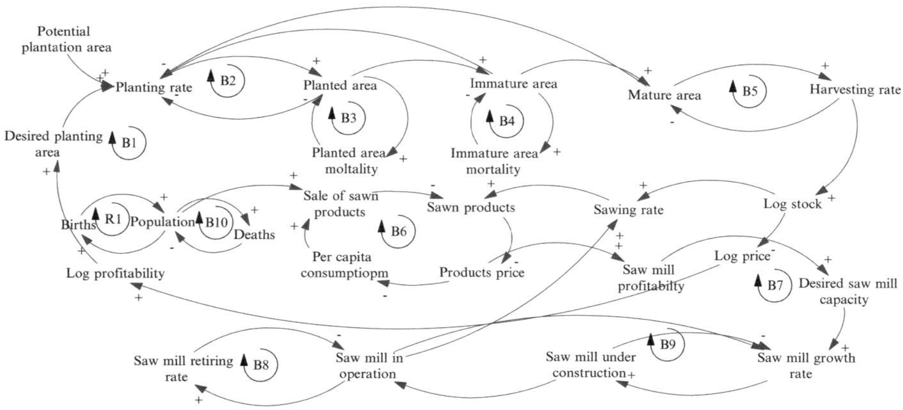

Fig. 3.10 Causal loop diagram of forest dynamics

timbers, and the expansion of the saw mills depends on profitability. Draw a causal loop diagram of the forest dynamics and management.

Solution

The causal loop diagram of forest dynamics is shown in Fig. 3.10. Planting rate increases planting and mature area and higher harvesting rate. The higher log stock would cause to lower the log price and the higher price resulting higher profitability would motivate to increase the planting area. This forms a negative feedback loop B1. The planting area forms negative feedback loop with planting rate (B2). Planted area and immature area are decreased by mortality, and the mature area is decreased by harvesting rate (B3, B4 and B5). Sawn products, products price and sales of sawn products form a negative feedback loop (B6). Increase in saw mill under construction would ultimately increase the saw mill under operation which results in higher sawing rate. This in turn would reduce product price, and the higher product price would result in higher profitability and saw mill growth rate. This forms a negative feedback loop B7. Saw mill under construction with saw mill growth rate and saw under operation with retiring rating rate balance the saw mill under construction and saw mill under operation, respectively (B8 and B9). Sales of sawn products increase with population and population comprises of one reinforcing loop (R1) and one balancing loop (B10).

3.4.7 Electricity Supply

Energy is needed for economic and social development (Bala 1997a, b, 1998). Per capita consumption of energy is a measure of physical quality of life (Bala 1998). Also per capita consumption of electrical energy is a measure of physical quality of life. Energy demands are increasing rapidly. Investments for power plants are triggered by the power supply gap, price and public pressure. Indicated price

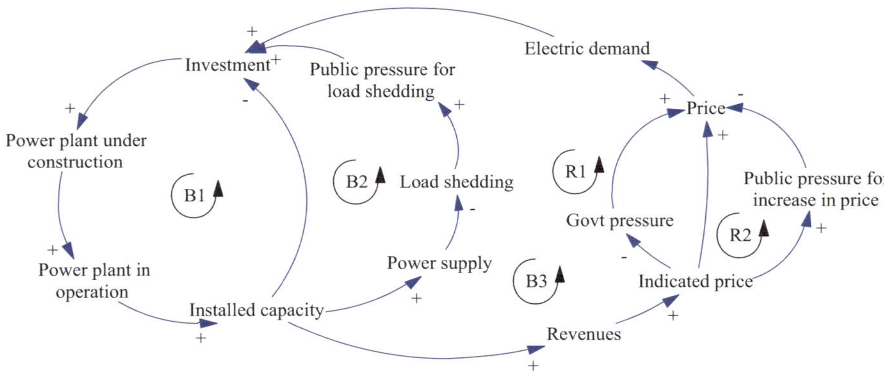

Fig. 3.11 Causal loop diagram of electricity supply system

depends on revenue received which creates govt pressure to increase the price as the indicated price increases while the public pressure for increase in price creates pressure to decrease the price. A power plant under construction becomes operational after some time delay and adds to the installed capacity to supply electricity and hence reduces the pressure on new investments. Draw the causal loop diagram of the simple electric supply system.

Solution
The causal loop diagram of the simple electricity supply system is shown in Fig. 3.11. Investment increases the installed capacity which in turn would invite less investment. This forms a negative feedback loop B1. The increase in installed capacity would increase power supply, and the increased power supply would reduce the load shedding. The load shedding increases public pressure which would increase the investment. The increased investment would result in increased installed capacity. This forms the negative feedback loop B2. Revenues, indicated price, government pressure, public pressure for increase in price, electric demand, investment and installed capacity form two positive feedback loops, and these are indicated R1 and R2.

3.4.8 Global Warming

Global warming refers to an increase in the average temperature of the atmospheric temperature resulting from greenhouse effect, and it is a prime concern in many developing countries. Earth's surface average temperature is increased by emissions of greenhouse gases and incoming solar radiation. The emissions are reduced by carbon sink of plant kingdom and algae in the sea. The water vapour increases the absorbed radiation. Draw the causal loop diagram of this simple global warming model.

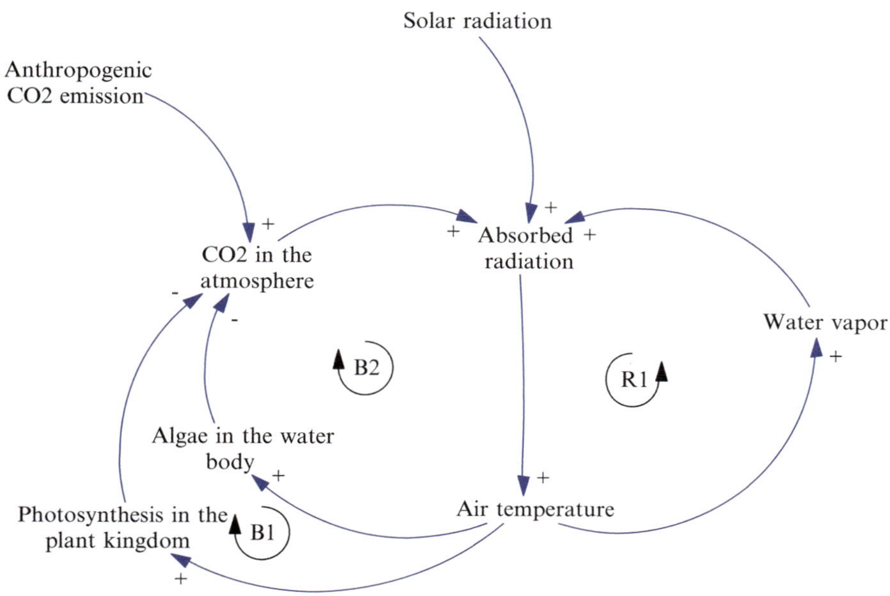

Fig. 3.12 Causal loop diagram of global warming

Solution

The causal loop diagram of the simple global warming model system is shown in Fig. 3.12. Absorbed radiation is increased by solar radiation and CO_2 in the atmosphere and water vapour. The increase in air temperature causes increase in photosynthesis in the plant kingdom and algae in the water body which decrease the CO_2 in the atmosphere (B1 and B2). Air temperature increases water vapour which increases the absorbed radiation. This forms a positive feedback loop (R1). Thus, there are one reinforcing loop and two distinct balancing feedback loops B1 and B2.

Exercises

Exercise 3.1 Rural and urban population

Population is divided into rural and urban. Population increases by birth rate and decreases by death rate. Both the rates are considered to be influenced by the availability of food, crowding and earning. Draw the causal loop diagram for the description of rural and urban population.

Exercise 3.2 Fish pond ecosystem

Fish pond ecosystem can be divided into two parts known as the biotic (living) and abiotic (nonliving) components. The abiotic substances are the water, the nutrients, oxygen, carbon dioxide, etc. The producers in the ecosystems are the large rated plants and the free-floating plants, usually algae, called phytoplankton. These store energy and liberate oxygen. The primary consumers are benthos or bottom form and zooplankton with little or no swimming ability. The phytoplanktons are consumed by zooplankton, which are in turn eaten by large aquatic life such as fish. Other consumers are the insects, frogs, man, etc., and a category called detritivores which live on the organic wastes. All of these forms produce organic waste and dead organisms. The decomposers, bacteria and fungi, utilise the organic carbon and generate CO_2 which in turn is used by the algae. Additional CO_2 is provided from the atmosphere and through the respiration of fish. In a healthy system, the availability of the nutrients, carbon, phosphorus and nitrogen is sufficiently small so as to limit the production of algae. Draw the causal loop diagram for fish pond ecosystem.

Exercise 3.3 Prawn production system
The prawn production system consists of two components: biological component and economic component. The biological component is a dynamic population model. Population of either sex at any time is determined from death rate, population level and harvest size. The number of deaths increases with an increase in population, and again the increase in number of deaths decreases the population. We can divide the population by length and size. Draw the causal loop diagram for prawn production system.

Exercise 3.4 Price forecasting of palm oil
Palm oil production is a highly complex system starting from plantation to export of palm oil. Our starting hypothesis about the market of palm oil should be something to do with the balance of supply and demand as in case with other agricultural commodities. The change in the palm oil price is a result of two major adjustments; change in palm oil price due to expectation adjustment and price adjustment. Short-term pressures arising from imbalance of supply and demand, changes in costs or competitors' prices will cause the traders to bid prices up or down relative to their expectation about equilibrium price. Other factors are such as substitutes, trade liberalisation and export restriction. Our initial dynamic hypothesis of palm oil market structure is based on standard assumptions how commodity markets typically work. If our hypothesis is correct, the model will be able to reproduce the general historical pattern followed by the palm oil price. Draw the causal loop diagram of price forecasting model of palm oil.

Exercise 3.5 Biofuel Promotion
Biofuel inventory and food inventory are key factors which regulate the prices and the profitability for farmers. The inventories act as buffers between production and supply and absorb variations on both sides. An increase in biofuel demand depletes biofuel inventory and causes biofuel price to increase and subsequently biofuel crop land to increase. This increase in land is at the expense of food crop land. The need for cultivable land is reinforced by the simultaneous increase in population. Government policies are represented through the variables such as incentives for biofuels and biofuel technology and management capabilities. The former is a

direct policy instrument, whereas the latter is an indirect one resulting in the funding of related R&D projects. These two instruments are competing for resources and have contrasting effects, as incentives increase land use, whereas technology increases the yield of the existing land. Draw causal the loop diagram.

References

Bala BK (1997a) Computer modelling of the rural energy system and of CO_2 emissions for Bangladesh. Energy 22: 999–1003

Bala BK (1997b) Computer modelling of energy and environment: the case of Bangladesh. In: Proceedings of 15th international system dynamics conference, Istanbul, Turkey, August 19–22, 1997

Bala BK (1998) Energy and environment: modelling and simulation. Nova, New York

Dudley RG (2008) A basis for understanding fishery management dynamics. Syst Dyn Rev 24 (10):1–29

FAO (Food and Agricultural Organization) (1996a) Implications of economic policy for food security: a training manual, Rome, 1996

FAO (Food and Agricultural Organization) (1996b) Technical background document prepared for the World Food Summit, Rome, 1996

FAO (2002) The state of food insecurity in the world: eradicating world hunger—taking stock ten years after the World Food Summit. Rome, 2002

Giraldo DP, Betancur MJ, Arango S (2008) Food security in developing countries: a systemic perspective. Paper presented at the international conference of the system dynamics society held on July 20–24 2008 at Athens, Greece, 2008

Graham M (1935) Modem theory of exploiting a fishery, and application to North Sea trawling. Journal du Conseil International Pour l'Exploration de la Mer 10(3):264–274

Meadows DL (1970) Dynamics of commodity production cycles. Wright Allen Press Inc., Cambridge, MA

Nerlove N (1958) Adaptive expectations and cobweb phenomena. Q J Econ 72:227–240

Rosser B Jr (2003) Complexities of dynamic forestry management policies. James Madison University Harrisonburg, USA, 2003

Schaefer MB (1954) Some aspects of the dynamics of populations important to the management of commercial marine fisheries. Inter-Am Trop Tuna Comm Bull 1:27–56

Schaefer MB (1957) A study of the dynamics of the fishery for yellow fin tuna in the eastern tropical Pacific Ocean. Inter-Am Trop Tuna Comm Bull 2:245–285

Sterman JD (2000) Business dynamics: systems thinking and modeling for a complex world. Irwin McGraw Hill, New York

Bibliography

Bala BK (1999) Principles of system dynamics, 1st edn. Agrotech Publishing Academy, Udaipur

Forrester JW (1968) Principles of systems. MIT Press, Cambridge, MA

Maani KE, Cavana RY (2000) Systems thinking and modelling: understanding change and complexity. Prentice Hall, Auckland

Mohapatra PKJ, Mandal P, Bora MC (1994) Introduction to system dynamics modelling. Universities Press, Hyderabad

Stock and Flow Diagram

4

The concept of stock–flow diagram was introduced in Chap. 2, and this chapter presents concepts, methodology and techniques of constructing stock–flow diagrams. The basic building blocks of the system structure: stock and flow are explained. Delays and graphical functions are highlighted. Function with and without integrations are presented. A good number of examples are included to demonstrate the techniques of constructing the system structures in terms of stock–flow diagrams and simulation using the software STELLA.

4.1 Introduction

One approach to study a system is verbal description, and another approach is causal loop diagram. Causal loops are wonderful to tell the story of the dynamic behaviour of complex and dynamic systems to others and represent the verbal and mental model more clearly using cause and effect relationships. But to show the relationship and to accentuate the loop structure of a system, the flow diagram is the best since causal loops suffer from their inability to capture the stock and flow of the systems. Essentially stock and flow represent systems of first-order difference equation central to the system dynamics theory.

Feedback loop is the basic structural element within the system boundary. The boundary separates the system from the surrounding environment. Dynamic behaviour is simulated by the feedback loops. The feedback loop is a path coupling decision, action, level (or condition) and information with the path returning to the decision point (Fig. 4.1). The decision process is one that controls action. It can be the governing process in biological development, or it may be the valve and actuator in a chemical plant. Whatever may be the decision process, it is embedded in a feedback loop. The decision is based on the available information; the decision controls an action that influences a system level and new information arises to modify the decision stream. A system may be a single loop or interconnected loops.

B.K. Bala et al., *System Dynamics*, Springer Texts in Business and Economics,
DOI 10.1007/978-981-10-2045-2_4

Fig. 4.1 Feedback loop

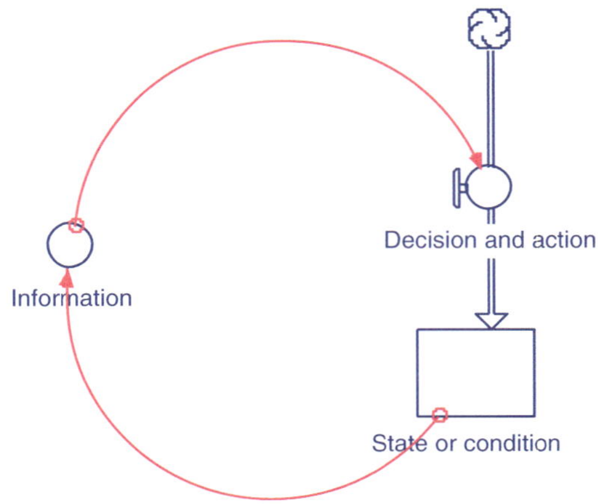

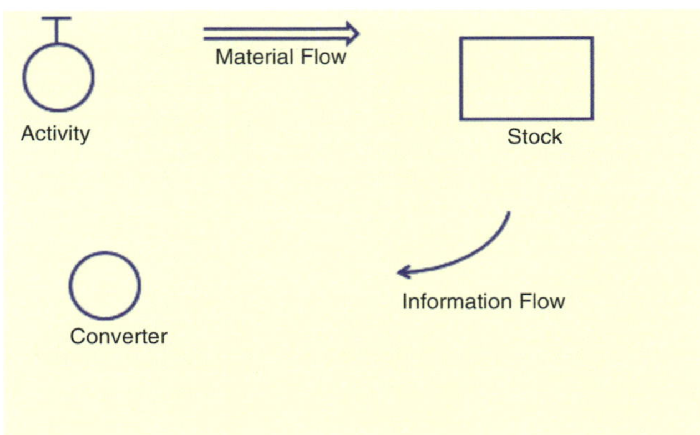

Fig. 4.2 Symbols of STELLA flow diagrams

The stocks and flows are two basic building blocks in system dynamics modelling. The equations of a system focus on the composition of each stock and flows. The stock–flow diagram should show how stocks and flows are interconnected to produce the feedback loops and how the feedback loops interlink to create the system. The symbols used in STELLA stock–flow diagrams are shown in Fig. 4.2. Stock–flow diagram with reference to STELLA stock–flow diagrams are described in the following subsections.

4.2 Stock

The first basic building block in system dynamics modelling is the stock, and this variable describes the condition or state of the system at any particular time. These variables are accumulations in the system. Examples of these variables are

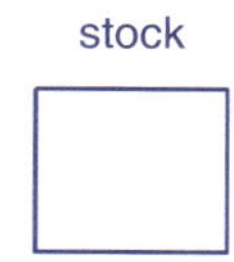

Fig. 4.3 Symbol of stock

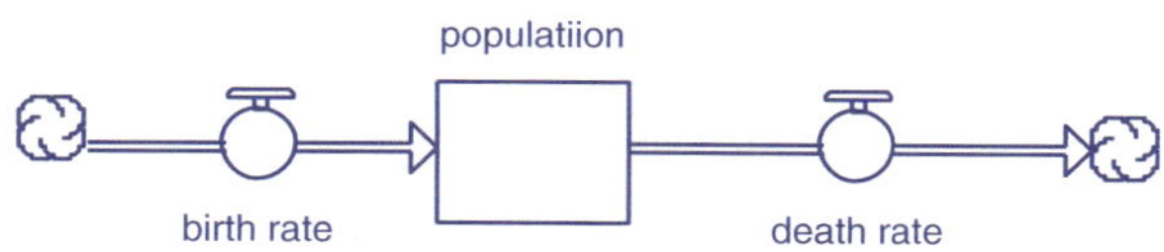

Fig. 4.4 Flow diagram of a stock of population

population or water in a reservoir. These are physical accumulations, while the examples of non-physical accumulations are public pressure, skill and knowledge.

All stock equations and any special function that represents integration are represented by a rectangle (Fig. 4.3). A simple single stock model of population in STELLA is shown in Fig. 4.4. Flow into the system is birth rate, and flow out of the system is death rate. The accumulation of the population is the stock of the model.

Stocks integrate flows into and out of the stock, and the net flow into the stock is the rate of change of the stock. The stock which is represented by a first-order integral equation is

Stock

$$\text{population}(t) = \text{population}(t - dt) + (\text{birth\_rate} \;-\; \text{death\_rate})^* dt \qquad (4.1)$$

Equation (4.1) is in the form of first-order finite difference equation. The simple population model shows that population increases by birth rate and decreases by death rate.

4.3 Flow

The second building block is the flow, and these variables tell how fast the stocks are changing. The flows are represented by valve symbols (Fig. 4.5). The examples of these variables are inflow (birth rate) and outflow (death rate to the stock of population), and these are shown in Fig. 4.6. The birth rate depends on the population size and birth fraction, while death rate depends on the population size and death fraction.

The inflow and outflow equations are

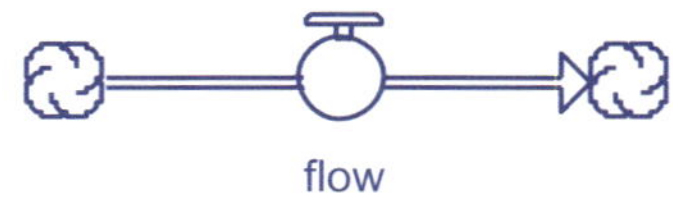

Fig. 4.5 Symbol of flow

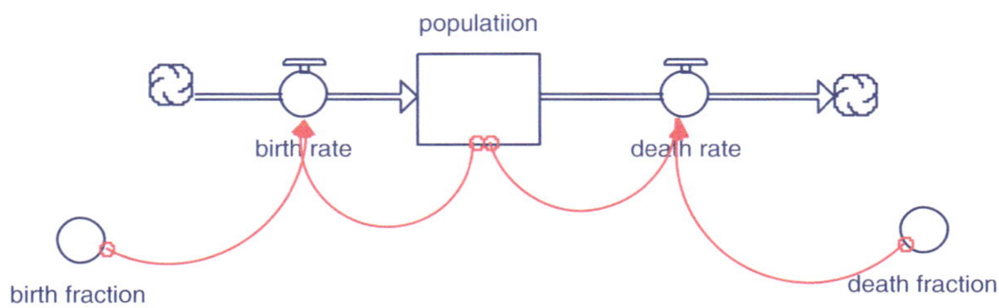

Fig. 4.6 Flow diagram showing flows to stock of population

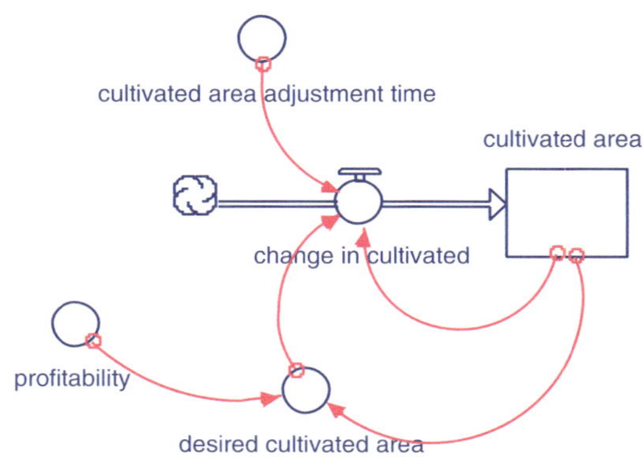

Fig. 4.7 Flow diagram illustrating converter

Inflow

$$\text{birth rate} = \text{population} * \text{birth fraction} \tag{4.2}$$

Outflow

$$\text{death rate} = \text{population} * \text{death fraction} \tag{4.3}$$

4.4 Converter

In STELLA the converter serves utilitarian role in the software. It holds the values for constants, defines external inputs to the model, calculates algebraic relationships and serves as the repository for graphical functions. In general, it converts inputs into outputs. Hence, the name is ‘converter’. In Fig. 4.7 desired cultivated area, profitability and cultivated area adjustment time are converters.

4.5 Delays

It's the time between the action and the result (consequence) of this action. Delay is a commonly observed phenomenon in material and information flow, and it takes time for the material to flow and information to process. Time delay is defined as the time required for the flow of material or information, and more precisely it is the time by which the output lags behind the input. Thus, the delay may be material delay or information delay. The material delay causes the delay in the supply of the material, and the information delay is the delay in processing the information.

In pipeline delay the delay time is constant, and the order of output is precisely same as the order of entry. The delay may be first order, second order, third order and even higher order. In first-order delay, the outflow is always proportional to the stock of the material in transit. The second-order delay consists of two first-order delays in series, with each of time delay of half of the original time delay, while a third-order delay consists of three first-order delays in series with each of the time delay of one third of the original time delay. Also higher-order delays approach the pipeline delay. Figure 4.8 shows the responses of different types of delay functions.

Delays

```
first_order_delay=SMTH1(input,2,0)
input=STEP(100,2)
pipeline_delay=DELAY(input,2,0)
third_order_delay=SMTH3(input,2,0)
```

4.5.1 Role of Delay

Delay plays an important role in information and material flow and hence the behavioural performance of the system and the key points which deserve special attentions are:

(a) Acknowledge delay as a *factor* in decision-making.
(b) Respect delay as an *element* to understand success or failure.
(c) Regard delay as a *force* in determining value of change.

4.5.2 Choice of Delay Function

The choice of delay function is quite important when inputs of sharp variations such as STEP function are used. The difference in the response of third- and higher-order delays is not quite large. For practical purposes three main possibilities are

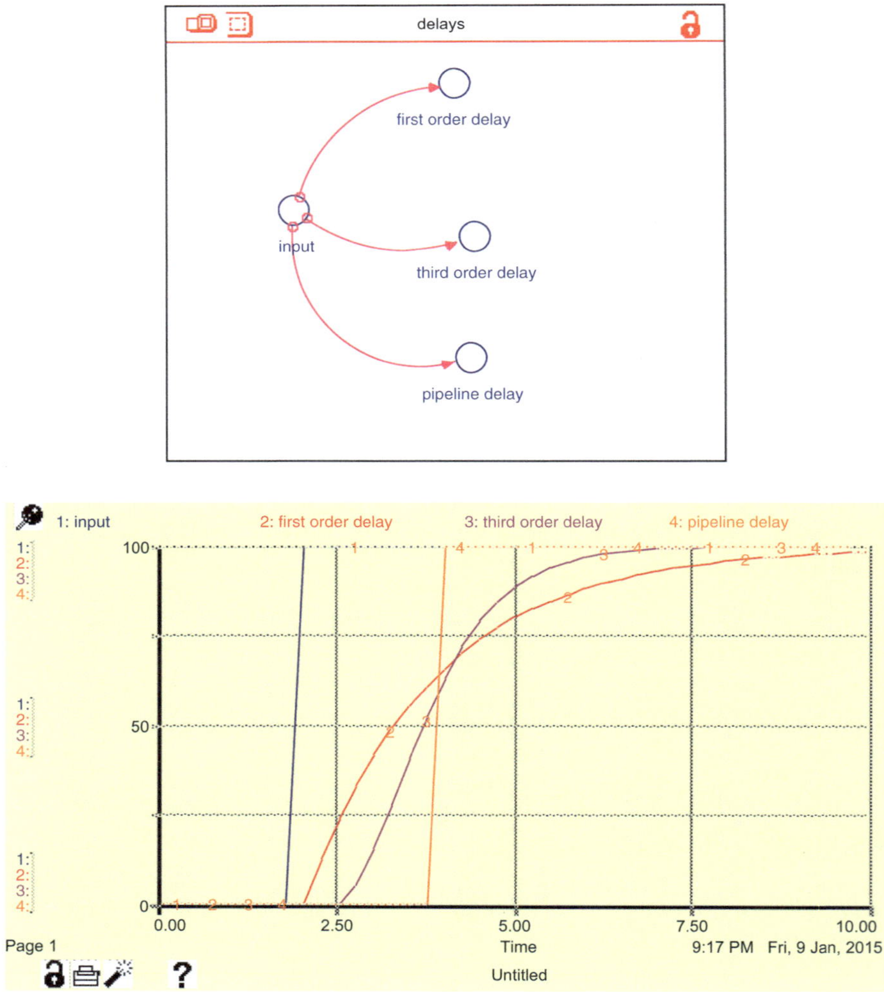

Fig. 4.8 STEP response of first-, third-order and pipeline delay

first-order delay, third-order delay and pure or pipeline delay. When the input has an immediate effect and then long tail, first-order delay is recommended. But when the output is the exact copy of the input, pipeline delay is appropriate. In intermediate case where the input has a delayed response and a spread of inflow, third-order delay is recommended. Third-order delay is usually a good compromise well within accuracy of the commonly available data, and that is why it is commonly used in system dynamics modelling.

4.6 Identification of Stock and Flow

In control theory stocks are known as state variable and flows are known as derivatives, while in system dynamics, these are also known as levels and rates, respectively. In system dynamics modelling first of all, the variables within the system boundary must be identified. Then identify which variables are affected or influenced by other variable and the affected variables which represent state of the system, i.e. quantities of material or other materials are stocks. The others having the unit per unit time of the unit of the stock variables are flows. It should be remembered that stocks are quantity and flows are quantity per unit time. The stocks are only changed by flows and the contents of stocks, and flow networks are conserved. Also converters (auxiliaries) are elaboration of flows, and hence these are affected by stocks and constants. For example, in a simple population model, the variables are birth rate, death rate and population. The population has a unit of quantity, while birth rate and death rate have a unit of quantity per unit time. Also population level is changed by birth rate and death as inflow and outflow, respectively. Hence, the population is a stock variable, and birth rate and death rate are flow variables.

4.7 Mathematical Representation of Stock and Flow

Integration or accumulation creates the dynamic behaviour of the systems, and integration occurs naturally both in physical and biological systems. Stock equation can be represented by first-order finite difference equation, and it can be expressed as

$$\mathrm{stock}(t) = \mathrm{stock}(t_0) + \int_{t_0}^{t} (\mathrm{inflow}(t) - \mathrm{outflow}(t))dt \tag{4.4}$$

And this integration equation in the differential equation form is

$$\frac{d(\mathrm{stock})}{dt} = \mathrm{inflow}(t) - \mathrm{outflow}(t) \tag{4.5}$$

Equation (4.5) is an initial value problem and several methods are available to solve this problem. STELLA provides three methods of solution of this initial value problem. These methods are (i) Euler's method, (ii) second-order Runge–Kutta method and (iii) fourth-order Runge–Kutta method. The numerical solution procedure of these methods is summarised below:

(i) Euler's Method

Denote the first-order problem in Eq. (4.5) as

$$\frac{d(\text{stock})}{dt} = \frac{dy}{dt} = y' = f(t, y), \qquad y(t_0) = y_0 \tag{4.6}$$

The derivative of y at i, $f(t,y) = f_i$ can be evaluated using Eq. (4.6) as soon as y_i is known. Once f_i is known, the approximate integration formula gives

$$\Delta y_i = y_{i+1} - y_i = \int_{t_i}^{t_{i+1}} f(t, y)\, dt = (t_{i+1} - t_i) f_i \tag{4.7}$$

And solving for y_{i+1}

$$y_{i+1} = y_i + (t_{i+1} - t_i) f_i \tag{4.8}$$

Equation (4.8) is known as Euler's formula, and it has a truncation error of order h^2.

(ii) Second-Order Runge–Kutta Method

Given the equation

$$\frac{dy}{dt} = y' = f(t, y), \tag{4.9}$$

the Runge–Kutta forward integration formulae are derived by determining a point

$$\bar{t} = t_i + \alpha h, \quad \bar{y} = y_i + \beta h, \quad (h = t_{i+1} - t_i) \tag{4.10}$$

and the increment is computed by the one-term formula as given by Eq. (4.11):

$$\overline{\Delta} y_i = y_{i+1} - y_i = f(\bar{t}, \bar{y}) h \tag{4.11}$$

The increment in Eq. (4.11) is identical with the increment computed by means of a predetermined number of terms of the Taylor series expansion in Eq. (4.12):

$$\Delta y_i = y_i' \, h + y_i'' \frac{h^2}{2} + y''' \frac{h^3}{6} + \cdots\cdots \tag{4.12}$$

The second-order Runge–Kutta formula is obtained by setting increment in Eq. (4.11) equal to the first two terms of the Eq. (4.12), and it is given by

$$f(\bar{t}, \bar{y}) = f(t_i + \alpha h, y_i + \beta h) = y_i' + y_i'' \frac{h}{2} + \ldots\ldots\ldots\ldots \tag{4.13}$$

When $f(\bar{t}, \bar{y})$ in Eq. (4.13) is expanded into a double Taylor series about (t, y) truncated after the first derivate terms, it can expressed as

$$f(t_i + \alpha h, y_i + \beta h) = f_i + \alpha h f_{t,i} + \beta h f_{y,i} \tag{4.14}$$

By Eq. (4.9), the right-hand side of Eq. (4.13) becomes

$$y'_i + y''_i \frac{h}{2} = \left[f + \frac{h}{2} \frac{df}{dt} \right]_{t=t_i} \tag{4.15}$$

where $f(t,y)$ is to be considered a function of t only, through substitution of $y(t)$ for y. Hence, the derivative of $f(t, y)$ with respect to t can be expressed as

$$\frac{df(t,y)}{dt} = \frac{\delta f}{\delta t} + \frac{\delta f}{\delta y}\frac{dy}{dt} = f_t + f_y f, \tag{4.16}$$

Substituting Eq. (4.16) into Eq. (4.15) can expressed as

$$y'_i + y''_i \frac{h}{2} = f_i + \frac{h}{2}\left(f_{t,i} + f_{y,i} f_i\right) \tag{4.17}$$

Substitution of Eq. (4.14) and Eq. (4.17) into Eq. (4.13) gives

$$f_i + \alpha h f_{t,i} + \beta h f_{y,i} = f_i + \frac{h}{2} f_{t,i} + \frac{h}{2} f_{y,i} \tag{4.18}$$

Equation (4.18) gives

$$\alpha = \frac{1}{2}, \quad \beta = \frac{1}{2} f_i, \quad \bar{t} = t + \frac{h}{2}, \quad y = y_i + \frac{h}{2} f_i \tag{4.19}$$

The evaluation of y_{i+1} by second-order Runge–Kutta formula requires the following three steps:

$$\Delta' y_i = h f_i = h f(t_i, y_i)$$

$$\overline{\Delta} y_i = h f(\bar{t}, \bar{y}) = h f\left(t_i + \frac{h}{2}, y_i + \Delta' y_i\right)$$

$$y_{i+1} = y_i + h f\left(t_i + \frac{h}{2}, y_i + \Delta' y_i\right) \tag{4.20}$$

Equation (4.20) is known as second-order Runge–Kutta formula, and it has truncation error of order h^3.

(iii) Fourth-Order Runge–Kutta Method

To obtain the fourth-order Runge–Kutta formula with error of order h^5, the following successive values of must be evaluated:

$$\Delta' y_i = hf(t_i, y_i)$$

$$\Delta'' y_i = hf\left(t_i + \frac{h}{2}, y_i + \frac{1}{2}\Delta' y_i\right)$$

$$\Delta''' y_i = hf(t_i + h, y_i + 2\Delta'' y_i - \Delta' y_i)$$

$$\overline{\Delta} y_i = \frac{1}{6}(\Delta' y_i + 4\Delta'' y_i + \Delta''' y_i)$$

$$y_{i+1} = y_i + \frac{1}{6}(\Delta' y_i + 4\Delta'' y_i + \Delta''' y_i) \quad (4.21)$$

Equation (4.21) is known as fourth-order Runge–Kutta formula, and it has truncation error of order h^5. Thus, for better prediction of the model the Runge–Kutta fourth-order method is recommended.

4.8 Solution Interval

The proper length of the solution interval is related to the shortest delays that are represented in the model. If the solution interval is too long, instability is generated which arises from the computing process, not from any inherent dynamic characteristic of the system itself. If the solution interval is too short, the equation will be evaluated unnecessarily often using extra computer time. As a practical rule of thumb, the solution interval should be half or less of the shortest first-order delay in the system.

A model should be always tested for sensitivity to step size. The model should be first run for full length of simulation. Some variables should be plotted and printed. Next, the step size should be cut half, and the simulation should be repeated. This process should be continued until the improvements in results become negligible.

4.9 Functions Without Integration

There are several types of special function having the nature of flows (rates) or converter (auxiliary). These are special computation procedures, table interpolation, test inputs, randomness and logical choices. These functions can be imposed in any flow channel, and they would not introduce any time delay or periodicity in the system. They alter only the instantaneous amplitude of the signals.

The group of special computation procedures includes the square root, exponential and logarithmic functions. In STELLA there is no special symbol for each of these functions. These functions in STELLA are represented by converters.

SQRT(<expression>)
The SQRT function gives the square root of *expression. Expression* can be variable or constant. For meaningful results, expression must be greater than or equal to 0.
Examples:
SQRT(400) returns 20

EXP(<expression>)
The EXP function gives e raised to the power of *expression. Expression* can be variable or constant. The base of the natural logarithm, e, is also known as Euler's number. e is equal to 2.7182818.... EXP is the inverse of the LOGN function (natural logarithm). To calculate powers of other bases, use the exponentiation operator (^).
Examples:
EXP(1) equals 2.7182818 (the value of Euler's constant)
EXP(LOGN(3)) equals 3

LOGN(<expression>)
The LOGN function calculates the natural logarithm of *expression. Expression* can be variable or constant. Natural logarithms use Euler's constant e, 2.7182818..., as a base. For LOGN to have meaningful results, expression must be positive. LOGN is the inverse of the EXP function, e raised to the power of *expression*.
Examples:
LOGN(2.7182818) equals 1
LOGN(EXP(3)) equals 3
LOGN(−10) returns ? (undefined for non-positive numbers)

LOG10(<expression>)
The LOG10 function gives the base 10 logarithm of *expression. Expression* can be variable or constant. For LOG10 to be meaningful, *expression* must evaluate to a positive value. LOG10 is the inverse of the letter E in scientific notation, or base 10 exponentiation (10 raised to the power of *expression*).
Examples:
LOG10(10) equals 1
LOG10(1E5) equals 5
LOG10(−10) returns ? (undefined for non-positive numbers)

Graphical Function
Second class of functions provides for interpolation in a table. Non-linear relationships appear repeatedly in system dynamics. Table functions locate by linear interpolation, the values intermediate between the points entered in a table. These functions in STELLA are represented by converters with a small shadow inside the converters, and the flow diagram symbol of graphical functions is shown in Fig. 4.9.

In STELLA graphical function is essentially a sketch that relates between some input and output. This can be represented as

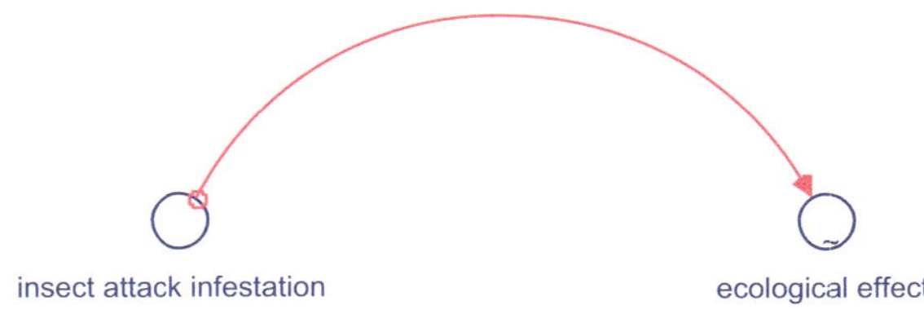

Fig. 4.9 Graphical function

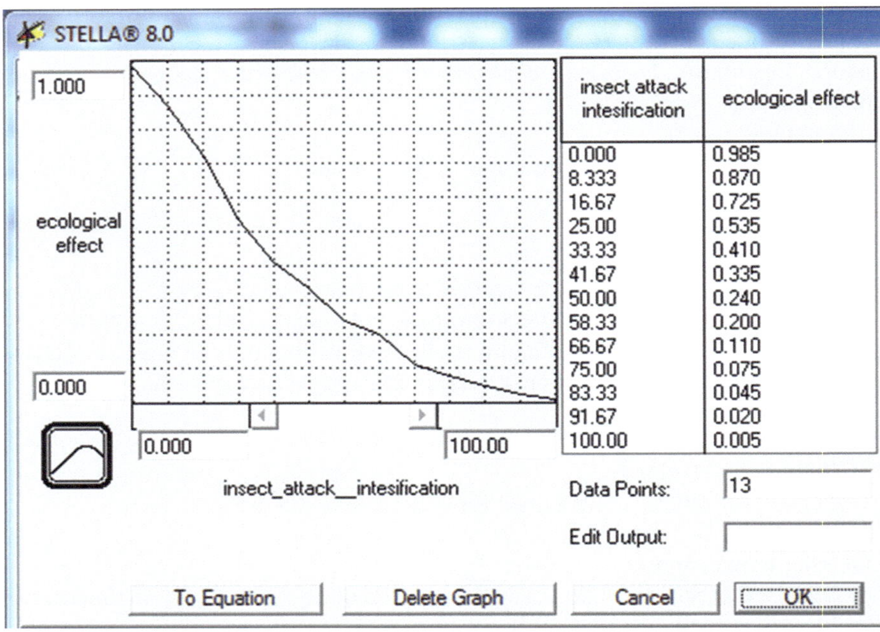

Fig. 4.10 Illustration of graphical function

Graphical_function_name GRAPH (input_ variable_ name) (x_1,y_1), (x_2, y_2), (x_3, y_3).........(x_n, y_n)
where

graphical_function_name is name of the graphical function.
Input_variable_name is the input variable for which corresponding graphical/table entry is made.

(x_1,y_1) is first graphical point with input variable x_1, and output variable y_1. (x_2,y_2) is the second graphical point with input variable x_2 and output variable y_2. Similarly (x_n,y_n) is the nth graphical point. n is the number of points in the graphical function. Figure 4.10 illustrates the graphical function of ecological effect as a function of insect attack infestation. We need to provide data on minimum and maximum values of the X-axis and Y-axis, number of data points and x values as well as y values of each point for the graphical function.

Test Input Functions

A group of function consisting of STEP, RAMP, SIN and COS are primarily used as test inputs.

STEP(<height>,<time>)

STEP function produces a specified constant height starting at a specified time, and it is frequently used as a shock excitation to determine the dynamics of the system recovery.

RAMP(<slope> [,<time>])

The RAMP function generates a linearly increasing or decreasing input over time with a specified slope (*slope*). Optionally, you may set the time at which the ramp begins. Slope and time can be either variable or constant.

PULSE(<volume> [,<first pulse>,<interval>])

The PULSE function generates a pulse input of a specified size (*volume*). In using the PULSE function, you have the option of setting the time at which the PULSE will first fire (*first pulse*), as well as the interval between subsequent PULSEs (*interval*). Each time that it fires a pulse, the software pulses the specified volume over a period of one time step (DT). Thus, the instantaneous value taken on by the PULSE function is volume/DT. Volume can be either variable or constant. First pulse and interval should be specified as constants.

SIN(<radians>)

SIN generates a sinusoidal fluctuation having a unit amplitude and specified period. The SIN function gives the sine of *radians*, where *radians* is an angle in radians. To convert measurement between degrees and radians, use the identity: pi (radians) = 180 (degrees). *Radians* can be constant or variable.

COS(<radians>)

COS in a similar fashion generates a cosine function of unit amplitude and specified period. The COS function gives the cosine of *radians*, where *radians* is an angle in radians. To convert measurement between degrees and radians, use the identity: pi (radians) = 180 (degrees). *Radians* can be constant or variable.

RANDOM(<min>,<max>, [<seed>])

The RANDOM function generates a series of uniformly distributed random numbers between *min* and *max*. RANDOM samples a new random number in each iteration of a model run. If you wish to replicate the stream of random numbers, specify *seed* as an integer between 1 and 32,767. To replicate a simulation, all variables that use statistical functions must have seeds specified. Each variable using a statistical function should use a separate seed value.

MAX(<expression>,<expression>,. . .)
The MAX function gives the maximum value among the expressions contained within parentheses.

MIN(<expression>,<expression>,. . .)
The MIN function gives the minimum value among the expressions contained within parentheses.

SWITCH(<Input1>,<Input2>)
The SWITCH function is equivalent to the following logic:
If *Input1* > *Input2* then 1 else 0.

Logical Function
The logical functions – IF, THEN, ELSE, AND, OR, NOT – are used to create expressions, and then give values based upon whether the resulting expressions are TRUE or FALSE. When you have multiple conditions, the expressions to be evaluated must be enclosed in parentheses ().

In STELLA the logical functions – IF, THEN, ELSE, AND, OR, NOT – are used to create expressions and give values based upon whether the resulting expressions are TRUE or FALSE. An example of bonus payments logic given below is essentially a logical function:

```
bonus_payments = 1F(TIME = 5) OR (sales > 5000) THEN Bonus ELSE 0
```

This statement sets bonus payments to the value of bonus at simulated time 5, or whenever the value of sales is greater than 5000. When neither condition is met, the statement gives the value, 0.

4.10 Functions Containing Integration

The functions SMOOTH, information delay and material delay contain integration. These are normally inserted in the flow channel or information channel in a model. Since these functions include stock, they change the time shape quantities moving between their inputs and outputs. These functions are assemblies of elementary stock and flow equations.

4.10.1 Smooth

SMOOTH produces first-order exponential average of a physical flow, and it is represented as shown in Fig. 4.11. Internally, the following equations will be generated and executed:

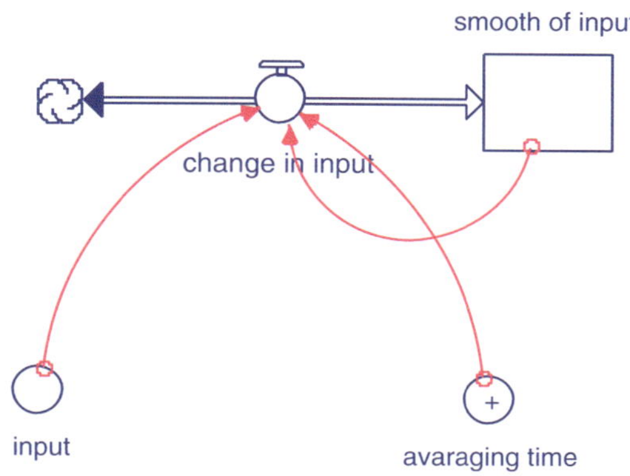

Fig. 4.11 Flow diagram of SMOOTH

```
smooth_of_input(t) = smooth_of_input(t - dt) + (change_in_input) * dt
INIT smooth_of_input = { Place initial value here... }
```

INFLOWS:

```
change_in_input = { Place right hand side of equation here... }
averaging_time = { Place right hand side of equation here... }
input = { Place right hand side of equation here... }
```

4.10.2 Information Delay

First-Order Information Delay

First-order information delay is used in the information channel to produce a first-order exponential delay. It represents the process of gradual, delayed adjustment of recognised information moving towards a value being supplied by a source. It is used to generate a delayed awareness of a changing situation. The flow diagram for first-order information delay is shown in Fig. 4.12. Internally, the following equations will be generated and executed. This function being first-order negative loop has a simple exponential response to the changes in its input:

```
output_with_information_delay1(t) = output_with_information_delay1
(t - dt) + (change_in_output) * dt
INIT output_with_information_delay1 = 0
```

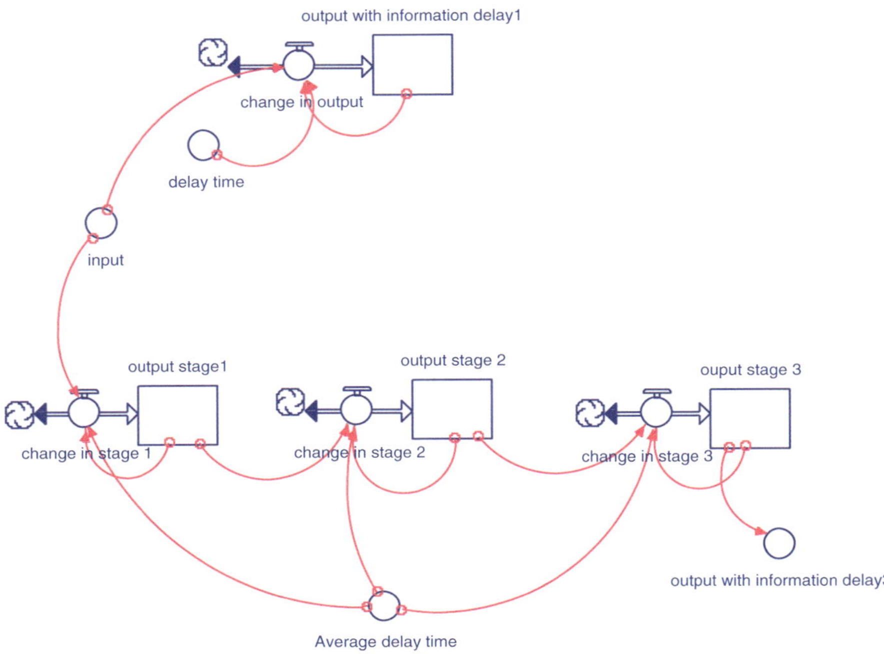

Fig. 4.12 Flow diagram of information delay

INFLOWS:

```
change_in_output = (input-output_with_information_delay1)/delay_time
delay_time = 2
input = STEP(100,2)
```

Third-Order Information Delay

Third-order information delay is a cascade of three first-order information delays as shown in Fig. 4.12. The equations and flow diagram show three sections like that for first-order information delay except in each section the delay is one third of the total. The following equations produce the third-order information delay in the information stream:

```
ouput_stage_3(t) = ouput_stage_3(t - dt) + (change_in_stage_3) * dt
INIT ouput_stage_3 = 0
```

INFLOWS:

```
change_in_stage_3 = (output_stage_2-ouput_stage_3)/
Average_delay_time


output_stage_2(t) = output_stage_2(t - dt) + (change_in_stage_2) * dt
INIT output_stage_2 = 0
```

INFLOWS:

```
change_in_stage_2 = (output_stage1-output_stage_2)/
Average_delay_time


output_stage1(t) = output_stage1(t - dt) + (change_in_stage_1) * dt
INIT output_stage1 = 0
```

INFLOWS:

```
change_in_stage_1 = (input-output_stage1)/Average_delay_time
Average_delay_time = 2/3
output_with_information_delay3 = ouput_stage_3
input = STEP(100,2)
```

The delay may be first order, second order, third order and even higher order. Third-order and first-order information delays are placed in the information channel. But third-order information delay produces a different time shape in output. Figure 4.13 shows the step responses of first-order and third-order information delay.

4.10.3 Material Delay

The outflow from a first-order material delay is always proportional to the stock of material in transit, and the flow diagram of a first-order material delay is shown in Fig. 4.14. Third-order material delay is similar to third-order information delay, and it is also the three cascaded first-order exponential loops. Therefore, it has the same response. It differs from third-order information delay being placed in flow channel and creates a delay in the transmission of quantity from the input to the output. Internally the following equations will be generated and executed. Figure 4.15 shows STEP responses of first- and third-order material delay.

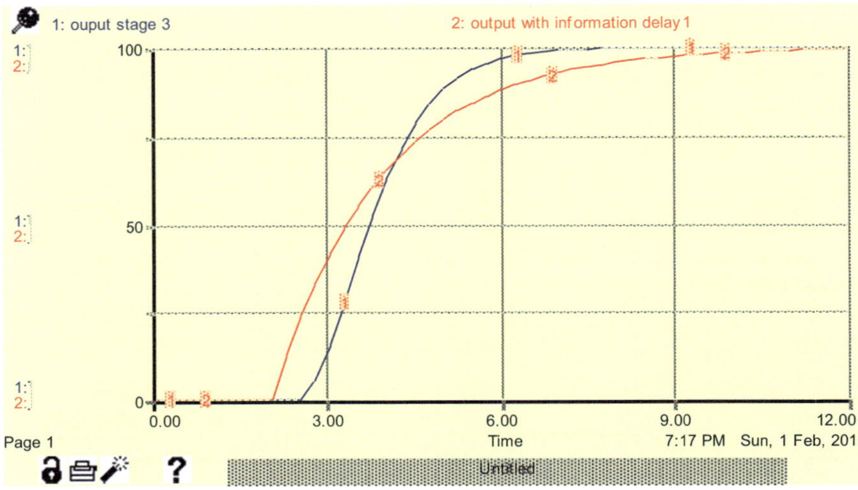

Fig. 4.13 STEP response of first-order information delay (output with information delay 1) and third-order information delay (output stage 3)

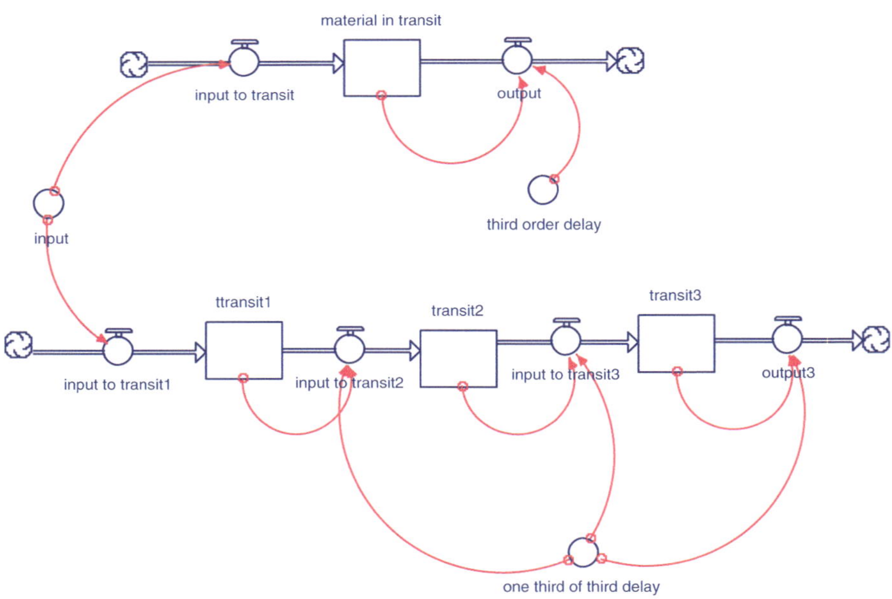

Fig. 4.14 Flow diagram of material delay

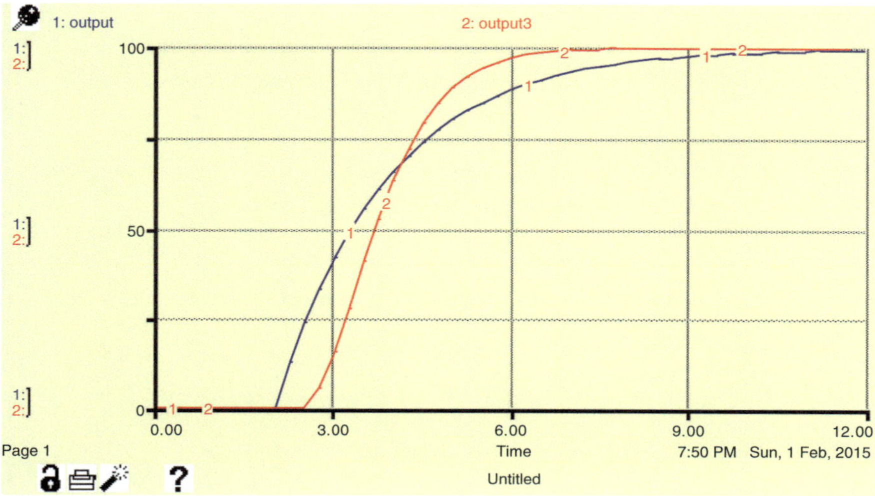

Fig. 4.15 STEP response of first-order material delay (output) and third-order material delay (output 3)

First-Order Material Delay

```
material_in_transit(t) = material_in_transit(t - dt) + (input_to_transit
 - output) * dt
INIT material_in_transit = 0
```

INFLOWS:

```
input_to_transit = input
```

OUTFLOWS:

```
output = material_in_transit/third_order_delay
third_order_delay = 2
```

Third-Order Material Delay

```
transit2(t) = transit2(t - dt) + (input_to_transit2 - input_to_
transit3) * dt
INIT transit2 = 0
```

INFLOWS:

```
input_to_transit2 = ttransit1/one_third_of_third_delay
```

OUTFLOWS:

```
input_to_transit3 = transit2/one_third_of_third_delay
```

```
transit3(t) = transit3(t - dt) + (input_to_transit3 - output3) * dt
INIT transit3 = 0
```

INFLOWS:

```
input_to_transit3 = transit2/one_third_of_third_delay
```

OUTFLOWS:

```
output3 = transit3/one_third_of_third_delay
```

```
ttransit1(t) = ttransit1(t - dt) + (input_to_transit1 - input_to_
transit2) * dt
INIT ttransit1 = 0
```

INFLOWS:

```
input_to_transit1 = input
```

OUTFLOWS:

```
input_to_transit2 = ttransit1/one_third_of_third_delay
one_third_of_third_delay = 2/3.
input = STEP(100,2)
```

4.11 Examples

4.11.1 Population Model

In this example the stock–flow diagram of population model is derived from the causal loop diagram of population model in example 3.4.1 in Chap. 3. In the causal loop diagram, population is a state variable (population level), and it is represented by a stock. The birth rate is an inflow (population/year) into the stock population, and the death rate is an outflow into stock population. The stock–flow diagram of a

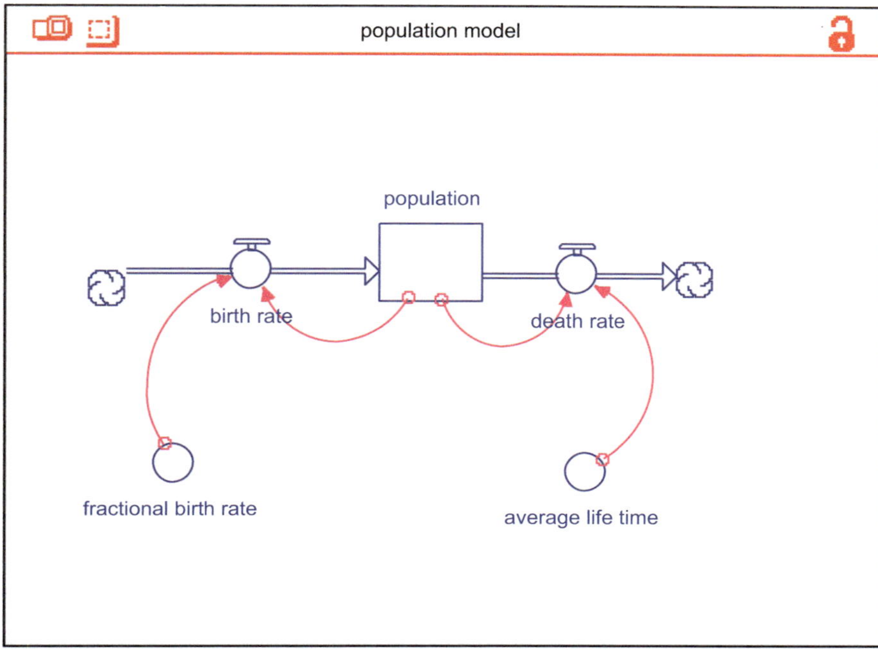

Fig. 4.16 Stock–flow diagram of a simple population model

simple population model derived from the causal loop diagram in example 3.4.1 in Chap. 3 is shown in Fig. 4.16, and the model consists of two feedback loops. The positive feedback loop consisting of population and birth generates population growth, and the negative loop consisting of population and death decreases the population.

Population is increased by births and decreased by deaths, and it is expressed as

```
population(t) = population(t - dt) + (birth_rate - death_rate) * dt
INIT population = 100
```

The birth is computed as a fixed percentage of population.

```
birth_rate = population*fractional_birth_rate
fractional_birth_rate = 0.03
```

The death is computed from population and average lifetime.

```
death_rate = population/average_life_time
average_life_time = 70
```

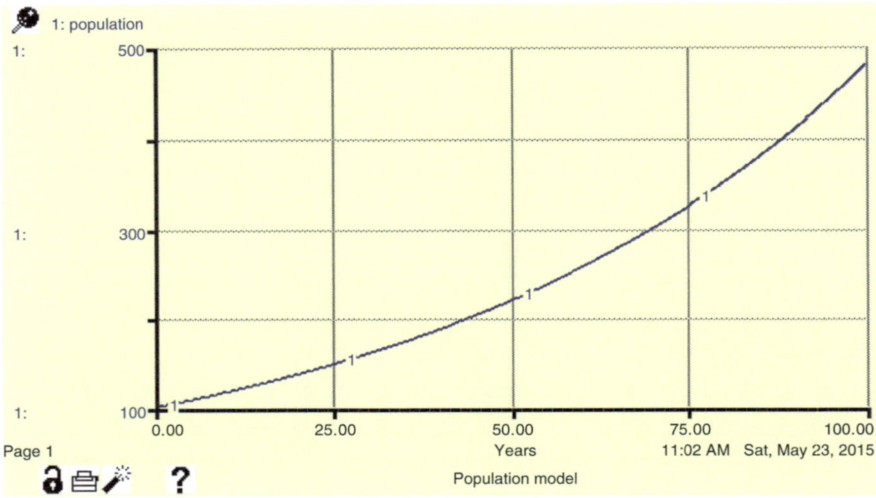

Fig. 4.17 Simulated population for a period of 100 years

Simulated population is shown in Fig. 4.17, and population grows by a fixed percentage exhibiting exponential growth. However, in practice the exponential growth cannot continue indefinitely. When the ultimate limit is approached, the negative loop becomes dominant.

STELLA model of the population is given below:

Population Model

```
population(t) = population(t - dt) + (birth_rate - death_rate) * dt
INIT population = 100
```

INFLOWS:

```
birth_rate = population*fractional_birth_rate
```

OUTFLOWS:

```
death_rate = population/average_life_time
average_life_time = 70
fractional_birth_rate = 0.03
```

4.11.2 Grain Storage System

Grains are basically stored to maintaining the quality for planting as a seed, continuity of supply and higher price. Here we consider the storage for grains at the retailer to iron out the price variations and provide consumer's welfare. The stock–flow diagram of simple grain storage system is shown in Fig. 4.18, and it consists of grain stock for consumers and grain on order to maintain desired supply of grain on order.

Grain stock is increased by receiving rate from grain on order and decreased by release rate from the grain stock.

```
grain_stock(t) = grain_stock(t - dt) + (receiving_rate - release_rate)
* dt
INIT grain_stock = 1000
```

Release rate depends on grain stock and release fraction.

```
release_rate = grain_stock*release_fraction
release_fraction = 0.025
```

Receiving rate depends on grain on order and receiving delay.

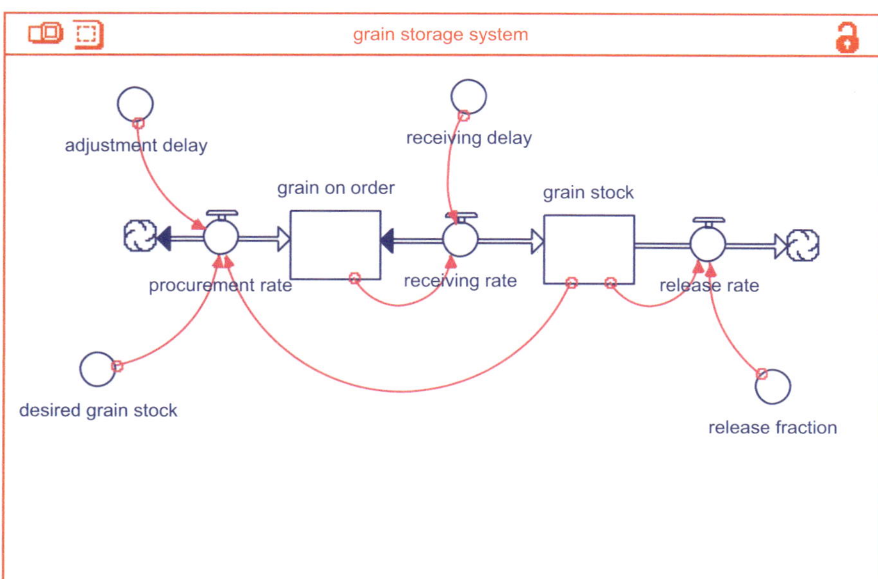

Fig. 4.18 Stock–flow diagram of grain storage system

```
receiving_rate = grain_on_order/receiving_delay
receiving_delay = 10
```

Grain on order is increased by procurement rate and decreased by receiving rate in the grain stock.

```
grain_on_order(t) = grain_on_order(t - dt) + (procurement_rate -
receiving_rate) * dt
INIT grain_on_order = 10000
```

The procurement rate depends on the difference between desired stock and current stock and procurement adjustment delay.

```
procurement_rate = (desired_grain_stock-grain_stock)/
adjustment_delay
adjustment_delay = 5
```

This model was simulated to see what happens to grain stock, receiving rate and procurement over time. Figure 4.19 shows the simulated grain stock, receiving rate and procurement over time. The grain stock, receiving rate and procurement rate oscillate over time due to addition of delay created by the supply line as represented by the grain on order and receiving rate.

STELLA model of the grain storage system is given below:

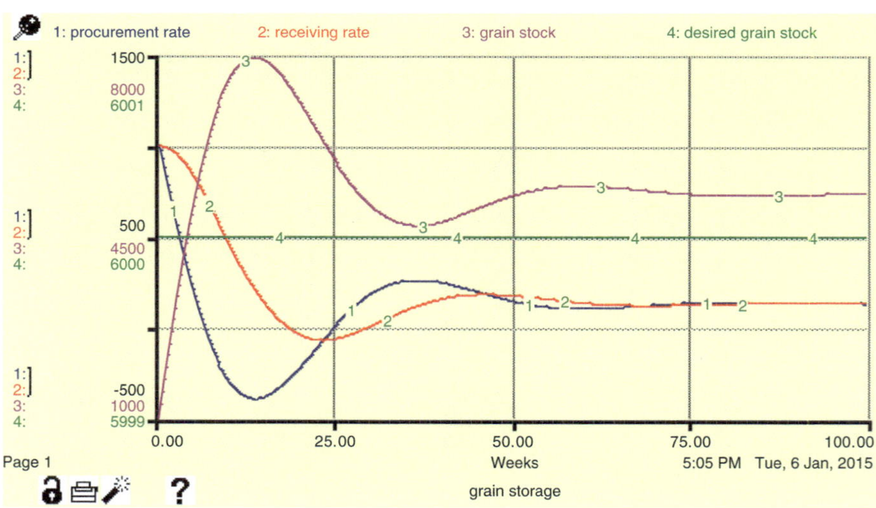

Fig. 4.19 Simulated grain stock, receiving rate and procurement over time

Grain Storage System

```
grain_on_order(t) = grain_on_order(t - dt) + (procurement_rate -
receiving_rate) * dt
INIT grain_on_order = 10000
```

INFLOWS:

```
procurement_rate = (desired_grain_stock-grain_stock) /
adjustment_delay
```

OUTFLOWS:

```
receiving_rate = grain_on_order/receiving_delay
```

```
grain_stock(t) = grain_stock(t - dt) + (receiving_rate - release_rate)
* dt
INIT grain_stock = 1000
```

INFLOWS:

```
receiving_rate = grain_on_order/receiving_delay
```

OUTFLOWS:

```
release_rate = grain_stock*release_fraction
```

```
adjustment_delay = 5
desired_grain_stock = 6000
receiving_delay = 10
release_fraction = 0.025
```

4.11.3 Food Security Model

Food security is a worldwide problem that has called the attention to governments and the scientific community. It particularly affects developing countries. The scientific community has had increasing concerns for strategic understanding and implementation of food security policies in developing countries, especially since the food crisis in the 1970s. The process of decision-making is becoming increasingly complex due to the interaction of multiple dimensions related to food security (Giraldo et al. 2008).

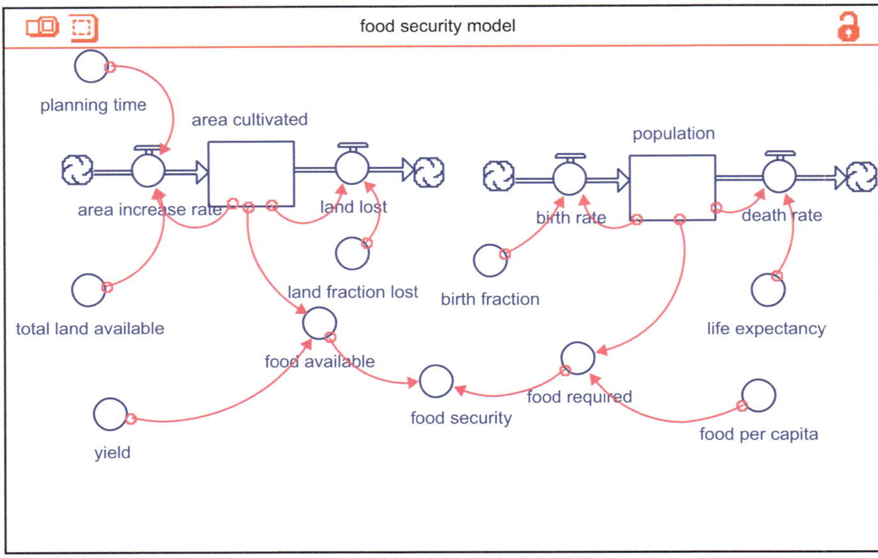

Fig. 4.20 Stock–flow diagram of food security model

In this model land suitable for rice production is a limited resource and the available land area suitable for rice production is converted into rice area. The rice area is also converted into industrial areas. The rice available is computed from area cultivated and yield. Population increases by birth and decreases by death. Food security defined as a ratio of rice available per capita and changes with both changes in food available and population. Figure 4.20 shows the stock–flow diagram of rice food security model.

Area cultivated is increased by transfer of land suitable for rice cultivation based on development policy of covering the land suitable for rice cultivation, and also the area cultivated is decreased by a discard rate of a fraction of rice area for housing, roads and highways and industries.

```
area_cultivated(t) = area_cultivated(t - dt) + (area_increase_rate -
land_lost) * dt
INIT area_cultivated = 68000
```

Area increase rate is computed from land available for cultivation and planning time horizon.

```
area_increase_rate = MAX((total_land_available-area_cultivated)/
planning_time,0)
total_land_available = 120000
planning_time = 15
```

Area lost rate is computed from area cultivated and lost fraction.

```
land_lost = area_cultivated*land_fraction_lost
land_fraction_lost = 0.00085
```

Food available depends on the yield of the crop rice (tons/ha) and area under rice cultivation.

```
food_available = area_cultivated*yield
yield = 3.2
```

Population level at any time t is computed from the population growth rate, and it is given by

```
population(t) = population(t - dt) + (birth_rate - death_rate) * dt
INIT population = 2800000
```

In this simple model, the birth rate is computed from population and birth fraction.

```
birth_rate = population*birth_fraction
birth_fraction = 0.025
```

And death rate is computed from population and life expectancy.

```
death_rate = population/life_expectancy
life_expectancy = 65
```

Food requirement depends on the per capita consumption of rice and population level, and it is computed as

```
food_required = population*food_per_capita
food_per_capita = 180/1000
```

Food security is defined as the ratio of rice production to rice requirement:

```
food_security = food_available/food_required
```

Food security level 1 means surplus food and less than 1 means shortage in food supply to lead healthy life.

Figure 4.21 shows the simulated area cultivated, area increase rate, land lost, population and food security in rice production for a simulation period of 100 years. The rice area initially increases till about 50 years and reaches the upper limit of available land suitable for rice and remains constant. The food security in rice increases till 25 years and then decreases due to increase in the population.

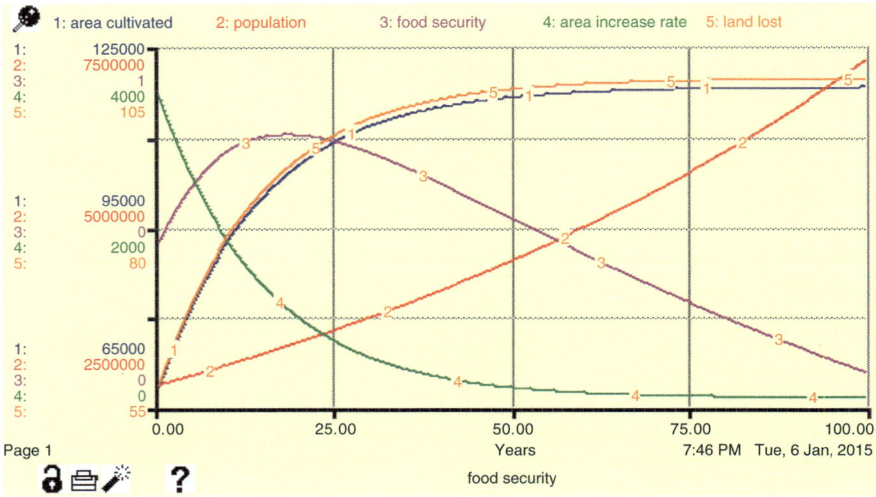

Fig. 4.21 Simulated area cultivated, area increase rate, land lost, population and food security

STELLA model of food security is given below:

Food Security Model

```
area_cultivated(t) = area_cultivated(t - dt) + (area_increase_rate -
land_lost) * dt
INIT area_cultivated = 68000
```

INFLOWS:

```
area_increase_rate = MAX((total_land_available-area_cultivated)/
planning_time,0)
```

OUTFLOWS:

```
land_lost = area_cultivated*land_fraction_lost
```

```
population(t) = population(t - dt) + (birth_rate - death_rate) * dt
INIT population = 2800000
```

INFLOWS:

```
birth_rate = population*birth_fraction
```

OUTFLOWS:

```
death_rate = population/life_expectancy

birth_fraction = 0.025
food_available = area_cultivated*yield
food_per_capita = 180/1000
food_required = population*food_per_capita
food_security = food_available/food_required
land_fraction_lost = 0.00085
life_expectancy = 65
planning_time = 15
total_land_available = 120000
yield = 3.2
```

4.11.4 Commodity Production Cycle Model

Agricultural commodities are sold in highly competitive markets whose prices change yearly, daily, hourly and even by the minute. Farmers are affected by fluctuations of price – good times bring him an increase in income, and depression causes his cash income to drop away to very little. The price received by the farmers is important not only to him; the rate of production depends on it to a large extent. In addition, pigs, and to some extent cattle and chickens, show fluctuations in price and supply of a persistent nature at a frequency consistent over long periods of time (Meadows 1970).

We formulate here a commodity production cycle model based on two feedback loops, a production and a consumption loop each trying to adjust inventory coverage to the desired level. Price links the supply and demand. Nerlove's formulation of expected price is incorporated (Nerlove 1958). If the price is high, this tends to reduce the consumption and increase the production after some time delay. This in turn, increases the inventory level, the price would fall. Similarly, if the price is too low, the consumption would increase to raise it back to the equilibrium. The stock–flow diagram of the commodity production cycle model is shown in Fig. 4.22.

In every commodity there will be stocks of products in physical forms and in many locations, and only production and consumption determine the inventory level.

```
distributor_inventory(t) = distributor_inventory(t - dt) + (supply_
rate - consumption_rate) * dt
INIT distributor_inventory = 6000
```

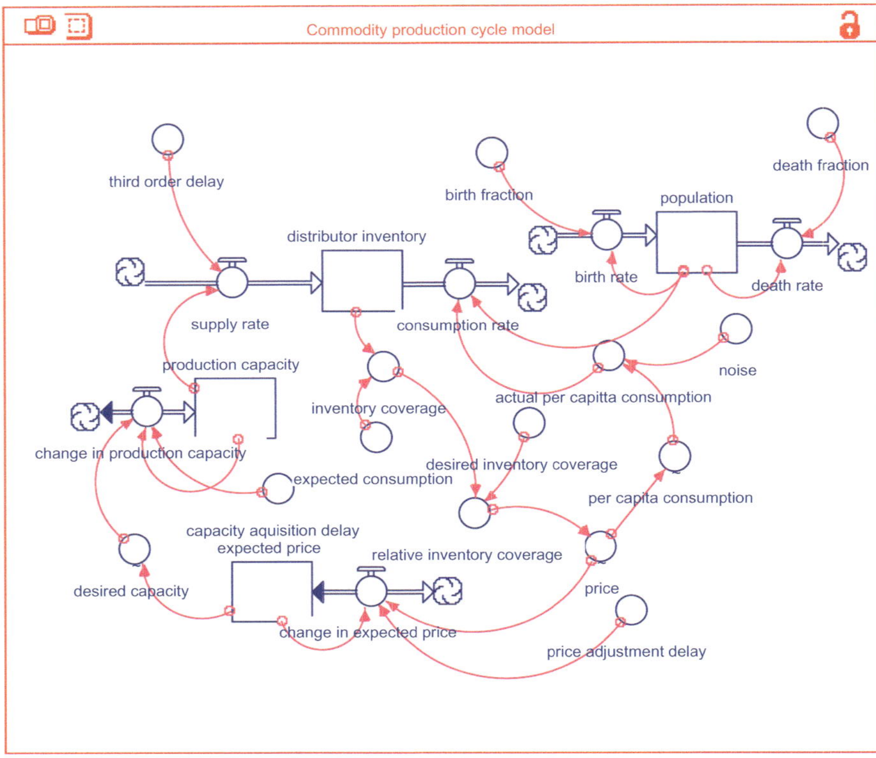

Fig. 4.22 Stock–flow diagram of the commodity production cycle model

The holdings of the inventory stocks of the commodity are concerned with length of time current inventory satisfy anticipated demand. The inventory coverage is simply the ratio of distributor inventory to expected consumption.

```
inventory_coverage = distributor_inventory/expected_consumption
```

Actual coverage is compared with desired coverage to determine the relative coverage as the ratio of inventory coverage to desired inventory coverage.

```
relative_inventory_coverage = inventory_coverage/
desired_inventory_coverage
desired_inventory_coverage = 1
```

When the relative inventory coverage increases, the price will fall, and conversely if the relative inventory coverage decreases, the price would increase. The relationship between price and relative inventory coverage is non-linear, and it is expressed by a graphical function.

```
price = GRAPH(relative_inventory_coverage)
(0.00, 100), (0.4, 94.0), (0.8, 80.0), (1.20, 50.0), (1.60, 20.0),
(2.00, 10.0)
```

We adopt the model of producer expectations as proposed by Nerlove (1958). The current rate of change in the expected price is proportional to the difference between recent expected price and recent actual price.

```
expected_price(t) = expected_price(t - dt) + (- change_in_expected_
price) * dt
INIT expected_price = 70

change_in_expected_price = (expected_price-price)/price_adjustment_
delay
```

Desired capacity is non-linearly related to expected price, and it is represented by a graphical function.

```
desired_capacity = GRAPH(expected_price)
(0.00, 0.00), (20.0, 40.0), (40.0, 200), (60.0, 1000), (80.0, 1200),
(100, 1280)
```

Production capacity is influenced by change in production capacity, and it is computed as

```
production_capacity(t) = production_capacity(t - dt) + (change_in_
production_capacity) * dt
INIT production_capacity = 600
```

Change in production capacity depends on desired capacity, present level of production capacity and capacity acquisition delay, and it is expressed as

```
change_in_production_capacity = (desired_capacity-production_
capacity)/capacity_aquisition_delay
capacity_aquisition_delay = 4
```

The model acknowledges a third-order delay due to biological, physical and meteorological factors in the supply of the commodity from initiation to final availability.

```
supply_rate = SMTH3(production_capacity,third_order_delay)
third_order_delay = 12
```

Consumption rate is the product of per capita consumption and population of consumers.

```
consumption_rate = population*actual_per_capita_consumption
```

Per capita consumption depends on the price, and it is non-linearly related to price. This relationship is given by graphical function.

```
per_capita_consumption = GRAPH(price)
(0.00, 7.00), (20.0, 6.00), (40.0, 4.00), (60.0, 2.00), (80.0, 1.00),
(100, 0.00)
```

Population is increased by birth and decreased by death:

```
population(t) = population(t - dt) + (birth_rate - death_rate) * dt
INIT population = 200
birth_rate = population*birth_fraction
death_rate = population*death_fraction

birth_fraction = 0.025/12
death_fraction = 0.008/12
```

The model was simulated to a step change in consumption with noise to address the stability of commodity production systems. Figure 4.23 shows the simulated production of price, supply rate and distributor inventory of the commodity

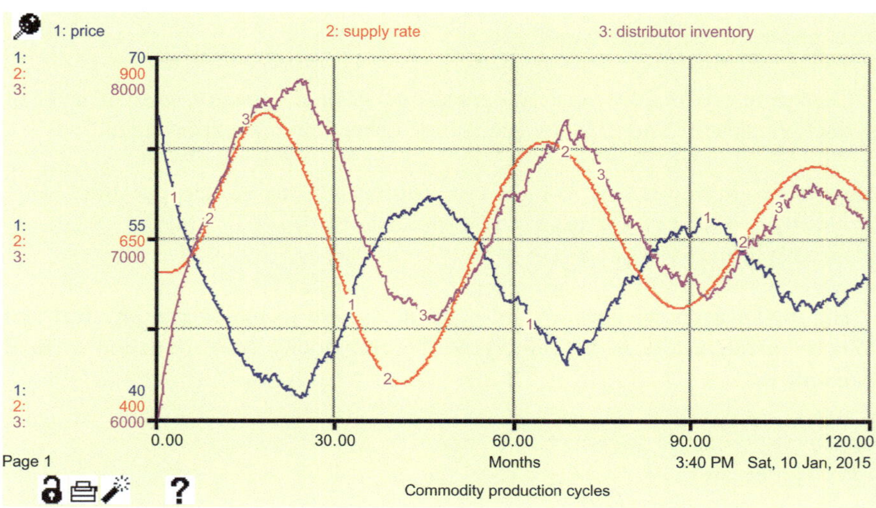

Fig. 4.23 Simulated price, supply rate and distributor inventory of the commodity production cycle model

production cycle model, and the simulated results exhibit periodicity and qualitatively reflect real-world behaviour.

STELLA model of the commodity production cycle model is given below:

Commodity Cycles

```
distributor_inventory(t) = distributor_inventory(t - dt) + (supply_
rate - consumption_rate) * dt
INIT distributor_inventory = 6000
```

INFLOWS:

```
supply_rate = SMTH3(production_capacity,third_order_delay)
```

OUTFLOWS:

```
consumption_rate = population*actual_per_capita_consumption
```

```
expected_price(t) = expected_price(t - dt) + (- change_in_expected_
price) * dt
INIT expected_price = 70
```

OUTFLOWS:

```
change_in_expected_price = (expected_price-price)/price_adjustment_
delay
```

```
population(t) = population(t - dt) + (birth_rate - death_rate) * dt
INIT population = 200
```

INFLOWS:

```
birth_rate = population*birth_fraction
```

OUTFLOWS:

```
death_rate = population*death_fraction
```

```
production_capacity(t) = production_capacity(t - dt) + (change_in_
production_capacity) * dt
INIT production_capacity = 600
```

INFLOWS:

```
change_in_production_capacity = (desired_capacity-
production_capacity)/capacity_aquisition_delay


actual_per_capita_consumption = per_capita_consumption*(1+noise)
birth_fraction = 0.025/12
capacity_aquisition_delay = 4
death_fraction = 0.008/12
desired_inventory_coverage = 1
expected_consumption = 6000
inventory_coverage = distributor_inventory/expected_consumption
noise = RANDOM(-0.5,0.5,2374)*STEP(0.5,10)
price_adjustment_delay = 5
relative_inventory_coverage = inventory_coverage/
desired_inventory_coverage
third_order_delay = 12


desired_capacity = GRAPH(expected_price)
(0.00, 0.00), (20.0, 40.0), (40.0, 200), (60.0, 1000), (80.0, 1200),
(100, 1280)


per_capita_consumption = GRAPH(price)
(0.00, 7.00), (20.0, 6.00), (40.0, 4.00), (60.0, 2.00), (80.0, 1.00),
(100, 0.00)


price = GRAPH(relative_inventory_coverage)
(0.00, 100), (0.4, 94.0), (0.8, 80.0), (1.20, 50.0), (1.60, 20.0),
(2.00, 10.0)
```

4.11.5 Food Relief Model

Food relief is provided for humanitarian ground. This model examines what will happen if suddenly food relief of a poor country having shortage of food and relying greatly on food relief is cut off. Stock–flow diagram of the food relief model is shown in Fig. 4.24.

Population is increased by birth rate and decreased by death rate.

```
population(t) = population(t - dt) + (birth_rate - death_rate) * dt
INIT population = 100E06
```

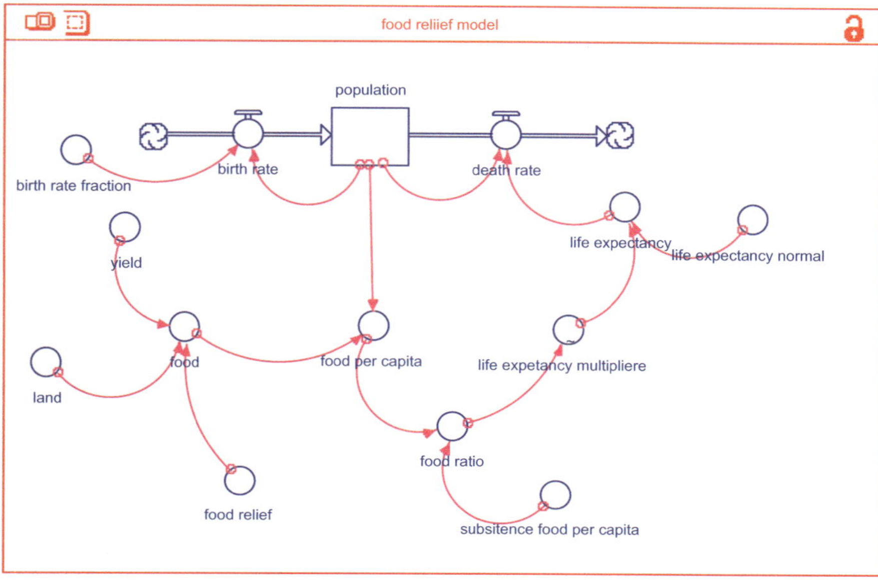

Fig. 4.24 Stock–flow diagram of the food relief model

Birth rate is assumed to depend on population and birth rate normal.

```
birth_rate=population*birth_rate_fraction
birth_rate_fraction=0.045
```

Death rate depends on population level, reference life expectancy and life expectancy multiplier.

```
death_rate=population/life_expectancy
```

Life expectancy is affected by food, and it is expressed in terms of life expectancy normal and life expectancy multiplier from food.

```
life_expectancy=life_expectancy_normal*life_expetancy_multipliere
life_expectancy_normal=45
```

Life expectancy multiplier is non-linearly related to food ratio, and it is expressed by graphical function.

```
life_expetancy_multiplier = GRAPH(food_ratio)
(0.00, 0.033), (0.313, 0.25), (0.625, 0.5), (0.938, 0.825), (1.25,
1.02), (1.56, 1.18), (1.88, 1.25), (2.19, 1.25), (2.50, 1.25)
```

Food ratio is defined as the ratio of food per capita to subsistence food per capita.

```
food_ratio = food_per_capita/subsitence_food_per_capita
subsitence_food_per_capita = 180
```

Food per capita is defined as the ratio of total food available to population.

```
food_per_capita = food/population
```

Food is the most crucial regulator of life expectancy, and it depends on total production of food and food relief available:

```
food = land*yield + food_relief
land = 7.4E06
yield = 2200
```

Food relief for humanitarian ground is considered to increase at the rate of 5000E5 kg/year, and it is interrupted after 50 years of continuous supply.

```
food_relief = IF(TIME) >= (50) THEN(0) ELSE(RAMP(5000E5,0))
```

RAMP function is used to provide continuous supply of relief.

Simulated population, birth rate, death rate and food per capita for continuous food relief for 50 years at a constant rate of 500E6 and then its interruption are shown in Fig. 4.25.

STELLA model of food relief is given below:

Food Relief Model

```
population(t) = population(t - dt) + (birth_rate - death_rate) * dt
INIT population = 100E06
```

INFLOWS:

```
birth_rate = population*birth_rate_fraction
```

OUTFLOWS:

```
death_rate = population/life_expectancy
```

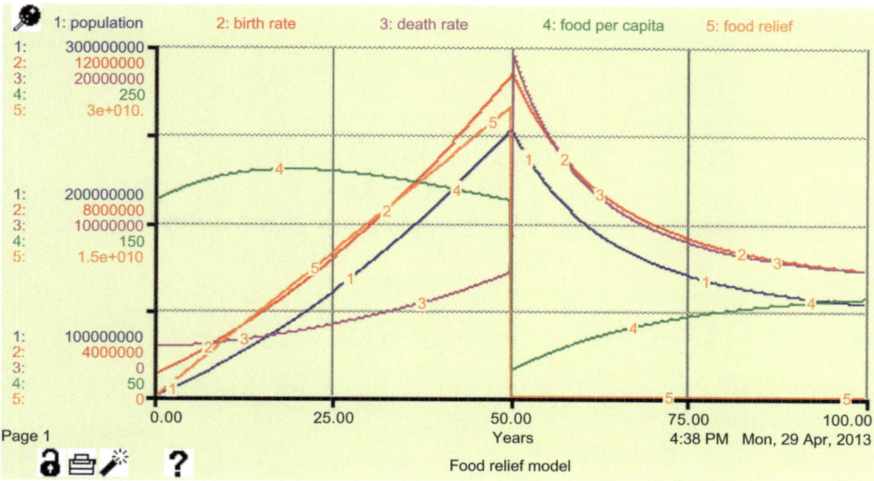

Fig. 4.25 Simulated population, birth rate, death rate and food per capita for continuous food relief for 50 years at a constant rate of 500E6 and then its interruption

```
birth_rate_fraction = 0.045
food = land*yield + food_relief
food_per_capita = food/population
food_ratio = food_per_capita/subsitence_food_per_capita
food_relief = IF(TIME)>=(50)THEN(0)ELSE(RAMP(5000E5,0))
land = 7.4E06
life_expectancy = life_expectancy_normal*life_expetancy_multipliere
life_expectancy_normal = 45
subsitence_food_per_capita = 180
yield = 2200


life_expetancy_multiplier = GRAPH(food_ratio)
(0.00, 0.033), (0.313, 0.25), (0.625, 0.5), (0.938, 0.825), (1.25, 1.02),
(1.56, 1.18), (1.88, 1.25), (2.19, 1.25), (2.50, 1.25)
```

4.11.6 Crop Livestock Model

Crop-based agriculture is the main source of livelihoods in low-income countries. When crop-based agriculture cannot support livelihoods, it is for the farmers to look for additional income-generating opportunities. In order to create livelihood, they must combine the capital endowments that they have access to and control, and livestock production is an important source of income in such situation to create livelihoods. The stock–flow diagram of the crop livestock model is shown in

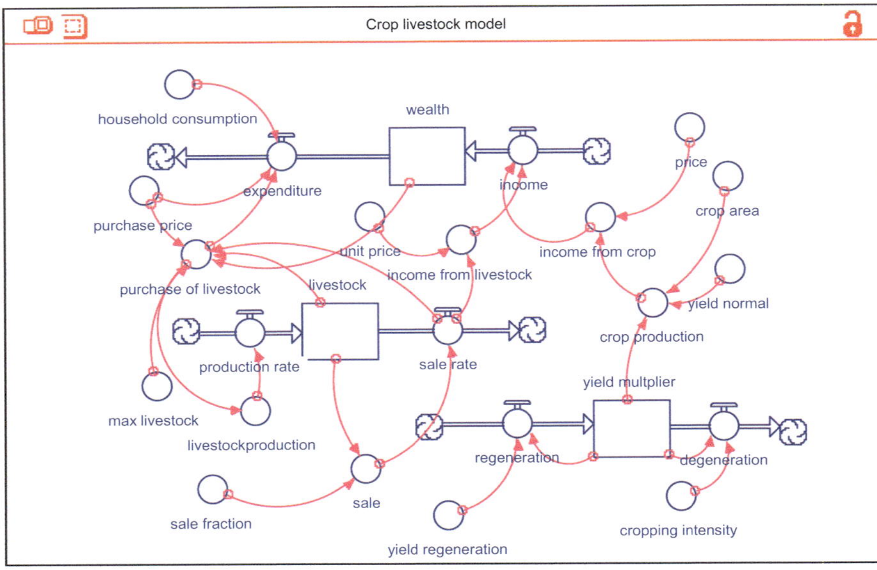

Fig. 4.26 The stock–flow diagram of the crop livestock model

Fig. 4.26. The model consists of a fixed amount of crop area and fixed number of family members for rearing livestock provided the family has some wealth as grant or bank loan to start the livestock husbandry.

As the crop area is limited, the crop production depends on the yield, and the crop production is computed from crop area, yield normal and yield multiplier. The effect of cropping intensity on yield, i.e. the effect of degeneration, and regeneration effect are assumed to be zero.

```
crop_production = crop_area*yield_normal*yield_multplier
crop_area = 0.5
yield_normal = 3.2
```

The wealth of the family increases from the income from crop production and also from the income from livestock and decreases by the expenditure on household consumption and the expenditure on purchase of livestock.

```
wealth(t) = wealth(t - dt) + (income - expenditure) * dt
INIT wealth = 10000
```

The income is derived from crop and from sale of livestock.

```
income = income_from_crop + income_from_livestock
income_from_crop = crop_production*price
income_from_livestock = sale_rate*unit_price
```

The expenditure comes from household consumption and from the purchase of livestock. The purchase of livestock is considered when it is less than maximum capacity of rearing the livestock.

```
expenditure = purchase_of_livestock*purchase_price
+ household_consumption
purchase_of_livestock = IF(INT(wealth/purchase_price)) <=
(max_livestock + sale_rate-livestock) THEN(INT(wealth/
purchase_price))ELSE(0)
purchase_price = 15000
household_consumption = 75000
```

The number of livestock is increased after certain maturation delay by addition from purchase and decreased from sale of the matured livestock.

```
livestock(t) = livestock(t - dt) + (production_rate - sale_rate) * dt
INIT livestock = 4
```

The model was simulated to see what happens to the wealth of the household when some livestock is provided to a small marginal farming family. Simulated wealth, livestock, crop production, income from crop production and income from livestock are shown in Fig. 4.27. The household can increase his income from livestock considerably, but it is after some time delay.

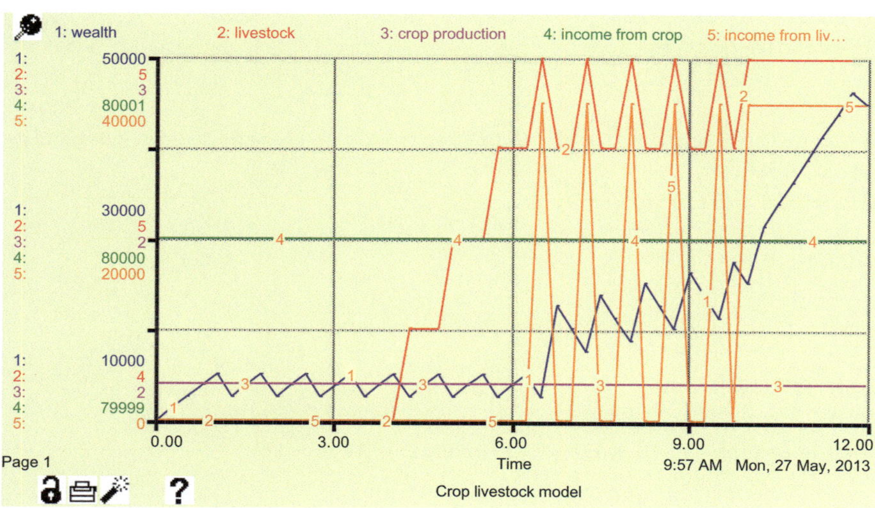

Fig. 4.27 Simulated wealth, livestock, crop production, income from crop production and income from livestock

STELLA model of the crop livestock system is given below:

Crop Livestock Model

```
livestock(t) = livestock(t - dt) + (production_rate - sale_rate) * dt
INIT livestock = 4
```

INFLOWS:

```
production_rate = livestockproduction
```

OUTFLOWS:

```
sale_rate = sale
```

```
wealth(t) = wealth(t - dt) + (income - expenditure) * dt
INIT wealth = 10000
```

INFLOWS:

```
income = income_from_crop + income_from_livestock
```

OUTFLOWS:

```
expenditure = purchase_of_livestock*purchase_price
+ household_consumption
```

```
yield_multplier(t) = yield_multplier(t - dt) + (regeneration - degener-
ation) * dt
INIT yield_multplier = 1
```

INFLOWS:

```
regeneration = yield_multplier*yield_regeneration
```

OUTFLOWS:

```
degeneration = yield_multplier*cropping_intensity
```

```
cropping_intensity = 0
crop_area = 0.5
```

```
crop_production = crop_area*yield_normal*yield_multplier
household_consumption = 75000
income_from_crop = crop_production*price
income_from_livestock = sale_rate*unit_price
livestockproduction = DELAY(purchase_of_livestock,3)
max_livestock = 8
price = 50000
purchase_of_livestock = IF(INT(wealth/purchase_price)) <=
(max_livestock + sale_rate-livestock)THEN(INT(wealth/
purchase_price))ELSE(0)
purchase_price = 15000
sale = INT(livestock*sale_fraction)
sale_fraction = 0.2
unit_price = 35000
yield_normal = 3.2
yield_regeneration = 0
```

4.11.7 Penaeid Shrimp Model

Shrimp farming plays a significant role in the economy of many developing countries in Asia. Due to the economic importance of penaeid shrimp worldwide, particularly in aquaculture, a great effort to understand the growth biology of *Penaeus* spp. has been made in recent years. This includes studies on the influence of environmental factors such as temperature (Wyban et al. 1995; Miao and Tu 1996; Ye et al. 2003; López-Martínez et al. 2003), salinity (Lemos et al. 2001) and lunar cycles (Griffith and Wigglesworth 1993) on shrimp growth. These valuable information may be synthesised and integrated into general models in order to simulate the growth of *Penaeus* individuals. Growth simulations for penaeid shrimp are based on single-equation models such as the von Bertalanffy (Wang 1998; Xiao 1999) and Gompertz growth equations (Jackson and Wang 1998; Xiao 1999), bioenergetic models (Mishra et al.2002) or the Fuzzy Inductive Reasoning (FIR) approach (Carvajal and Nebot 1998). This example presents a system dynamics computer model of penaeid shrimp in STELLA adapted from a computer model developed by Franco et al. (2006).

An individual-based model is presented to simulate the growth of a penaeid shrimp from the juvenile stage to the end of the life cycle. Shrimp larval stages (nauplius, zoea and mysis) have a very short duration (less than 3 weeks) (Haywood et al. 1995) and are not included in the model. The growth model includes five physiological processes, ingestion, assimilation, elimination, respiration and female reproduction, and is forced by water temperature and food availability. The stock–flow diagram of the model is shown in Fig. 4.28.

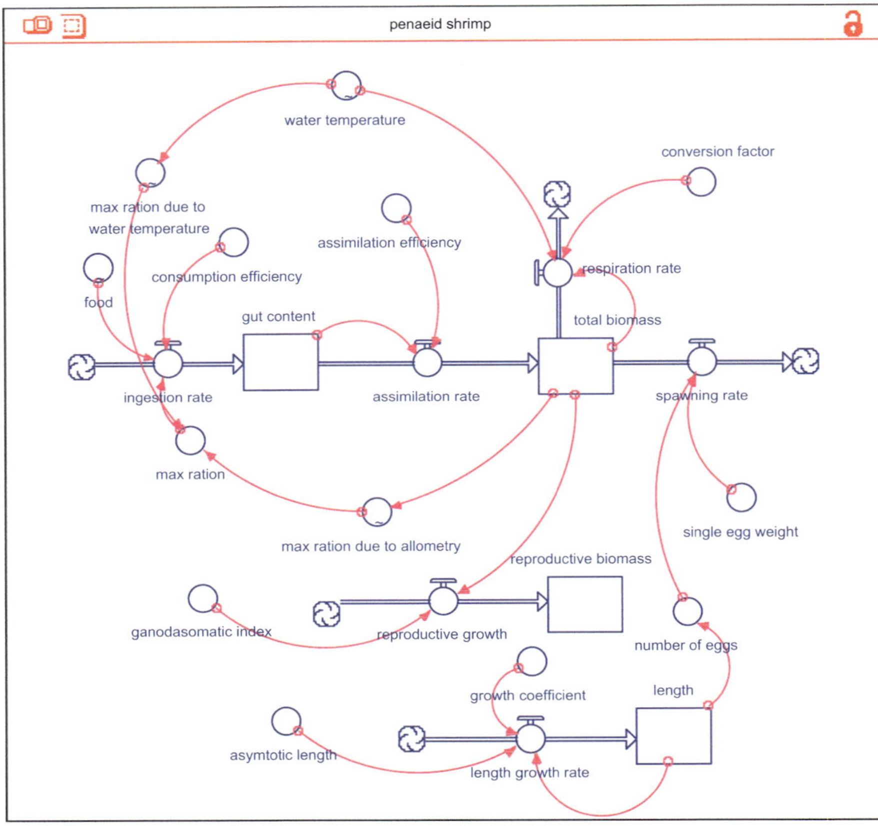

Fig. 4.28 Stock–flow diagram of the penaeid shrimp model

Ingestion

Daily ingestion is simulated after Ivlev (1945) (Eq. (4.23)). In this equation the quantity of food ingested by shrimp increases with the concentration of food available (expresses as dry weight: DW), up to a maximum ration, $R_{\max}$.

$$I = R_{\max}\left(1 - e^{-kF}\right) \tag{4.23}$$

where

$I =$ Ingestion (gDW day^{-1})
$R_{\max} =$ Maximum ration (gDW day^{-1})
$k =$ Consumption efficiency (mgDW^{-1} m^2)
$F =$ Food (mgDW m^{-2})

The food intake as a function of time is expressed as

```
food=GRAPH(TIME)
(0.00, 695), (29.2, 750), (58.3, 750), (87.5, 730), (117, 720),
(146, 710), (175, 710), (204, 735), (233, 685), (263, 630), (292, 665),
(321, 720), (350, 710)
```

R_{max} is a function of both allometry and water temperature (Eq. (4.24)):

$$R_{max} = f(B_t)f(T) \tag{4.24}$$

where

B_t = Total biomass (gDW)
T = Temperature (°C)
R_{max} is computed from both allometry and water temperature as

```
max_ration=max_ration_due_to_allometry*max_ration_due_to_water_
temperature*2.5
```

Water temperature varies with time and these variations with time are expressed as

```
water_temperature=GRAPH(TIME)
(0.00, 27.3), (29.2, 29.5), (58.3, 30.5), (87.5, 26.5), (117, 27.3),
(146, 28.8), (175, 31.5), (204, 29.5), (233, 25.8), (263, 27.5),
(292, 27.8), (321, 29.5), (350, 29.8)
```

The maximum ration due to allometry was calculated using Eq. (4.24), obtained for *P. subtilis* (Nunes and Parsons 2000):

$$f(B_t) = 0.09(\Phi B_t)^{0.62} \tag{4.25}$$

where

Φ = Conversion factor from dry weight to fresh weight (dimensionless)

The relationship in Eq. (4.25) is simplified, and the maximum ration due to allometry expressed is as a function of total biomass.

```
max_ration_due_to_allometry=GRAPH(total_biomass)
(1.00, 0.15), (1.90, 0.243), (2.80, 0.327), (3.70, 0.393), (4.60, 0.474),
(5.50, 0.543), (6.40, 0.615), (7.30, 0.663), (8.20, 0.708), (9.10, 0.735),
(10.0, 0.75)
```

The effect of water temperature on food consumption was simulated as a positive relationship over a given temperature range, followed by a decrease of food intake beyond the optimal temperature (Niu et al. 2003; Wasielesky et al. 2003; Kumlu and Kir 2005). Wasielesky et al. (2003) found that food consumption increased with temperature but stabilised from 26 °C to 32 °C. Additionally, Hewitt and Duncan (2001) observed that the mean daily consumption of *P. japonicus* was highest at 32 °C (range 28–36 °C). A second-order polynomial was used to describe this relationship in the model, based on data from Wyban et al. (1995) and on the assumption that the shrimp ration is maximal at 32 °C:

$$f(T) = \frac{-0.02T^2 + 1.44T - 17.41}{R_{32}} \tag{4.26}$$

where

R_{32} = Ration at 32 °C

The relationship in Eq. (4.24) is as a function of water temperature.

```
max_ration_due_to_water_temperature = GRAPH(water_temperature)
(25.0, 0.902), (25.7, 0.92), (26.4, 0.951), (27.1, 0.99), (27.8, 1.02),
(28.5, 1.06), (29.2, 1.08), (29.9, 1.08), (30.6, 1.09), (31.3, 1.10),
(32.0, 1.10)
```

Several authors refer that larger sizes and faster growth rates are observed in females (Dredge 1990; Chow and Sandifer 1991; Buckworth 1992); therefore, the model also takes sexual dimorphism into account. This difference was represented as a lower food intake for males than females (Hansford and Hewitt 1994). Growth of male animals (expressed as carapace length) was estimated using the von Bertalanffy equation, parameterised from Le Reste and Marcille (1976):

$$L_t = L_\infty \left[1 - e^{K(t - t_0)}\right] \tag{4.27}$$

where

L_t = Carapace length at time t (mm)
L_∞ = Asymptotic carapace length (mm)
K = Growth coefficient (day^{-1})
t = Age (day)
t_0 = Starting age (day)

The Eq. (4.27) can be expressed in the form of differentia equation as

$$\frac{dL_t}{dt} = K(L_\alpha - L_t) \tag{4.28}$$

In STELLA this equation is

```
length(t) = length(t - dt) + (length_growth_rate) * dt
INIT length = 0.0
```

And length growth rate is expressed as

```
length_growth_rate = MAX(0, (growth_coefficient*(asymptotic_length-
length)))
asymptotic_length = 29.9
growth_coefficient = 0.013
```

The growth curve expressed as carapace length was converted to fresh weight and used to parameterise the maximum ration equation for male penaeids by tuning the coefficients to optimally correlate model biomass outputs to the growth curve:

$$f(B_t) = 0.06(\Phi B_t)^{0.68} \tag{4.29}$$

Assimilation/Faeces Production

Part of the food ingested is assimilated in the gut, and the remaining fraction is eliminated as faeces. The amount of food assimilated is dependent on the gut content and assimilation efficiency (Eq. 4.30).

$$A = GA_e \tag{4.30}$$

where

A = Assimilation (g DW day^{-1})
G = Gut content (g DW)
A_e = Assimilation efficiency (day^{-1})

Respiration

Shrimp metabolic losses are assumed to be due to oxygen consumption (respiration) and dependent on the individual biomass and temperature. Respiration was simulated following the equation derived by Dall (1986) for *P. esculentus* (Eq. 4.31), which gives the weight-specific respiration rate:

$$M_w = 0.05e^{0.07T}(\Phi B_t)^{-.0.185} \tag{4.31}$$

where

M_w = Weight-specific oxygen consumption (day^{-1})

The weight-specific respiration was then multiplied by the dry weight of the animal to give the total respiration, M, expressed as g DW day^{-1}.

Gut Content and Total Biomass

The balance between ingestion and assimilation determines the gut content (Eq. 4.32.):

$$\frac{dG}{dt} = R_{\max}\left(1 - e^{-kF}\right) - GA_{\mathrm{e}} \tag{4.32}$$

where the first term on the right-hand side corresponds to the ingestion and the second to the assimilation.

Equation (4.32) in STELLA is

```
gut_content(t) = gut_content(t - dt) + (ingestion_rate - assimilation_
rate) * dt
INIT gut_content = 0.5
The ingestion rate is
ingestion_rate = max_ration* (1.0-EXP
(−consumption_efficiency*food))
consumption_efficiency = 0.12
```

The assimilate rate is

```
assimilation_rate = gut_content*assimilation_efficiency
assimilation_efficiency = 0.8
```

The change in total individual biomass (Eq. 4.33) is thus

$$\frac{dB_{\mathrm{t}}}{dt} = GA_{\mathrm{e}} - M - WS \tag{4.33}$$

where

M = Respiration (gDW day^{-1})
S = Spawning (gDW day^{-1})

Equation (4.33) in STELLA is

```
total_biomass(t) = total_biomass(t - dt) + (assimilation_rate - respiration_
rate - spawning_rate) * dt
INIT total_biomass = 1
```

The respiration rate is

```
respiration_rate = 0.05*EXP(0.07*water_temperature)*
((total_biomass*conversion_factor)^(−0.185))*total_biomass
```

The spawning rate is

```
spawning_rate = number_of_eggs*single_egg_weight
single_egg_weight = 2.0E-6
```

Reproduction and Non-somatic Biomass

To describe female reproduction, two processes were considered: maturation and spawning. The gonadosomatic index is employed to simulate gonad growth as it is widely used to assess the maturation state in these animals (Cha et al. 2002). A constant rate of gonad growth over the maturation period was considered. The non-somatic biomass may be described according to Eq. (4.34):

$$\frac{dB_{\mathrm{g}}}{dt} = B_{\mathrm{t}} G_{\mathrm{SI}} \tag{4.34}$$

where

B_{g} = Reproductive (non-somatic) biomass (gDW)
G_{SI} = Gonadosomatic index (day^{-1})

Equation (4.34) in integral form in STELLA is

```
reproductive_biomass(t) = reproductive_biomass(t - dt) + (reproductive_
growth) * dt
INIT reproductive_biomass = .0001
```

The reproductive growth is

```
reproductive_growth = total_biomass*ganodasomatic_index
ganodasomatic_index = .05
```

The somatic tissue biomass $\mathrm{B_s}$ may then be calculated as $B_{\mathrm{t}} - B_{\mathrm{g}}$. The loss of biomass due to spawning (S) was calculated by multiplying the number of eggs produced by the mean weight of a single egg (Eq. 4.35):

$$N = 9.3 C_{\mathrm{L}}{}^{2.8} \tag{4.35}$$

where

N = Number of eggs produced (day^{-1})
C_{L} = Carapace length (mm)

This relation (4.35) in STELLA is
number_of_eggs = 9.3*(length^2.8)

$$S = N E_{\mathrm{w}}$$

where

E_{w} = Single egg weight (gDW)

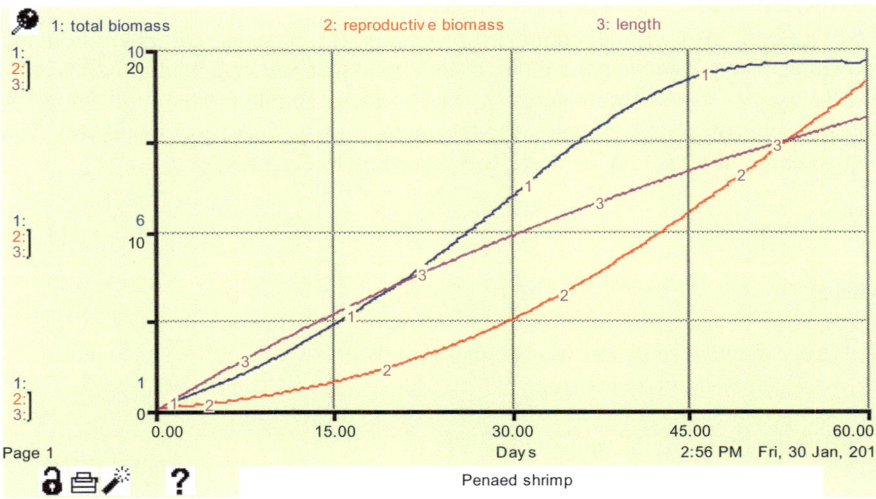

Fig. 4.29 Simulated total biomass, reproductive biomass and length of penaeid shrimp

Figure 4.29 shows the simulated shrimp growth curve, expressed as fresh weight. The model reproduces the typical pattern of growth of a penaeid shrimp, characterised by a rapid weight gain during the juvenile stage followed by the achievement of asymptotic length of adults (Dall et al. 1990). The individual reaches an asymptotic length after 6–8 months, which is in accordance with data from Rothlisberg (1998).The sexual dimorphism is confirmed by a larger size and a faster growth rate of females than males.

STELLA model of the penaeid shrimp growth is given below:

Penaeid Shrimp

```
gut_content(t) = gut_content(t - dt) + (ingestion_rate - assimilation_
rate) * dt
INIT gut_content = 0.5
```

INFLOWS:

```
ingestion_rate = max_ration*(1.0-EXP(−consumption_efficiency*food))
```

OUTFLOWS:

```
assimilation_rate = gut_content*assimilation_efficiency
```

```
length(t) = length(t - dt) + (length_growth_rate) * dt
INIT length = 0.0
```

INFLOWS:

```
length_growth_rate = MAX(0, (growth_coefficient * (asymptotic_length-
length)))

reproductive_biomass(t) = reproductive_biomass(t - dt) + (reproductive_
growth) * dt
INIT reproductive_biomass = .0001
```

INFLOWS:

```
reproductive_growth = total_biomass*ganodasomatic_index

total_biomass(t) = total_biomass(t  -  dt) + (assimilation_rate  -
respiration_rate - spawning_rate) * dt
INIT total_biomass = 1
```

INFLOWS:

```
assimilation_rate = gut_content*assimilation_efficiency
```

OUTFLOWS:

```
respiration_rate = 0.05*EXP(0.07*water_temperature)*
((total_biomass*conversion_factor)^(-0.185))*total_biomass
spawning_rate = number_of_eggs*single_egg_weight

assimilation_efficiency = 0.8
asymptotic_length = 29.9
consumption_efficiency = 0.12
conversion_factor = 5
ganodasomatic_index = .05
growth_coefficient = 0.013
max_ration = max_ration_due_to_allometry*max_ration_due_to_water_
temperature*2.5
number_of_eggs = 9.3*(length^2.8)
single_egg_weight = 2.0E-6

food = GRAPH(TIME)
(0.00, 695), (29.2, 750), (58.3, 750), (87.5, 730), (117, 720),
(146, 710), (175, 710), (204, 735), (233, 685), (263, 630), (292, 665),
(321, 720), (350, 710)
```

```
max_ration_due_to_allometry = GRAPH(total_biomass)
(1.00, 0.15), (1.90, 0.243), (2.80, 0.327), (3.70, 0.393), (4.60, 0.474),
(5.50, 0.543), (6.40, 0.615), (7.30, 0.663), (8.20, 0.708), (9.10, 0.735),
(10.0, 0.75)
```

```
max_ration_due_to_water_temperature = GRAPH(water_temperature)
(25.0, 0.902), (25.7, 0.92), (26.4, 0.951), (27.1, 0.99), (27.8, 1.02),
(28.5, 1.06), (29.2, 1.08), (29.9, 1.08), (30.6, 1.09), (31.3, 1.10),
(32.0, 1.10)
```

```
water_temperature = GRAPH(TIME)
(0.00, 27.3), (29.2, 29.5), (58.3, 30.5), (87.5, 26.5), (117, 27.3),
(146, 28.8), (175, 31.5), (204, 29.5), (233, 25.8), (263, 27.5),
(292, 27.8), (321, 29.5), (350, 29.8)
```

4.11.8 Crop Irrigation Model

Planning and management of crop irrigation systems require development of analytical models in order to facilitate analysis. The problem is a complex bioeconomic system. System dynamics model of crop irrigation system can be used to determine the quantity and timing of irrigation water supply. Figure 4.30 shows the stock–flow diagram of a simple crop irrigation model. This model provides a framework in which demand and benefits of irrigation may be quantified to aid in irrigation policy planning. Any plant simulation model should include (1) a method of estimating evapotranspiration, (2) a method of estimating soil moisture levels in the root zone and (3) a provision of a means of relating crop growth in each time period to the incidence and severity of moisture stress in that period.

The variations of the reference evaporation with development stage of the crop are a non-linear relationship, and it is expressed by a graphical function.

```
evaporation = GRAPH(TIME)
(0.00, 3.80), (11.0, 4.20), (22.0, 4.60), (33.0, 1.10), (44.0, 2.40),
(55.0, 2.30), (66.0, 2.10), (77.0, 2.30), (88.0, 2.70), (99.0, 3.40),
(110, 3.00), (121, 3.20)
```

Potential evapotranspiration is computed from reference evaporation and crop factor.

```
potential_evapotranspiration = evaporation*crop_factor
```

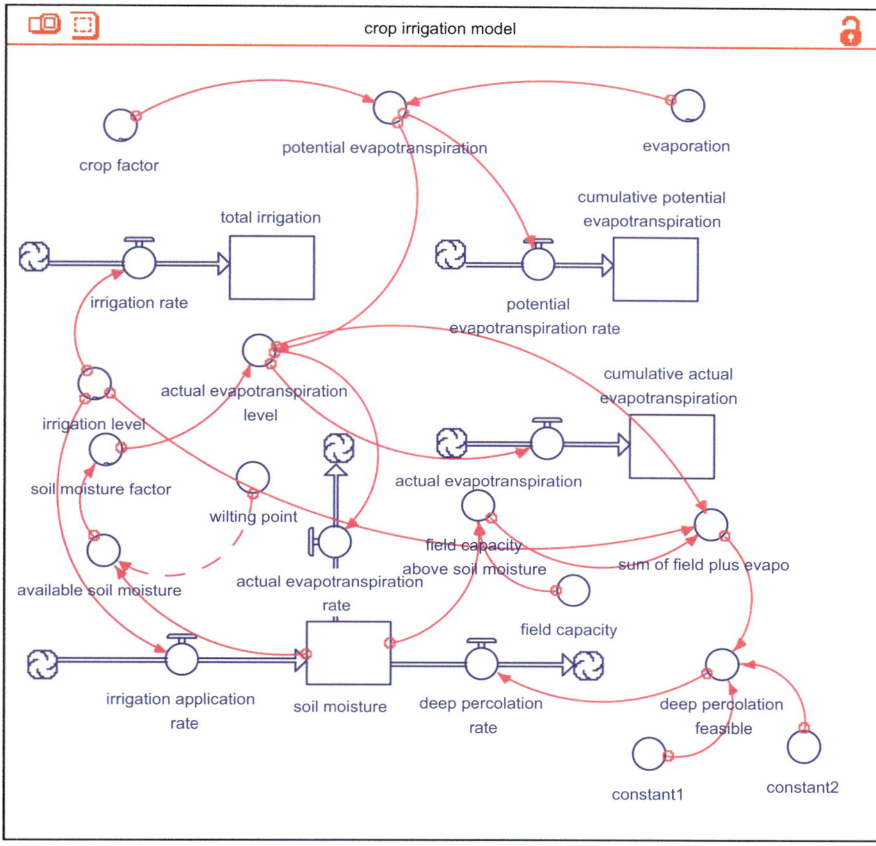

Fig. 4.30 Stock–flow diagram of crop irrigation model

The crop factor with the development stage of the crop is a non-linear relationship, and it is expressed by a graphical function.

```
crop_factor = GRAPH(TIME)
(0.00, 0.3), (12.7, 0.35), (25.5, 0.4), (38.2, 0.55), (50.9, 0.65),
(63.6, 0.85), (76.4, 1.10), (89.1, 1.20), (102, 1.10), (115, 1.05),
(127, 0.9), (140, 0.75)
```

Cumulative potential evapotranspiration is the accumulation of the potential evapotranspiration during the crop production period.

```
cumulative_potential_evapotranspiration(t) =
cumulative_potential_evapotranspiration(t - dt) + (potential_
evapotranspiration_rate) * dt
INIT cumulative_potential_evapotranspiration = 0.0
```

Actual evapotranspiration is evaluated from potential evapotranspiration and soil moisture factor.

```
actual_evapotranspiration = actual_evapotranspiration__level
actual_evapotranspiration__level = potential_evapotranspiration*
soil_moisture_factor
```

Cumulative actual evapotranspiration is the accumulation of the actual evapotranspiration during the crop production period.

```
cumulative_actual_evapotranspiration(t) =
cumulative_actual_evapotranspiration(t - dt)
+ (actual_evapotranspiration) * dt
INIT cumulative_actual_evapotranspiration = 0.0
```

The soil moisture factor is computed from the relation between soil moisture factor and available soil moisture, and it is expressed by a graphical function.

```
soil_moisture_factor = GRAPH(available_soil_moisture)
(0.00, 0.00), (9.00, 0.03), (18.0, 0.06), (27.0, 0.09), (36.0, 0.12),
(45.0, 0.16), (54.0, 0.19), (63.0, 0.23), (72.0, 0.26), (81.0, 0.32),
(90.0, 0.38), (99.0, 0.45), (108, 0.53), (117, 0.62), (126, 0.72),
(135, 0.86), (144, 0.95), (153, 1.00), (162, 1.00), (171, 1.00),
(180, 1.00)
```

The irrigation provided during the development stage of the crop is expressed by a graphical function.

```
irrigation_level = GRAPH(TIME)
(7.00, 0.00), (8.00, 26.8), (9.00, 0.00), (10.0, 0.00), (11.0, 0.00),
(12.0, 0.00), (13.0, 0.00), (14.0, 0.00), (15.0, 27.0), (16.0, 0.00),
(17.0, 0.00), (18.0, 0.00), (19.0, 0.00), (20.0, 0.00), (21.0, 0.00),
(22.0, 27.4), (23.0, 0.00), (24.0, 0.00), (25.0, 0.00), (26.0, 0.00),
(27.0, 0.00), (28.0, 0.00), (29.0, 0.00), (30.0, 0.00), (31.0, 13.8),
(32.0, 0.00)
```

Total irrigation is the accumulation of the irrigation during the crop production period.

```
total_irrigation(t) = total_irrigation(t - dt) + (irrigation_rate) * dt
INIT total_irrigation = 0.0
```

The deep percolation is computed from the consideration that depletion must not take place in unfeasible values, and this is implemented using logical function.

```
deep_percolation__feasible = IF(sum_of_field_plus_evapo = constant2)
THEN(constant1)ELSE(−sum_of_field_plus_evapo)
```

```
sum_of_field_plus_evapo = field_capacity_above_soil_moisture
+ actual_evapotranspiration__level-irrigation_level
field_capacity_above_soil_moisture = field_capacity-soil_moisture
field_capacity = 260
constant1 = 0.0
constant2 = 0.0
```

Soil moisture level at any time is computed from the soil moisture level at previous time, plus irrigation and minus actual evapotranspiration and deep percolation

```
soil_moisture(t) = soil_moisture(t - dt) + (irrigation_application__
rate - actual_evapotranspiration__rate - deep_percolation__rate) * dt
INIT soil_moisture = 240
```

Simulated soil moisture and cumulative actual evapotranspiration are shown in Fig. 4.31. The data and parameters of this model were adopted from Bala et al. (1988). Simulated soil moisture and cumulative actual evapotranspiration agree well with the values of the field level study reported by Bala et al. (1988).

STELLA model of the crop irrigation model is given below:

Crop Irrigation Model

```
cumulative_actual_evapotranspiration(t) = cumulative_actual_
evapotranspiration(t - dt) + (actual_evapotranspiration) * dt
INIT cumulative_actual_evapotranspiration = 0.0
```

INFLOWS:

```
actual_evapotranspiration = actual_evapotranspiration__level
```

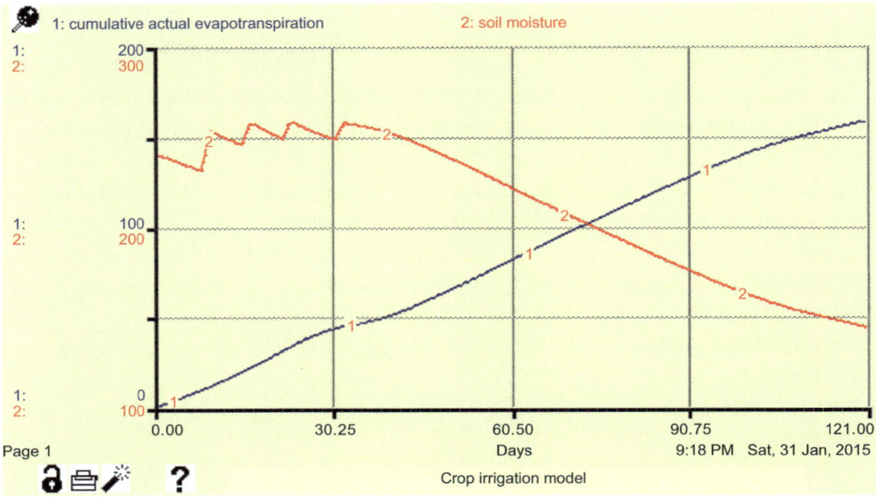

Fig. 4.31 Simulated soil moisture and cumulative actual evapotranspiration

```
cumulative_potential_evapotranspiration(t) =
cumulative_potential_evapotranspiration(t - dt) + (potential_
evapotranspiration_rate) * dt
INIT cumulative_potential_evapotranspiration = 0.0
```

INFLOWS:

```
potential_evapotranspiration_rate = potential_evapotranspiration
```

```
soil_moisture(t) = soil_moisture(t - dt) + (irrigation_application__
rate - actual_evapotranspiration__rate - deep_percolation__rate) * dt
INIT soil_moisture = 240
```

INFLOWS:

```
irrigation_application__rate = irrigation_level
```

OUTFLOWS:

```
actual_evapotranspiration__rate = actual_evapotranspiration__level
deep_percolation__rate = deep_percolation__feasible
```

```
total_irrigation(t) = total_irrigation(t - dt) + (irrigation_rate) * dt
INIT total_irrigation = 0.0
```

INFLOWS:

```
irrigation_rate = irrigation_level

actual_evapotranspiration__level = potential_evapotranspiration*soil_
moisture_factor
available_soil_moisture = soil_moisture-wilting_point
constant1 = 0.0
constant2 = 0.0
deep_percolation__feasible = IF(sum_of_field_plus_evapo = constant2)
THEN(constant1)ELSE(-sum_of_field_plus_evapo)
field_capacity = 260
field_capacity_above_soil_moisture = field_capacity-soil_moisture
potential_evapotranspiration = evaporation*crop_factor
sum_of_field_plus_evapo = field_capacity_above_soil_moisture
+actual_evapotranspiration__level-irrigation_level
wilting_point = 80

crop_factor = GRAPH(TIME)
(0.00, 0.3), (12.7, 0.35), (25.5, 0.4), (38.2, 0.55), (50.9, 0.65),
(63.6, 0.85), (76.4, 1.10), (89.1, 1.20), (102, 1.10), (115, 1.05),
(127, 0.9), (140, 0.75)

evaporation = GRAPH(TIME)
(0.00, 3.80), (11.0, 4.20), (22.0, 4.60), (33.0, 1.10), (44.0, 2.40),
(55.0, 2.30), (66.0, 2.10), (77.0, 2.30), (88.0, 2.70), (99.0, 3.40),
(110, 3.00), (121, 3.20)

irrigation_level = GRAPH(TIME)
(7.00, 0.00), (8.00, 26.8), (9.00, 0.00), (10.0, 0.00), (11.0, 0.00),
(12.0, 0.00), (13.0, 0.00), (14.0, 0.00), (15.0, 27.0), (16.0, 0.00),
(17.0, 0.00), (18.0, 0.00), (19.0, 0.00), (20.0, 0.00), (21.0, 0.00),
(22.0, 27.4), (23.0, 0.00), (24.0, 0.00), (25.0, 0.00), (26.0, 0.00),
(27.0, 0.00), (28.0, 0.00), (29.0, 0.00), (30.0, 0.00), (31.0, 13.8),
(32.0, 0.00)

soil_moisture_factor = GRAPH(available_soil_moisture)
(0.00, 0.00), (9.00, 0.03), (18.0, 0.06), (27.0, 0.09), (36.0, 0.12),
(45.0, 0.16), (54.0, 0.19), (63.0, 0.23), (72.0, 0.26), (81.0, 0.32),
(90.0, 0.38), (99.0, 0.45), (108, 0.53), (117, 0.62), (126, 0.72),
(135, 0.86), (144, 0.95), (153, 1.00), (162, 1.00), (171, 1.00),
(180, 1.00)
```

4.11.9 Pollution Model

Pollution is the action that causes environmental degradation. Yes, pollution is one of the causes of degradation and also the pollution generated affects the quality of life. Environmental degradation due to pollution caused by human actions can be viewed in terms of population and pollution normal due to human action. Environmental considerations are now increasingly becoming important components of planning and computer modelling can provide greater insight of the systems and better understanding of the control measures to be adopted. Stock–flow diagram of the pollution model is shown Fig. 4.32.

Pollution level increases by human activities and decreases by absorption.

```
pollution(t) = pollution(t - dt) + (genration_rate - absorption_rate) * dt
INIT pollution = 10000000 Pollution generation is taken as population
multiplied by pollution normal.
```

```
genration_rate = population*pollution_normal
pollution_normal = 0.25
```

Pollution absorption depends on the existence of pollution to be absorbed. It depends on the natural processes that will determine the amount absorbed in a specified time. Pollution absorption is determined by the amount of pollution divided by absorption time.

```
absorption_rate = pollution/pollution_absorption_time
```

The pollution absorption time depends on the pollution ratio, and it is represented as a graphical function.

```
pollution_absorption_time = GRAPH(pollution_ratio)
(0.00, 0.6), (10.0, 2.50), (20.0, 5.00), (30.0, 8.00), (40.0, 11.5),
(50.0, 15.5), (60.0, 20.0)
```

Pollution ratio is defined as the ratio of pollution level to pollution standard.

```
pollution_ratio = pollution/pollution_standard
pollution_standard = 15000000
```

Population is increased by birth rate and decreased by death rate.

```
population(t) = population(t - dt) + (births - deaths) * dt
INIT population = 100000000
```

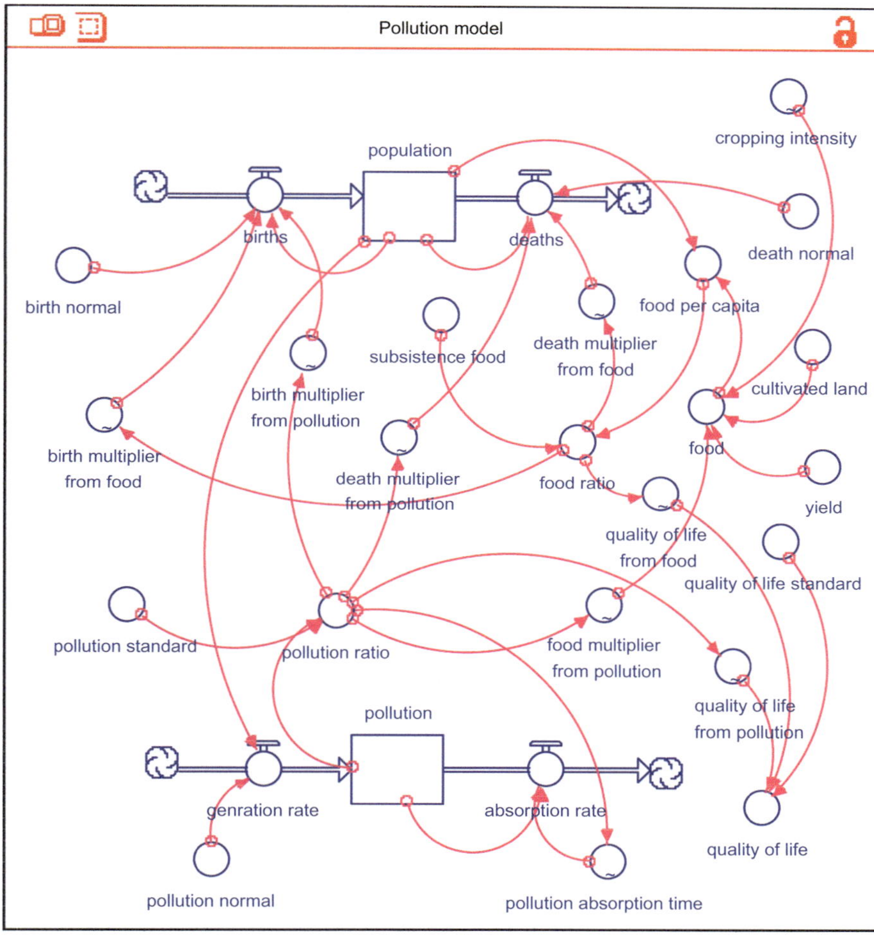

Fig. 4.32 Stock–flow diagram of the pollution model

Actual birth rate is computed from population level, birth rate normal, birth rate from food multiplier and birth rate from pollution multiplier.

```
births = population*birth_normal*birth_multiplier__from_food*birt-
h_multiplier__from_pollution
birth_normal = 0.04
```

Birth multiplier from food is non-linearly related to food ratio, and it is presented graphically. The food ratio is defined as the ratio of food per capita to food per capita standard.

```
birth_multiplier__from_food = GRAPH(food_ratio)
(0.00, 0.00), (1.00, 1.00), (2.00, 1.60), (3.00, 1.90), (4.00, 2.00)

food_ratio = food_per_capita/subsistence_food
subsistence_food = 180
```

Birth multiplier from pollution is non-linearly related to pollution ratio, and it is presented graphically.

```
birth_multiplier__from_pollution = GRAPH(pollution_ratio)
(0.00, 1.02), (10.0, 0.9), (20.0, 0.7), (30.0, 0.4), (40.0, 0.25),
(50.0, 0.15), (60.0, 0.1)

food_ratio = food_per_capita/subsistence_food
subsistence_food = 180
```

Death rate depends upon population level, death rate normal, death rate from food multiplier and death rate from pollution multiplier.

```
deaths = population*death_normal*death_multiplier__from_food*death_
multiplier__from_pollution
death_normal = .028
```

Food can be a powerful indicator of population. The non-linear relationship between death rate multiplier from food and food ratio is represented graphically.

```
death_multiplier__from_food = GRAPH(food_ratio)
(0.00, 30.0), (0.25, 3.00), (0.5, 2.00), (0.75, 1.40), (1.00, 1.00),
(1.25, 0.7), (1.50, 0.6), (1.75, 0.5), (2.00, 0.5)
```

Death multiplier from pollution is non-linearly related to pollution, and this relationship is expressed graphically.

```
death_multiplier__from_pollution = GRAPH(pollution_ratio)
(0.00, 0.92), (10.0, 1.30), (20.0, 3.30), (30.0, 4.80), (40.0, 6.80),
(50.0, 9.20), (60.0, 9.50)
```

Food per capita is defined as the ratio of food available to population.

```
food_per_capita = food/population
```

The food availability depends on cultivated land, yield, cropping intensity and food from pollution multiplier.

```
food=cultivated_land*yield*cropping_intensity*food_multiplier__f-
rom_pollution
cultivated_land=7400000
yield=2200
```

Food multiplier pollution is non-linearly related to pollution, and it is expressed graphically.

```
food_multiplier__from_pollution=GRAPH(pollution_ratio)
(0.00, 1.02), (10.0, 0.9), (20.0, 0.65), (30.0, 0.35), (40.0, 0.2),
(50.0, 0.1), (60.0, 0.05)
```

The cropping intensity changes with time, and it is expressed graphically as a function of time.

```
cropping_intensity=GRAPH(TIME)
(0.00, 1.75), (10.0, 1.80), (20.0, 1.90), (30.0, 2.00), (40.0, 2.30),
(50.0, 2.50)
```

Food has the most powerful influence on quality of life. The quality of life multiplier from food depends on food, and it is expressed graphically.

```
quality_of_life_from_food=GRAPH(food_ratio)
(0.00, 0.00), (1.00, 1.00), (2.00, 1.80), (3.00, 2.40), (4.00, 2.70)
```

Pollution affects the quality of life and the quality of life multiplier from pollution depends on pollution, and it is expressed graphically.

```
quality_of_life__from_pollution=GRAPH(pollution_ratio)
(0.00, 1.05), (10.0, 0.85), (20.0, 0.6), (30.0, 0.3), (40.0, 0.15),
(50.0, 0.05), (60.0, 0.02)
```

Quality of life is used here as the measure of performance of the system. It is computed as a multiplication of quality of life standard multiplied by multipliers derived from food and pollution.

```
quality_of _life=quality_of_life_standard*quality_of_life_from_
food*quality_of_life__from_pollution
quality_of_life_standard=1
```

Simulated population, food per capita, pollution and quality of life are shown in Fig. 4.33.

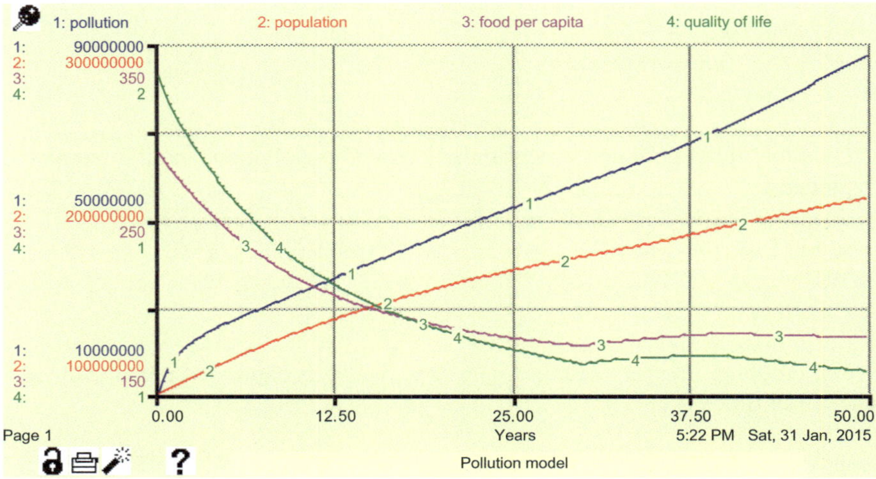

Fig. 4.33 Simulated population, food per capita, pollution and quality of life

Simulated results show that pollution increases as the population increases which is the observed phenomenon. Also food per capita decreases with time. As a result quality of life decreases with time.

STELLA model of pollution is given below:

Pollution Model

```
pollution(t) = pollution(t - dt) + (genration_rate - absorption_rate) * dt
INIT pollution = 10000000
```

INFLOWS:

```
genration_rate = population*pollution_normal
```

OUTFLOWS:

```
absorption_rate = pollution/pollution_absorption_time
```

```
population(t) = population(t - dt) + (births - deaths) * dt
INIT population = 100000000
```

INFLOWS:

```
births = population*birth_normal*birth_multiplier__from_food*birth_
multiplier__from_pollution
```

OUTFLOWS:

```
deaths = population*death_normal*death_multiplier__from_food*death_
multiplier__from_pollution

birth_normal = 0.04
cultivated_land = 7400000
death_normal = .028
food = cultivated_land*yield*cropping_intensity*food_multiplier__
from_pollution
food_per_capita = food/population
food_ratio = food_per_capita/subsistence_food
pollution_normal = 0.25
pollution_ratio = pollution/pollution_standard
pollution_standard = 15000000
quality_of_life = quality_of_life_standard*quality_of_life_from_
food*quality_of_life__from_pollution
quality_of_life_standard = 1
subsistence_food = 180
yield = 2200

birth_multiplier__from_food = GRAPH(food_ratio)
(0.00, 0.00), (1.00, 1.00), (2.00, 1.60), (3.00, 1.90), (4.00, 2.00)

birth_multiplier__from_pollution = GRAPH(pollution_ratio)
(0.00, 1.02), (10.0, 0.9), (20.0, 0.7), (30.0, 0.4), (40.0, 0.25),
(50.0, 0.15), (60.0, 0.1)

cropping_intensity = GRAPH(TIME)
(0.00, 1.75), (10.0, 1.80), (20.0, 1.90), (30.0, 2.00), (40.0, 2.30),
(50.0, 2.50)

death_multiplier__from_food = GRAPH(food_ratio)
(0.00, 30.0), (0.25, 3.00), (0.5, 2.00), (0.75, 1.40), (1.00, 1.00),
(1.25, 0.7), (1.50, 0.6), (1.75, 0.5), (2.00, 0.5)

death_multiplier__from_pollution = GRAPH(pollution_ratio)
(0.00, 0.92), (10.0, 1.30), (20.0, 3.30), (30.0, 4.80), (40.0, 6.80),
(50.0, 9.20), (60.0, 9.50)
```

```
food_multiplier__from_pollution=GRAPH(pollution_ratio)
(0.00, 1.02), (10.0, 0.9), (20.0, 0.65), (30.0, 0.35), (40.0, 0.2),
(50.0, 0.1), (60.0, 0.05)

pollution_absorption_time=GRAPH(pollution_ratio)
(0.00, 0.6), (10.0, 2.50), (20.0, 5.00), (30.0, 8.00), (40.0, 11.5),
(50.0, 15.5), (60.0, 20.0)

quality_of_life_from_food=GRAPH(food_ratio)
(0.00, 0.00), (1.00, 1.00), (2.00, 1.80), (3.00, 2.40), (4.00, 2.70)

quality_of_life__from_pollution=GRAPH(pollution_ratio)
(0.00, 1.05), (10.0, 0.85), (20.0, 0.6), (30.0, 0.3), (40.0, 0.15),
(50.0, 0.05), (60.0, 0.02)
```

Exercises

Exercises 4.1 Draw the stock–flow diagram for the simple biological model of a fish shown in Fig. 4.34.

Exercises 4.2 The birth rate of population of a country is 0.045 for adults, and the average life expectancy is 50 years. After a delay of 5 years, they reach school age, and their education takes 10 years. Assume initial number of babies, school children and adults to be 2500, 30,000 and 90,000, respectively. Draw the causal loop and stock–flow diagram of the system and plot the dynamic responses of the model.

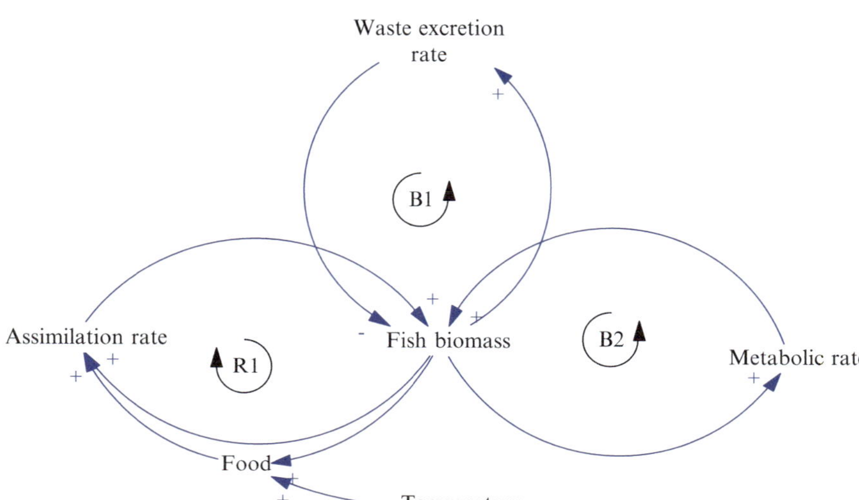

Fig. 4.34 Causal loop diagram of a simple biological model of fish

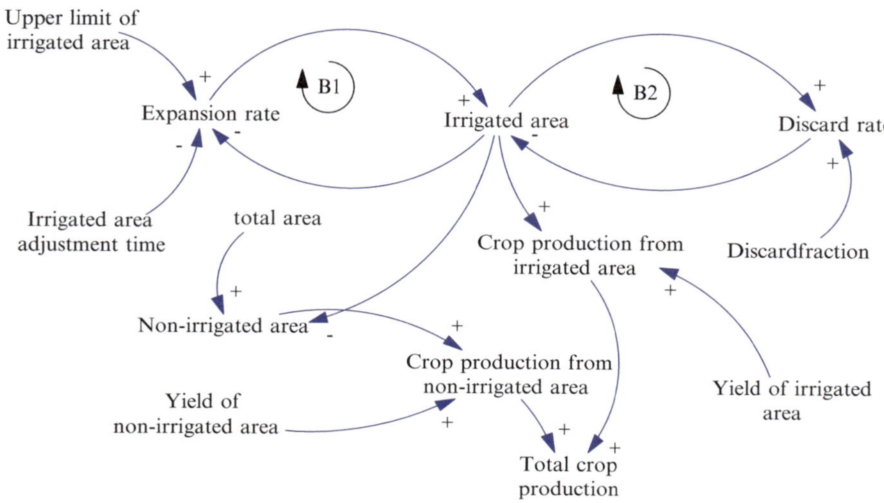

Fig. 4.35 Causal loop diagram of crop production model

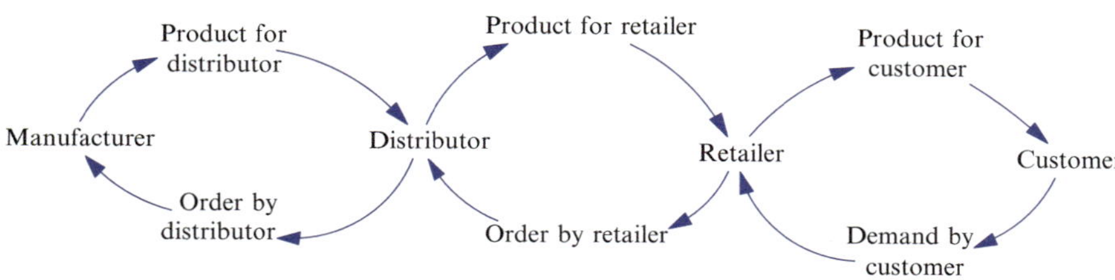

Fig. 4.36 Causal loop diagram of a simple food supply chain

Exercises 4.3 Draw stock–flow diagram and plot dynamic response of the crop production model shown in Fig. 4.35. Assume initial irrigated area = 1.22E6 ha, irrigated area adjustment time = 10 years, upper limit of irrigated area =7.4E6 ha, total area = 7.4E6 ha, yield of irrigated area = 2800 kg/ha, yield of nonirrigated area = 1300 kg/ha and average life time of irrigation device =15 years.

Exercises 4.4 Causal loop diagram of simple supply chain of food product is shown in Fig. 4.36. Elaborate the causal loop diagram and convert it into stock–flow diagram and plot dynamic response of the supply chain for a product of your locality.

Exercises 4.5 Identify the causal loops and draw stock–flow diagram, and plot dynamic response of food grain storage management system as shown in Fig. 4.37. Use food supply = rural food stock + released stock and food supply = 1.275E6 tons, rural storage, rural stock = 1E6 tons, govt. storage = 0.725E6 tons, grain on order = 0.05E6 tons, security reserve =

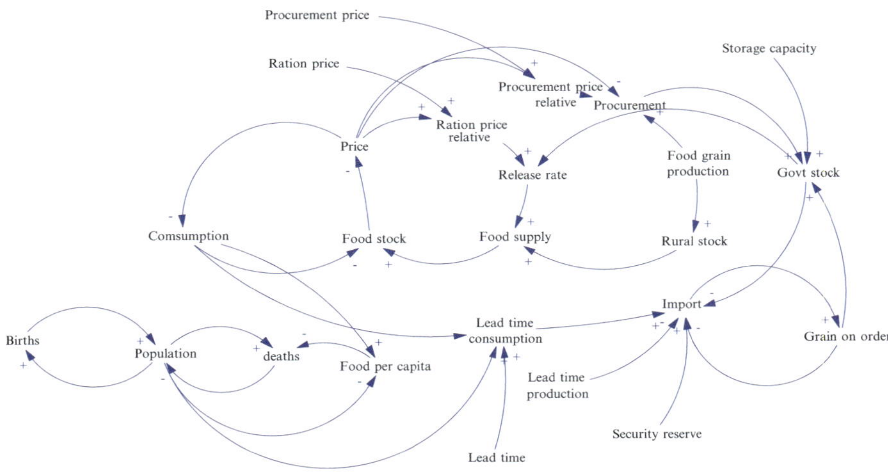

Fig. 4.37 Food grain storage management system

1.5E6 tons, delay in ordering grain =2 days and delay in receiving grain = 3 days, population = 1E8, birth rate normal = 0.003, death rate normal = 0.00125. Lead time = 3 days, ration price = 0.80 $/kg and procurement price = 1.20 $/kg.

References

Bala BK, Satter MA, Halim MA, Talukdar MSU (1988) Simulation of crop irrigation systems. Agric Syst 27:51–65

Buckworth RC (1992) Movements and growth of tagged blue endeavour prawns, Metapenaeus endeavouri (Schmitt 1926), in the western Gulf of Carpentaria Australia. Aust J Mar Freshwat Res 43:1283–1299

Carvajal R, Nebot A (1998) Growth model for white shrimp in semi-intensive farming using inductive reasoning methodology. Comput Electron Agric 19:187–210

Cha HK, Oh C-W, Hong SY, Park KY (2002) Reproduction and population dynamics of Penaeus chinensis (Decapoda: Penaeidae) on the western coast of Korea, Yellow Sea. Fish Res 56:25–36

Chow S, Sandifer PA (1991) Differences in growth, morphometric traits, and male sexual maturity among Pacific white shrimp, Penaeus vannamei, from different commercial hatcheries. Aquaculture 92:165–178

Dall W (1986) Estimation of routine metabolic rate in a penaeid prawn. J Exp Mar Biol Ecol 96:57–74

Dall W, Hill BJ, Rothlisberg PC, Staples DJ (1990) Biology of the Penaeidae. In: Blaxter JHS, Southward AJ (eds) Advances in marine biology, vol 27. Academic, London, p 489

Dredge MCL (1990) Movement, growth and natural mortality rate of the red spot king prawn, Penaeus longistylus Kubo, from the Great Barrier Reef lagoon. Aust J Mar Freshwat Res 41:399–410

Franco AR, Ferreira JG, Nobre AM (2006) Development of a growth model for penaeid shrimp. Aquaculture 259:268–277

Giraldo DP, Betancur MJ, Arango S (2008) Food security in developing countries: a systemic perspective. Paper presented at the international conference of the System Dynamics Society held on July 20–24 2008 at Athens, Greece, 2008

Griffith DRW, Wigglesworth JM (1993) Growth rhythms in the shrimp Penaeus vannamei and P. schmitti. Mar Biol 115:295–299

Hansford SW, Hewitt DR (1994) Growth and nutrient digestibility by male and female Penaeus monodon: evidence of sexual dimorphism. Aquaculture 125(1/2):147–154

Haywood MDE, Vance DJ, Loneragan NR (1995) Seagrass and algal beds as nursery habitats for tiger prawns (Penaeus semisulcatus and P. esculentus) in a tropical estuary. Mar Biol 122:213–223

Hewitt DR, Duncan PF (2001) Effect of high water temperature on the survival, moulting and food consumption of Penaeus (Marsupenaeus) japonicus (Bate, 1888). Aquac Res 32(4):305–313

Ivlev VS (1945) The biological productivity of waters. Uspekhi Sovremenoi Biologii 19 (1):98–120

Jackson CJ, Wang YG (1998) Modelling growth rate of Penaeus monodon Fabricius in intensively managed ponds: effects of temperature, pond age and stocking density. Aquac Res 29:27–36

Kumlu M, Kir M (2005) Food consumption, moulting and survival of Penaeus semisulcatus during over-wintering. Aquac Res 36(2):137–143

Le Reste L, Marcille J (1976) Biologie de la crevette Penaeus indicus H. Milne Edwards a Madagascar: croissance, recruitement, migrations, reproduction, mortalité. Contribuition à l' étude d'une baie eutrophique tropicale. Travaux et documents de l'ORSTOM 14 (2):109–127 (in French)

Lemos D, Phan VN, Alvarez G (2001) Growth, oxygen consumption, ammonia-N excretion, biochemical composition and energy content of Farfantepenaeus paulensis Pérez-Farfante (Crustacea, Decapoda, Penaeidae) early postlarvae in different salinities. J Exp Mar Biol Ecol 261:55–74

López-Martínez J, Arreguín-Sánchez F, Hernández-Vázquez S, García-Juárez AR, Valenzuela-Quiñonez W (2003) Interannual variation of growth of the brown shrimp Farfantepenaeus californiensis and its relation to temperature. Fish Res 61:95–105

Meadows DL (1970) Dynamics of commodity production cycles. Wright Allen Press Inc., Cambridge, MA

Miao S, Tu S (1996) Modelling the effect of thermic amplitude on growing Chinese shrimp, Penaeus chinensis (Osbeck). Ecol Model 88:93–100

Mishra AK, Verdegem M, van Dam A (2002) A dynamic simulation model for growth of penaeid shrimps. In: Cruz-Suárez LE, Ricque-Marie D, Tapia-Salazar M, Gaxiola-Cortés MG, Simoes N (eds) Avances en Nutrición Acuícola VI. Memorias del VI Simposium Internacional de Nutricón Acuícola. 3–6 Septiembre. Cancún, Quintana Roo, Mexico, pp 448–470

Nerlove N (1958) Adaptive expectations and cobweb phenomena. Q J Econ 72:227–240

Niu C, Zhang T, Sun R (2003) Effects of temperature on food consumption, growth and oxygen consumption of freshwater prawn Macrobrachium rosenbergii (de Man 1879) postlarvae. Aquac Res 34(6):501–506

Nunes AJP, Parsons GJ (2000) Size-related feeding and gastric evacuation measurements for the southern brown shrimp Penaeus subtilis. Aquaculture 187:133–151

Rothlisberg P (1998) Aspects of penaeid biology and ecology of relevance to aquaculture: a review. Aquaculture 164:49–65

Wang YG (1998) Growth curves with explanatory variables and estimation of the effect of tagging. Aust N Z J Stat 40(3):299–304

Wasielesky W, Bianchini A, Sanchez CC, Poersh LH (2003) The effect of temperature, salinity and nitrogen products on food consumption of pink shrimp Farfantepenaeus paulensis. Braz Arch Biol Technol 46(1):135–141

Wyban J, Walsh WA, Godin DM (1995) Temperature effects on growth, feeding rate and feed conversion of the pacific white shrimp (Penaeus vannamei). Aquaculture 138:267–279
Xiao Y (1999) General age- and time-dependent growth models for animals. Fish Bull US 97:690–701
Ye Y, Bishop JM, Fetta N, Abdulqader E, Al-Mohammadi J, Alsaffar AH, Almatar S (2003) Spatial variation in growth of the green tiger prawn (Penaeus semisulcatus) along the coastal waters of Kuwait, eastern Saudi Arabia, Bahrain, and Qatar. ICES J Mar Sci 60:806–817

Bibliography

Bala BK (1997a) Computer modelling of the rural energy system and of CO_2 emissions for Bangladesh. Energy 22:999–1003
Bala BK (1997b) Computer modelling of energy and environment: the case of Bangladesh. In: Proceedings of 15th international system dynamics conference, Istanbul, Turkey, 19–22 August 1997
Bala BK (1998) Energy and environment: modelling and simulation. Nova Science Publishers, New York
Bala BK (1999) Principles of system dynamics, 1st edn. Agrotech Publishing Academy, Udaipur
Cavana RY, Maani KE (2000) A methodological framework for systems thinking and modelling (ST&M) interventions. Ist international conference on systems thinking in management. Internet
Dudley RG (2008) A basis for understanding fishery management dynamics. Syst Dyn Rev 24 (10):1–29
FAO (Food and Agricultural Organization) (1996) Implications of economic policy for food security: a training manual. Food and Agriculture Organization of the United Nations, Rome
Forrester JW (1968) Principles of systems. MIT Press, Cambridge, MA
Graham M (1935) Modem theory of exploiting a fishery, and application to North Sea trawling. Journal du Conseil International Pour l'Exploration de la Mer 10(3):264–274
Rosser B Jr (2003) Complexities of dynamic forestry management policies. James Madison University, Harrisonburg
Schaefer MB (1954) Some aspects of the dynamics of populations important to the management of commercial marine fisheries. Inter Am Trop Tuna Comm Bull 1:27–56
Schaefer MB (1957) A study of the dynamics of the fishery for yellow fin tuna in the eastern tropical Pacific Ocean. Inter Am Trop Tuna Comm Bull 2:245–285
Sterman JD (2000) Business dynamics: systems thinking and modeling for a complex world. Irwin McGraw Hill, New York

Parameter Estimation and Sensitivity Analysis

5

The computer model based on stock–flow diagram discussed in Chap. 4 needs estimation of parameters and sensitivity analysis of the parameters. This chapter presents parameter estimation and sensitivity analysis. Parameter estimation using aggregated and disaggregated data are discussed. Estimation of table function, conversion factor and normal fractional flow rate is highlighted with examples. Sensitivity analysis and size of solution interval have been explained, and some examples of sensitivity analysis have been presented.

5.1 Introduction

System dynamics modelling offers an attractive tool for policy testing and evaluation. Policy alternatives can be simulated by computer. But simulation requires a model, and model building is still an art in many respects. After formulation of dynamic hypothesis, including mapping stock and flow diagram of a model, the parameters of the model should be either assigned numerical values or expressed in terms of equations with parameter values to simulate the model. The estimation of parameters is one of the important steps in model building. The correct estimation of parameters is required to provide plausible behaviours of the system over time. Also sensitivity analysis of the parameters is important to assess how the parameter values affect the system behaviour and how important it is to determine the parameters accurately. This chapter discusses parameter estimation techniques and sensitivity analysis.

5.2 Parameter Estimation Techniques

Parameter estimation techniques can be classified on the basis of assumption and data. The data may be categorised in two categories which are disaggregated and aggregated data.

B.K. Bala et al., *System Dynamics*, Springer Texts in Business and Economics,
DOI 10.1007/978-981-10-2045-2_5

5.3 Estimation Using Disaggregated Data

Many system dynamics studies including the studies of Forrester use parameters determined on the basis of descriptive information obtained from the participants in the systems modelling. Such information is disaggregated data.

For an example of parameters that can be estimated with disaggregated data, consider an equation below which represents death rate of the population model:

$$\text{Death rate}(\text{person/year}) = \frac{\text{Population (person)}}{\text{Average life expectancy (year)}} \tag{5.1}$$

In STELLA

death_rate = population/average_life_expectancy

The model assumes a constant average life expectancy, so that every year, 1/average life expectancy of the population is dead. Average life expectancy can be estimated many ways from disaggregate data, and the equation is not used for computation of average life expectancy. The equation is only needed to define the function in the model. There are many ways to find the average life expectancy. One time-consuming approach is to make a survey to find the number of people dead and their ages at the time of death. A second approach is to examine the ages of the people alive and observe the age at which very few people are alive. The third approach is to ask the oldest people who have observed a lot of deaths, about the ages at which most of the old people died. The fourth approach is to examine the health conditions and extrapolate life time expectancy. The fifth approach is to study the descriptive history of population, food and society, etc., and can have some idea of life expectancy. There may be some other means also to find the life expectancy. The data discussed above come from records, a history book, expert testimony and the modeller's own day-to-day experience. Thus, disaggregated data are by far the most abundant source of knowledge about real systems.

Another example is organic degradation equation in paddy production model as represented below:

$$\text{Organic degradation (tonne/year)} = \frac{\text{Cumilative organic (tonne)}}{\text{Organic degradation time (year)}} \tag{5.2}$$

The model estimates a constant degradation time. Every year, 1/organic degradation time of cumulative organic is degraded. As discussed above, the same approach also can be used to estimate the degradation time either by survey, experiment, historical data or any other sources of information.

5.3.1 Table Functions

Many of the relationships between the variables in complex systems are non-linear especially in socio-economic systems, and such non-linear relationships are expressed in the form of table functions. This section describes a less straightforward but a common technique to estimate the table function. The table function seems to be a formidable estimation problem, but it can be broken into subproblems, which are:

1. Estimate the value and the slope of the function at (a) one extreme, (b) the normal value, and (c) the other extreme.
2. Connect these known values and slopes with a smooth curve.

Once these subproblems are solved, the table function is known within a narrow range of values.

For example, consider market price and inventory relationship of the commodity production cycle model.

The non-linear relationships between the price and inventory in STELLA are:

```
price = GRAPH(inventory)
(0.00, 100), (1000, 96.5), (2000, 94.0), (3000, 90.0), (4000, 80.0),
(5000, 70.0), (6000, 50.0), (7000, 30.5), (8000, 20.0), (9000, 16.0),
(10000, 10.0)
```

To estimate table function for price, first consider the extreme condition of zero inventory, where the price would be maximum. After that, consider the normal condition when there is some moderate quantity of inventory and the table function has a negative slope. At the other extreme, the inventory is large, and the price is very low. Now draw a smooth curve through the estimated points (Fig. 5.1). Thus, solving the subproblems of extreme conditions, and connecting known points with smooth curves, allows the modeller to estimate a non-linear table function with confidence.

For another example, consider the relationship between soil moisture factor and available soil moisture of the crop irrigation model.

The non-linear relationships between soil moisture factor and available soil moisture in STELLA are:

```
soil_moisture_factor = GRAPH(available_soil_moisture)
(0.00, 0.00), (9.00, 0.03), (18.0, 0.06), (27.0, 0.09), (36.0, 0.12),
(45.0, 0.16), (54.0, 0.19), (63.0, 0.23), (72.0, 0.26), (81.0, 0.32),
(90.0, 0.38), (99.0, 0.45), (108, 0.53), (117, 0.62), (126, 0.72),
(135, 0.86), (144, 0.95), (153, 1.00), (162, 1.00), (171, 1.00), (180, 1.00)
```

To estimate table function for soil moisture factor, first consider the extreme condition of zero available soil moisture, where the soil moisture factor would be zero. Now consider the normal conditions when there are some moderate values

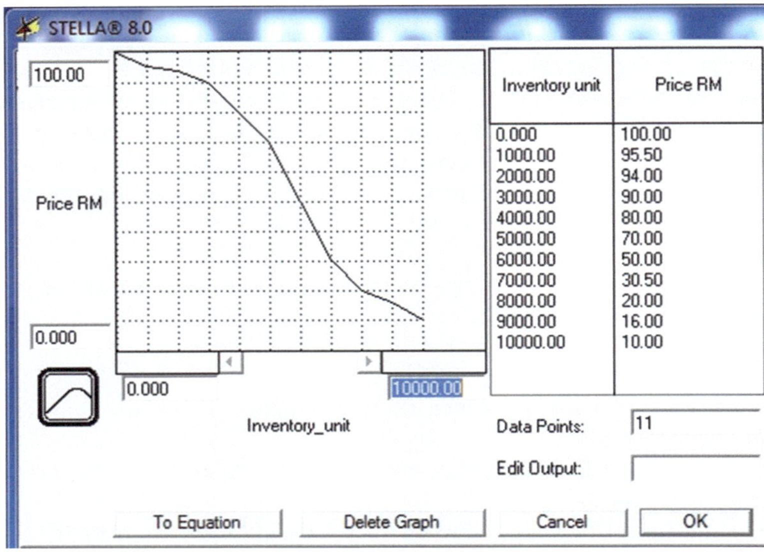

Fig. 5.1 Price versus inventory relationship

and the table function has a positive slope. At the other extreme, the available soil moisture is large and the soil moisture factor is unity. Now draw a smooth curve through the estimated points (Fig. 5.2). Thus, solving the subproblems of extreme conditions, and connecting known points with smooth curves, allows the modeller to estimate a non-linear table function with confidence.

The non-linear relationships between rice per capita consumption and gross domestic product per capita can be expressed in the form of table function. This relationship is illustrated in Fig. 5.3.

```
rice_per_capita_consumption = GRAPH(GDP_per_capita) (4000, 83.7),
(4500, 84.8), (5000, 85.6), (5500, 85.8), (6000, 85.7), (6500, 85.0),
(7000, 84.0), (7500, 82.4), (8000, 80.5), (8500, 78.0), (9000, 75.2)
```

Table functions can also be estimated in normalised form. Consider a cumulative input such as chemical and organic fertilisers in food production. In the long term, the cumulative application of these inputs will affect land fertility, and the fertility will change over time non-linearly. The use of inputs depends on technology and variety of the crop. To develop meaningful relationship between the land fertility effect and cumulative input, it is better to normalise the relationship. In order to normalise the cumulative input, the cumulative input should be divided by its initial value and the land fertility by initial land fertility to determine the fertility effect. After that, the effect of cumulative input on land fertility can be expressed in normalised form as in Fig. 5.4.

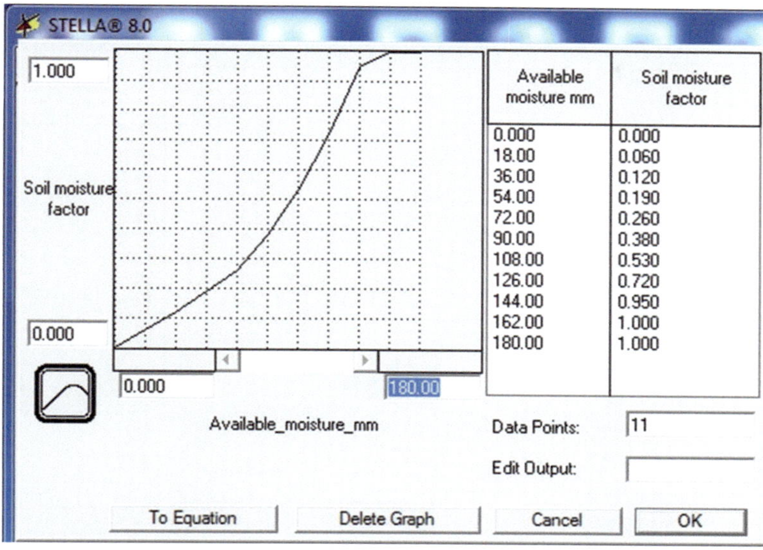

Fig. 5.2 Soil moisture factor versus available soil moisture relationship

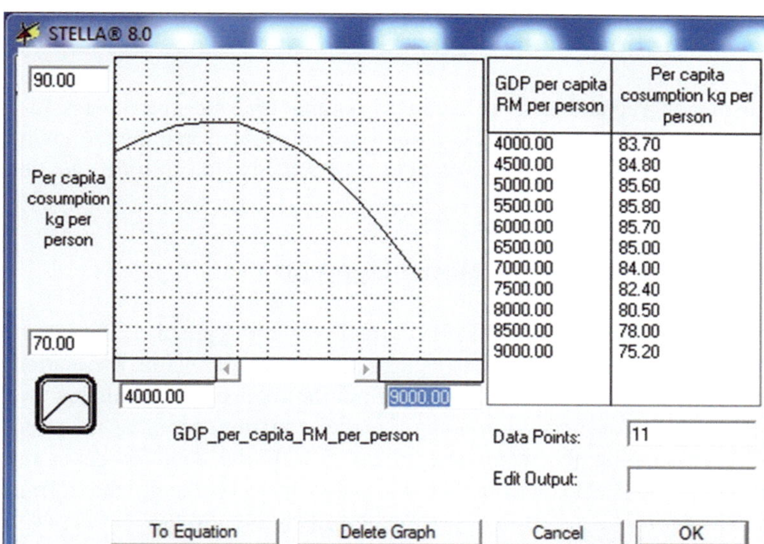

Fig. 5.3 Rice per capita consumption (kg/person) versus GDP per capita (RM/person)

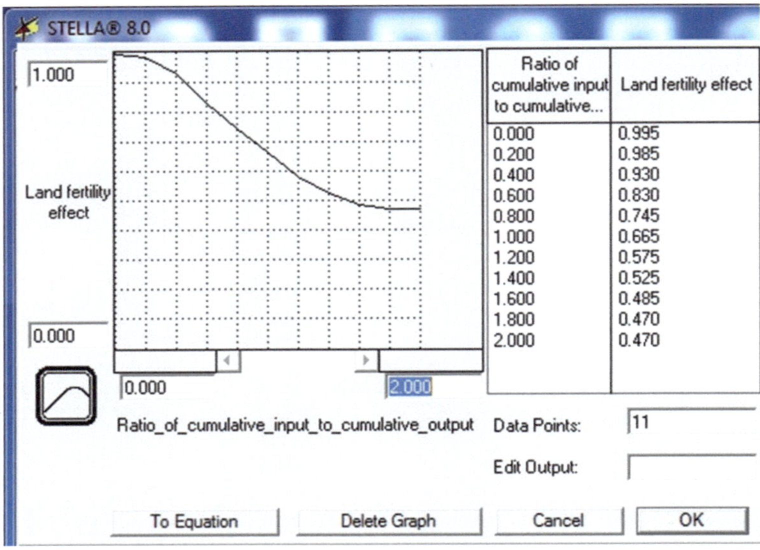

Fig. 5.4 Effect of cumulative input on land fertility

5.4 Estimation Using Aggregated Data

When using aggregated data to estimate parameters, the correctness of model equation(s) is (are) assumed. Two main techniques of parameter estimations using aggregated data are using (a) single equation and (b) multiple equations.

5.4.1 Estimation Using a Model Equation

The model equation is assumed to be correct and the aggregated data that correspond to model variables are used to estimate the parameter value. Estimation using a model equation encompasses all single equation regression techniques.

For an example of estimation using a model equation, consider the population model for calculating the death rate in Eq. 5.1. If the data for death rate and population are available, LE can be estimated by rearranging the formulae in Eq. 5.1. Therefore, the equation for estimated LE is as follows:

$$DR = \frac{P}{LE} \tag{5.3}$$

$$LE = \frac{P}{DR} \tag{5.4}$$

DR Death rate (person/year)
P Population (person)
LE Average life expectancy (year)

Parameters in biological and aquacultural models are essentially estimated using regression techniques. For example, the relation between fasting metabolic rate and body weight of a fish can be expressed as follows:

$$\text{FMR} = \alpha \times (\text{WoF}^{\gamma}) \tag{5.5}$$

FMR Fasting metabolic rate (Mole ATP/day)
WoF Weight of fish (g)

α and γ are the parameters estimated regressing the experimental data on fasting metabolic rate and body weight of fish. For catfish, these parameters are $\alpha = 0.0012$ and $\gamma = 0.8$.

Estimation using a model equation is less frequent in socio-economic system dynamics studies. However, two forms of model equation estimation—one involving conversion factors and the other fractional rates of flow—have been useful in many studies and are discussed below.

Conversion Factors

Many model parameters are conversion factors, and they convert quantities from one dimension to another. For example, consider the housing sector of the residential community model. Land per house (LPH) converts housing units (H) to an equivalent number of hectares:

$$\text{LFO} = \frac{(H \times \text{LPH})}{A} \tag{5.6}$$

LFO Land fraction occupied (dimensionless)
H Housing unit (unit)
LPH Land per house (ha/unit)
A Area (ha)

Such equation can be manipulated to compute the parameter as function of the real data, in this case, the real LFO, H and A:

$$\text{LPH} = \frac{(\text{LFO} \times A)}{H} \tag{5.7}$$

Consider another example. Here, straw feed per cattle converts cattle population to an equivalent amount of straw feed. The following equation uses straw feed per cattle in the definition of fraction of available straw used as cattle feed (FoS):

$$\mathrm{FoS} = \frac{(\mathrm{SFC} \times CP \times DY)}{\mathrm{ASF}} \tag{5.8}$$

FoS Fraction of Straw (dimensionless)
SFC Straw feed per cattle (kg/day)
CP Cattle population (head)
DY 365 days (day)
ASF Available straw feed (kg/year)

This equation can be used to compute the parameter as a function of the real data in this case FoS, ASF, *CP* and *DY*:

$$\mathrm{SFC} = \frac{(\mathrm{FoS} \times \mathrm{ASF})}{(CP \times DY)} \tag{5.9}$$

Estimating conversion factors offers a straightforward means of ensuring that the absolute magnitudes of model variables are realistic.

Normal Fractional Rates of Flow

Equation below defines the death rate in terms of a level population, a normal fractional rate of death (DRN) and dimensionless multipliers:

$$DR = \mathrm{POP} \times \mathrm{DRN} \times \mathrm{DRFM} \times \mathrm{DRPM} \tag{5.10}$$

DR Death rate
POP Population
DRN Death rate normal
DRFM Death rate multiplier from food
DRPM Death rate multiplier from population

The format for above equation is:

$$\mathrm{Rate} = \mathrm{Level} \times \mathrm{Normal\ fraction} \times \mathrm{Multipliers} \tag{5.11}$$

This format is widely used and the multiplier can be easily estimated when it is normalised around 1.0. It also facilitates the estimation of the normal fraction, and it can be represented by:

$$\mathrm{Normal\ fraction} = \frac{\mathrm{Rate}}{(\mathrm{Level} \times \mathrm{Multipliers})} \tag{5.12}$$

Under normal conditions the multipliers assume a value of 1.0, so the fractional flow rate can be obtained by dividing the observed rate by the observed level. For example, suppose 1970 is the normal period for the population model. Then the birth rate normal is obtained from the number of births during 1970 divided by the number of population in 1970. Choosing a normal condition does not bias behaviour of the model. A normal condition is merely a system state about which parameters are defined.

5.5 Estimation Using Multiple Equations

Estimation using several multiple equations consists of using several equations to compute a parameter value. For example, birth rate normal (BRN) can be estimated by finding the value of BRN that causes population growth rate to fit the observed rate of growth. This estimation will use all the equations. The fitting could be performed either with repeated simulations or, if possible, by computation. For an example of such a computation, the population growth rate is 2.5 %. Also from observations of the dead, the life expectancy (LE) is estimated to be 66 years, that is, 1/66 of the population are dead per year. Therefore the BRN should be $1/66 + 0.025 = 0.04$.

5.6 Sensitivity Analysis

Here we consider the parameter sensitivity analysis, and it is conducted to assess how sensitive is the model behaviours to the changes in the values of the parameters. Parameter sensitivity analysis is usually conducted by setting different values such as ±15 % changes in the parameter to study the changes in the model behaviours. Sensitivity analysis provides an opportunity to determine the level of accuracy needed in the estimation of the parameter to make the model valid and useful. If the parameter is insensitive to the model behaviour, it may be used in the policy analysis and design. On the other hand, parameters significantly affecting the model behaviour should be chosen as candidates for additional data collection (Sterman 2000).

The parameters of a system dynamics models are subject to uncertainty. So, sensitivity analysis is an important task for reliability of the simulated results and checking the robustness of the model behaviour for the changes in parameter values. The sensitivity of the important parameters should be estimated. The productivity is one of the most important parameters affecting the model behaviour, i.e. self-sufficiency ratio which is the focus of the food security model (Bala et al. 2014). Figure 5.5 shows the changes in food self-sufficiency ratio for rice productivity for changes of 2.50 t/ha (curve 1), 3.50 t/ha (curve 2) and 4.50 t/ha (curve 3). The food self-sufficiency ratio changes from decrease to increase in values for the changes in the rice productivity from 3.50 t/ha to 2.50 t/ha and 4.50 t/ ha, and this corresponds with real-world situation.

Another example of sensitivity analysis is the sensitivity of cocoa production systems in Malaysia to the changes in the subsidy level (Fatimah et al. 2015). Figure 5.6 shows the simulated cocoa plantation area for full subsidy (curve 3),

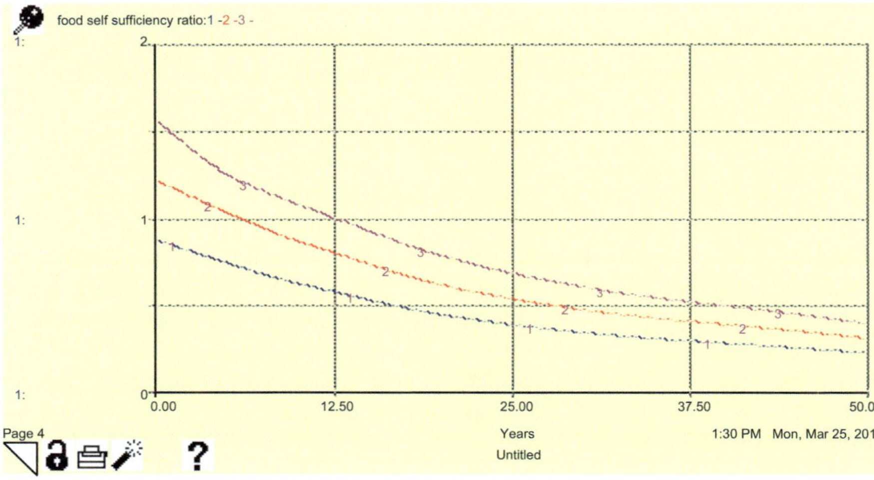

Fig. 5.5 Sensitivity of food self-sufficiency ratio to rice productivity (2.50 t/ha, 3.50 t/ha and 4.50 t/ha)

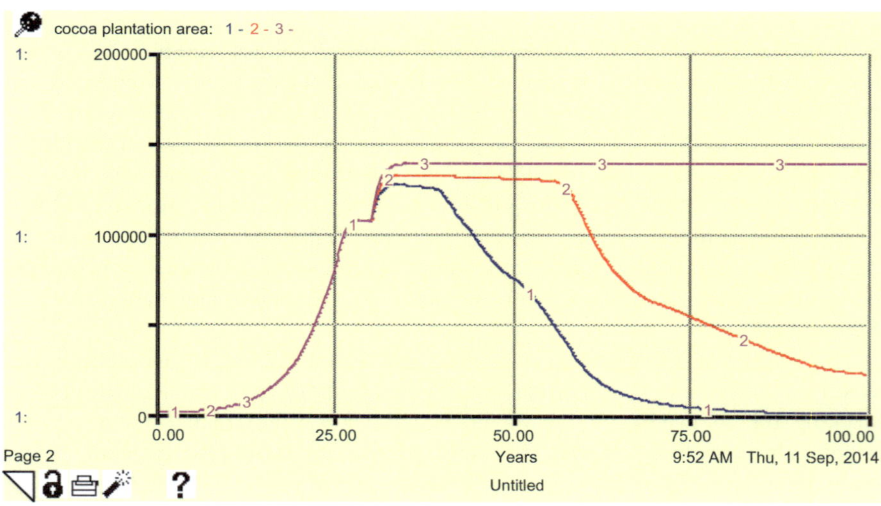

Fig. 5.6 Sensitivity analysis of subsidy for cocoa production systems in Malaysia

80 % subsidy (curve 2) and 60 % subsidy (curve 1) of the cost to cover conservation practices and insect control. The system is sustainable for full subsidy since financial support is provided for joint trade-off of biodiversity and yield along with integrated pest management (IPM) pest control through farmer field schools (FFS), but the sustainability decreases with the degree of reduction of subsidy as lesser opportunities available to maintain biodiversity and insect control in terms of financial support.

5.7 Size of Solution Interval (DT)

It has been mentioned in Chap. 4 that solution interval should be half or less than of the shortest time delay in the model. Let us now consider a simple model of grain storage system as shown in Fig. 5.7 to address the effects of the size of solution interval, and the STELLA equations of the model are given below:

STELLA Equations

```
grain_stock(t) = grain_stock(t - dt) + (receiving_rate) * dt
NIT grain_stock = 50
```

INFLOWS:

```
receiving_rate = (desired_grain_stock-grain_stock)/receeiving_delay
desired_grain_stock = 100
receeiving_delay = 0.5
```

Here in this model, the initial grain stock is 50 tons and desired stock is 100 tons, and receiving time delay is 0.5 months. Let us see what happens to the simulated behaviour of this simple system if solution intervals are 0.20, 0.50 and 1.0. Figures. 5.8, 5.9 and 5.10 show the simulated responses of the system for the solution intervals of 0.20, 0.50 and 1.0, respectively. The simulated response of the system in Fig. 5.8 is the inherent system behaviour for a solution interval of less

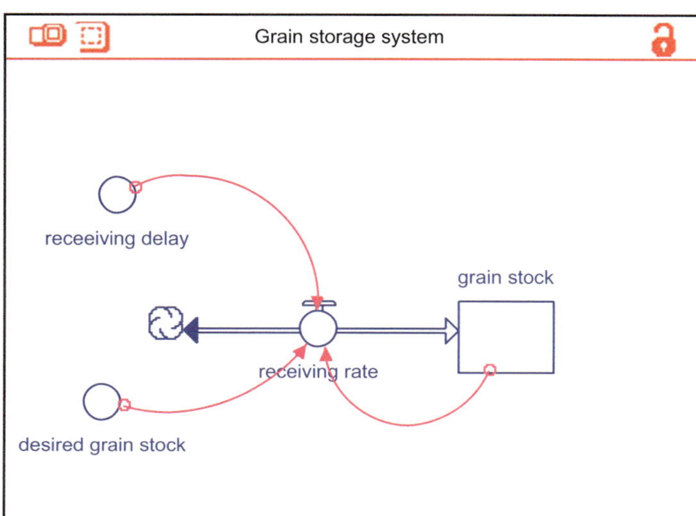

Fig. 5.7 Stock–flow diagram of a simple grain storage system

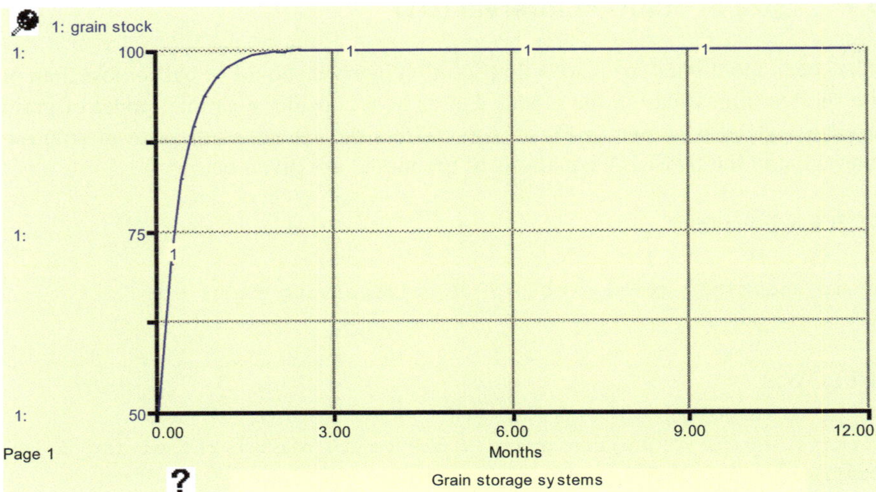

Fig. 5.8 Grain stock response for solution interval of 0.20

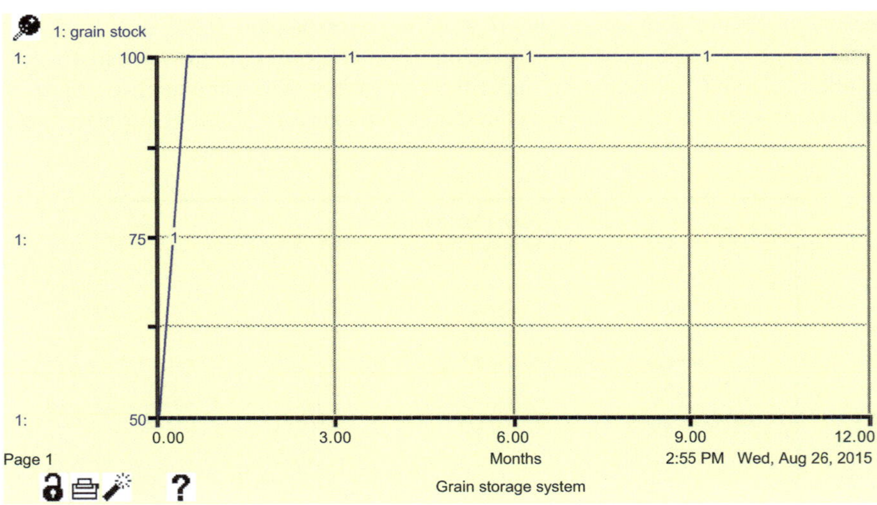

Fig. 5.9 Grain stock response for solution interval of 0.50

than half of the time delay. The simulated response in Fig. 5.9 is poor for a solution interval equal to time delay, and the simulated response in Fig. 5.10 for a solution interval of double of the time delay is the worst, and these responses are not the inherent system behaviour, but these are the results of improper size of solution interval. The solution interval is not part of the real system, but the improper size of it affects the simulated results seriously.

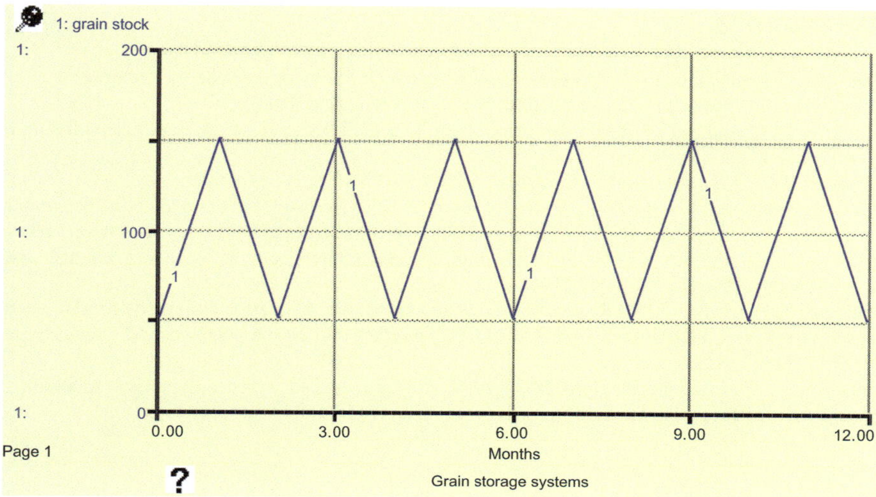

Fig. 5.10 Grain stock response for solution interval of 1.0

Exercises

Exercise 5.1 What is meant by parameter estimation? Describe briefly the techniques of the parameter estimation.

Exercise 5.2 What is table function? Show with examples how table functions are estimated.

Exercise 5.3 What are the main techniques of parameter estimation using aggregated data? Describe how multipliers are estimated?

Exercise 5.4 What is meant by sensitivity analysis? Describe with examples how the sensitivity of a parameter analysed.

Exercise 5.5 What is meant by solution interval? Show that improper use of solution interval can provide explosive behaviour which is not the behaviour of the system.

References

Bala BK, Alias EF, Fatimah MA, Noh KM, Hadi AHA (2014) Modeling of food security in Malaysia. Simul Model Pract Theory 47:154–164

Fatimah MA, Bala BK, Alias EF, Abdulla I (2015) Modeling of boom and bust of cocoa production systems in Malaysia. Ecol Model 309–310:22–32

Sterman JD (2000) Business dynamics: systems thinking and modeling for a complex world. McGraw-Hill Higher Education, Boston

Bibliography

Bala BK (1999) Principles of system dynamics. Agrotech Publishing Academy, Udaipur

Bossel H (1994) Modeling and simulation. K.Peters, Wellesley, 484 pp

Breierova L, Choudhuri M (2001) An introduction to sensitivity analysis. Massachusetts Institute of Technology, Cambridge, MA

Forrester JW (1961) Industrial dynamics. M.I.T. Press, Cambridge, MA

Hekimuglu M, Barlas Y (2015) Sensitivity analysis of system dynamics models by behaviour pattern measures. Department of Industrial Engineering. Bogazici University, Istanbul, Turkey

Maani KE, Cavana RY (2000) Systems thinking and modelling: understanding change and complexity. Prentice Hall, Auckland

Manetsch TJ, Park GL (1982) Systems analysis and simulation with applications to economic and social systems. Department of Electrical Engineering and System Science, Michigan State University, East Lansing

Mohapatra PKJ, Mandal P, Bora MC (1994) Introduction to system dynamics modelling. Universities Press, Hyderabad

Tests for Confidence Building 6

The models developed based on the concepts, methodology and techniques of system dynamics discussed in the earlier chapters must be tested to build up confidence in the models. This chapter presents the tests for confidence building in the system dynamics models. These tests are discussed under the broad heading of tests for structure, tests for behaviour and tests for policy implications. The logical sequences of conducting these testes are also presented.

6.1 Introduction

Once we have developed a model, how can we develop our confidence in the use of the model for developing scenarios and management strategies? How can we make others trust our model? This chapter deals with the tests for confidence building in system dynamics models. Model validation to develop confidence in the model is important, but it is a controversial aspect of any process-based model in general and system dynamics (Barlas 1996). The validity and usefulness of dynamic models should be judged, not against an imaginary perfection but in comparison with the mental and descriptive models which we would otherwise use (Forrester 1968). Models should be judged, not on absolute scale but on relative scale. If they succeed in clarifying our knowledge and insights into the systems for better understanding and management, the model should be accepted.

Tests for building confidence in system dynamics models essentially consist of validation, sensitivity analysis and policy analysis of the system dynamics models. The two important notions of building confidence in system dynamics models are testing and validation of system dynamics models. Testing means the comparison of a model to empirical reality for the accepting or rejecting the model, and validation means the process of establishing confidence in the soundness and usefulness of the model.

In testing mode, the model structures are compared directly to descriptive knowledge of real system structures, and model behaviour may be compared to

B.K. Bala et al., *System Dynamics*, Springer Texts in Business and Economics,
DOI 10.1007/978-981-10-2045-2_6

observed real system behaviour. In validation mode, the model behaves plausively and generates problem symptoms or modes of behaviour observed in the real world. The modeller's confidence needs to be transferred to the target audience.

Validation is complicated by many relevant audiences. For a scientist, a model is useful if it generates insight into the structure of real system, makes correct prediction and stimulates meaningful questions for future research. For the public and political leaders, a model is useful if it explains the causes of important problems and provides a basis for designing policy to improve the behaviour of the system. Validity meaning confidence in a model's usefulness is inherently relative concepts. One must choose between competing models.

In system dynamics model, behaviour testing is very common. This is because very often system dynamics models incorporate such variables for which no real-life data is available. Under such circumstances, assumed relationships are based on available literatures, and the appropriateness of such relationships is justified in the overall context of the generated model behaviour. Though system dynamics modellers try to make best use of the available data for parameter estimation, sometimes by statistical methods, vigorous use of statistical tests is rare.

The main reason why system dynamics methods are not statistically tested is that the system models include variables for which no real-life data exists. Since system dynamics models include both statistically valid variables and assumed variables, it cannot be claimed that the model behaviour would deteriorate because of the use of statistically non-tested variables. In fact, the use of the assumed variables would improve the behaviour generation capacity of the model, and it may be claimed that the model has passed improvement test. In situation where the modeller has clear idea and supporting data of the mode of behaviour of the system, the model building and validation becomes relatively an easy task.

The ultimate objective of model validation is to develop confidence in predictions. The first step of the ultimate objective of system dynamics model validation is to establish the validity of the model structure. The next step is the behaviour reproduction of the model compared to the real behaviour of the system, and it is meaningful only if we have sufficient confidence in the model structure (Barlas 1996). The logical sequences of model validation are shown in Fig. 6.1. The model behaviour changes to the change in parameter values, and policy issues are also important to understand how the model will behave under changing conditions. Keeping these philosophical aspects in mind, the tests for confidence building in system dynamics models are to be designed. The tests for building confidence in system dynamics models may be broadly classified as:

1. Tests for structure
2. Tests for behaviour
3. Tests for policy implications

One must realise that not all the tests are to be considered for validation of a model, but structure and behaviour pattern tests are essential, and policy

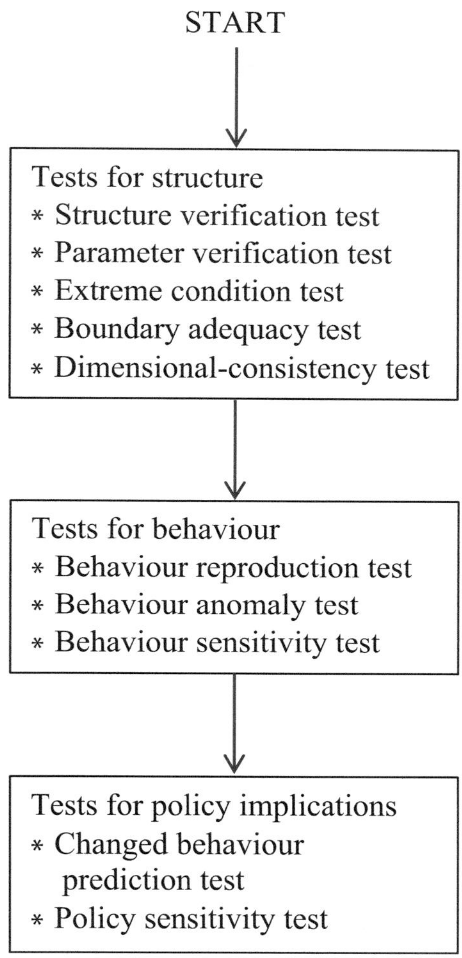

Fig. 6.1 Logical sequences of the tests for model validation

implication test makes the validation sufficient. It is rather those tests which are essential for establishing the creditability of the model are to be included.

6.2 Tests of Model Structure

The first step in the validation of system dynamics models is the structure validity tests, and this can be further classified as direct structure tests and structure-oriented behaviour tests. In direct validation structure tests, the validity of the model structure is assessed by direct comparison of the model structure with the knowledge of the real systems. It is accomplished by comparing mathematical equations and logical relationship with the available knowledge of the real systems. No simulation of the model is needed. Direct structure tests can be classified as empirical tests and theoretical tests. Empirical tests are conducted by comparing the model structure with the information (quantitative and qualitative) obtained about the system, while theoretical tests are conducted by comparing the model structure with the generalised knowledge of the system from literature such as research reports and studies. Structure confirmation tests are the toughest tasks to do as we need to compare the equations of the model directly with the knowledge of the real systems.

In structure-oriented behaviour tests, the validity of the model structure is assessed by the comparison of the behaviour of model predicted with the knowledge of behaviour of the real systems expected and usually observed in reality. It is a qualitative validation of the model.

Broadly speaking, tests of model structure may be classified as:

1. Structure verification test
2. Parameter verification test
3. Extreme condition test
4. Boundary adequacy test
5. Dimensional consistency test

6.2.1 Structure Verification Test

The structure verification test applies as empirical means comparing the form of equations of the model with relationships that exist in the real systems. The structure of the model that is the relationships in the equations should be in line with the descriptive knowledge of the system. It may also be conducted as theoretical tests by comparing the model equations with the generalised knowledge of the systems in the literatures. All the equations should be well argued and based on available information. The structure of the model should match observable goals, pressures and constraints of the real systems. Verifying a model structure is an easier task and takes less skill than some other tests.

6.2.2 Parameter Verification Test

The second structure verification test is the parameter confirmation test, and it means evaluating the constant parameters against the knowledge of the real systems both conceptually and numerically. Every constant (and variable) should have a clear real-life meaning. The basic choice is formal statistical estimation or judgemental estimation. Econometrics or other methods may be used to estimate the parameters.

The choice of appropriate initial values for stock equations, values of constants and table functions is directly related to the model description, and the values should be based on the published data from various sources. Computer software packages are now available to estimate and justify the exact values of the parameters to produce the expected behaviour of the system. Structure verification and parameter verification are interrelated, and both tests have the same basic objective.

6.2.3 Extreme Condition Test

The model should be robust under extreme conditions. There is an important direct structure test to the robustness of the model under direct extreme conditions, and it evaluates the validity of the equations under extreme conditions by assessing the plausibility of the resulting values against knowledge/anticipation of what would happen under similar conditions in real life. It is relatively easy to anticipate which variables and what values would these variables take under extreme condition in real systems.

The model must be capable to cope with external conditions. If the extreme conditions are incorporated in the model, the result is an improved model in the normal operating region. System dynamics model structure permits extreme combinations of socks in the system under study. A model should be questioned if extreme condition test is not met. It is not acceptable that extreme condition is not necessary on the plea that it does not occur in real life. Extreme condition test is effective for two reasons: (a) it is a powerful tool to detect the defect in model structure, and (b) it enhances the usefulness of the model for analysing policies that may force a system to operate outside historical regions of behaviour. Hence the extreme condition test is a strong test.

Let us consider that the supply chain model of rice milling systems is to be tested for extreme conditions to detect the defect in the model structure and enhance the usefulness of the model for policy analysis. Figure 6.2 shows simulated milling inventory, wholesale inventory and retail inventory under extreme condition of crop failure, i.e. zero crop production. Under this condition, the milling inventory and then wholesale inventory and retail inventory are reduced to zero since the rice production is zero. The model results confirmed to the expected patterns of results and realities. This model complied with the basic principles of supply chain management and was consistent with supply chain theory and research results. Thus, the model is able to provide qualitative and quantitative understanding of the supply chain performances of rice milling systems. Hence the model is reliable and validated under extreme conditions.

6.2.4 Boundary Adequacy Test

Boundary adequacy test considers structural relationships necessary to satisfy the model's purpose. Boundary adequacy test asks whether or not model aggregation is appropriate and if a model includes all relevant structure.

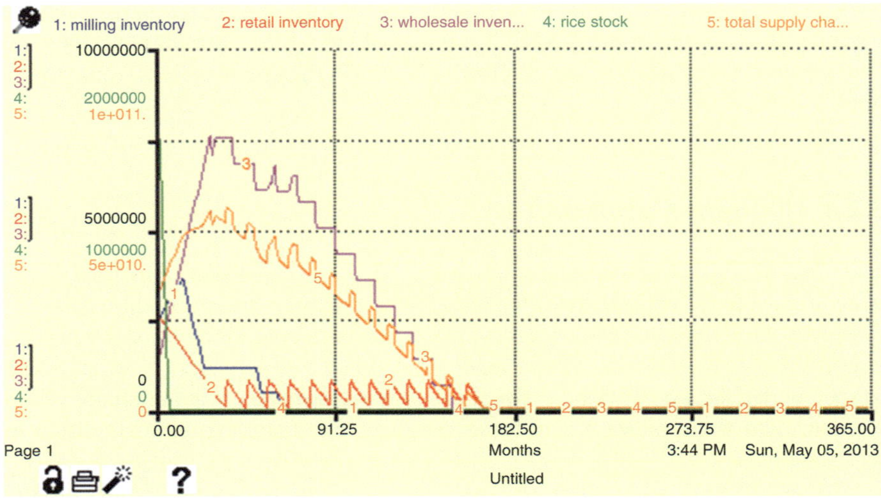

Fig. 6.2 Simulated milling inventory, wholesale inventory and retail inventory under extreme condition of crop failure, i.e. zero crop production

Once the model boundary is established, it is necessary to check whether any additional feedback loop has been omitted or not. If the additional feedback has any significant impact, it must be included. In essence, the model must include all variables and feedback loops encompassing the entire system under study which affect the dynamics or policy implications of the model.

6.2.5 Dimensional Consistency Test

Dimensional consistency test is one of the basic tests, and it must be carried out during the construction of the model. Dimensional consistency test involves checking the right-hand side and left-hand side of each of the equations of the model for dimensional consistency. It is better to specify the units of measure of each variable during construction of the model. The dimension of the left-hand and right-hand side of an equation should be the same for correct model formulation. Moreover, the dimensions of the variables should be close to their physical meaning; none of the dimension should be divorced from its actual meanings. Dimensional consistency test entails the dimensional analysis of a model's rate equation. Surprisingly many models fail this simple test. Hence, dimensional consistency test is the most powerful when applied in conjunction with the parameter verification test.

6.3 Tests of Model Behaviour

The second important step in the validation of system dynamics models is the model behaviour validity tests, and the tests for model behaviour should be conducted once the structural validation tests are completed successfully. The core tests of model behaviour may be classified as:

1. Behaviour reproduction test
2. Behaviour anomaly test
3. Behaviour sensitivity test

6.3.1 Behaviour Reproduction Test

Once the structure confirmation tests are completed successfully, the next test is the behaviour pattern tests to measure how accurately the model can reproduce the dynamic behaviour of the real systems. Behaviour reproduction tests compare how well the model-generated behaviour matches model-observed behaviour of the real system. Behaviour reproduction tests include symptom generation, frequency generation, relative phasing, multiple mode and behaviour characteristic. Behaviour reproduction tests become much more convincing when one can show why the tests are passed.

Many tools are available to assess the model behaviour to reproduce the system behaviour. Most common techniques are descriptive statistics to measure point by point fit. The most commonly used measure of the fit is the coefficient of determination (R^2), and it measures the fraction of variance explained by the model. The mean absolute error (MAE), mean absolute percent error (MAPE) and root mean square error (RMSE) all provide measures of the average error between the simulated and actual values. However, the emphasis should be on pattern rather than point prediction.

In the literature on modelling and simulation, there are a wide range of tests involving point by point comparisons of model-generated and model-observed behaviour. Despite widespread acceptance, such tests involving point by point measures of goodness of fit are generally less appropriate for socio-economic system dynamics models.

The reproduction of historical behaviour is the single most important test to build up confidence in models. Figure 6.3a shows the comparison between the predicted and historical behaviour of food self-sufficiency ratio in Malaysia. The model-simulated food self-sufficiency ratio agrees reasonably well with historical behaviour, and the model is reliable. Figure 6.3b shows the comparison of simulated and reported changes in wholesale price of rice in 2011 in Bangladesh. The model can simulate the actual behaviour of the system closely and can be used for policy analysis.

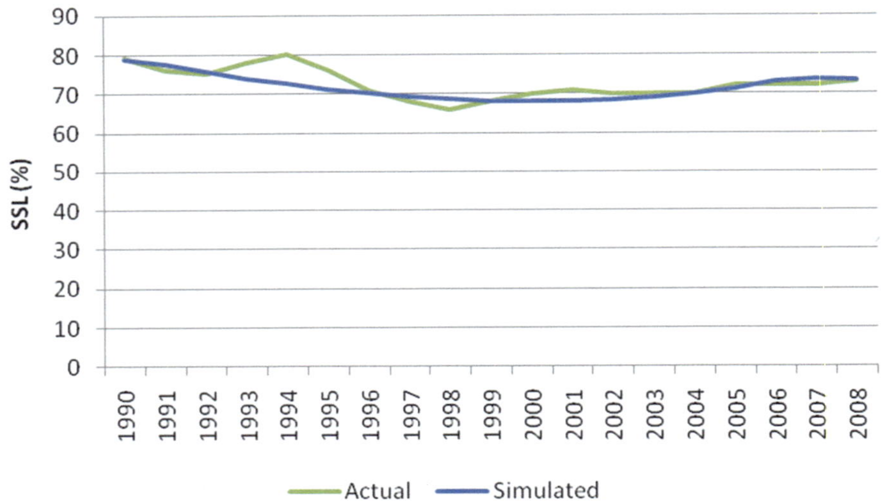

Fig. 6.3a Simulated and historical data of food self-sufficiency ratio in Malaysia

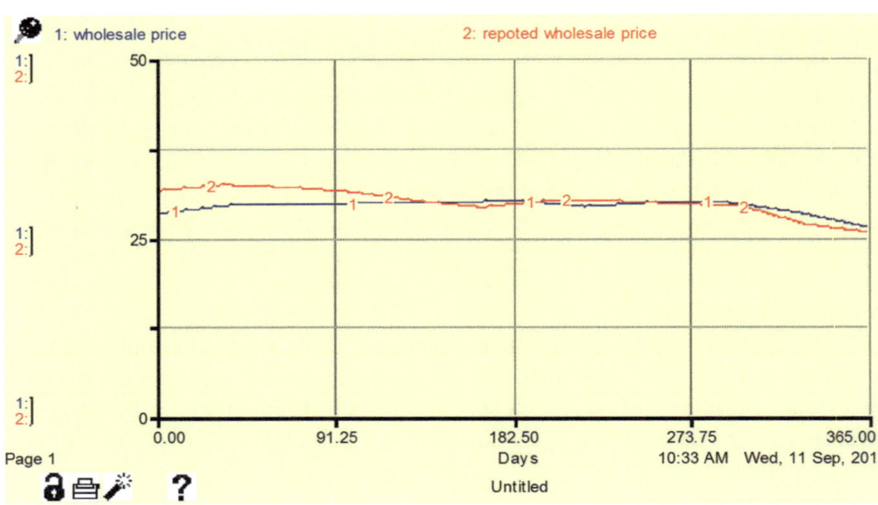

Fig. 6.3b Comparison of simulated and reported changes in wholesale price of rice in 2011 in Bangladesh

6.3.2 Behaviour Anomaly Test

While simulating a system dynamics model, one expects it to behave like real system under study, but frequently model builders face anomalous features of model behaviour, and these contradict the behaviour of the real system. Whenever there is anomaly in model behaviour, there may be defect in model assumptions. One can often defend particular assumptions by showing how implausible behaviour arises if the assumption is altered. Loop knockout analysis is a common method to search for behaviour anomalies. Anomalous behaviour resulting from knockout test suggests the importance of the loop and may help to establish the plausibility of system behaviour.

6.3.3 Behaviour Sensitivity Test

The behaviour sensitivity test shows the sensitivity of the model behaviour to changes in parameter values. The parameter sensitivity test ascertains whether or not plausible shifts in parameter values cause a model to fail behaviour tests previously passed.

The behaviour sensitivity test is typically conducted by experimenting with different parameter values and analysing their impacts on behaviour. Typically, the behaviour of system dynamics models is insensitive to plausible changes in most parameter values. It appears that systems are insensitive. On the other hand, both real systems and models or real systems are sensitive to a few parameters. Finding a sensitive parameter does not necessarily invalidate the model. Even though it has a substantial effect on behaviour, plausible variations may not lead to failure of other behaviour tests.

The model of supply chain of rice milling systems was simulated to address the impacts of rice productivity on the supply chain performances. Here the rice productivity is the yield of rice per ha. Rice productivity may be reduced from crop damage due to floods or pest infestation, and also it can be increased by development of higher yield hybrid rice through research and development. Rice productivity for this policy is defined as

$$\text{rice production rate} = \text{area under rice} \times \text{yield of rice} \tag{6.1}$$

$$\text{yield of rice} = 1.81, 2.81 \text{ and } 3.81 \text{ tons/ha} \tag{6.2}$$

Simulated milling inventory, wholesale inventory, retail inventory and total supply chain cost for rice productivity of 1.81 tons per ha, 2.81 tons per ha (present average rice productivity) and 3.81 tons per ha are shown in Figs. 6.4a, 6.4b, 6.4c and 6.4d, respectively. Milling inventory is reduced to zero for most of the period of the reduced productivity of rice (1.81 tons per ha), while it is high for bumper production of rice (3.81 tons per ha) (Fig. 6.4a). Wholesale inventory is reduced significantly in the fourth quarter of the year for reduced rice productivity (1.81 tons per ha) (Fig. 6.4b). Total supply chain cost is also reduced in the third and fourth quarter of the year for reduced rice productivity (Fig. 6.4d). However, in all the cases, the retail inventory is almost the same except towards the end of the year when both milling and wholesale inventories are empty for reduced productivity

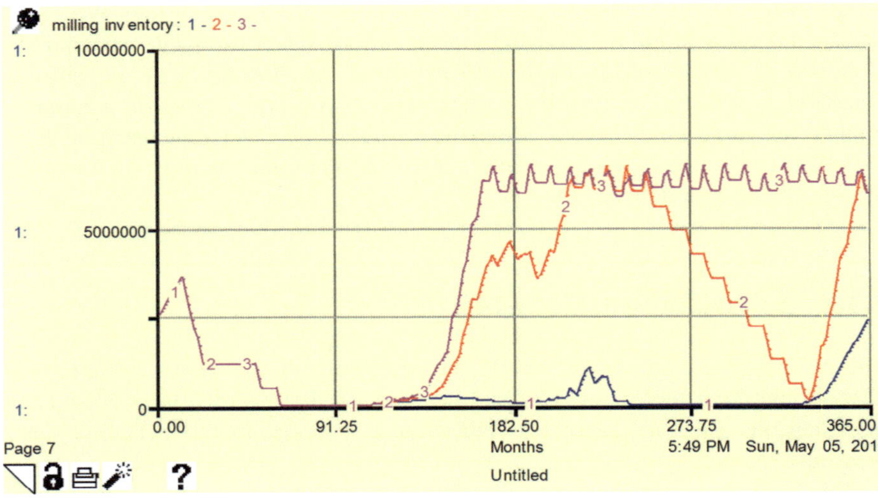

Fig. 6.4a Simulated milling inventory for rice productivity of 1.81 tons per ha, 2.81 tons per ha and 3.81 tons per ha

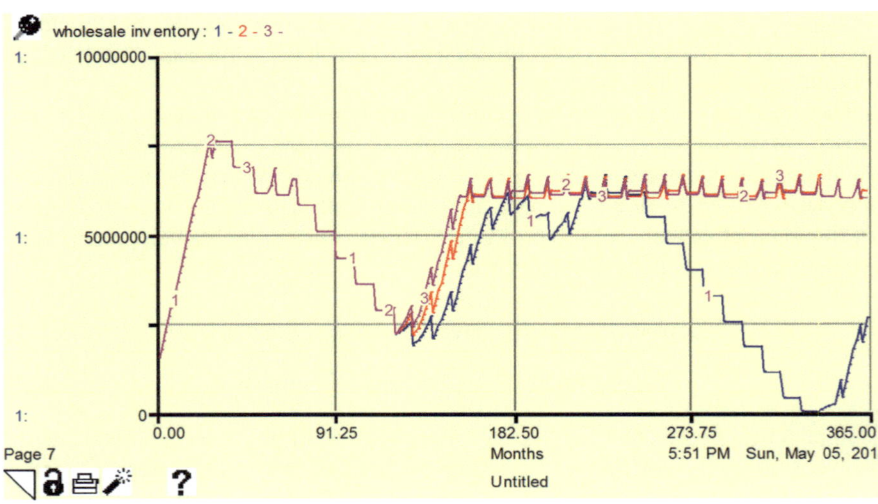

Fig. 6.4b Simulated wholesale inventory for rice productivity of 1.81 tons per ha, 2.81 tons per ha and 3.81 tons per ha

(Fig. 6.4c). Thus, increased wholesale inventory is a possible solution for the retail inventory to face the shortage of rice during the off-peak harvesting season of rice production. This implies that as long as wholesale inventory is available, the retail inventory is stabilised based on economic order quantity and reordering point

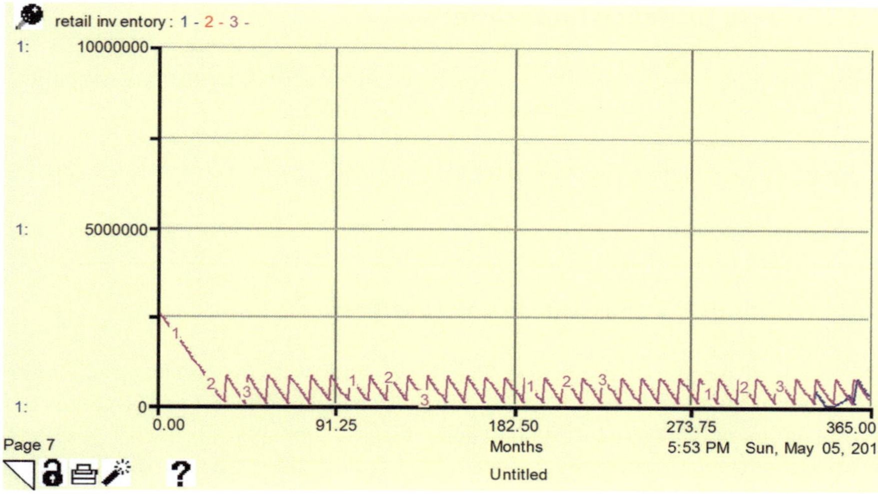

Fig. 6.4c Simulated retail inventory for rice productivity of 1.81 tons per ha, 2.81 tons per ha and 3.81 tons per ha

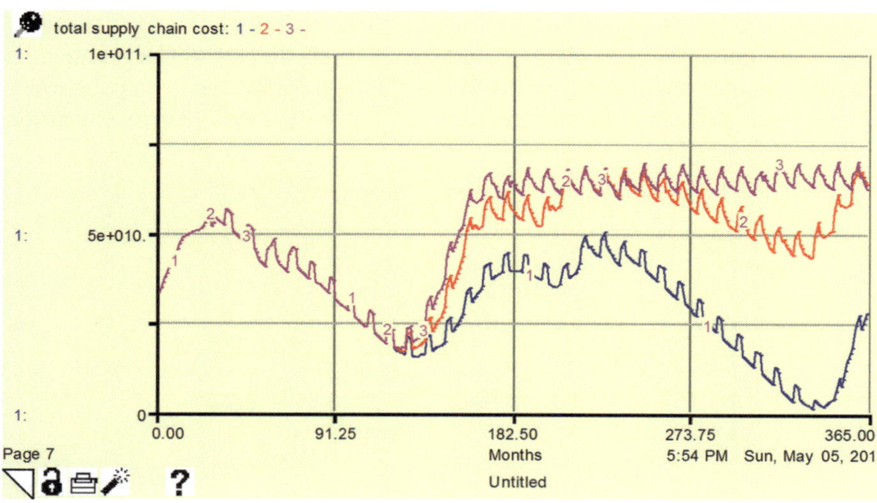

Fig. 6.4d Simulated total supply chain cost for rice productivity of 1.81 tons per ha, 2.81 tons per ha and 3.81 tons per ha

operation of milling, wholesale and retail inventory. This demonstrates that the policy based on economic order quality and reordering point can ensure the availability of rice even under reduced production of rice, i.e. during crop damage/failure unless both wholesale and milling inventories are empty.

6.4 Tests of Policy Implications

Tests should be conducted to build confidence in model's implications for policy. The core tests of policy implications may be classified as:

1. Changed behaviour prediction test
2. Policy sensitivity test

6.4.1 Changed Behaviour Prediction Test

The changed behaviour prediction test shows how well the model predicts the behaviour of the system if a governing policy is changed. The test can be conducted by changing policies in the model and verifying the plausibility of resulting behaviour changes. Alternatively one can examine the response of the policy already pursued to see how well model response agrees with the real system response. This test essentially shows the impacts of exogenous policies on the model behaviour.

Figure 6.5 shows the comparison of food self-sufficiency ratios for basic run and IPCC climate scenario for base year yield and yield increase of 6 tons per ha within next 50 years. The climate change impacts on food self-sufficiency level are small for all these runs. However, food self-sufficiency levels for both the base year yield and the yield increase of 6 tons per ha follow similar patterns, and in both the cases, the food self-sufficiency levels increase for about 10 years due to expansion

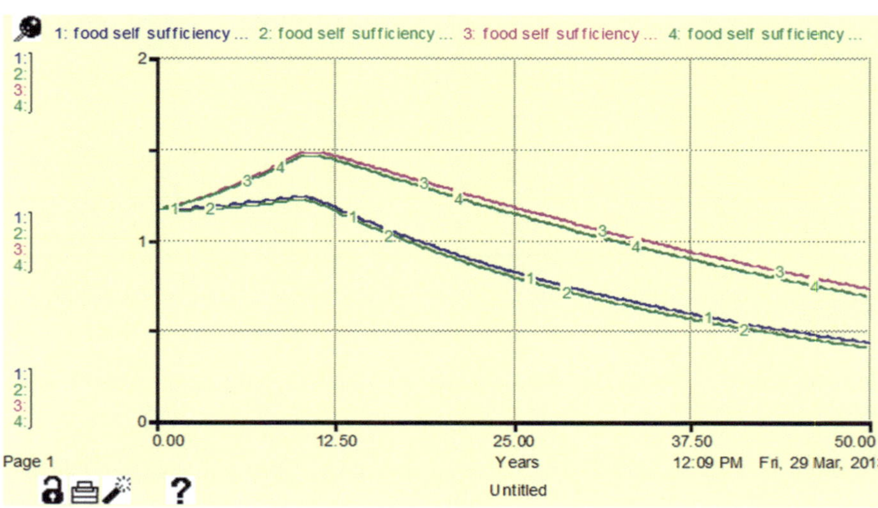

Fig. 6.5 Comparison of food self-sufficiency ratios for basic run and IPCC climate scenario for base year yield and yield increase of 6 tons per ha withinTests of policy implications:Changed behavior prediction test next 50 years

irrigation, and then it decreases as a result of the discard of irrigated area for its use for infrastructure development. The improved productivity increases the self-sufficiency level for about 12.5 years ahead, and the food self-sufficiency at the end of 50 years of simulation period increases from 43 to 73 %. Food self-sufficiency level is more seriously challenged by the decreasing cultivable land and growing population, and these are essentially demanding more increase in the productivity in the vertical direction due to the constraints of non-availability of additional cultivable irrigable land and more attention to control the growing population to improve rice self-sufficiency in the long run.

6.4.2 Policy Sensitivity Test

Researchers and decision makers have to make up their minds about where to concentrate their efforts to improve policies. When policy improvement is the desired objective, the question of policy sensitivity arises, although it may not be recognised as such. What kind of researchers should be involved? What mechanisms should be included or left out of formal and mental models? For which relationships and parameters should one seek better data and higher-quality estimates? These questions are best answered by policy sensitivity analysis.

The traditional and frequently used form of sensitivity analysis in system dynamics is to vary model assumptions and to observe how behaviour changes. In the branch of operations research using optimisation, sensitivity analysis is to vary model assumptions and to observe how optimal policies change. To avoid confusion between these two types of sensitivity, the terms behaviour sensitivity and policy sensitivity are used, of which the latter is the focus here. Policy sensitivity exists when a change in assumptions reverses the impacts or desirability of a proposed policy (Sterman 2000). For example, when one set of assumptions causes sustainable supply of palm oil, while another does not, the model exhibits policy sensitivity. If a particular policy change always produces improvement, regardless of changes in a sensitive parameter, then the policy recommendation is not affected (Forrester 1969). For example, when both sets of policy assumptions produce improvement of food security, the model exhibits policy insensitivity. These two statements clearly support and emphasise the sensitivity of the outcome of particular policies to uncertain parameters. To distinguish the two types of sensitivity analysis, we denote the latter policy outcome sensitivity.

Policy outcome sensitivity analysis is the most appropriate type of analysis when there remains uncertainty about parameter assumptions. This type of analysis can be expanded to include risk. The policy sensitivity we focus on here is the most appropriate for the purpose of model building. What parameters are most important for policy recommendations and require thorough analysis? What simplifications and aggregations are important for policies?

From a more practical viewpoint, if the simplified policy is the best one can do, or it is the type of policy that will be used in practice, then the bias is of less concern. Then it is interesting in itself to see how the optimised practical policy varies with changes in model assumptions. As a problem is demonstrated not to be very sensitive to uncertain parameters, decision makers' confidence in the problem formulation increases and so does the likelihood that some first policy measures will be implemented. That this type of analysis is needed in renewable resources management, e.g. fisheries, is indicated by the long delays in implementing appropriate policies and by laboratory experiments showing tendencies towards misperception.

We start with the problem of model validation, where a major challenge is to choose between potentially large numbers of models that pass tests of falsification. Policy sensitivity analysis is of no direct help in this choice. However, it can be used to find out if the choice of model has policy implications. If it has not, the remaining uncertainty is mostly of academic interest.

This is a potentially important insight since a heated 'academic' debate about the correctness of models can confuse the policy debate. In principle, policy sensitivity analysis could even be used to investigate the importance of the 'unavoidable a priori' assumptions of different disciplines (Meadows 1980). If different disciplines prescribe different policies, policy sensitivity analyses could be used to identify the causes of disagreement and to direct further validation effort towards the identified causes.

Next, we consider model aggregation. In principle, aggregation is a very challenging problem. Most system dynamics models resort to aggregation of people, resources, perceptions, etc. On the other hand, in non-linear dynamic models, aggregation leads to model errors except in some very special cases (Krysiak and Krysiak 2002). For this reason, Krysiak and Krysiak argue that 'environmental economics as well as other fields of economics may benefit from using more complex models'. On the other hand, increased complexity has a cost in terms of efforts to validate, analyse and explain models. Policy sensitivity analysis can be used to identify the appropriate aggregation level for policies.

Finally, policy sensitivity test shows the degree of robustness of model behaviour and policy recommendations. Such testing can help to show the uncertainty in parameter values. In the worst case, the parameter change can invalidate the recommended policies that were given. However, the policy recommendations are not likely to be affected by uncertainties in parameters.

In summary, tests for building confidence in system dynamics models should be conducted in some logical sequences, and we should move one step to the next step only if we are able to establish sufficient confidence in the current step. The logical sequences of formal steps of model validation as suggested by Barlas (1996) capture the essentials of the validation of system dynamics models to build up confidence in system dynamics models. The logical sequences of formal steps of model validation as suggested by Barlas (1996) are shown in Fig. 6.6. Once the model has passed through structural tests, we should start behaviour pattern tests. If the model passes through both direct and indirect structural tests, but fails the behaviour pattern tests, then we should re-estimate certain parameter and/or input function.

Next it would be logical to skip the structural tests and apply behaviour pattern tests. Once we have reached the final step of the behaviour pattern test, we should

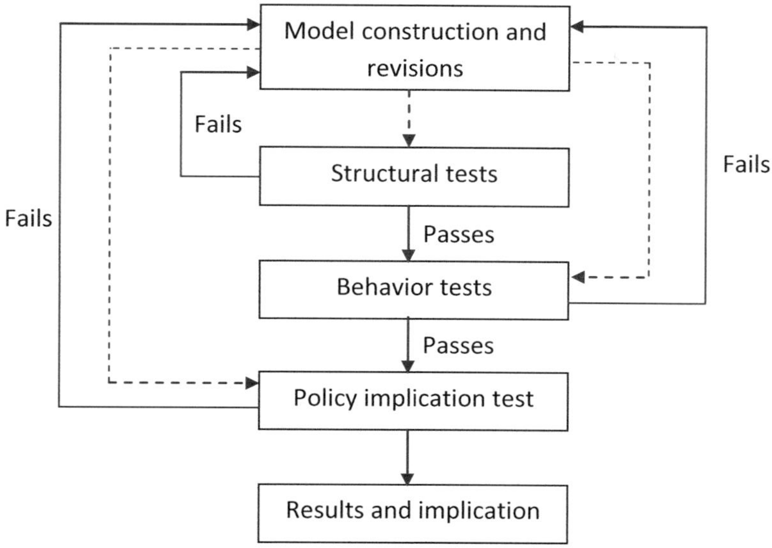

Fig. 6.6 Logical sequences of formal steps of model validation as suggested by Barlas (1996)

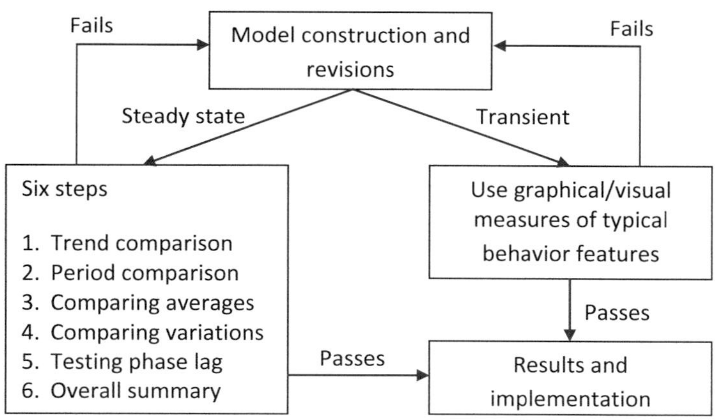

Fig. 6.7 Logical sequence of behaviour pattern tests

give emphasis on the accuracy of the pattern predictions. However, behaviour pattern tests are weak tests that provide no information on the validity of the structure of the model. Behaviour pattern tests must be carried out in logical order too. Figure 6.7 shows the logical sequence of behaviour pattern tests. If the problem involves transient and highly nonstationary behaviour, it is not possible to apply any standard statistical measure. On the other hand, if the problem involves a long-term simulation, it is possible to apply standard statistical measures and tests. Note that if the model is considered to fail the behaviour pattern tests, once again we should go back to model revisions, and the model revisions involve parameter/ input changes rather any other tests.

Exercises

Exercise 6.1 What are the tests for confidence building in system dynamics models? Describe the logical sequences of the tests of model validation.
Exercise 6.2 Describe tests of model structure with examples. Discuss the importance of verification tests. Explain why extreme condition test is important with examples.
Exercise 6.3 Describe tests of model behaviour with examples. Discuss the importance of behaviour reproduction tests with examples and also show logical sequence of behaviour pattern tests.
Exercise 6.4 Describe tests of policy implications with examples. Discuss the importance of policy sensitivity test.
Exercise 6.5 Describe the logical sequences of formal steps of model validation suggested by Barlas (1996).

References

Barlas Y (1996) Formal aspects of model validity and validation in system dynamics. Syst Dyn Rev 12(3):183–210
Forrester JW (1968) Principles of systems. MIT Press, Cambridge, MA
Forrester JW (1969) Urban dynamics. MIT Press, Cambridge, MA
Krysiak FC, Krysiak D (2002) Aggregation of dynamic systems and the existence of a regeneration function. J Environ Econ Manag 44:517–539
Meadows DH (1980) The unavoidable a prior. In: Randers J (ed) Elements of system dynamics method, Productivity Press, Portland
Sterman JD (2000) Business dynamics: systems thinking and modelling for a complex world. Irwin/Macgraw Hill, Boston

Bibliography

Bala BK (1999) Principles of system dynamics, First editionth edn. Agrotech Publishing Academy, Udaipur
Maani KE, Cavana RY (2000) Systems thinking and modelling: understanding change and complexity. Prentice Hall, Auckland
Mohapatra PKJ, Mandal P, Bora MC (1994) Introduction to system dynamics modelling. Universities Press, Hyderabad
Moxnes E (2005) Policy sensitivity analysis: simple versus complex fishery models. Syst Dyn Rev 21(2):123–145

Scenario Planning and Modelling 7

The models tested for building up confidence (as discussed in Chap. 6) should be used for scenario planning and modelling. This chapter presents scenario planning and modelling. Participatory system dynamics modelling and scenario planning are broadly discussed and the steps in scenario planning are highlighted. Also, policy planning for development strategy and implementation of development strategy and modelling are discussed. Some examples of scenario planning and modelling are presented.

7.1 Introduction

Once the model has been validated, the next steps are policy design and evaluation. Policy design includes management strategies and different policy options for improvement of the system performances. Policy design must involve all the stakeholders of the system being modelled not only in the simulation at the final stage but also during the development of the model. This will assist in minimising the gap between the model builders and the stakeholders. Also, this will make the stakeholders feel to be part of policy design using computer models.

Scenario planning is an approach for exploring the different possible futures and it relies on an understanding of today to explore what might happen in the future. This methodology also includes both simulation using system dynamics model and conversations with stakeholders related to scenario planning. It is to be noted that system dynamics is an approach for exploring the current structure of a system and the reasons for its behaviour, while scenario planning methodology generates scenarios as the solution derived from the simulation results of the system dynamics model.

B.K. Bala et al., *System Dynamics*, Springer Texts in Business and Economics,
DOI 10.1007/978-981-10-2045-2_7

7.2 Scenario Planning and Modelling

Scenarios are descriptions of different possible futures and scenario planning is an approach designed to develop scenarios. It considers the decision-maker's current mental models and aims to question their assumptions and their model's limitations. The aim of scenario planning and scenarios is to change the mental models of decision-makers to consider new possibilities and create common language and common mental models from which to begin discussing strategic options (Featherston and Doolan 2013). Commonly cited requirements of scenarios are that they be believable, reasonable and realistic.

The difference between scenario planning and system dynamics is the different time frames upon which they focus. System dynamics focuses on present systems and the reasons for their behaviour, whereas scenario planning focuses on what may happen in the future, often between 5 and 25 years. However, scenario planning relies on an understanding of today in order to be able to map out the space into which the future might fall. Furthermore, system dynamics and scenario planning are not static approaches; as a system might evolve, they can be used to understand the causes for systemic behaviour in the evolved system and explore the spaces it might proceed to in the future.

The system dynamics approach addresses a question that remains unanswered by the scenario planning approach. Why there is gap between perceived and actual behaviour of the system is unanswered by scenario planning. But system dynamics is able to address such an issue. System dynamics also offers a more formal approach to addressing mental models. While scenario planning's focus is on informing mental models, the approaches' literature provides little besides information collection and advocacy and inquiry as means to achieving that. However, system dynamics procedural approach of forming a hypothesis, understanding a system structure and testing the hypothesis provides a guide to identifying and addressing a specific, systemic problem. The specificity of the system dynamics approach is one glaring difference between these approaches. Scenario planning is used to explore the entire system and understand how it might evolve. System dynamics, however, despite being capable of exploring such broad system applications is applied to a very specific problem with limited system scale. This perhaps limits the conclusions that can be drawn from this type of study. Perhaps as a consequence of this, the information that comes from the system dynamics process does not encourage a reassessment of the scenarios. This indicates that despite the information from the system dynamics approach providing useful information for policy development, it does not cause a change in the scenarios, but instead helps to understand why the behaviour is occurring and what could be done about it in the future, essentially informing the mental models that fill in the gaps left by the scenario narratives. The learning that occurs during the scenario planning phase is fed into the system diagrams and in effect acts as a problem structuring method to identify where a system dynamics approach is needed. System dynamics then addresses a problem that is not addressed by scenario planning. It was then the combination of these results depicts a situation where the organisation is in a stalemate and begins to provide a framework from which policies could be addressed to influence this.

Both system dynamics modelling and system dynamics-based scenario planning are advocated for involving all the stakeholders in modelling and scenario planning. We present here first participatory system dynamics modelling and then participatory system dynamics-based scenario planning.

7.3 Participatory System Dynamics Modelling

System dynamics uses simulation models to support policy and policy planning. More specifically, system dynamics models are based on causal loop diagrams, stock and flow diagrams and non-linear finite difference integral equations, and the stakeholders form an important part of the methodology (Forrester 1961; Gardiner and Ford 1980; Vennix 1996; Hsiao 1998; Elias et al. 2000; Maani and Cavana 2000). Group model building is a system dynamics model-building process which involves a client group deeply in the process of modelling (Vennix 1996, 1999; Andersen and Richardson 1997; Rouwette et al. 2011).

Many reasons can be found in the literature to take stakeholders into account in the process of model building (de Gooyert et al. 2012), and the participatory model-building interventions can be classified into three distinct types: modelling stakeholder behaviour, modelling with stakeholders and modelling stakeholder behaviour with stakeholders. When the stakeholders are involved in the client group as well as modelling the behaviour of those stakeholders, the intervention should include several iterations. The intervention will support learning for the stakeholders, and learning will change the behaviour and this will invalidate the simulation model used to base on a decision. The simulation model should be updated after each iteration to take the new stakeholder behaviour into account.

Participatory modelling is an approach that includes a broad group of stakeholders in the process of formal decision analysis. It generally includes the stakeholders in the development and/or use of a computer model, although some soft approaches, such as cognitive mapping, does not (Mendoza and Prabhu 2006). Voinov and Gaddis (2008) describe participatory modelling as the process of incorporating stakeholders, often including the public, and decision-makers into an otherwise purely analytic modelling process to support decisions. It involves all the stakeholders to a greater or lesser degree in the process. A fully participatory process is one in which participants help structure the problem, describe the system, create an operational computer model of the system, use the model to identify and test policy interventions and choose one or more solutions based on the model analysis. A minimally participatory modelling process is one in which a model is used to help stakeholders understand the basis for an already selected decision.

Involving the stakeholders in the development process helps them understand a system's interactions and behaviour and can help design management strategy relevant to local concerns. It provides an opportunity to integrate scientific knowledge with local knowledge and builds a shared representation of the problem. When involving stakeholders in the setting of goals, make sure that the model assumptions are appropriate and develop politically feasible solution scenarios to build trust among stakeholders (Tàbara and Pahl-Wostl 2007). Simulation models also allow experimentation for policy design and policy analysis.

Participatory system dynamics modelling uses system dynamics perspective, and the stakeholders or clients participate to some degree in different stages of the modelling process: problem definition, system description, identification of policy levers, model development and/or policy analysis. The steps of the participatory system dynamics modelling and simulation are illustrated in Fig. 7.1. Participatory system dynamics modelling is more than simply eliciting knowledge from clients about the problem and the system. It involves building shared ownership of the analysis, problem, system description and solutions or a shared understanding of the trade-offs among different decisions. Steps in participatory system dynamics modelling are summarised as:

1. Define the problem and develop dynamic hypothesis.
2. Identify and recruit the 'right' stakeholders.
3. Elicit collective understanding of the 'system'.
4. Triangulate collective theory with data and literature.
5. Simulation modelling.

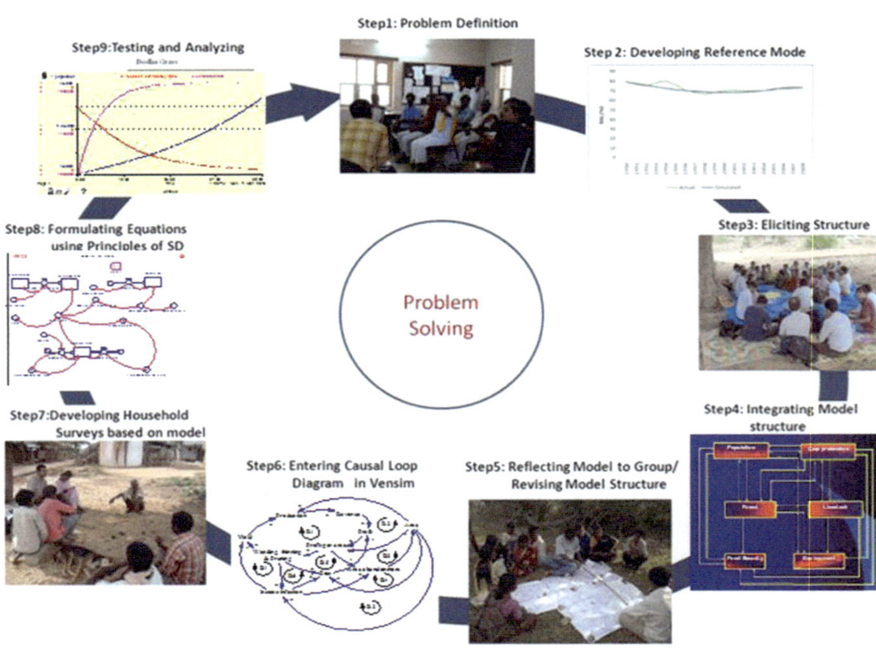

Fig. 7.1 Steps of the participatory system dynamics modelling and simulation (Source: Yadama et al. 2010)

6. Policy levers.
7. More simulation modelling.
8. Policy insights.

The goal of a system dynamics approach is to understand how a dynamic pattern of behaviour is generated by a system and to find leverage points within the system structure that have the potential to change the problematic trend to a more desirable one. The key steps in a system dynamics approach are identifying one or more trends that characterise the problem, describing the structure of the system generating the behaviour and finding and testing leverage points in the system to change the problematic behaviour. System dynamics is an appropriate modelling approach for sustainability questions because of the long-term perspective and feedback dynamics inherent in such questions. One of the key benefits of participatory system dynamics modelling is participant learning about system connections and feedback, both about the system and about other participants.

System dynamics modelling has been used in a number of applications with stakeholder groups examining difficult environmental issues (van den Belt 2004; Mendoza and Prabhu 2006; Langsdale et al. 2009; Beall and Zeoli 2008; Beall and Ford 2010). More recently, Beall and Ford (2010) compared nine cases that used a variety of techniques for engaging stakeholders and problem-solving. These case analyses illustrate the range of issues and settings in which participatory system dynamics modelling can be used and represent the beginnings of efforts to develop best principles and practices.

In participatory model building, the core modelling team mainly builds the model. Multiple participatory model-building sessions are conducted. The structures developed are later integrated to form a larger model. During this process, the community team takes the models to the community and bring back vital information and feedback as given to them by the community members. This iterative process provides significant insights from the community members. These insights are incorporated into the model and are again taken back to the community for reflection. This process is resource intensive, but is imperative to build confidence in the system dynamics modelling. The modelling workshop/focus group discussion is a significant part of the project for many reasons. From the model point of view, the system dynamics workshop/focus group discussion holds special significance because it ensures that the model was built based on key stakeholders embedded in the system and experiencing the dynamic problem and the experts who have been working with the system for many years.

Yadama et al. (2010) reported a participatory system dynamics approach of engaging poor communities and households to model the interactions between household livelihood strategies and natural resource dependence, with a particular focus on forest resources. Drivers of socio-economic and ecological systems and feedback mechanisms between the two are multiple, difficult to generalise and hard to reduce to a core representative set. The methodological strategy of using participatory rural appraisal techniques in combination with participatory system dynamics model building was used to elicit data on a key dynamic problem over time in a forest-dependent community. The goal was to develop dynamic models based on the knowledge and behaviour of actors most directly embedded in particular social–ecological systems. It was emphasised that people, dependent on natural resources for their living, are the real experts to help develop dynamic models of human and natural system interaction and outlined the four phases in which participatory approaches were used to work with a community to identify a dynamic problem that concerns forest resource dependence and their livelihoods, the associated reference modes that portray the dynamic model. In this way confidence in the initial models that emerge from the community were built up and reflections on community-driven participatory modelling were developed.

7.4 Participatory System Dynamics-Based Scenario Planning

Scenario planning attempts to capture the richness and range of possibilities, stimulating the decision-makers and managers to consider changes they would otherwise ignore. At the same time, it organises those possibilities into stories that are easier to grasp than huge volume of data. Above all, however, scenarios are aimed at challenging the manager's mental models and their prevailing mindsets (Cavana 2010). A scenario is not a forecast or an intention to describe a future state, but it is intended to provide a possible set of future conditions. However, scenarios must help decision-makers develop their own feel for the future of the system, the forces at work within it, the uncertainties that underlie the alternative scenarios and the concepts useful for interpreting key data (Wack 1985).

There are several alternative ways of constructing scenarios, but a method most consistent with system dynamics modelling and simulation proposed by Nakajima et al. (2014) is presented here (Fig. 7.2). This method consists of creating policy options based on system dynamics modelling and scenario repeatedly and then better scenario is achieved, and in this method the simulation results play the role to system dynamics and scenario planning.

We can easily perform the simulation using causal loop diagrams and stock–flow diagrams and system dynamics model strongly related to causal loop diagram and stock–flow diagram. The model can be used to get simulation results. Conversation

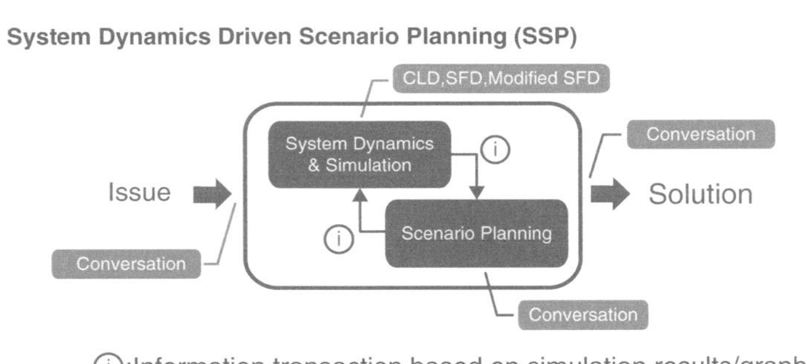

Fig. 7.2 Method of scenario planning

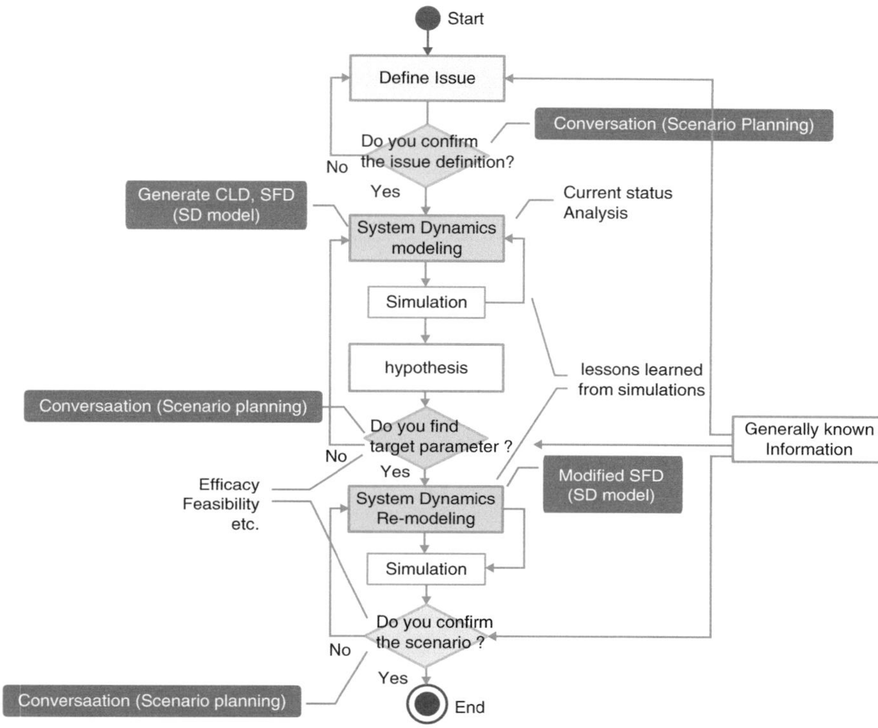

Fig. 7.3 Process of scenario planning

is an effective way in the process of scenario planning. The scenario is a kind of hypothesis. If it is validated, it becomes a theory. Thus, scenario is a speculative idea. There are merits and demerits of conversation in scenario generation. But there are more merits than demerits to improve the accuracy of the generated scenarios as a solution through the conversation. Of course, conversation is a kind of human interface to confirm the important decision among the members.

Now participatory process of scenario planning is discussed (Fig. 7.3). First of all, the issue, i.e. the problem, is defined involving the stakeholders through conversation as an approach of scenario planning. The next step is to develop causal loop diagrams and stock–flow diagrams. After this, the model is simulated to generate the dynamic behaviour. Once the model is simulated, a hypothesis is made and target parameters are identified having conversation with the stakeholders. Finally, policy scenarios are developed based on the simulation results having conversation with the stakeholders.

In order to understand the current status of rice self-sufficiency in Malaysia, we simulated the food security model of Malaysia. We used the simulation model to derive a policy scenario. Though it is known that a simulation is useful for discussion regarding rice food security in Malaysia, the discussion process which is based on simulation result is tied to a policy scenario.

The system dynamics model is not to predict point by point. Rather, it is intended to understand the behaviour of the system and to confirm how the behaviour of the system changes as a result of setting different values of parameters. In addition, changing the values of the parameters or changing the model, we can learn the behaviour of the system from the simulation results. However, to undertake scenario analysis, we cannot consider every possible change that may take place in the future. Instead, we need to identify the changes that are likely to happen or could have the largest impact on the system performance. Also, we need to consider the variables, parameters and factors that are uncertain or are most important to the system.

7.4.1 Simulation

The modelling of food security in Malaysia was simulated and it confirmed the behaviour of the food security in Malaysia. The purpose of this simulation was to make policy scenarios for food security in Malaysia.

Firstly, the model was simulated to know the possibility of occurrence of rice self-sufficiency in Malaysia. Next, the model was simulated to know how much food self-sufficiency in rice in Malaysia can be achieved. The factors which have the possibility to control were identified and simulations were conducted. From the simulation it was realised that a rice self-sufficiency of 73% can be achieved in 2025 which then gradually decreases to 22% in 2050.

7.4.2 Scenario

Having identified a range of uncertainties and factors that could have an impact on the system performance, scenarios are developed. These scenarios are then needed to be checked for internal consistency, and some additional research is needed to determine the realistic boundary values for the parameters and graphical relationships so that learning scenarios can be developed.

7.5 Steps in Scenario Planning

The steps in the scenario planning using system dynamics are as follows:

1. Define and confirm the issue through conversations.
2. Recognise current status using simulation results.
3. Make a hypothesis to solve the problem.
4. Figure out the parameters which change the behaviour of the system using causal loop diagram, stock–flow diagram and simulation results.
5. Learn the effectiveness of the parameter which influences the system behaviour using remodelling stock–flow diagram and simulation.

6. Verify the feasibility of the parameter using generally known information and generate scenario.
7. Confirm the scenarios through conversation.

7.6 Policy Planning for Different Development Strategies and Modelling

Policy planning for the strategic management process allows top managers to find answers to questions related to strategy effectiveness under a variety of alternative policy scenarios. Particularly, specific initiatives for awakening the awareness of the target stakeholders could be identified, ideas could be generated of how to stir up with the different policies, discovery could also occur of what would be the best strategies to stimulate growth and so on. In turn validated simulation models can verify answers to the above questions with a specific degree of certainty. Further, sensitivity analysis can provide a deep understanding of the structure and behaviour of future scenarios and may be modelled. Based on the investigative research method of computer simulation and experimentation, plausible strategies can be tested for their effectiveness prior to their implementation. In this way possible mistakes which can prove detrimental can be avoided.

7.7 Policy Planning for Implementation of the Development Strategy and Modelling

Once confidence in the structure and behaviour of the model has been established and policy design has been completed involving all the stakeholders, the model can be considered for implementation. For successful implementation, emphasis should be given on involving the target stakeholders, communicating modelling insights and demonstrating different development scenarios and outcomes of the short-run and long-run policy implication.

7.8 Policy Design and Evaluation

Once system dynamics models are developed and tested for building confidence, these can be used for policy design and evaluation. Policy design is more than changing the values of the parameters. Policy design includes creating new strategies, structures and decision rules (Sterman 2000). Policy design should be based on the interventions needed to achieve the desired goals from the dynamics of policy scenarios developed involving all the stakeholders. Of course, the policy options should be selected based on the dynamics of some selected indicators of the simulated scenarios. Also, the robustness of the policies and their sensitivities to the uncertainties of the model parameters and structures must be assessed. Policy evaluation should be conducted before and during implementation phases involving all the stakeholders to design corrective measures needed.

7.9 Some Examples of Scenario Planning and Modelling

System dynamics is an effective systematic method for making policy scenario. An example of the system dynamics model used for scenario planning is shown in Fig. 7.4, and from simulation results, the parameter that impacted the behaviour of the system is recognised. Our main focus is scenario planning and modelling. In this phase we formulate and test different policies and strategies of food security in Malaysia. Here policy refers to changes to a single internal variable such as

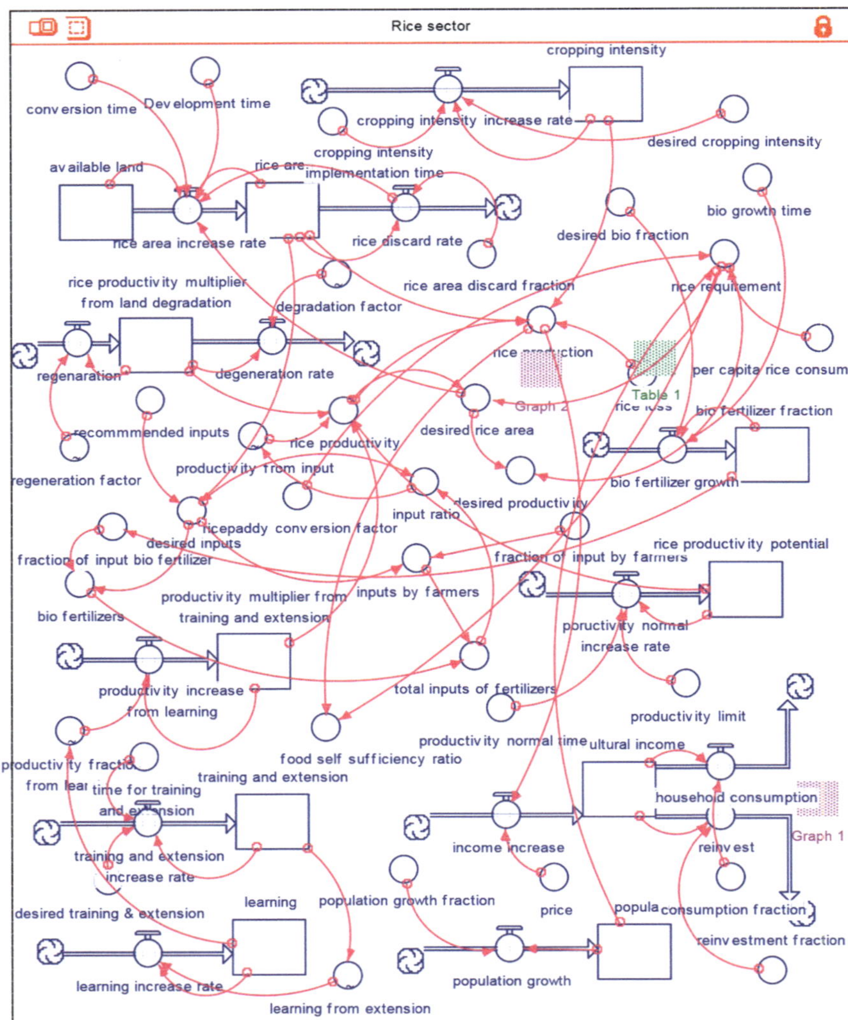

Fig. 7.4 Flow diagram of the system dynamics model of rice production systems

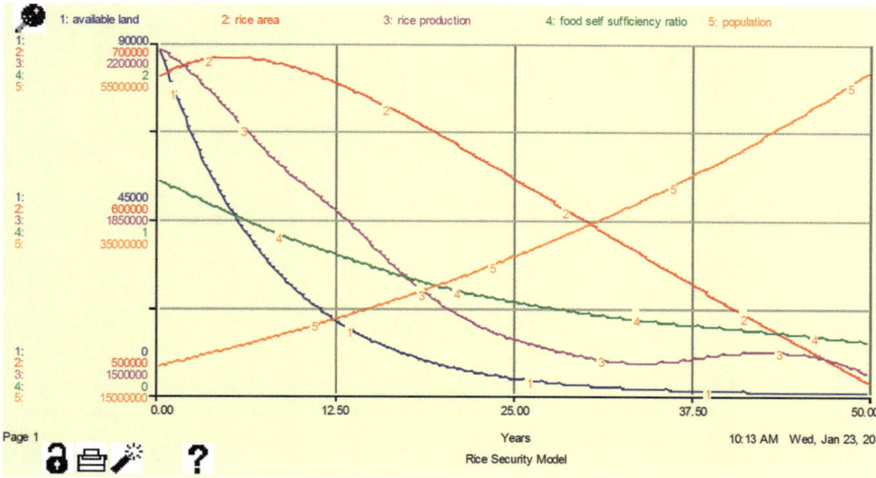

Fig. 7.5 Simulated rice production and rice self-sufficiency level for agricultural subsidies in inputs

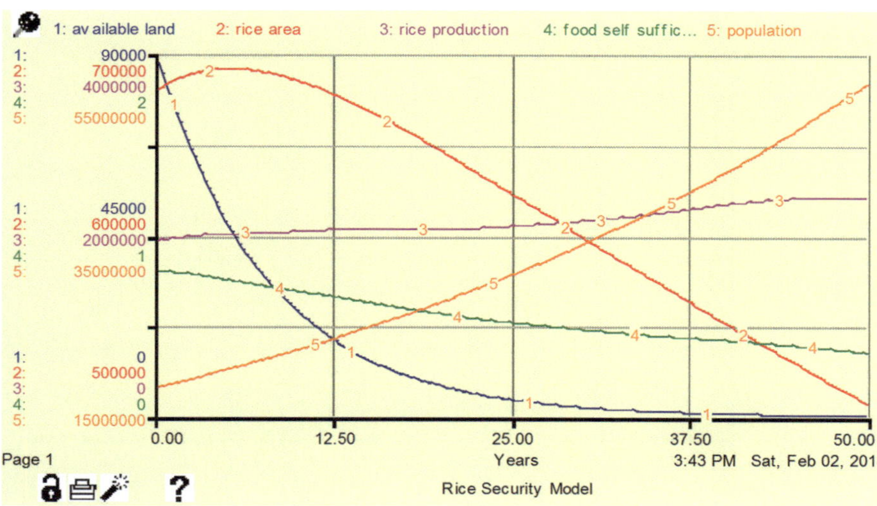

Fig. 7.6 Simulated rice production and rice self-sufficiency level for gradual transition to bio-fertilisers (50 %) and R & D

potential yield and extension services. Strategy is the combination of a set of policies and as such deals with internal or controllable changes. When these strategies are tested under varying external conditions, this is termed as a scenario modelling. This involves working closely with all major stakeholders. Figure. 7.5 shows the behaviour of food security in Malaysia under business as usual condition. Figures 7.6, 7.7 and 7.8 show three scenarios for sustainable development of rice

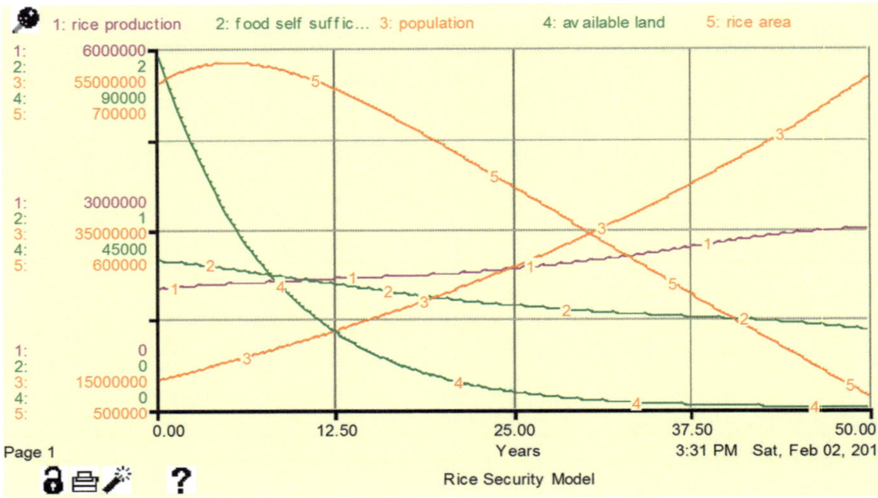

Fig. 7.7 Simulated rice production and rice self-sufficiency level for gradual transition to bio-fertilisers (50 %), R &D and training and extension

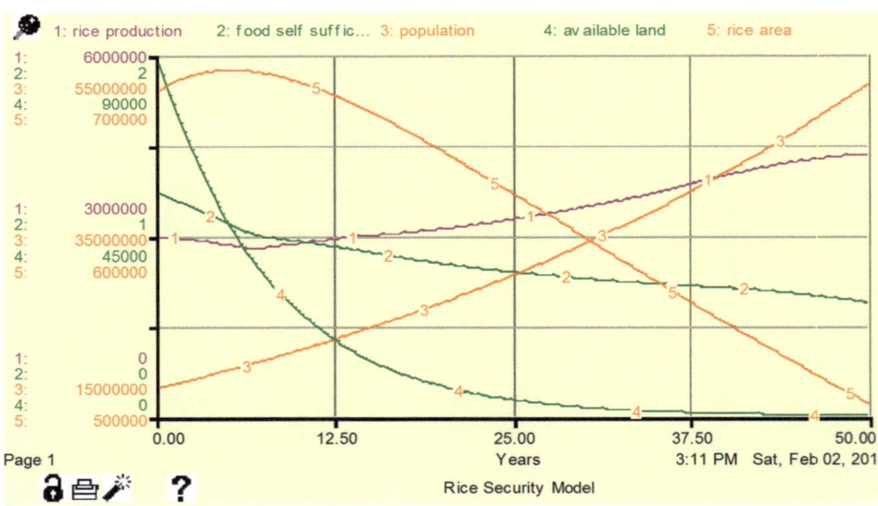

Fig. 7.8 Simulated rice production and rice self-sufficiency level for gradual transition to bio-fertilisers (50 %), R & D, training and extension and cropping intensity

self-sufficiency in Malaysia. These scenarios as the solution of rice self-sufficiency in Malaysia are derived from system dynamics simulation results and scenario making through conversations of rice self-sufficiency in Malaysia. Finally, we have developed scenario planning based on the results of the simulations.

Exercises

Exercise 7.1 What is scenario planning and modelling? How stakeholders can be involved in scenario planning and modelling?

Exercise 7.2 What is participatory system dynamics modelling? Describe steps in participatory system dynamics modelling.

Exercise 7.3 What is participatory system dynamics scenario planning? Describe the method of system dynamics-based scenario planning.

Exercise 7.4 What is scenario planning? Describe the process of scenario planning. What are the steps in scenario planning?

Exercise 7.5 How policy planning for different development strategies and modelling can aid the managers to avoid the possible mistakes? For successful implementation, what should be emphasised?

References

Andersen DF, Richardson GP (1997) Scripts for group model building. Syst Dyn Rev 13 (2):107–129

Beall A, Ford A (2010) Reports from the field: assessing the art and science of participatory environmental modeling. Int J Syst Soc Change 1:72–89

Beall A, Zeoli L (2008) Participatory modeling of endangered wildlife systems: simulating the sage-grouse and land use in Central Washington. Ecol Econ 68:24–33

Cavana RY (2010) Scenario modelling for managers: a system dynamics approach. Proceedings of the 45th annual conference of the ORSNZ, November, 2010

de Gooyert V, Rouwette FAJA, van Kranenburg HI (2012) A typology of stakeholder involvement in group model building. In: Proceeding of the 30th international conference of the System Dynamics Society

Elias AA, Cavana RY, Jackson LS (2000) Linking stakeholder literature and system dynamics: opportunities for research. In: 1st international conference on systems thinking in management, pp 174–179

Featherston Charles, Doolan Mathew (2013) Using system dynamics to inform scenario planning: a case study. The 31st international conference of the System Dynamics Society, Cambridge, MA

Forrester JW (1961) Industrial dynamics. M.I.T. Press, Cambridge, MA

Gardiner PC, Ford A (1980) Which policy run is best, and who says so. TIMS Stud Manage Stud 14:241–257

Hsiao N (1998) Conflict analysis of public policy stakeholders combining judgment analysis and system dynamics modeling. In: Proceeding of the 16th international conference of the System Dynamics Society

Langsdale SM, Beall A, Carmichael J, Cohen SJ, Forster CB, Neale T (2009) Exploring the implications of climate change on water resources through participatory modeling: case study of the Okangan Basin, British Columbia. J Water Res. Pl-ASCE 1116

Maani KE, Cavana RY (2000) Systems thinking and modelling: understanding change and complexity. Prentice Hall, Auckland

Mendoza GA, Prabhu R (2006) Participatory modeling and analysis for sustainable forest management: overview of soft system dynamics models and applications. Forest Policy Econ 9:179–196

Nakajima Y, Yasui T, Ohkami Y, Ktake N (2014) Modelinga policy for managing polio vaccine in Japan: scenario planning based on system dynamics. Internet.

Rouwette EAJA, Korzilius H, Vennix JAM, Jacobs E (2011) Modeling as persuasion: the impact of group model building on attitudes and behavior. Syst Dyn Rev 27(1):1–21

Sterman JD (2000) Business dynamics: systems thinking for a complex world. Irwin McGraw Hill, Boston

Tàbara D, Pahl-Wostl C (2007) Sustainability learning in natural resource use and management. Ecol Soc 12(3). Available online: http://www.ecologyandsociety.org/vol12/iss2/art12/. Accessed 17 Apr 2010

van den Belt M (2004) Mediated modeling: a system dynamics approach to environmental consensus building. Island Press, Washington, DC

Vennix J (1996) Group model building. Irwin McGraw-Hill, New York

Vennix JAM (1999) Group model-building: tackling messy problems. Syst Dyn Rev 15 (4):379–402

Voinov A, Gaddis EJB (2008) Lessons for successful participatory watershed modeling: a perspective from modeling practitioners. Ecol Model 216:197–207

Wack P (1985) Scenarios: shooting the rapids. Harv Bus Rev 63:131–142

Yadama GN, Hovmand PS, Chalise N (2010) Community driven modeling of social-ecological systems: lessons from Andhra Pradesh, India. Proceedings of the 28th international conference of the System Dynamic Society, Seoul

Part II

Cases and Applications

Modelling of Boom and Bust of Cocoa Production Systems in Malaysia

8

In the previous chapters in Part I, the concepts, methodology and techniques of system dynamics modelling and simulation in the areas of agricultural, biological, environmental and socio-economic systems are presented. This chapter presents applications of system dynamics modelling in ecological systems of boom and bust of cocoa production systems in Malaysia to demonstrate how to construct a system dynamics model and simulate it for policy planning and design. The model presented is an illustration of modelling and simulation of practical problems, and such an experience is essential to face the challenge of modelling and simulation of dynamic systems. To achieve this goal, the model of this case study is organised as follows: (1) introduction, (2) dynamic hypothesis, (3) causal loop diagram, (4) stock–flow diagram, (5) model validation, (6) simulation and policy analysis and (7) conclusion. The simulated results indicate that the collapse of cocoa production systems can be avoided through biodiversity conservation and insect control resulting from sustainable production systems and implementation of such policy demands adequate subsidy to retain high biodiversity, control pest and disease and attain acceptable yields through extension services through farmer field schools.

8.1 Introduction

The cocoa tree (*Theobroma cacao L.*) is an understorey forest species which evolved in the Amazon (Motamayor et al. 2008), and it is currently grown in many countries of the humid tropics. The largest cocoa-producing countries are Cote d'Ivoire, Ghana, Indonesia, Nigeria, Cameroon, Brazil, Ecuador and Malaysia, and these contribute 90 % of world production (Latiff 2007). Cocoa beans are primarily exported to Europe and North America to be processed to produce cocoa and chocolate.

Cocoa was introduced into Malaysia for commercial cultivation in 1950, became the third major commodity product in Malaysia after palm oil and rubber and was

B.K. Bala et al., *System Dynamics*, Springer Texts in Business and Economics,
DOI 10.1007/978-981-10-2045-2_8

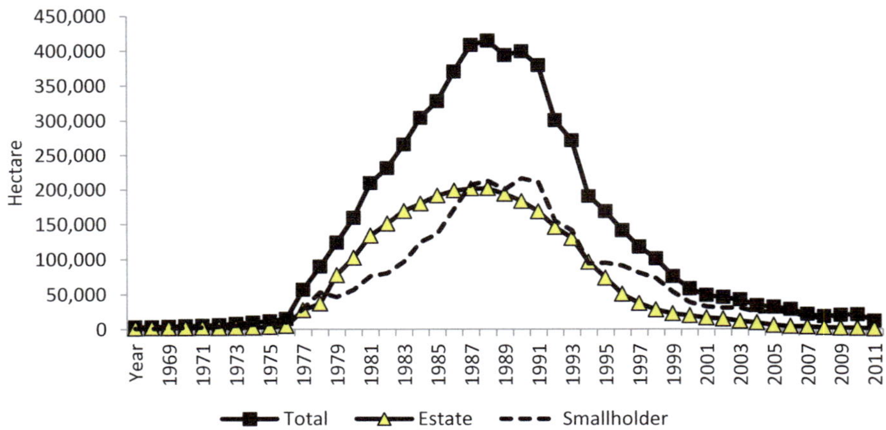

Fig. 8.1 Cocoa cultivation by type of holdings in Malaysia (Source: Department of Statistics, Malaysia 2014)

considered to be a crop for agricultural diversification in the Second Malaysia Plan, 1971–1975. The availability of superior planting materials and planting technology and the implementation of the government policy to encourage the growing of cocoa as an intercrop with coconuts coupled with the high favourable prices led to the rapid expansion of the cocoa planting industry in Malaysia (Fig. 8.1). The area planted increased to 123,855 ha in 1980 and 414, 236 ha in 1989. The high plantation rate is attributed to the unprecedented high cocoa bean prices in the 1970s and 1980s. (Lee 2013). But post-1980s marked a decrease in cocoa planting area with the decline of cocoa production due to poor world cocoa prices, labour constraints, competition for land use from oil palm cultivation and the severe spread and infestation of the cocoa pod borer (Lee 2013). The cultivated area decreased sharply from 393,465 ha in 1990 to 190,127 in 1995, and it continued till 2005 and the area was reduced to 33,398 ha due to severe pest infestation. By 2013, the cocoa planted area was reduced to only 13,728 ha. Fluctuations with small decline in cocoa areas were noted from 2005 (33,398 ha) to 2013 (13,728 ha), and during this period, strong government support was provided for cocoa planting especially in the rural and outlying areas to improve livelihood and elevate income with poverty reduction with targeted area of 40,000 ha by 2020 (MPIC 2011). However, there is an apparent uptrend of cocoa dried bean prices throughout the three regions of Malaysia as that of world cocoa prices. Malaysia has been dropped to the 12th position from 4th position in the world cocoa production (Lee 2013).

Questions remain to answer: What caused the boom and bust of cocoa plantation in Malaysia in particular? Why do cocoa plantations exhibit an extremely unstable pattern of development with ecological damages of biodiversity resulting from outbreak of severe pest infestation and diseases? What should be the policy for sustainable development of cocoa production systems in Malaysia?

Several studies have reported on the boom and bust of agricultural commodities like the shrimp aquaculture industry (Arquitt et al. 2005; Bala and Hossain 2010; Prusty et al. 2011). In these cases, from a systems perspective, we can conclude that when the industry is prone to exceed and consume its environmental carrying capacity, a boom and bust type of development results. Clough et al. (2009) reported a qualitative model of the boom and bust of the cocoa production systems. Franzen and Mulder (2007) highlighted the important ecological, economic and social considerations for sustainable cocoa production. This clearly indicates that there is a large research gap to understand the boom and bust and to search for policies for sustainable development of cocoa production and marketing in Malaysia.

Cocoa trees can be planted in the forest or under planted shade, but most cocoa plantations are planted into thinned forests. Shade removal increases the yield in the short run which damages ecosystems and reduces the biodiversity. As the boom busts, the plantation area falls sharply to a very low level due to extremely low productivity. Underlying production busts are failures of the industry participants particularly the policy makers to understand the problem and take effective measures. However, the boom and bust of cocoa production in Malaysia has been well documented (Lee 2013). Current cocoa production systems are not sustainable because of non-eco-friendly production of cocoa beans, although there is a demand of cocoa in chocolate industries. Sustainable development of cocoa production and marketing is a major concern for the policy makers and authorities who are searching for a sustainable planning to accomplish the targeted goals. Although boom and bust is the major concern, in addition, the sustainable development aims to achieve social, economic and ecological success in the cocoa production. However, sustainability cannot be achieved unless the ecological imbalance is rectified. The sustainable production should increase profit within the framework of ecological conservation of biodiversity. To cover this gap of understanding of boom and bust and designing sustainable production of cocoa beans and propose a realistic model which can add not only knowledge of boom and bust but also the implementation knowledge of sustainable production of cocoa in Malaysia, Fatimah et al. (2015) developed a system dynamics model to examine the underlying causes of the boom and bust of cocoa production systems in Malaysia and developed policies for sustainable development of production of cocoa in Malaysia. The modelling of the boom and bust of cocoa production systems presented here is adopted from Fatimah et al. (2015).

8.2 Dynamic Hypothesis

The dynamic hypothesis is a conceptual model typically consisting of a causal loop diagram, a stock–flow diagram or their combination. The dynamic hypothesis seeks to define the critical feedback loops that drive the system's behaviour. When the model based on the feedback concept is simulated, the endogenous structure of the model should generate the reference mode behaviour of the system, and thus, the endogenous structure causes the changes in dynamic behaviour of the system (Sterman 2000). The boom and bust of cocoa production systems can be represented by causal loop diagram and stock–flow diagram, and the simulation model based on the causal loop diagram and stock–flow diagram can generate dynamic behaviour of the cocoa production systems. The cocoa production systems in the form of causal loop diagram and stock–flow diagram are hypothesised to generate the observed boom and bust of cocoa production systems in the reference mode. In essence, the degradation of the biodiversity resulting from the reduction of shade level and subsequent large-scale insect infestation caused the boom and bust of cocoa production systems in Malaysia (Fig. 8.1), and this dynamics resulted from the endogenous consequences of the feedback structure (Sterman 2000).

8.3 Causal Loop Diagrams

The key factors influencing cocoa production are yield, thinning of shading trees and area under cultivation. The boom and bust of cocoa production in Malaysia is described by two reinforcing loops and six balancing loops. When production of cocoa and thinning generate profit for the producers, these motivate them to continue cultivation generating a feedback loop to work (R_1). Also when production of cocoa generates profit per ha, the producers are motivated to increase cultivated area (R_2). Area under cultivation and thinning of shading trees which creates ecological imbalance invites insect infestation. When insect infestation becomes intolerable and cocoa beans are damaged significantly, the cocoa production is affected negatively (B_1), and also cultivation is abandoned due to severe insect infestation (B_2). Profit per ha also is deciding factor for abandonment of cultivation area (B_3), and the cultivation cost affects the profit per ha (B_4). Thinning cost and cost incurred in insect infestation also affect profit per ha (B_5 and B_6). The dynamic interaction between cocoa production and ecology gives rise to the feedback loops. The causal loop diagrams of cocoa production systems are shown in Fig. 8.2 which shows the initial dynamic hypothesis of the boom and bust of cocoa plantation, and it is based on standard assumptions of how the cocoa production system typically works. If our hypothesis is correct, the model will be able to reproduce the general historical pattern of boom and bust followed by the simulated boom and bust of cocoa plantation.

8.4 Stock–Flow Model

Figure 8.3 shows the stock–flow diagram of the boom and bust of cocoa production systems in Malaysia. Fundamental equations that correspond to major state variables shown in Fig. 8.3 are as follows:

The cocoa plantation area is increased by the cocoa plantation rate based on the profitability of the cocoa plantation, and also the cocoa plantation area is abandoned based on yield and profit. This is expressed as:

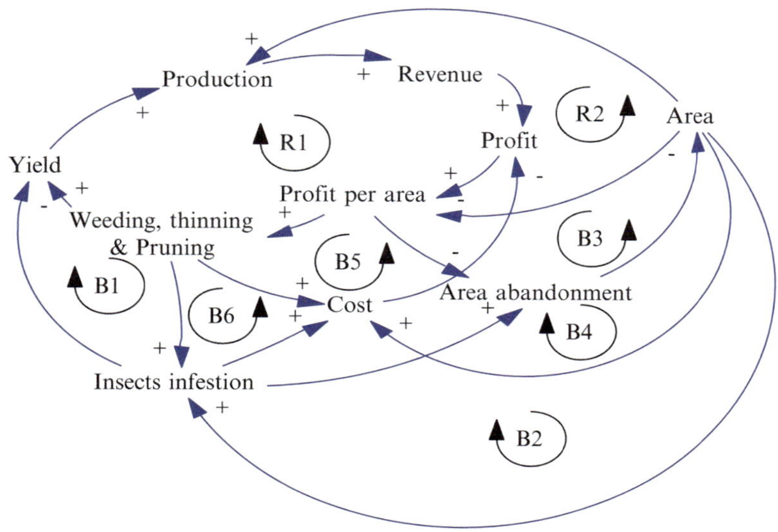

Fig. 8.2 Causal loop diagrams of cocoa production system

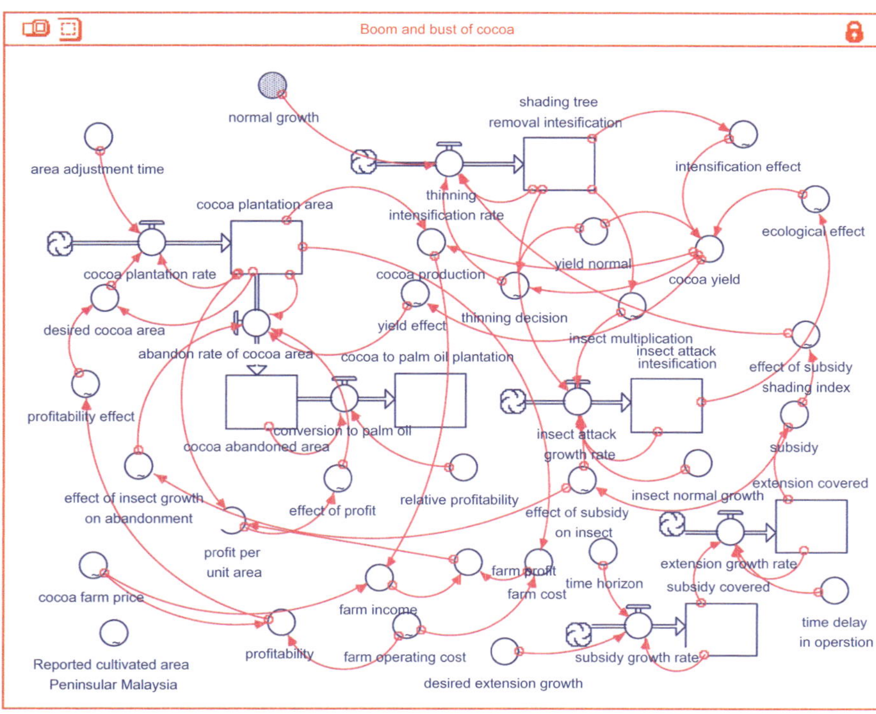

Fig. 8.3 Stock–flow diagram of boom and bust of cocoa production in Malaysia

$$\text{cocoa plantation area}\,(t) = \text{cocoa plantation area}\,(t - \Delta t) + \text{cocoa plantation rate} \times \Delta t - \text{abandon rate of cocoa} \times \Delta t \tag{8.1}$$

The cocoa plantation rate depends on the present level of the cocoa plantation area, the desired cocoa area and the time delay to reduce the gap between the desired cocoa area and level of cocoa plantation area, and it is expressed as:

$$\text{cocoa plantation rate} = \text{MAX}\big(0, (\text{desired cocoa area} - \text{cocoa plantation area}) / \text{area adjustment time}\big) \tag{8.2}$$

Desired area is computed from the level of the cocoa plantation area and profitability effect as:

$$\text{desired cocoa area} = \text{cocoa plantation area} \times \text{profitabilty effect} \tag{8.3}$$

The expansion of the cocoa area depends on the profitability of cocoa production and the non-linear relationship between profitability effect and profitability is expressed as:

profitability_effect = GRAPH(profitability)
(0.00, 0.02), (0.1, 0.738), (0.2, 1.16), (0.3, 1.50), (0.4, 1.76), (0.5, 2.00), (0.6, 2.14), (0.7, 2.21), (0.8, 2.24), (0.9, 2.24), (1, 2.27)

Abandon rate of the cocoa area depends on the level of the cocoa plantation area, yield effect and effect of profit, and it is expressed as:

$$\text{abandon rate of cocoa area} = \text{cocoa plantation area} \times \text{yield effect} \times \text{effect of profit} \tag{8.4}$$

The abandonment of cocoa area under cultivation depends on the yield of cocoa and the non-linear relationship between yield effect on the abandonment of cocoa cultivation area and yield of cocoa is expressed as:

yield_effect = GRAPH(cocoa_yield)
(0.00, 0.645), (1.00, 0.555), (2.00, 0.47), (3.00, 0.41), (4.00, 0.37), (5.00, 0.32), (6.00, 0.29), (7.00, 0.26), (8.00, 0.23), (9.00, 0.215), (10.0, 0.21)

Also the abandonment of cocoa area under cultivation depends on the profit per unit area and the non-linear relationship between effect of profit on the abandonment of cocoa cultivation area and profit per unit area is expressed as:

effect_of_profit = GRAPH(profit_per__unit_area)

(0.00, 1.11), (10.0, 1.09), (20.0, 1.08), (30.0, 1.07), (40.0, 1.06), (50.0, 1.05), (60.0, 1.04), (70.0, 1.03), (80.0, 1.03), (90.0, 1.03), (100, 1.02)

Cocoa yield is increased by the development of new hybrid varieties of cocoa through research and development, and also it depends on the ecological effect resulting from insect infestation due to limiting the shading index and intensity of shading index. This is described as:

$$\text{cocoa yield} = \text{yield normal} \times \text{ecological effect} \times \text{intensification effect} \tag{8.5}$$

The ecological effect resulting from the degradation of biodiversity and hence insect infestation affects the cocoa yield and the non-linear relationship between ecological effect and insect attack intensification, and it is expressed as:

ecological_effect = GRAPH(insect_attack__intesification)
(0.00, 1.00), (8.33, 0.87), (16.7, 0.725), (25.0, 0.535), (33.3, 0.41), (41.7, 0.335), (50.0, 0.24), (58.3, 0.2), (66.7, 0.11), (75.0, 0.075), (83.3, 0.045), (91.7, 0.02), (100.0, 0.005)

The shading tree removal enhances not only the cocoa yield, but it also invites insect infestation and the non-linear relationship between intensification effect and shading tree removal intensification is expressed as:

intensification_effect = GRAPH(shading_tree__removal_intesification)
(0.00, 1.04), (8.33, 1.18), (16.7, 1.24), (25.0, 1.29), (33.3, 1.34), (41.7, 1.37), (50.0, 1.41), (58.3, 1.45), (66.7, 1.46), (75.0, 1.48), (83.3, 1.48), (91.7, 1.48), (100.0, 1.48)

Yield normal is increased by development of new high-yielding/hybrid varieties through research and development.

Shading index reduction increases the yield in the short run, but large-scale reduction of shading index invites insect infestation which results in ecological degradation in the long run. Shading index is reduced by thinning the trees for higher yields in case of cocoa plantation under forest trees, and also shading index can be adjusted by changing the plant-to-plant distance of the coconut tree in case of cocoa plantation under coconut plantation, and the shading removal intensification is expressed as:

$$\begin{aligned}\text{shading tree removal intensification}\,(t) = {} & \text{shading tree removal intensification}\,(t - \Delta t) \\ & + \text{thinning intensification rate} \times \Delta t\end{aligned} \tag{8.6}$$

The thinning intensification rate depends on the present level of shading tree removal intensification, cocoa yield and subsidy for shading.

Shading tree and cocoa plants invite the insects, and the severity of insect damage depends on the intensity of insect attack, and the insect attack intensification is expressed as:

$$\begin{aligned}\text{insect attack intensification}\,(t) = {} & \text{insect attack intensification}\,(t - \Delta t) \\ & + \text{insect attack growth rate} \times \Delta t\end{aligned} \tag{8.7}$$

The insect attack growth rate depends on the present level of insect population, insect multiplication, shading tree removal intensification level and subsidy to control insects.

Cocoa production in Fig. 8.3 depends on cocoa yield (tons/ha) as well as on area under cocoa plantation, and it is computed as:

$$\text{cocoa production} = \text{cocoa yield} \times \text{cocoa area} \tag{8.8}$$

The coverage of the subsidy and extension through farmer field schools is expanded with a broad policy of high biodiversity and acceptable yields for sustainable development. These are described as:

$$\begin{aligned}\text{extension covered}\,(t) = {} & \text{extension covered}\,(t - \Delta t) \\ & + \text{extension growth rate} \times \Delta t\end{aligned} \tag{8.9}$$

and

$$\text{subsidy covered}\,(t) = \text{subsidy covered}\,(t - \Delta t) + \text{subsidy growth rate} \times \Delta t \tag{8.10}$$

The extension growth rate depends on the subsidy covered, present level of subsidy and time horizon in operation.

These systems of equation are solved by Runge–Kutta fourth-order method using STELLA software.

8.5 Model Validation

Initial values and the parameters were estimated from the primary and secondary data collected from different research reports, statistical year books of Malaysia and field visits. Tests were also conducted to build up confidence in the model. Tests for building confidence in system dynamics models essentially consist of validation, sensitivity analysis and policy analysis (Bala 1999). The two important notions of the building confidence in the system dynamics models are testing and validation of the system dynamics models. Testing means the comparison of a model to empirical reality for accepting or rejecting the model, and validation means the process of establishing confidence in the soundness and usefulness of the model. In the

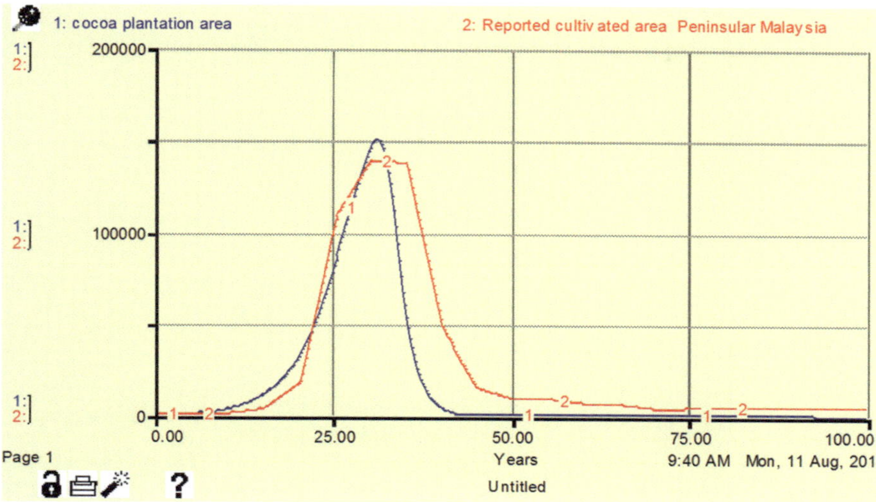

Fig. 8.4 Simulated and historical patterns of boom and bust of cocoa production systems in Peninsular Malaysia

behaviour validity tests, emphasis should be on the behavioural patterns rather than on point prediction (Barlas 1996).

To build up confidence in the predictions of the model, various ways of validating a model such as model structures, comparing the model predictions with historical data, checking whether the model generates plausible behaviour and checking the quality of the parameter values were considered.

Figures 8.4, 8.5 and 8.6 show the comparisons of simulated behaviours of plantation area of cocoa with the historical data. The historical data of cocoa plantation area in Peninsular Malaysia, Sabah and Sarawak show the boom and bust of cocoa production area in Malaysia. Simulated behaviours are numerically sensitive to parameters and shapes of the table functions. However, the basic patterns of the historical and simulated behaviours agree adequately, and model predictions represent reality.

8.6 Simulation and Policy Analysis

Figures 8.4, 8.5 and 8.6 show the collapse of cocoa production systems which may be attributed to mainly the reduction of the shade level which reduces the biodiversity and resulting large-scale insect infestation. The model was simulated to test how cocoa plantation changes with the changes in subsidy for maintaining high biodiversity attain acceptable yields and IPM-FFS (integrated pest management-farmer field schools) for pest management. The subsidy coverage (0–100 %) was gradually increased to cover the extension services (0–100 %) through farmer field schools for maintaining the recommended shading index and hence the biodiversity

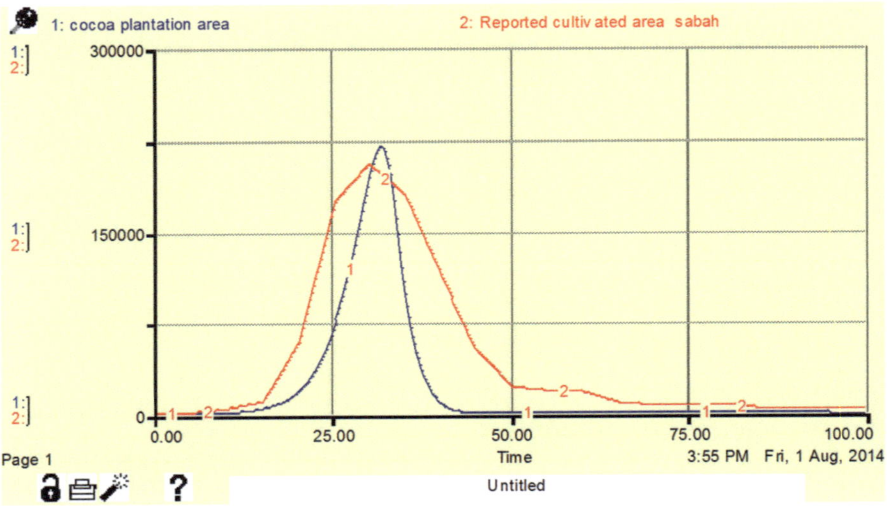

Fig. 8.5 Simulated and historical patterns of boom and bust of cocoa production systems in Sabah, Malaysia

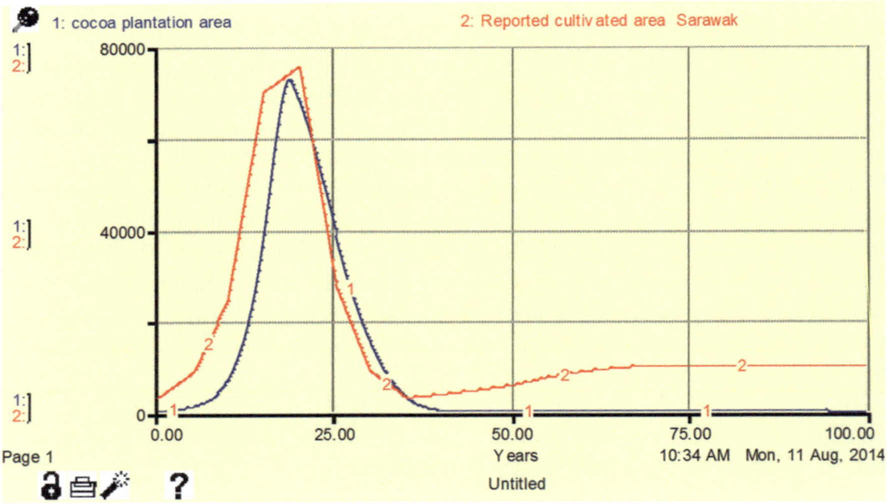

Fig. 8.6 Simulated and historical patterns of boom and bust of cocoa production systems in Sarawak, Malaysia

and providing pest control including IPM. Figure 8.7 shows the simulated cocoa plantation area for full subsidy (curve 3), 80 % subsidy (curve 2) and 60 % subsidy (curve 1) of the cost to cover conservation practices and insect control. The system is sustainable for full subsidy since financial support is provided for joint trade-off

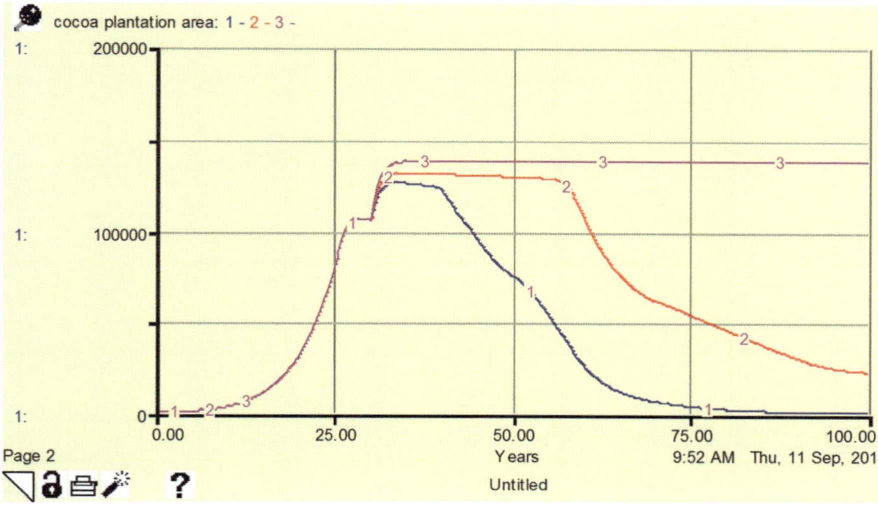

Fig. 8.7 Sensitivity analysis of subsidy for cocoa production systems in Malaysia

of biodiversity and yield along with IPM pest control through FFS, but the sustainability decreases with the degree of reduction of subsidy as there are lesser opportunities available to maintain biodiversity and insect control in terms of financial support. Waldron et al. (2012) also suggest that a simple development help specifically targeted at cocoa smallholders as the best short-term means to improve the long-term stability of the production in a sustainable environment, together with cocoa smallholders' economic status.

The model was also simulated for integration of subsidy with extension through farmer field schools to retain high biodiversity and attain acceptable yields along with proper integrated pest management for sustainable development of cocoa production systems in Malaysia. The model was simulated for two options: (1) First option is starting the joint programme of subsidy and extension at the peak of the boom. The subsidy coverage (100 %) at the peak of the boom covers the extension services (100 %) through farmer field schools for maintaining the recommended shading index and hence the biodiversity and providing pest control including IPM and (2) Second option is starting the joint programme of subsidy and extension at the beginning of the boom of cocoa. The subsidy coverage (0–100 %) was gradually increased from the beginning of the simulation period to cover the extension services (0–100 %) through farmer field schools for maintaining the recommended shading index and hence the biodiversity and providing pest control including IPM. Figure 8.8 shows reported results and simulated cocoa plantation area with subsidy and extension through farmer field schools at the peak of the boom, while Fig. 8.9 shows simulated cocoa plantation area with subsidy and extension through farmer field schools for coverage of the cocoa plantation with subsidy and extension through farmer field schools with 5 years of time horizon for

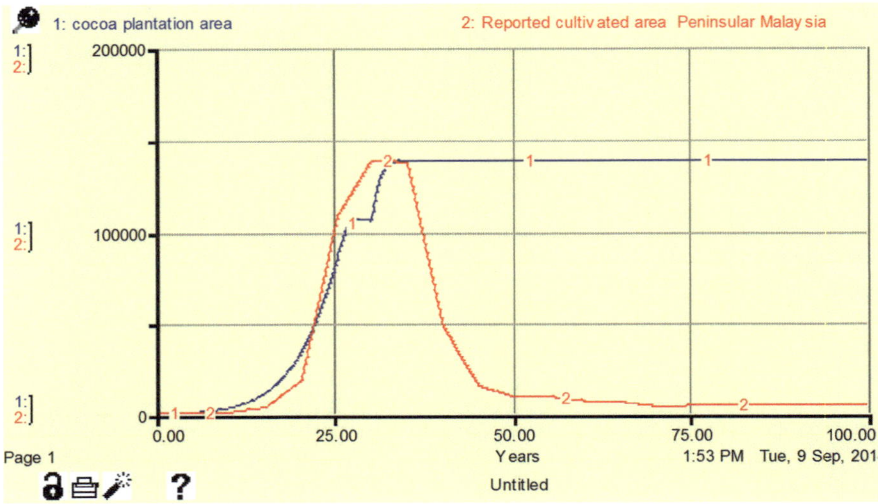

Fig. 8.8 Simulated results for subsidies for thinning and insect control

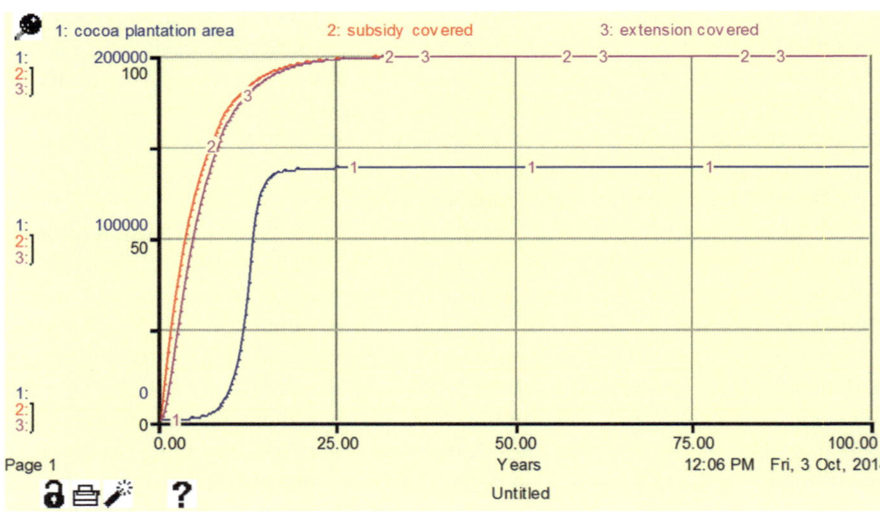

Fig. 8.9 Simulated cocoa plantation area, subsidy covered and extension covered for cocoa production systems in Malaysia

implementation of the subsidy and extension programme. In both the cases, the cocoa plantation area becomes stabilised. The cocoa production with subsidy and extension through farmer field schools at beginning of the production cycle not only stabilises the system but also returns to the stable condition much more quickly. The achievement of the stable cocoa production system largely depends on the success of integrated subsidy and extension programme to attain high biodiversity and acceptable yields through farmer field schools.

8.7 Conclusion

The boom and bust of cocoa production is one of the consequences of the short-term benefit of higher yield at the expense of reducing the biodiversity, thereby decreasing local habitat diversity and subsequently inviting pest and diseases. The subsidy policy proposed limits cocoa production to ecological carrying capacity which can be achieved by maintaining a joint trade-off between the shading index and proper actions for insect control of the cocoa production systems and acceptable yields. Simulated results support that suggested policy can lead to a sustainable production system. The Malaysian case study of the boom and bust of the cocoa production system gives an opportunity to develop a hypothesis of cocoa boom and bust and provided structure for policy simulation for sustainable development. Of course, adequate subsidy with effective extension of technologies from R&D for policy implementation is needed.

Exercises

Exercise 8.1 What do you mean by boom and bust? What caused the boom and bust of cocoa production systems in Malaysia?

Exercise 8.2 Draw the causal loop diagrams of cocoa production systems and then include insect infestation and finally shading of top covers.

Exercise 8.3 Draw a stock–flow diagram of cocoa production systems and explain how a causal loop diagram and stock–flow diagram represent the dynamic hypothesis of cocoa production systems. Also simulate the model.

Exercise 8.4 How can we make somebody to build up confidence in the system dynamics model of cocoa production systems? Also discuss the usefulness of the cocoa production system model.

Exercise 8.5 Discuss the policy issues to stabilise the boom and bust of cocoa production systems in Malaysia. How can the policy of stabilisation of cocoa production systems be implemented using the participatory systems approach?

References

Arquitt S, Hongang X, Johnstone R (2005) A system dynamics analysis of boom and bust in the shrimp aquaculture industry. Syst Dyn Rev 21:305–324

Bala BK (1999) Principles of system dynamics. Agrotech Publishing Academy, Udaipur

Bala BK, Hossain MA (2010) Modeling of food security and ecological footprint of coastal zone of Bangladesh. Environ Dev Sustain 12:511–529

Barlas Y (1996) Formal aspects of model validity and validation in system dynamics. Syst Dyn Rev 12:183–210

Clough Y, Faust H, Tscharntke T (2009) Cacao boom and bust: sustainability of agroforests and opportunities for biodiversity conservation. Conserv Lett 2:197–205

Department of Statistics (2014) Malaysia economic statistics – time series 2013. Department of Statistics, Malaysia, Putrajaya

Fatimah MA, Bala BK, Alias EF, Abdulla I (2015) Modeling of boom and bust of cocoa production systems in Malaysia. Ecol Model 309–310:22–32

Franzen M, Mulder MB (2007) Ecological, economic and social perspectives on cocoa production worldwide. Biodivers Conserv 16:3835–3849

Latiff IA (2007) Cocoa industry of Malaysia. In: 50 years of Malaysian agriculture: transformational issues, challenges and direction edited by Fatimah Mohamed Arshad, Nik Mustapha R Abdullah, Bisant Kaur and Amin Mahir Abdullah, Universiti Putra Malaysia, Serdang

Lee CH (2013) Planting cocoa – challenges and realities. Received from www.ipicex.com/doc1/2012/onral/planting cocoa – challenges and reality in malaysia.pdf

Motamayor JC, Lachenaud P, da Silva E Mota JW (2008) Geographic and genetic population differentiation of the Amazonian chocolate tree (*Theobroma cacao L.*). PLoS One 3, e3311

MPIC (Ministry of Plantation Industries and Commodities) (2011) National commodity policy (2011–2020). Ministry of Plantation Industries and Commodities, Putrajaya

Prusty SK, Mohapatra PKJ, Mukherjee CK (2011) Sustainable growth strategies for Indian shrimp industry: a multiple perspective. Proceedings of the 29th international conference of the system dynamics society, held on 24–28 July, Washington, DC

Sterman JD (2000) Business dynamics: systems thinking and modeling for a complex world. Irwin-McGraw-Hill, Boston

Waldron A, Justicia R, Smith L, Sanchez M (2012) Conservation through chocolate: a win-win for biodiversity and farmers in Ecuador's lowland tropics. Conserv Lett 5(3):213–221

Bibliography

Forrester JW (2007) System dynamics – the next fifty years. Syst Dyn Rev 23(2–3):359–370

Modelling of Hilsa Fish (*Tenualosa ilisha*) Population in Bangladesh

9

In the previous chapter, the modelling and simulation of boom and bust of cocoa production systems in Malaysia have been presented. This chapter demonstrates how to construct a system dynamics model of hilsa fish and simulate the model for policy planning and design for sustainable development. The model presented in this chapter is organised as follows: (1) introduction, (2) dynamic hypothesis, (3) causal loop diagram, (4) stock–flow diagram, (5) model validation, (6) simulation and policy analysis and (7) conclusion to illustrate the system dynamics applications based on systems thinking. Such experiences are essential to face the challenge of modelling and simulation of dynamic systems. The model predicts the long-term trends in the hilsa population over several decades and assesses the impacts of harvesting the juveniles and spawning adults. Simulated results show that increased harvesting of the adults entering the rivers and the juveniles in the rivers cause gradual decline in hilsa fish population and even may cause this valuable resource to disappear within a short period of time. Also the optimal strategies for sustainable development of hilsa fish have been addressed.

9.1 Introduction

Hilsa (*Tenualosa ilisha*) is the national fish of Bangladesh. It is the largest single species fishery in Bangladesh especially during the monsoon and contributes about 22–25 % of the total fish production of the country. The average annual production of hilsa is 200,000 tons, and its value is Tk 20,000 million (1 US dollar = taka 70.00). The contribution of this fish in the GDP of Bangladesh is 1 %, and annually taka 500–600 million is earned as foreign currency from the export of hilsa. It is harvested mainly by gill net, and about 2 % of the total population of Bangladesh is directly or indirectly involved in this fishery.

There is a decline of fisheries worldwide due to overexploitations of the stocks (FAO 1995), and there is also a decrease of about 19 % in harvesting of hilsa in the

B.K. Bala et al., *System Dynamics*, Springer Texts in Business and Economics,
DOI 10.1007/978-981-10-2045-2_9

recent years. If this trend continues, the existence of this fish will be at risk, and eventually it will become a rare species.

The sustainable management of marine renewable resources requires integration of biological, economic and social aspects (Nunes et al. 2004; Bald et al. 2009). Considering the importance of hilsa fish in nutrition, diet, employment and economy, the sustainable management of hilsa fishery is very important. For sustainable production of hilsa fish as well as increase in production, sound development policy and management strategies are urgently needed.

Several studies have been reported on modelling of population dynamics of fish (Tian et al. 2004; Robadue and Simanek 2007) and other species such as gooseneck barnacle (Bald et al. 2006). Few studies have been reported on the contributions of overfishing on the decline/collapse of the fish stock (Lillegard et al. 2005; Kaitala et al. 2003). Limited studies have been reported on parameters such as mortality, catch rate and maximum sustainable yield of hilsa using FISAT software (Miah et al. 1997; Amin et al. 2000; Haldar and Amin 2005; Hashemi et al. 2010). Mome and Arnason (2007) reported a simple bioeconomic model similar to FISAT software, and it is not a true model; rather, it is a supplementary model.

The system dynamics simulation model of population dynamics of hilsa fish presented here to assess the policy options and management strategies for sustainable development of hilsa fish is adopted from Bala et al. (2014). To understand and model the population dynamics of hilsa fish, a knowledge of the reproduction system of hilsa fish is essential. To bridge this gap, we present a brief and comprehensive description of the reproduction system of hilsa fish relevant to system dynamics modelling of population dynamics of hilsa fish. Then we focus on the system dynamics modelling of hilsa fish population and examine the harvesting policies of juveniles and spawning adults and the policies for sustainable exploitation of hilsa fish.

There are five species of hilsa under *Tenualosa* genus found in the world. The two species, *Tenualosa ilisha* and *Tenualosa toli* are found in Bangladesh. The availability of hilsa is 50–60 % in Bangladesh, 20–25 % in Myanmar and 10–15 % in India and the rest 5–10 % all over the world.

The species of hilsa is heterosexual, and the females due to faster growth rate attain larger sizes than the males. Shafi et al. (1977a, b) observed that the body of females is broader and the girth is comparatively larger. Opinions differ whether one sex predominates over the other in different phases of the life history of hilsa or whether the distribution conforms to the normal 1:1 ratio. Quereshi (1968) observed that although the sex ratio was 1 during the monsoon, the females dominated in October.

There is some evidence to suggest that there are two different and distinct spawning stocks or varieties. Quddus (1983) reported that the stocks responsible for the two spawnings are different—the 'broad' variety taking part in monsoon spawning and the 'slender' variety in the winter spawning. Comparative studies by Quddus (1982) of the 'broad' and 'slender' types of hilsa indicate a highly significant difference between the two types in the fecundity estimates. In the 'broad' type, the fecundity estimates range from 0.6 to 1.5 million for the fish in the size range of 33–51 cm and for the 'slender' type from 0.4 to 0.6 million for fish ranging in size from 32 to 49 cm. Hatching takes place in about 23–26 h at an average temperature of 23 °C (Jones and Menon 1951), about 18–26 h at 28.0 °C to 28.5 °C (Kulkarni 1950) and at about 24–28 h (Motwani et al. 1957). The length of the newly hatched larvae is recorded as 2.3 mm by Jones and Menon (1951), 3.1 mm by Kulkarni (1950), 2.50–2.55 mm by Motwani et al. (1957) and 2.4–3.0 mm by Karamchandani (1961).

The juveniles remain in the rivers/estuaries till they probably reach a length of 150–160 mm which is believed to be attained in about 5 months time. It is possible that the seaward migration is undertaken at this size/age. It has been reported that the operation of small-mesh nets in the period immediately succeeding the breeding season results in a large-scale destruction of young fish in Bangladesh waters (Quereshi 1968).

Conflicting views have been expressed on the minimum size of hilsa at first maturity. Hilsa may attain first maturity at the end of the first year or at the beginning of the second year. In Bangladesh waters (Meghna), Shafi et al. (1978) observed a size of 21 cm in the case of males and 32 cm in the case of females, as the size at first maturity.

Hilsa is essentially a plankton feeder, and generally, the items which are preponderant are crustaceans (particularly copepods), diatoms and green and blue algae; organic detritus, mud and sand have also been recorded. Pillay (1958) has reported no evidence of cessation or appreciable decrease in the feeding activity during the spawning period, but some workers reported that during spawning migration, the intensity of feeding decreases or ceases altogether (Pillay and Rosa 1963).

To model the various phases of the life cycle of hilsa fish, the information on the reproduction system such as sex ratio, eggs per nest, larva, juvenile and maturation of hilsa fish from different sources has been carefully examined and incorporated in the model to make it to represent the real-world situations of the hilsa fish in Bangladesh.

9.2 Dynamic Hypothesis

The dynamic hypothesis is a conceptual model, and it hypothesises to define the critical feedback loops to construct stock–flow diagrams that drive the system's behaviour in the reference mode. When the model is simulated, the endogenous structure of the model generates the reference mode behaviour of the system, and thus, the endogenous structure causes the changes in dynamic behaviour of the system. In system dynamics modelling, causal loop diagrams identify the principal feedback loops of a system, and the stock–flow diagrams that describe the structure of the systems are hypothesised to generate the reference mode of the behaviour over time (Stermant 2000). Hilsa population can be represented by causal loop diagrams and stock–flow diagrams, and the simulation model based on the causal loop diagrams and stock–flow diagrams would generate the dynamic behaviour of the hilsa population. The population dynamics of hilsa in the form of causal loop diagrams and stock–flow diagrams are hypothesised to generate the observed hilsa population dynamics in the reference mode. In essence the decline in hilsa population results from the overharvesting of juveniles and spawning adults, and this dynamics results from the endogenous consequences of the feedback structure.

9.3 Causal Loop Diagram

The causal loop diagram of the population dynamics model of hilsa fish is shown in Fig. 9.1. The population dynamics model of hilsa fish is dominated by one positive and four negative loops. The positive loop shows the reproductive and maturation process, ultimately producing more adult hilsa. The number of spawning adults of hilsa fish will determine the number of larvae production and leads to the production of matured adults of hilsa fish. Juvenile rate, prematuration rate, maturation rate and spawning rate have positive effect on the production of hilsa fish. In the absence of any stabilisation, this loop would result in the exponential growth of the hilsa population. The negative feedback loops are balancing loops of the different stages of the life cycle of hilsa fish, and these are deaths of larvae, juvenile, premature adults and matured adults. An increase in the number of hilsa fish population at any stage of the life cycle will increase the number of deaths, and this in turn will decrease the number of hilsa population. Also harvesting at any stage of the life cycle of hilsa fish has a negative effect on the hilsa population. The negative loops cause the stabilisation at the different stages of the life of hilsa fish due to death and harvesting of the hilsa fish.

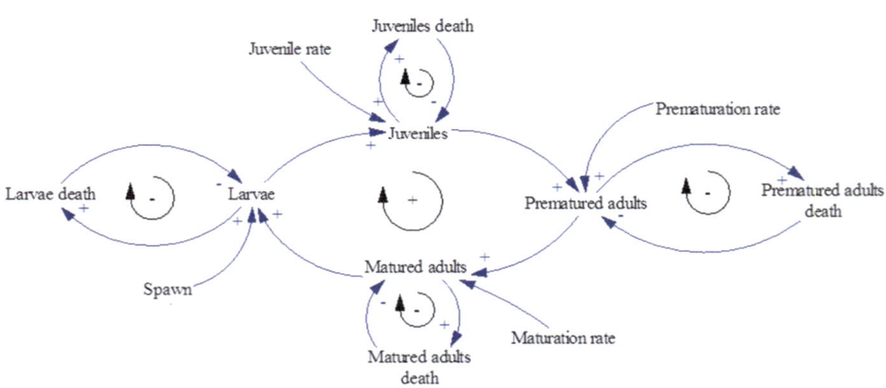

Fig. 9.1 Causal loop diagram of the population dynamics of hilsa fish

9.4 Stock–Flow Diagram

Figure 9.2 shows the stock–flow diagram of the population dynamics of hilsa fish. The flow diagram shows how stocks and flows are interconnected to produce the feedback loops and how the feedback loops interlink to create the system. The model is designed with seven stocks to keep track of the population of hilsa at the various phases of the life cycle is shown in Fig. 9.2. The life cycle begins when the adults of hilsa are about to spawn in rainy season during monsoon for laying eggs. The adults about to spawn in the rainy season for laying eggs can be described as:

$$\text{adults\_about\_to\_spawn}(t) = \text{adults\_about\_to\_spawn}(t - dt) + (\text{adults\_leaving} - \text{adults\_at\_spawning} - \text{harvesting}) * dt \quad (9.1)$$

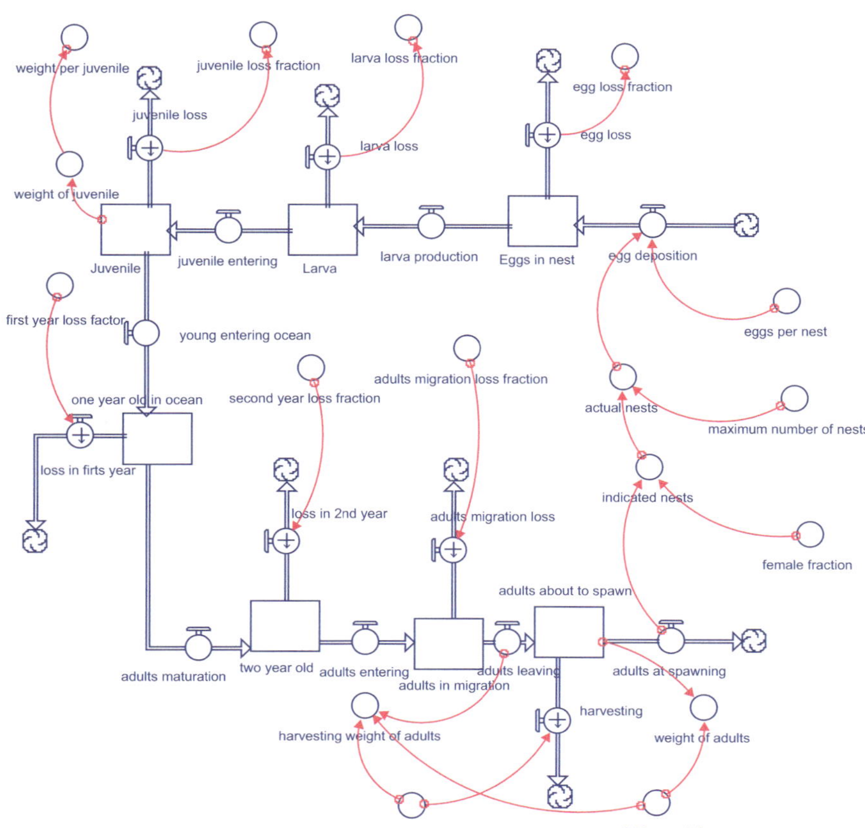

Fig. 9.2 Flow diagram of the population dynamics of hilsa fish

where adults_about_to_spawn(t) is adults about to spawn at time (t), adults_leaving is the 2-year-old adults which have left sea and joined spawning at time (t-dt), adults_at_spawning is the adults which have started spawning at time (t-dt) and harvesting is harvesting rate of adults about to spawn at time (t-dt).

The adults about to spawn lay eggs in the spawning grounds in the rivers and the eggs in the nest can be expressed as:

$$\begin{aligned}\text{Eggs\_in\_nest}(t) = {} & \text{Eggs\_in\_nest}(t - dt) \\ & + (\text{egg\_deposition} - \text{larva\_production} - \text{egg\_loss}) * dt\end{aligned} \tag{9.2}$$

where Eggs_in_nest(t) is eggs_layed in the nest at time (t), egg_deposition is egg deposited in the nest at time (t-dt), larva_production is the larva produced from eggs at time (t-dt) and egg_loss is egg lost in the nest at time (t-dt).

These eggs hatch 1 month later and larvae are produced from the eggs. The production of larvae can be expressed as:

$$\begin{aligned}\text{larva}(t) = {} & \text{larva}(t - dt) \\ & + (\text{larva\_production} - \text{juvenile\_entering} - \text{larva\_loss}) * dt\end{aligned} \tag{9.3}$$

where larva(t) is larva at time (t), larva_production is the larva produced at time (t-dt), juvenile_entering is the juvenile emerged from larva at time (dt-t) and larva_loss is larva lost at time (t-dt).

Larva takes 2 months to become juvenile. The juveniles are the recruiting phase of hilsa and the recruitment occurs more or less throughout the year with a major peak in June and July (Milton 2010). The juvenile population can described as:

$$\begin{aligned}\text{Juvenile}(t) = {} & \text{Juvenile}(t - dt) \\ & + (\text{juvenile\_entering} - \text{young\_entering\_ocean} \\ & - \text{juvenile\_loss}) * dt\end{aligned} \tag{9.4}$$

where Juvenile(t) is the Juvenile population at time (t), juvenile_entering is larva emerged into juvenile at time (t-dt), young_entering ocean is the young juvenile leaving then river at time (t-dt) and juvenile_loss is the juvenile lost in the river at time (t-dt).

The Juveniles remain in the rivers/estuaries for 8 months till they probably reach a length of 150–160 mm and then make the journey for sea. The 1-year-old in the ocean can be expressed as:

$$\begin{aligned}\text{one\_year\_old\_in\_ocean}(t) = {} & \text{one\_year\_old\_in\_ocean}(t - dt) \\ & + (\text{young\_entering\_ocean} - \text{adults\_maturation} \\ & - \text{loss\_in\_first\_year}) * dt\end{aligned} \tag{9.5}$$

where one_year_old_in_ocean(t) is the 1-year-old hilsa in ocean at time (t), young_entering_ocean is the young hilsa entering in the ocean at time (t-dt), adults_maturation is the matured hilsa at time (t-dt) and loss_ in_the_first_year is the hilsa lost in the first year in the sea at time (t-dt).

The juvenile that left the river develops maturity within 2 years in the sea and the 2-year-old matured hilsa in the sea can be described as:

$$\begin{aligned}\text{two\_year\_old}(t) &= \text{two\_year\_old}(t - dt) \\ &\quad + (\text{adults\_maturation} - \text{adults\_entering} - \text{loss\_in\_2nd\_year}) * dt\end{aligned} \tag{9.6}$$

where 2_year_old(t) is the 2-year-old hilsa in the sea at time (t), adults_maturation is the young 1-year-old hilsa maturating into 2-year-old hilsa in the ocean at time (t-dt), adults_entering is the 2-year-old hilsa leaving the sea and entering the rivers/ estuaries at time (t-dt) and loss_in_2nd_year is the loss of 2-year-old hilsa in the sea at time (t-dt).

The adult fish (2-year-old) from the sea return to rivers/estuaries where they lay eggs and this category of 2-year-old adults in rivers/estuaries is described as:

$$\begin{aligned}\text{adults\_in\_migration}(t) &= \text{adults\_in\_migration}(t - dt) \\ &\quad + (\text{adults\_entering} - \text{adults\_leaving} - \text{adults\_migration\_loss}) * dt\end{aligned} \tag{9.7}$$

where adults_in_migration(t) is 2-year-old adult hilsa migrating from sea to river at time t, adults_entering is the adults joining migration for journey to river at time (t-dt), adults_leaving is the adults left sea and entered the river at time (t-dt) and adults_migration_loss is the loss of adult hilsa during migration at time (t-dt).

Equations 9.1, 9.2, 9.3, 9.4, 9.5, 9.6, and 9.7 are solved using Runge–Kutta fourth-order method using STELLA software. The parameters of the model are estimated from studies, reports and publications. The parameters of the model are shown in Table 9.1.

In this part of modelling, we present a simple dynamic model for optimal harvesting strategy of adult hilsa for sustainable development of hilsa fishery using the population model. The optimisation problem is defined as the problem of maximisation of the total harvest of adult hilsa for the entire simulation period under conditions that the catch of adult hilsa fish is maintained at the maximum sustainable yield, i.e. spawning biomass is conserved and the harvesting juvenile is not allowed to exceed the present level of loss fraction of juvenile for sustainable harvest for all cycles of hilsa production. The objective function is to maximise the total adult harvested biomass during the whole year for sustainable development of hilsa fishery using the population model. The harvesting of adults of hilsa was optimised for sustainable development of hilsa fishery in Bangladesh. The model was simulated to:

Maximise

$$Z = \int_0^{T_t} HB(t)\,dt \tag{9.8}$$

Subject to the constraints

Table 9.1 Parameters used in the model

Parameter	Units	Adult		Egg	Larva	Juvenile	Mature
		M	F				
Average length*	mm	400	430	0.8	17.15	180	340
Average weight*	G	810	830	–	–	90	530
Female fraction****	Decimal	–	0.5	–	–	–	–
Eggs per ground**	Million	–	–	0.20	–	–	–
Stock*	Million	187.17	187.17	7,486,800	74,868	14973.6	898.42
Loss fraction***	Decimal	–	–	0.99	0.8	0.4	0.3
Stock after loss*	Million	37.43	37.43	74,868	14973.6	8984.2	628.89
Harvesting fraction***	Decimal	0.8	0.8	–	–	0.95	–
Second year loss fraction*	Decimal	–	–	–	–	–	0.5
Stock at second year*	Million	–	–	–	–	–	314.45

Sources: Mome and Arnason* (2007), Milton** (2010), Amin et al*** (2000, 2002) and Halder and Amin**** (2005)

$$C_p = MSY$$

$$L_f \leq L_{fe}$$

$$0.0 \leq H_f \leq 1.0$$

$$\int_0^{T_{p1}} HB(t)\,dt = \int_0^{T_{p2}} HB(t)\,dt = \int_0^{T_{p3}} HB(t)\,dt = \ldots$$

where HB(t) is the harvested weight of adult hilsa at time (t), tons/month; C_p the catch (yield), tons; MSY the maximum sustainable yield, tons; L_f the loss fraction of juvenile, decimal; L_{fe} the present loss fraction of juvenile, decimal; H_{fe} the harvesting fraction of adult hilsa, decimal; T_p the period of cycle, month; and T_t the total simulation period, month.

An adaptive pattern search was used for the solution of the constrained optimisation problem stated in Equation 9.8. This method does not require derivatives or approximation to derivatives to solve the problem, and it has been found to be very effective in solving non-linear constrained optimisation problems in system dynamics (Keloharju and Wolstenholme 1989; Dangerfield and Roberts 1996). The technique consists essentially of an exploratory search and a pattern search under constrained conditions.

9.5 Model Validation

The tests for building confidence in system dynamics models include (1) tests of structure, (2) tests for behaviour and (3) tests for policy implications (Bala 1999). Various ways of validating a system dynamics model have been considered such as comparing the model results with collected data; checking whether relations among the variables are logical and real, whether the model generates plausible behaviour and whether the model can generate anticipated behaviour under extreme condition; and checking the quality of parameter's value. Parameters have been derived from studies, other reports and publications to judge the plausibility of the model.

The behaviours of the key variables in the base run were examined. However, the management decisions have not been tested with time series field data since time series data are not available but have been compared with the reported values available. Forrester and Senge (1980) pointed out that for the public and political leaders, a useful model should explain causes of important problems and provide a basis for designing policies that can improve behaviour in the future. In the behaviour validity tests, emphasis should be on the behavioural patterns rather than on point prediction (Barlas 1996). The model presented generated plausible behaviours. An extreme condition test was conducted to check whether the model is capable to cope with extreme conditions and can provide the anticipated behaviour. One such extreme condition is the destruction of the spawning grounds, and it is anticipated that first juvenile and then adults should disappear within a very short

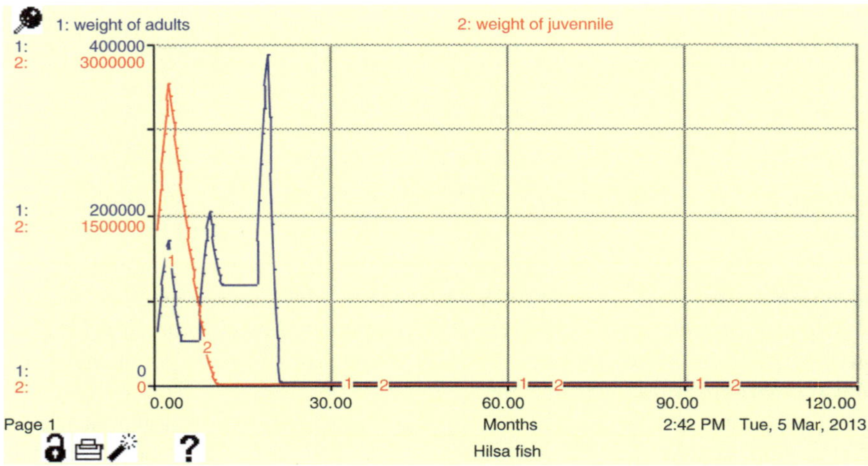

Fig. 9.3 Weights (tons) of adults about to spawn for total destruction of the habitats, i.e. eggs per nest = 0.0

time if the spawning grounds are destroyed. Figure 9.3 shows the simulated results for such a condition, and the simulated results show that first juvenile and then adults will disappear within 10 and 20 years, respectively. The model predictions are exactly what were anticipated.

The sensitivity of the important parameters was also estimated. In this study, behaviour sensitivity analysis of total weights of hilsa fish about to spawn to eggs per nest (150,000, 2,000,000 and 250,000) and adult migration loss (0.15, 0.25 and 0.35) was conducted. Figures 9.4 and 9.5 show the changes in total weights of adults for changes of eggs per nest (150,000, 2,000,000 and 250,000) and adult migration loss (0.15, 0.25 and 0.35), respectively. The total weights of adults increase with the increase of eggs per nests, while the total weights of adults decrease with the increase in values of adult migration loss, and these correspond with real-world situations.

The model is able to provide qualitative and quantitative understanding of the impacts of harvesting of juvenile and spawning adult hilsa fish, and the model represented the perceptions of reality and could be a good communication tool to advocate actions towards better management strategies of the harvesting of juvenile and spawning adult hilsa fish for sustainable exploitation of the hilsa fishery.

9.6 Simulation and Policy Analysis

System dynamics model is essentially developed to predict policy scenarios and applied to the understanding and modelling of complex systems. Because of the difficulty of documenting the effects of juvenile and adult hilsa fish, questions about

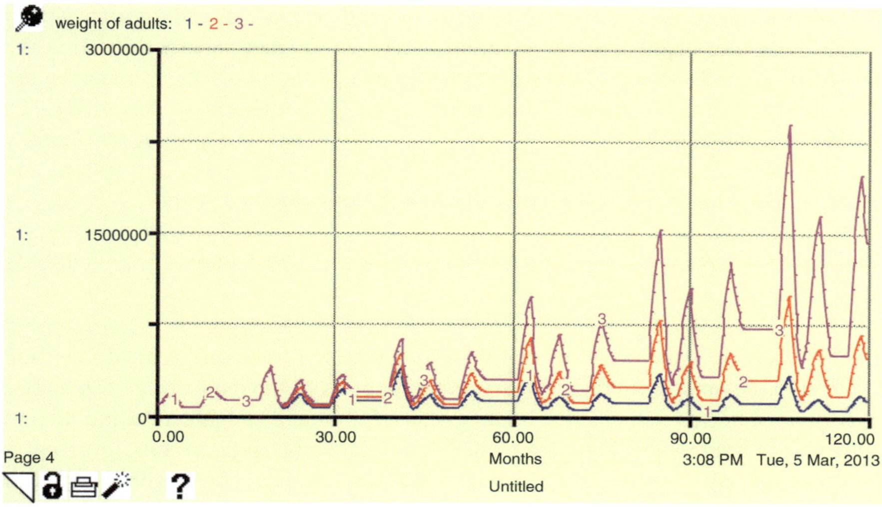

Fig. 9.4 Sensitivity analysis of the total weights of adults about to spawn to eggs per nest (150,000, 200,000 and 250,000)

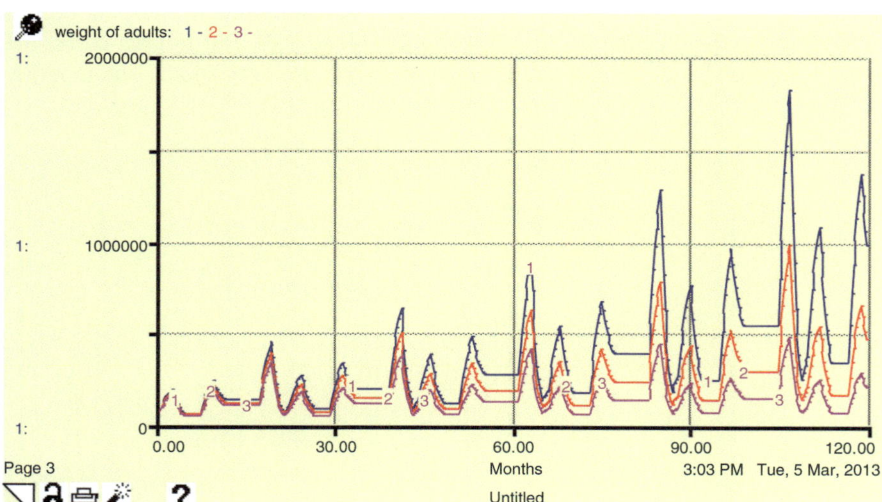

Fig. 9.5 Sensitivity analysis of the total weights of adults about to spawn to adult migration loss (0.15, 0.25 and 0.35)

the effectiveness of hilsa fishery management have been addressed through system dynamics model of hilsa fish population dynamics. In Bangladesh restricting the harvest of juvenile hilsa fish is extensively advocated for hilsa fishery management for sustainable development. This question is also addressed in this study. To examine the potentiality of the model as a tool for policy analysis, several policy options and management strategies are designed and tested. The policy options and management strategies are simulated for a time horizon of 10 years.

Scenario 1: Present practice of harvesting of juveniles and adults continued (base case)

This scenario is based on the information collected from different sources and updated to the present practices to assess the impacts of the harvesting of juveniles and adults of hilsa as usual in the coming years. Simulated weight and population of hilsa fish are shown in Figs. 9.6 and 9.7, respectively, and the time horizon is 10 years starting 2004. Figure 9.6 shows that the monthly changes in weights of adult and juvenile are stabilising with time for the juvenile loss fraction of 0.95 and harvesting loss fraction of 0.8. And also Fig. 9.7 shows that the monthly changes in population of 2-year-old fish, adults in migration, eggs in nest, juvenile and adults about to spawn are stabilising with time. The juvenile loss fraction and harvesting fraction include both the natural mortality and fishing mortality of juvenile population and adult hilsa fish population. The present harvesting practice does not result in any progressive decline of hilsa fish (Figs. 9.6 and 9.7). But in both cases, the growth rates of juveniles and adults about to spawn are very small. The monthly

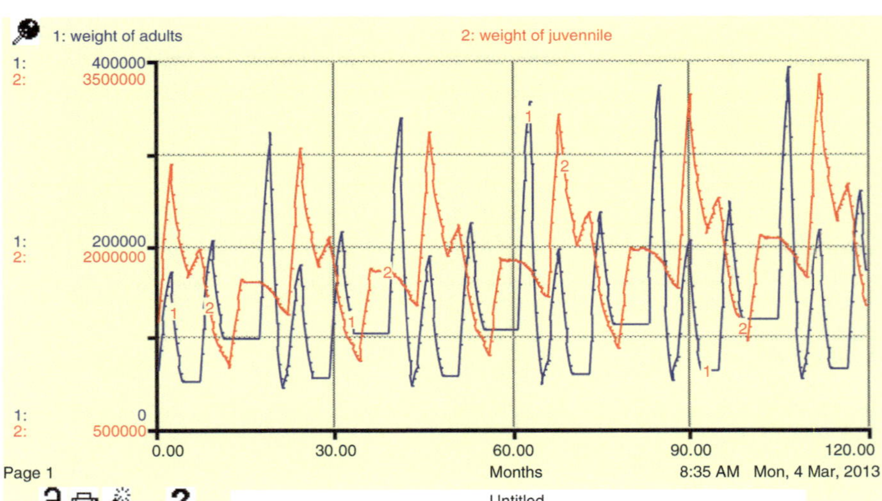

Fig. 9.6 Weights (tons) of juvenile and adult population of hilsa for the present practice of harvesting of juveniles and adults

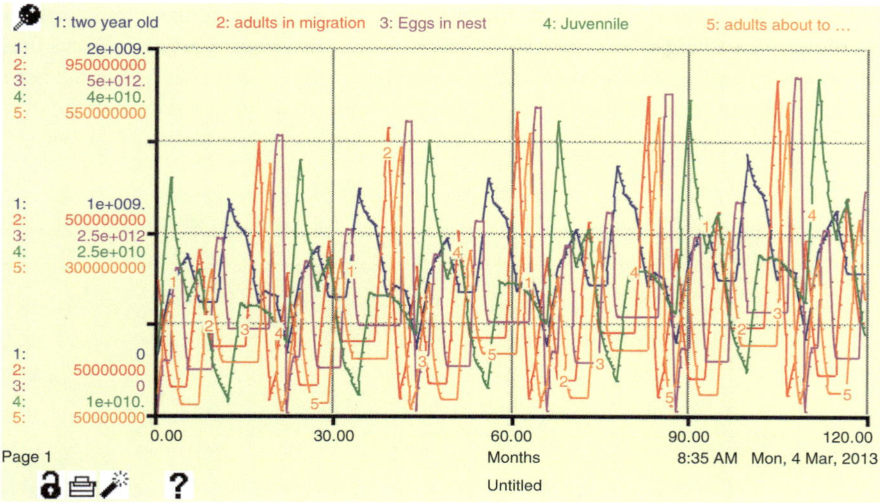

Fig. 9.7 Population of 2-year-old, adult in migration, egg and juvenile of hilsa for the present practice of harvesting of juveniles and adults

changes in the weights of adult hilsa of the standing stock during harvesting season in Fig. 9.6 show that the weights of standing stock of adult hilsa fish change from 290,000 to 3 80,000 tons which compare well with the reported value of 354,000 tons (Mome and Arnason 2007). The simulated results predicted the maximum sustainable yield to be 268, 000 tons. But Mome and Amarson (2007) reported that hilsa catch ranged between 194,981 and 280,328 tons from 1987 to 2007 and BBS (2012) reported a total catch of 313,753 tons of hilsa in Bangladesh which clearly indicates overfishing of hilsa in Bangladesh. Carrying capacity has been defined in many different ways, and it is not a constant number, and changes in environmental conditions may reduce the carrying capacity (Ayllon et al. 2012). The carrying capacity of hilsa fish was computed from the simulated results based on Gordon–Schaefer model (Siddique 2011), and it was found to be 670,000 tons. The monthly changes in the weights of juvenile hilsa during harvesting season also follow the patterns of the monthly changes in the weights of adult hilsa. The monthly changes in the population of 2-year-old, adult in migration, juvenile and egg of hilsa fish in Fig. 9.7 show similar patterns of changes with different magnitudes and time lags. No time series data are available to compare the simulated population dynamics of the hilsa fish.

Scenario 2: Harvesting mortality of juveniles is increased from 0.95 to 0.98

Scenario 2 is designed to assess the impacts of increasing the harvested amounts of juveniles, i.e. 3 % increase in harvesting mortality on the standing stock of hilsa in the future years. Figures 9.8 and 9.9 show the impacts of 3 % increase in harvesting mortality of the juvenile on the standing stock of hilsa in the future years. Figure 9.8 shows that the weights of adult hilsa fish during harvesting season decrease from 153,000 to 75,000 tons within 30 months. Figure 9.8 also shows that the weights of adult and juvenile are decreasing rapidly, and it will be almost zero after 90 months of simulation time. Figure 9.9 shows that the population of 2-year-old fish, adults in migration, eggs in nest, juvenile and adult about to spawn are

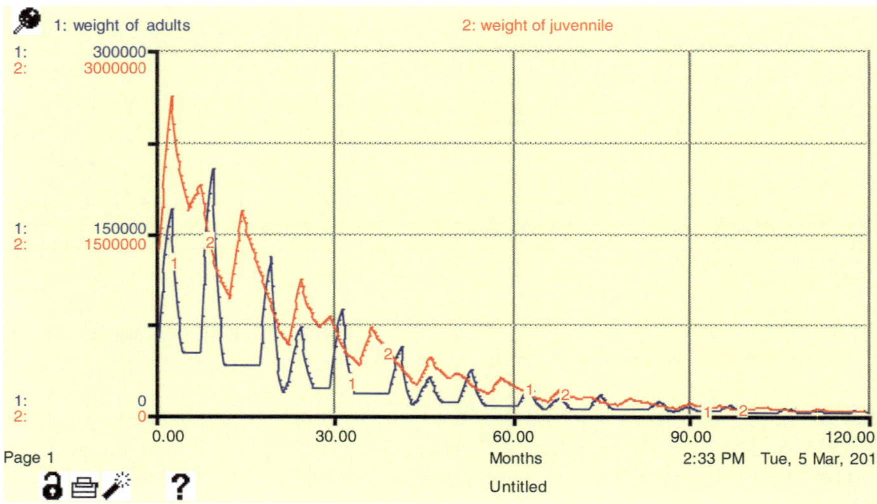

Fig. 9.8 Weights (tons) of juvenile and adult population of hilsa for the increase in harvesting mortality of juveniles from 0.95 to 0.98

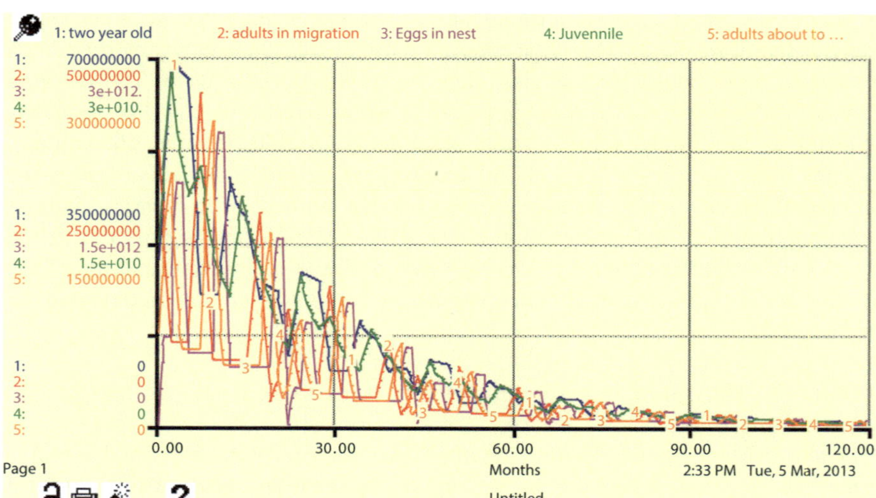

Fig. 9.9 Population of 2-year-old, adult in migration, egg and juvenile of hilsa for the increase in harvesting mortality of juveniles from 0.95 to 0.98

decreasing and ultimately become almost zero after 90 months of simulation time. Thus, increased juvenile harvesting has a negative impact on hilsa fish to disappear this valuable fish resource within a short period of time. This progressive decline of hilsa fish population for increased juvenile harvesting can be explained by the fact that the increased juvenile harvesting reduces the juvenile entering the ocean which ultimately reduces the adults in migration. This also reduces adults about to spawn and the larva production. As a result the juvenile population is further reduced resulting in progressive decline of hilsa fish population. Several researchers also reported similar findings (Amin et al. 2000; Haldar and Amin 2005; Hashemi et al. 2010). Amin et al. (2000) also reported overfishing of juvenile of hilsa for a loss fraction of 0.948 which is within the range of loss fraction of the overfishing of juvenile of hilsa in this study. To stabilise the hilsa fishery system, the juvenile standing stock/population should not exceed the level presented in Scenario 1.

Scenario 3: Harvesting of adults about to spawn is increased from 0.80 to 0.95

Scenario 3 is designed to assess the impacts of increasing the harvested amounts of adult hilsa, i.e. 15 % increase in the harvesting mortality of adults on the standing stock of hilsa in the future years. The effects of increase of the harvesting loss fraction to 0.95 on hilsa fish population are shown Figs. 9.10 and 9.11 respectively. Figure 9.10 shows that the weights of adult hilsa fish during harvesting season decrease from 153,000 to 37,500 tons within 30 months. Figure 9.10 also shows that the weights of adult and juvenile are decreasing rapidly, i.e. the production of hilsa will decrease to almost zero within 60–70 months. Figure 9.11 also gives similar trends for 2-year-old, adults in migration, eggs in nest, juvenile and adults about to spawn. It also shows that the existence of hilsa will be at risk, i.e. it will be almost zero after 70 months. This progressive decline of hilsa fish population for increased adult harvesting can be explained by the fact that the increased adult harvesting reduces the adults about to spawn which ultimately reduces the larva production. This also reduces juvenile production and entering the ocean and the adults leaving migration. As a result, the adults about to spawn are further reduced resulting in progressive decline of hilsa fish population.

Thus, the simulated results show that for increasing harvesting fraction, i.e. harvesting mortality of adult about to spawn is more effective to cause disappear hilsa species within several years. To avoid such a collapse of hilsa fishery in Bangladesh, the harvesting strategies should be not to exceed the maximum sustainable yield. Haldar and Amin (2005) also reported that fishes are harvested at a level than optimum fishing mortality and suggested some management policy to reduce the fishing mortality.

Scenario 4: Sustainable harvesting strategy of adults for the present juvenile loss fraction of 0.95

Scenario 4 is designed to develop sustainable harvesting strategy of the adults for the existing practices of overfishing of juvenile (0.95). For this policy, the

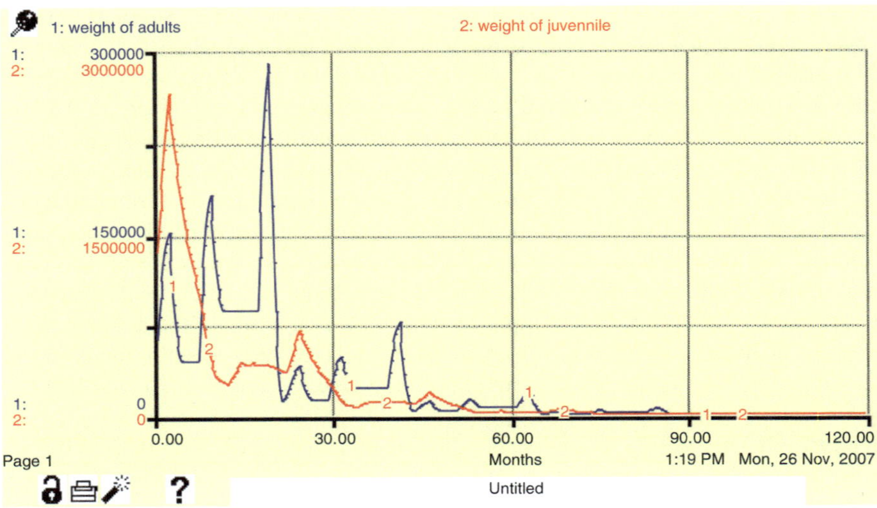

Fig. 9.10 Weights (tons) of juvenile and adult population of hilsa for the increase in harvesting mortality of adult from 0.80 to 0.95

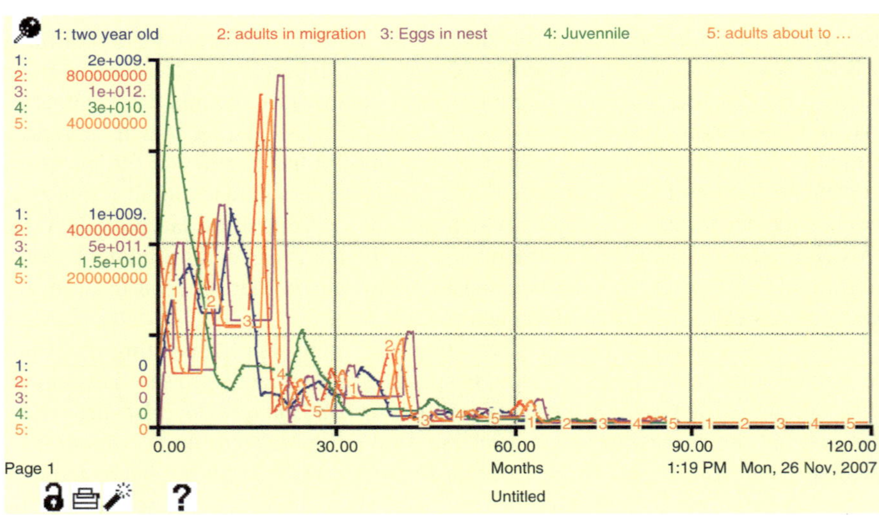

Fig. 9.11 Population of 2-year-old, adult in migration, egg and juvenile of hilsa for the increase in harvesting mortality of adult from 0.80 to 0.95

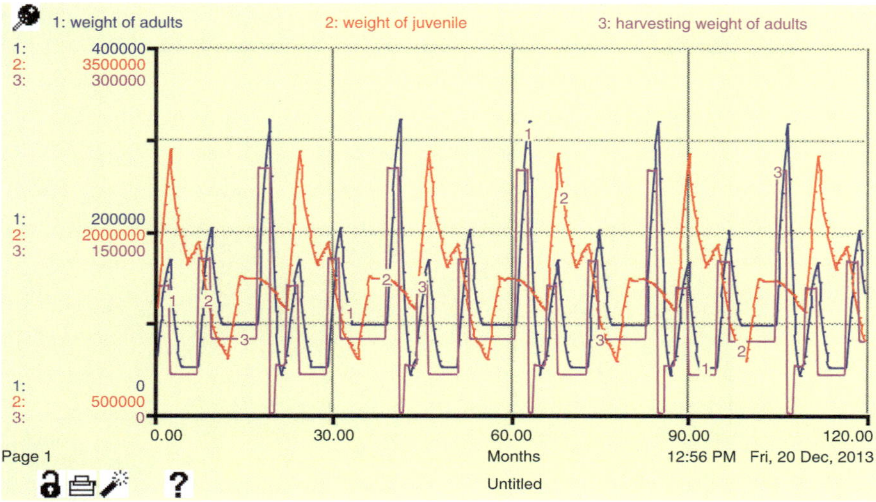

Fig. 9.12 Weights (tons) of juvenile, adult population and harvesting adults of hilsa for optimal harvesting of adult hilsa for the present loss fraction of the juveniles

restriction on maximum sustainable is relaxed. This policy considers proportional harvesting of the adults since this will ensure round-the-year fishing and hence livelihood of the fishermen. Figure 9.12 shows that simulated weights of the juvenile, adult hilsa and harvested adult hilsa for optimal harvesting strategy for sustainable development of hilsa fishery are stabilised. The optimal harvesting fraction of the adults for the existing juvenile harvesting practices of 0.95 was found to be 0.81 for sustainable development of the hilsa fishery. This will result in the stock of the adult hilsa in the peak season to be 320,137 tons and the highest peak season harvest of 200,821 tons.

Scenario 5: Sustainable harvesting strategy for sustainable development

Scenario 5 is designed to develop sustainable harvesting strategy of the adults and the reduction of the loss fraction of juvenile needed to maintain the carrying capacity of hilsa fishery. This policy also considers proportional harvesting of the adults since this will ensure round-the-year fishing and hence livelihood of the fishermen. Figure 9.13 shows that simulated weights of the juvenile, adult hilsa and harvested hilsa for sustainable development of hilsa fishery are stabilised. The optimal harvesting fraction of the adults and loss fraction of the juvenile harvesting fraction were found to be 0.85 and 0.935, respectively, for sustainable development of the hilsa fishery. This will cause to increase the stock of the adult hilsa and harvesting adults in the peak season 1.35 and 1.24 times higher than those of the previous policy (Scenario 4). Thus, this policy not only stabilises the hilsa fishery but also enhances the harvested adult hilsa and maintains the maximum sustainable yield.

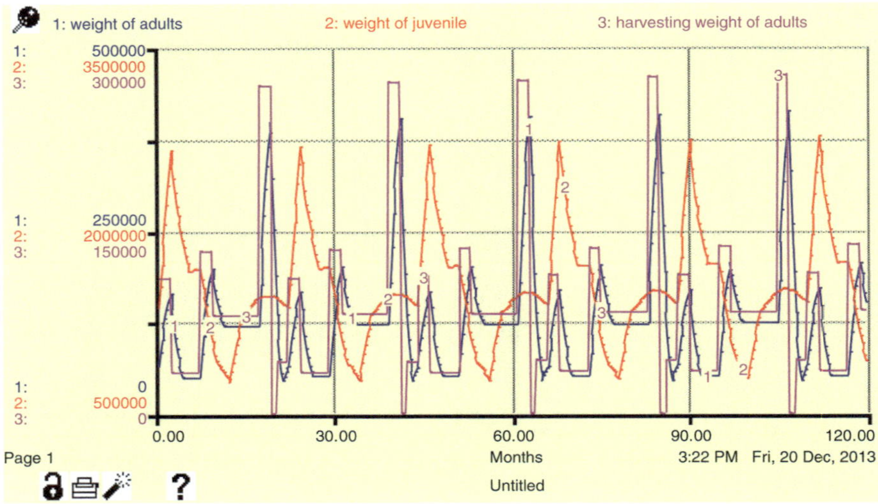

Fig. 9.13 Weights (tons) of juvenile and adult population of hilsa for optimal harvesting of adult hilsa for sustainable development of hilsa fishery

9.7 Conclusions

A computer model for hilsa fish population dynamics using system dynamics technique has been developed, and the model was simulated for the time horizon of 10 years. Two policy options such as juvenile harvesting mortality and increased level of harvesting of the adults are considered. Juvenile stage is the most critical stage for hilsa. Simulated results show that juvenile harvesting mortality and increased level of harvesting of the adults are the two sensitive policy issues for sustainable production of hilsa in Bangladesh. If appropriate policy options and management strategy for juvenile harvesting mortality and harvesting of the spawning adults are implemented, sustainable production of hilsa fish in Bangladesh could be maintained. Otherwise, this valuable resource of hilsa fish in Bangladesh will disappear within a short period of time. To address this problem, optimal harvesting strategy for sustainable hilsa fishery has been developed, and it not only enhances but also stabilises the hilsa fishery system. This model can be used to design policy options and management strategies for sustainable production of hilsa fish. However, further refinement of the model and updating of the data are needed.

Exercises

Exercise 9.1 Why is there a decline of fisheries worldwide? Discuss the importance of hilsa fishery in the context of economy and livelihood of hilsa fishing community in Bangladesh.

Exercise 9.2 Draw the causal loop diagrams of hilsa fishery which includes local price of hilsa fish and livelihood of hilsa fishing community.

Exercise 9.3 Draw stock–flow diagram of the causal loop diagram in exercise 9.2 and simulate hilsa production, local price of hilsa and livelihood of the hilsa fishing community.

Exercise 9.4 What are the two main types of harvesting practices of hilsa population in Bangladesh which are common scenario? Discuss the relative importance and trade-off for sustainable development.

Exercise 9.5 Discuss what should be the sustainable harvesting strategy for sustainable development. How can this be implemented using participatory systems approach?

References

Amin SMN, Rahman MA, Haldar GC, Nahar S, Dewan S, Mazid MA (2000) Population dynamics of jatka (juvenile hilsa) in the Meghna river, Bangladesh. Asian Fish Sci 13:383–389

Allyon D, Almodovar A, Nicola GG, Parra I, Elvira B (2012) Modeling carryingcapacity dynamics for the conservation and management of territorial salmonids. Fish Res 134–136:95–103

Bala BK (1999) Principles of system dynamics. Agrotech Publishing Academy, Udaipur

Bala BK, Fatimah MA, Alias EF, Sidique SF, Noh KM, Rowshon MK (2014) Modeling of hilsa fish (Tenualosa ilisha) population in Bangladesh: modeling and policy implications. Ecol Model 283:19–30

Bald J, Borja A, Muxika I (2006) A system dynamics model for the management of the gooseneck barnacle (*Pollicipes pollicipes*) in the marine reserve of Gaztelugatxe (Northern Spain). Ecol Model 194:306–315

Bald J, Sinquin A, Borja I, Caill-Milly N, Duclrcq B, Dang C, Montaudouin X (2009) A system dynamics model for the management of the Manila clam, *Ruditapes philippinarum* (Adams and Reeve, 1850) in the Bay of Arcachon (France). Ecol Model 220:2828–2837

Bangladesh Bureau of Statistics (BBS) (2012) Statistical pocket yearbook of Bangladesh 2011. Statistics and Informatics Division, Ministry of Planning, Dhaka, Bangladesh

Barlas Y (1996) Formal aspects of model validity and validation in system dynamics. Syst Dyn Rev 12:183–210

Dangerfield B, Roberts C (1996) An overview of strategy and tactics in system dynamics optimization. J Oper Res Soc 47(3):405–423

FAO (1995) The state of world fisheries and aquaculture. Food and Agriculture Organization of the United Nations, Fisheries Department, Rome

Forrester JW, Senge PM (1980) Tests for building confidence in system dynamics models. In: Legato Jr AA, Forrester JW, Lyneis JM (eds) System dynamics, TIMS studies in the management sciences, vol 14. North-Holland, Amsterdam, pp 209–228

Haldar GC, Amin SMN (2005) Population dynamics of male and female hilsa, *Tenualosa ilisha* of Bangladesh. Pak J Biol Sci 8(2):307–313

Hashemi S, Mohammadi G, Eskandary G (2010) Population dynamics and stock assessment of hilsa shad (*Tenualosa ilisha* Hamilton-Buchanan, 1822) in coastal waters of Iran (Northwest of Persian Gulf). Aust J Basic Appl Sci 4(12):5780–5786

Jones S, Menon PMG (1951) Observations on the life history of the Indian shad, Hilsa ilisha (Hamilton). Proc Indian Acad Sci 33(3):101–125

Kaitala V, Jonzen N, Emberg K (2003) Harvesting strategies in a fish stock dominated by low frequency variability: the Norwegian spring-spawning herring. Mar Resour Econ 18:263–278

Karamchandani SJ (1961) On the location of spawning grounds of Indian shad, Hilsa *ilisha* (Hamilton) in freshwater regions of the Narbada river. J Zool Soc India 30(10):373–375
Keloharju R, Wolstenholme EE (1989) A case study in system dynamics optimization. J Oper Res Soc 40(3):221–230
Kulkarni CV (1950) Breeding habits, eggs and early life history of Indian shad, *Hilsa ilisha* (Hamilton) in the Narbada river. Proc Natl Inst Sci India B15(2):169–176
Lillegard M, Engen S, Saether B, Toresen R (2005) Harvesting strategies for Norwegian spring-spawning herring. OIKOS 110:567–577
Miah MS, Haldar GC, Rahman MA (1997) Estimation of growth and mortality parameters of hilsa *tenualosa ilisha* (HAM.) population in the Meghna river of Bangladesh. Indian J Fish 44 (2):111–131
Milton DA (2010) Status of hilsa (*Tenualosa ilisha*) management in the Bay of Bengal: assessment of production risk and data gaps for more effective regional management. FAO report
Mome MA, Arnason R (2007) The potential of artisanal hilsa fishery in Bangladesh: An economically efficient fisheries policy. Final project 2007, The United NationsUniversity, Reykjavik, Iceland
Motwani MP, Jhingran VG, Karamchandani SJ (1957) On the breeding of the Indian shad, *Hilsa ilisha* (Hamilton) in freshwaters. Sci Cult 23:47–48
Nunes PALD, Rossetto L, Bleij A (2004) Measuring the economic value alternative clam fishing management practices in the Venice Lagoon: results from a conjoint valuation application. J Mar Syst 51(1–5):309–320
Pillay TVR (1958) Biology of the hilsa, Hilsa *ilisha* (Hamilton) of the river Hooghly. Indian J Fish 5:201–257
Pillay SR, Rosa Jr H (1963) Synopsis of biological data on hilsa, *Hilsa ilisha* (Hamilton) 1882. FAQ Fish B/al. Synops (25): 1: 1–6 :38
Quddus MMA (1982) Two types of Hilsa ilisha and their population biology from Bangladesh waters. Doctoral thesis, the University of Tokyo
Quddus MMA (1983) Hilsa shad of Bangladesh. Bangladesh Portrait 26:42–44
Quereshi MR (1968) Hilsa fishery in East Pakistan. Pak J Sci Ind Res 11(95–1):03
Robadue DD, Simanek RM (2007) A system dynamics perspective on a global fishing enterprise: the case of tuna ranching in Mexico. In: Proceedings of the 25th international conference of the System Dynamics Society, July 29 to August 2, Boston
Shafi M, Quddus MMA, Hossain H (1977a) Observations on the food and feeding habits of young *Hilsa ilisha* (Hamilton-Buchanan) from the river Dhaleswari. In: Proceedings, 2nd Bangladesh science conference, A-40
Shafi M, Quddus MMA, Islam N (1977b) Studies of gonad weight, sex-ratio and fecundity of *Hilsa ilisha* (Hamilton-Buchanan) from the river Meghna. J Asiat Soc Bangladesh (Sc.) 2(2):51–58
Shafi M, Quddus MMA, Islam N (1978) Maturation and spawning of *Hilsa ilisha* (Hamilton-Buchanan) of the river Meghna. Dacca Univ Stud B26(2):63–71
Siddique AB (2011) Harvest production and economic efficiency of Bangladeshi industrial trawlers in the Bay of Bengal. MS thesis, Norwegian College of Fishery Science, University of Tromso, Norway
Stermant JD (2000) Business dynamics: system thinking and modeling for a complex world. Irwin/ McGraw-Hill, Boston
Tian Y, Akamine T, Suda M (2004) Modeling the influence of oceanic-climatic changes on the dynamics of pacific saury in the north western pacific using a life cycle model. Fish Oceanogr 13(1):125–137

Bibliography

Forrester JW (1968) Principles of systems. MIT Press, Cambridge, MA

Modelling of Food Security in Malaysia 10

In the previous chapter, a case study on modelling and policy implication of hilsa fish population in Bangladesh has been presented. This chapter presents modelling of food security in Malaysia as a case study and application of system dynamics. The model is organised as (1) introduction, (2) dynamic hypothesis, (3) causal loop diagram, (4) stock–flow diagram, (5) model validation, (6) simulation and policy analysis and (7) conclusion. The findings suggest that gradual transition to bio-fertiliser, funding for R&D for development of high-yielding hybrid rice and increasing the cropping intensity hold big promise towards productivity enhancement. The promotion training and extension services using participatory approach of learning by doing or the farmer field schools are desirable to reduce the productivity gap.

10.1 Introduction

Global food security is a serious challenge to the mankind and it has emerged more seriously since the food price volatility in 2007–2008. Food security is rising to the centre of global discourse and has become an issue of national policy as well as public concern. Also, we often find food insecurity in the developing countries to mention some of the problems of complex and dynamic systems. The food crisis in 2007–2008 has affected particularly the developing and rice-importing countries in Asia. These prompted many South Asian countries to increase the domestic production and hence reduce the imports. The Malaysian government also adopted this approach and introduced the national food security policy. Malaysia has translated the food security concerns into three-pronged policy objectives and these are to ensure high price to paddy farmers to produce rice, to achieve a certain level of self-sufficiency in rice and to ensure a stable and high quality of rice to the consumers. It was perceived that under an open market, producers are susceptible to price volatility and consumers are subjected to supply instability in the world market. With these premises, the government has insulated the paddy and rice industry by

B.K. Bala et al., *System Dynamics*, Springer Texts in Business and Economics,
DOI 10.1007/978-981-10-2045-2_10

instituting a number of market interventions, from farm to import level. At the farm level, farmers are provided with input and cash subsidies and subsidised irrigation rates, and the price is guaranteed at the RM 750 per ton (one RM = 0.31 US dollar). The import of rice is entrusted to BERNAS, a corporate entity. This is the depth and width of the government's involvement and commitment in ensuring rice security to the country.

The policy objectives have been met as proven by the achievement of the SSL (self-sufficiency level), farmers received a relatively higher income and consumers are insulated from the vagaries of the world market. While these intended outcomes are satisfactory, the unintended consequences are showing symptoms of inefficiencies which may contribute partly to the slowing down of the productivity growth in the later years. These include heavy input subsidies (petroleum-based fertiliser) which result in the reduction of soil fertility and water quality. This in turn leads to more usage of this input to replenish the soil and hence continuous dependency on the fertiliser. While this strategy is effective in the short term, its long-term effect has been proven damaging. The recent increase in petroleum price pushes the input cost by more than 70 %. The government has responded to these problems by adding more fertiliser subsidies to help farmers with the soil problem as well as to cushion the impact of price hikes of inputs. The regional divide particularly in productivity between farmers in the granary and non-granary areas suggests poor transfer of technology. This is not surprising as the current extension system has not moved in parallel to the advancement in farming practices. These issues motivate us to explore the potential of R&D (varietal improvement), bio-organic fertiliser and new extension methods (field farmer schools) in enhancing the productivity and production of rice in Malaysia. This in turn helps Malaysia to achieve the rice availability which is one of the major pillars of food security.

Food security is a social sustainability indicator, and the most commonly used indicators in the assessment of food security conditions are food production, food self- sufficiency level, income, total expenditure, food expenditure, share of expenditure of food, calorie consumption, nutritional status, etc. (Riely et al. 1999). These studies give a descriptive statistics of the food security. Accounting tools for quantifying food security are essential for assessment of food security status and also for policy planning for sustainable development. Several studies have been reported to identify the determinants of food security at the household level using logistic models (Babatunde et al. 2007; Faridi and Wadood 2010; Haile et al. 2005; Mahajan and Joshi 2011; Sikwela 2008). Recently, Bala and Hossain (2010, 2013) reported a quantitative method of food security and also reported food security at regional levels, but this study does not consider subsidies, R&D and extension service. Also more recently, Anderson et al. (2013) reported an empirical basis for re-examining the effectiveness and efficiency of various policy options of ensuring food security in Asia and elsewhere.

Food security is a worldwide problem that has called the attention to governments and the scientific community. It particularly affects developing countries. The scientific community has had increasing concerns for strategic understanding and implementation of food security policies in developing countries, especially since the food crisis in the 1970s. The process of decision-making is becoming increasingly complex due to the interaction of multiple dimensions related to food security (Giraldo et al. 2008).

Food security is a situation in which people do not live in hunger or fear of starvation. FAO (1996a, b) defined the objective of food security as assuring to all human beings the physical and economic access to the basic food they need. This implies three different aspects: availability, stability and access. Food security exists when all people at all times have access to sufficient, safe and nutritional food to meet their dietary needs and food preferences for an active and healthy life (FAO 2002). Food security includes at a minimum (1) the ready availability of nutritionally adequate and safe foods and (2) an assured ability to acquire acceptable foods in socially acceptable ways (USDA 1999). The term food security has been used over time to mean different things, and it can be a useful measure of household and individual welfare, particularly if combined with estimates of household food acquisition and allocation behaviour (Per Pinstrup-Anderson 2009).

Historically, national food self-sufficiency has been equated with food security, but the definition could also be expanded to include agricultural self-reliance (Panagariya 2002). National food security is achieved when all citizens are individually food secure; but this is less straightforward than it appears. Nations often talk self-sufficiency ratios (the proportion of domestic consumption) as a measure of food security. A ratio of 1 or more would imply that the country's production satisfies all its food consumption needs. While this may occur for particular commodities, it is almost never possible for a country to be fully self-sufficient (Majumder et al. 2012).

In the wake of recent food price spikes and growing demands for food and also the growing concerns of climate change impacts in emerging Asia and for biofuels in Europe and the United States of America, the governments are re-examining their strategies for dealing both short-term and long-term food security, and the options include boosting agricultural production rates to deal with long-term concerns and using more appropriate domestic policy measures rather than trade policy to cope with price volatility (Anderson et al. 2013).

Considering the above notions of food security and the worst experiences with food crisis following the price volatility in 2007 and 2008, understanding the impacts of the recent competition between food and biofuel for land and climatic change impacts on food production and also realising the central issue of increasing local productivity and slowing population growth, Bala et al. (2014) were motivated to study the dynamics of food security in Malaysia to understand and design policy for increasing local productivity and reducing field level productivity gaps and address the questions of withdrawal of fertiliser subsidies and gradual transition to bio-fertilisers using systems approach. Modelling of food security in Malaysia presented here is adopted from Bala et al. (2014).

10.2 Dynamic Hypothesis

The dynamic hypothesis is a conceptual model typically consisting of a causal loop diagram, stock–flow diagram or their combination which drives the system's behaviour. When the model based on feedback concept is simulated, the endogenous structure of the model generates the reference mode behaviour of the system which results from the endogenous structure of the system. In system dynamics modelling, causal loop diagrams and stock–flow diagrams are used to describe the basic cause–effect relationships hypothesised to generate the reference mode of the behaviour over time (Sterman 2000). Food security can be represented by causal loop diagrams and stock–flow diagrams, and the simulation model based on the causal loop diagrams and stock–flow diagrams can generate the dynamic behaviour of the food security. Rice production and its self-sufficiency in the form of causal loop diagrams and stock–flow diagrams are hypothesised to generate the observed rice production and food security in the reference mode. In essence the decline in food security results from both the production and population, and this dynamics resulted from the endogenous consequences of the feedback structure.

10.3 Causal Loop Diagram

The causal loop diagram of rice production system in Malaysia is shown in Fig. 10.1. Causal loop diagrams identify the principal feedback loops of the systems. The causal loop diagrams have been used to describe basic causal mechanisms hypothesised to generate the reference mode of behaviour over time (Sterman 2000). A feedback loop contains two or more casualty related variables that close back on themselves. The relationship between one variable and next in the loop can be either positive or negative. A positive relationship means that if one increases, the other also increases. In a negative relationship, the two variables change inversely. If there are an even numbers of negative relationships in total, then the loop is positive; if there are odd numbers of negative relationships, the loop is negative. Positive feedback loops generate growth, i.e. re-enforcing, and negative feedback loops are goal seeking. There are eight main feedback loops in rice production system in Malaysia, of which two are positive, i.e. re-enforcing, and six are negative, i.e. goal seeking.

10.4 Stock–Flow Diagram

The flow diagram of the system dynamics model of rice production system in Malaysia is shown in Fig. 10.2. Fundamental equations that correspond to major state variables shown in Fig. 10.2 are as follows:

Rice production in Fig. 10.2 depends on rice productivity (tons/ha) as well as on the area under rice cultivation and cropping intensity, and it is computed as

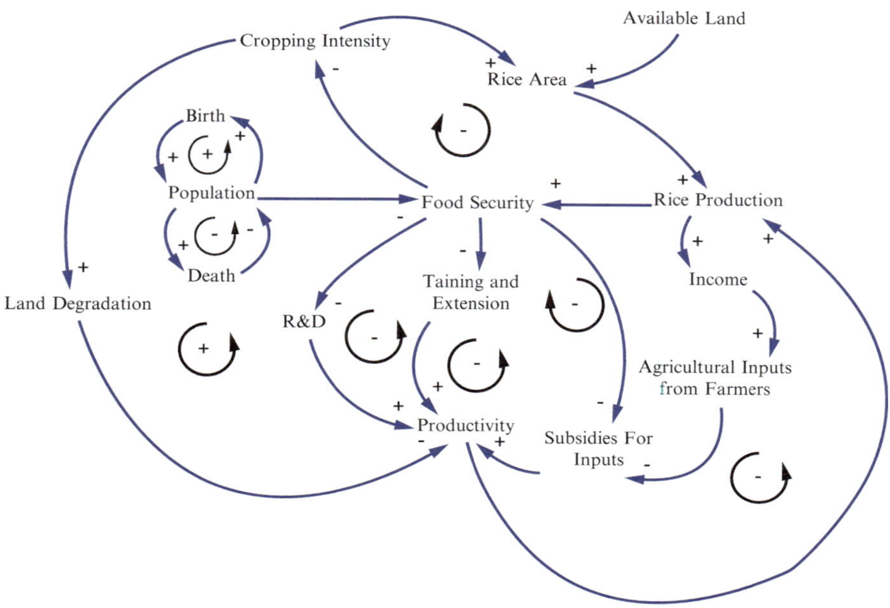

Fig. 10.1 Causal loop diagram of system dynamics model for rice production system in Malaysia

$$\text{rice production}\,(t) = \text{rice productivity}(t) \times \text{rice area}\,(t) \times \text{cropping intensity}(t) \tag{10.1}$$

The rice area is increased by transfer of land suitable for rice cultivation based on development policy of covering the land suitable for rice cultivation within 8 years, and also the rice area is decreased by a discard rate of 0.0085 fraction of rice area for housing, roads and highways and industries. This is expressed as

$$\text{rice area}\,(t) = \text{rice area}\,(t - \Delta t) + \text{rice area increase rate} \times \Delta t - \text{rice discard rate} \times \Delta t \tag{10.2}$$

Cropping intensity increases with development policy and programmes and also depends on potentiality of multiple crops and crop diversification, and it is expressed as

$$\text{cropping intensity}\,(t) = \text{cropping intensity}\,(t - \Delta t) + \text{cropping intensity increase rate} \times \Delta t \tag{10.3}$$

Rice productivity is increased by development of new hybrid varieties of rice through research and development and reducing productivity gaps through learning through training and extension. In addition, it is affected by agricultural inputs and land degradation. This is described as

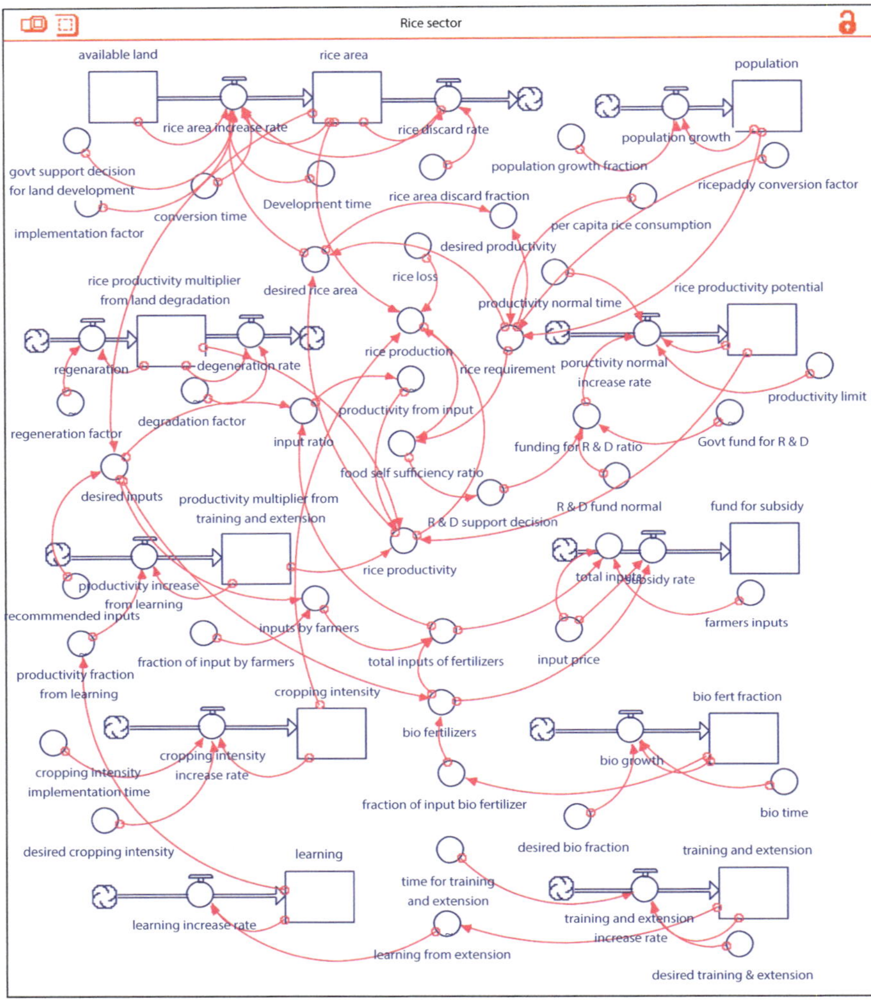

Fig. 10.2 Flow diagram of the system dynamics model of rice production systems in Malaysia

$$\begin{aligned}\text{rice productivity}\,(t) = {} & \text{rice productivity potential}(t) \times \text{productivity from input}(t)\\ & \times \text{rice productivity multiplier from land degradation}(t)\\ & \times \text{productivity multiplier from training and extension}\,(t)\end{aligned} \tag{10.4}$$

Rice productivity potential is increased by development of new high-yielding/hybrid varieties through research and development, and it is expressed as

$$\begin{aligned}\text{rice productivity potential}\,(t) = {} & \text{rice productivity potential}\,(t - \Delta t)\\ & + \text{productivity normal increase rate} \times \Delta t\end{aligned} \tag{10.5}$$

Rice productivity, i.e. yield/ha, is influenced by land degradation due to cropping intensity and can be increased by keeping the land fallow for a long time by regeneration, and it is described as a multiplier:

$$\text{rice productivity multiplier from land degradation } (t) = \text{ rice productivity multiplier from land degradation } (t-\Delta t) + \text{regeneration} \times \Delta t - \text{degeneration rate} \times \Delta t \tag{10.6}$$

The published research indicates that training and extension through farmer field schools is having a substantial impact in terms of increase in farm productivity, reduction in farmers' use of pesticides and improved farming knowledge (Davis et al. 2012; Larsen and Lilleor 2012; Mwagi et al. 2003; Praneetvatakul and Waibel 2003; Rola et al. 2002). Rice productivity can also be increased by training and extension through farmer field schools which provides an opportunity of learning by doing for the farmers to reduce the productivity gaps. Farmer field school is based on experiential learning techniques and participatory methods, and here, learning means to work based on experiences and knowledge, and it is run by training and extension service department/agency (Bala 2012). The contribution of the training and extension reduces the yield gap and it depends on learning. The productivity multiplier from training and extension is given by

$$\text{productivity multiplier from training and extension } (t) = \text{productivity multiplier from t raining and extension } (t-\Delta t) + \text{productivity increase from learning} \times \Delta t \tag{10.7}$$

Productivity multiplier from training and extension is related to learning level through multiplier. The growth rate fraction of productivity multiplier from training and extension is expressed as a non-linear function of learning level.

Learning level influences the productivity and the learning rate depends on the training extension activities, and the learning level is expressed as

$$\text{learning } (t) = \text{learning } (t-\Delta t) + \text{learning increase rate} \times \Delta t \tag{10.8}$$

Learning from training and extension services increases with the farmers under training and extension services. Learning from training and extension is expressed as a non-linear function of the farmers under training and extension services.

The training and extension service provides the knowledge, i.e. learning and the coverage of the training, and extension activities are given by

$$\begin{aligned}\text{training and extension } (t) = {} & \text{t raining and extension } (t-\Delta t) \\ & + \text{training and extension increase rate} \times \Delta t\end{aligned} \tag{10.9}$$

Commercial fertilisers create environmental degradation and also affect the soil structure and health. Gradual transition to bio-fertilisers based on a target policy is expressed as

$$\text{bio fertilizer fraction}\,(t) = \text{bio fertilizer fraction}\,(t - \Delta t) + \text{bio fertilizer growth} \times \Delta t \qquad (10.10)$$

Rice requirement depends on the per capita consumption of rice and population level and it is computed as

$$\text{rice requirement}\,(t) = \text{per cpita consumption}\,(t) \times \text{population}(t) \qquad (10.11)$$

Population level at any time t is computed from the population growth rate and it is given by:

$$\text{population}\,(t) = \text{population}\,(t - \Delta t) + \text{population growth rate} \qquad (10.12)$$

Self-sufficiency level (SSL) of rice is defined as the ratio of rice production to rice requirement:

$$\text{Self sufficiency Level (SSL)}\,(t) = \frac{\text{rice production}(t)}{\text{rice requirement}(t)} \qquad (10.13)$$

SSL above 1 means surplus food and SSL less than 1 means shortage in food supply to lead healthy life.

10.5 Model Validation

Initial values and parameters were estimated from the primary and secondary data collected from different research reports, statistical year books of Malaysia and field visits, and these are shown in Table 10.1.

To build up confidence in the predictions of the model, various ways of validating a model such as model structures, comparing the model predictions with historic data, checking whether the model generates plausible behaviour and

Table 10.1 Data on initial values and parameters

Name of the variable and parameter	Unit	Value
Rice area	Ha	680,647
Available land	Ha	88,000
Conversion time	Year	8
Discard rate	Per year	0.0085
Population	Million	18.1024
Population growth	Fraction per year	0.021
Per capita rice consumption	Kg per year	80
Rice productivity potential	Tons per ha	6
Rice productivity multiplier from land degradation	Dimensionless	0.99
Productivity multiplier from training land extension	Dimensionless	0.61
Bio-fertiliser	Fraction of total inputs	0.0

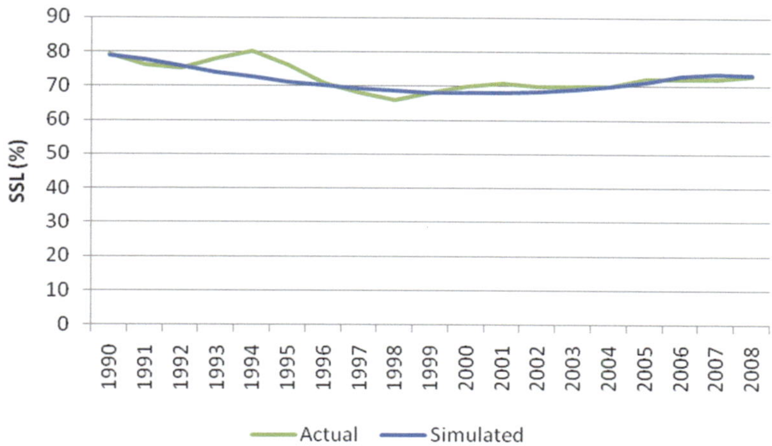

Fig. 10.3 Simulated and historical data of food self-sufficiency ratio in Malaysia

checking the quality of the parameter values were considered. Figure 10.3 shows the comparison between the predicted and historical behaviour of food self-sufficiency ratio in Malaysia. The model-simulated food self-sufficiency ratio agrees reasonably well with historical behaviour and the model is reliable. The validated model was used for baseline scenario and policy analysis.

The parameters of a system dynamics models are subject to uncertainty. So, sensitivity analysis is an important task for reliability of the simulated results and checking the robustness of the model behaviour for the changes in parameter values. The sensitivity of the important parameters was also estimated. The productivity is one of the most important parameters affecting the model behaviour, i.e. self-sufficiency ratio which is the focus of this study. In this study behaviour, sensitivity analysis of food self-sufficiency ratio to rice productivity was conducted, and Fig. 10.4 shows the changes in food self-sufficiency ratio for rice productivity for changes of 2.50 t/ha (curve 1), 3.50 t/ha (curve 2) and 4.50 t/ha (curve 3). The food self-sufficiency ratio changes from decrease to increase in values for the changes in the rice productivity from 3.50 t/ha to 2.50 t/ha and 4.50 t/ha, and this corresponds with real-world situation.

10.6 Simulation and Policy Analysis

A system dynamics model was developed to address the important policy options of rice food security in Malaysia, and the model was simulated to analyse policy issues of input subsidies, withdrawal of input subsidies, gradual transition to bio-fertilisers, R&D for increased productivity, extension services through farmer field schools (FFS) and increase in cropping intensity to address how it behaves in the long run to achieve self-sufficiency level in rice production in Malaysia. The

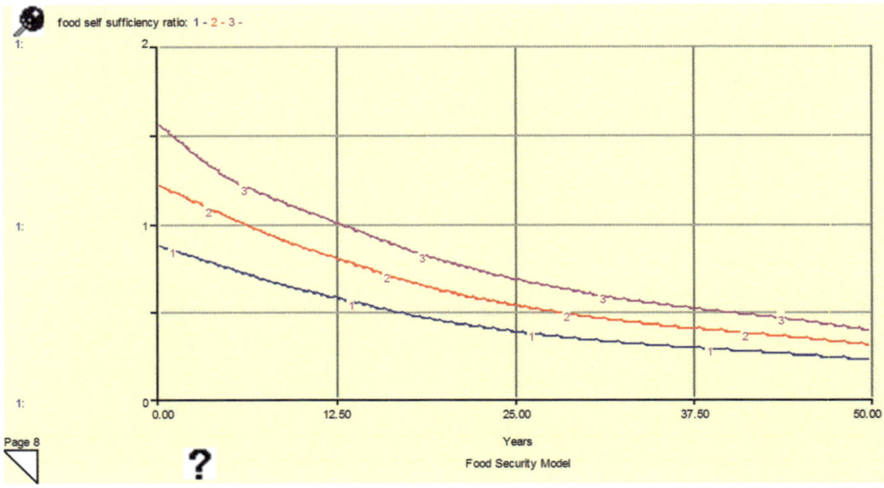

Fig. 10.4 Sensitivity of food self-sufficiency ratio to rice productivity (2.50 t/ha, 3.50 t/ha and 4.50 t/ha)

simulated results for these important policy options for a simulation period of 50 years starting from 2011 are presented and discussed here.

10.6.1 Subsidies for Agricultural Inputs

Figure 10.5 shows the simulated available land, rice area, rice production and self-sufficiency level in rice production for subsidies on agricultural inputs (67 %) and population for a simulation period of 50 years. The rice area initially increases till about 8 years and reaches the upper limit of available land suitable for rice and then decreases it due to the discard of rice area for housing and industrial enterprises. This results in the decrease in the rice production to 1.54 million tons at the end of the simulation period of 50 years. The self-sufficiency in rice decreases to 22 % due to the decrease in rice production and increase in the population. This fact implies that although subsidies on agricultural inputs have positive effect on productivity, only the subsidies on agricultural inputs are not sufficient to improve the SSL in rice production in Malaysia at the desired level, and the rice production in Malaysia is not sustainable at the desired level in the long run. However, it is good for short run. Other opportunities such as hybrid varieties for higher yields and better management practices to reduce the productivity gaps in the fields through training and extension services and increase in cropping intensity are needed to be searched.

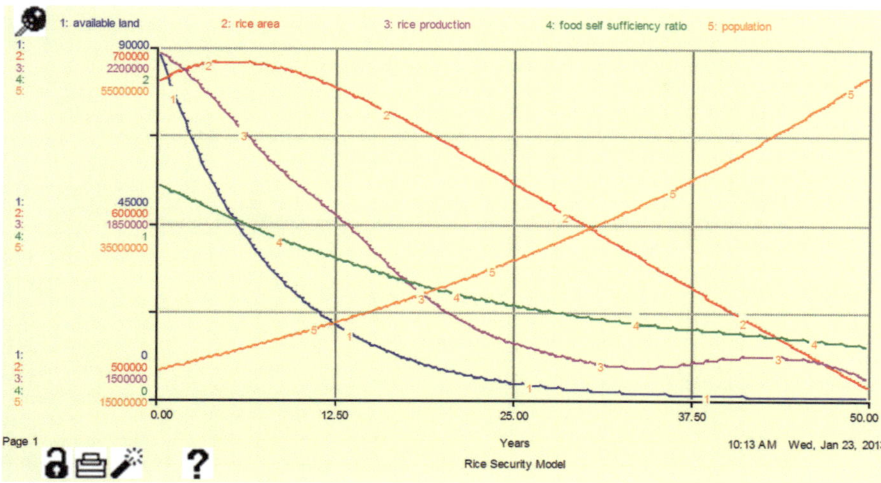

Fig. 10.5 Simulated available land, rice area, rice production and rice self-sufficiency level for agricultural subsidies in inputs and population

10.6.2 Withdrawal of Input Subsidies

Figure 10.6 shows the simulated available land, rice area, rice production and self-sufficiency level in rice production for withdrawal of subsidies on agricultural inputs and population for a simulation period of 50 years. The rice production and hence the self-sufficiency in rice fall drastically as a result of withdrawal of subsidies on agricultural inputs. This causes a further decrease in rice production to 1.33 million tons at the end of the simulation period of 50 years and hence the self-sufficiency in rice decreases to 19 %, i.e. less than 3 % of the basic as usual in Fig. 10.5. Ramli et al. (2012) also reported that the removal of fertiliser subsidy will decrease the paddy production and, consequently, decrease the self-sufficiency level (SSL) in Malaysia. This implies that the withdrawal of subsidies on agricultural inputs has a negative effect on productivity since rice production in Malaysia does not have comparative advantage in terms of the income derived. It is not a wise decision to withdraw the subsidies unless an alternative arrangement/mechanism to apply optimum level of agricultural inputs such as gradual transition to bio-fertilisers as low-cost nutrients to increase the productivity and reduce the environmental degradation and introduction of hybrid seeds and better management practices to increase the productivity of rice is ensured.

10.6.3 Gradual Transition to Bio-fertilisers (50 %)

Figure 10.7 shows the simulated available land, rice area, rice production and self-sufficiency level in rice production for gradual transition to bio-fertilisers and

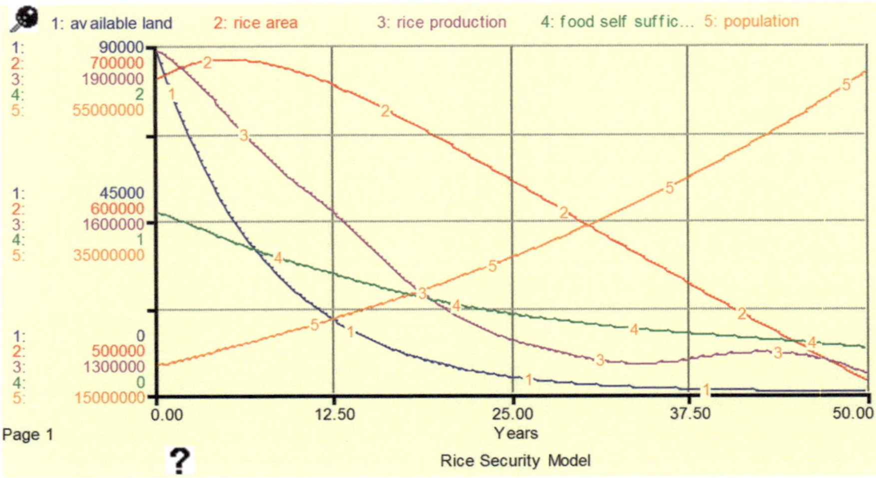

Fig. 10.6 Simulated available land, rice area, rice production and rice self-sufficiency level for withdrawal of agricultural subsidies in inputs and population

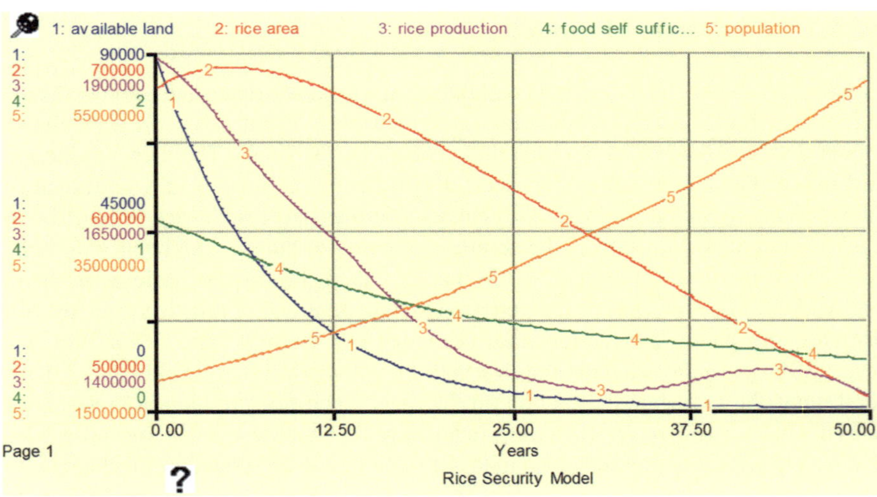

Fig. 10.7 Simulated available land, rice area, rice production and rice self-sufficiency level for gradual transition to bio-fertilisers and population

population for a simulation period of 50 years. The policy is to withdraw 50 % of the subsidy for agricultural inputs immediately and another 50 % of the agricultural input, i.e. commercial fertiliser is replaced by bio-fertilisers within next 50 years. The rice production in this policy increases from 1.33 million tons in the case of withdrawal of input subsidies to 1.42 million tons at the end of the simulation period of

50 years, and the rice self-sufficiency level settles to 21 % from 19 % for withdrawal of the subsidies. The rice self-sufficiency ratio (21 %) is comparable with the reference run in Fig. 10.5 which is 22 %. Still policy mechanisms are needed to increase the productivity which can be achieved through better management practices using farmer field schools, using hybrid variety and encouraging the farmers to use bio-fertiliser. Bio-fertilisers are eco-friendly and are environmentally safe. They form not only part of integrated nutrients but are of low cost which is of immense help to the farming community. This could increase the productivity to improve the SSL level and reduce the agricultural input costs of fertiliser subsidies. Lakshmi et al. (2012) reported that combined application of organics and blue-green algae (BGA) not only recorded higher yield but found to emit less methane in paddy cultivation. Application of BGA (blue-green algae) and *Azolla* reduced methane flux without reducing rice yields and can be used as a practical mitigation option for minimising the global warming potential of rice ecosystem and enhancement by nitrogen fixation.

10.6.4 Gradual Transition to Bio-fertilisers (50 %) and R&D

Figure 10.8 shows the simulated available land, rice area, rice production and self-sufficiency level in rice production for gradual transition to bio-fertilisers, R&D and population for a simulation period of 50 years. The R&D is designed to increase the potential rice productivity from 6 tons/ha to 12 tons/ha within the simulation period of 50 years. The rice production increases to 2.42 million tons and then becomes almost constant, but the rice self-sufficiency ratio decreases to 35 % due to the

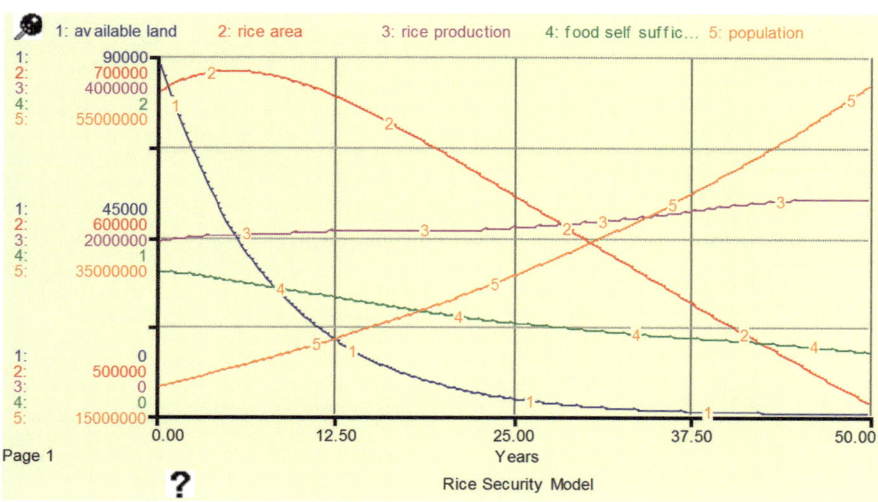

Fig. 10.8 Simulated available land, rice area, rice production and rice self-sufficiency level for gradual transition to bio-fertilisers (50 %), R&D and population

increasing population at the end of the simulation period of 50 years. Although the self-sufficiency level decreases, it rises at a much higher level in comparison to all the previous runs. Thus, this policy option is superior to the policy option illustrated in Fig. 10.7 due to R&D resulting in hybrid rice variety with much higher productivity. Anderson et al. (2013) reported that the best long-term option is investing more in agricultural R&D, and this has the potential to boost per capita consumption in both food-exporting and food-importing countries. There is no alternative to R&D to sustain the production and improve the SSL in rice production in Malaysia (Fatimah et al. 2011). The R&D policy incorporating bio-fertilisers not only improves the SSL but also minimises the global warming.

10.6.5 Gradual Transition to Bio-fertilisers (50 %), R&D and Training and Extension Services

Figure 10.9 shows the simulated available land, rice area, rice production and self-sufficiency level in rice production for gradual transition to bio-fertilisers, R&D, training and extension services and population for a simulation period of 50 years. The training and extension services for learning by doing through farmer field schools are expected to reduce the productivity gap at field levels, and these services will be provided to all of the farming communities within next 50 years. Rice production increases to about 2.99 million tons and the self-sufficiency level becomes 43 % towards the end of 50 years of the simulation period. This policy option further increases the rice production from 2.42 million tons to 2.99 million tons and the rice self-sufficiency level from 35 to 43 % from the policy illustrated in

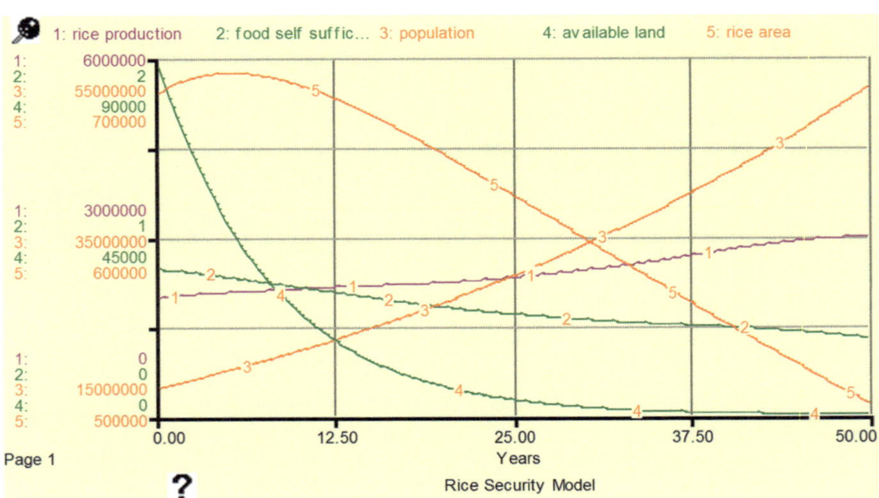

Fig. 10.9 Simulated available land, rice area, rice production and rice self-sufficiency level for gradual transition to bio-fertilisers (50 %), R&D and training and extension and population

Fig. 10.8, and thus it is superior to the policy option illustrated in Fig. 10.8 due to incorporation of R&D resulting in higher productivity and training and extension services to reduce the productivity gaps in the fields. There exists a large yield gap at the field levels in the rice production of Malaysia. Yield gap is manageable and can be narrowed down by deploying more efforts in training and extension services as well as governments' appropriate intervention particularly on the institutional issues (Duwayri and Tran 2013). Several studies have been reported on the positive impacts of agricultural training and extension on agricultural production and food security (Bala 2012; Davis et al. 2012; Larsen and Lilleor 2012; Majumder et al. 2012).

10.6.6 Gradual Transition to Bio-fertilisers (50 %), R&D, Training and Extension Services and Cropping Intensity (150 %)

Figure 10.10 shows the simulated available land, rice area, rice production and self-sufficiency level in rice production for gradual transition to bio-fertilisers, R&D, training and extension services, cropping intensity and population for a simulation period of 50 years. Cropping intensity is expected to increase from the present level to 150 % of the present level. Rice production increases but becomes almost stagnant at 4.34 million tons towards the end of 50 years with a self-sufficiency level of 63 %. This increase of 63 % in rice self-sufficiency is due to R&D resulting higher productivity, the extension and training services to reduce productivity gaps in the fields and increasing cropping intensity to increase the area under rice crop by vertical expansion. The rice production increases from 2.99 million tons to

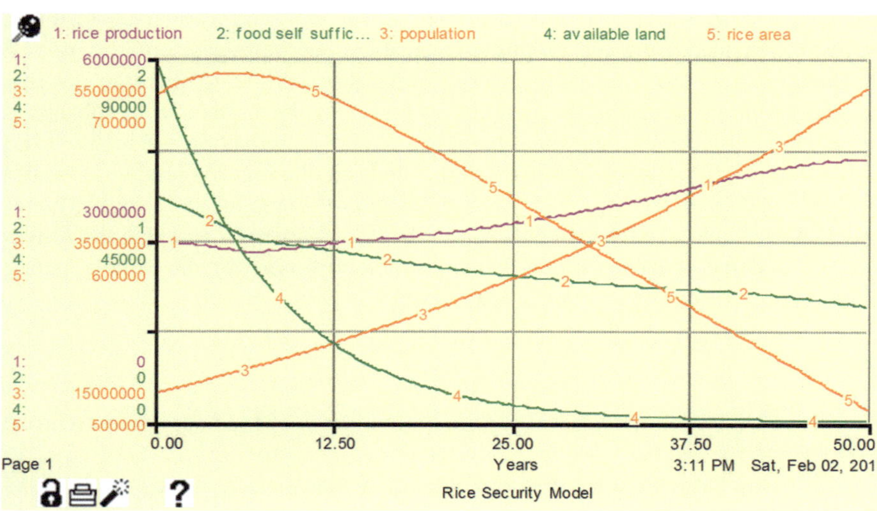

Fig. 10.10 Simulated rice production and rice self-sufficiency level for gradual transition to bio-fertilisers (50 %), R&D, training and extension and cropping intensity and population

4.34 million tons, and the food self-sufficiency ratio increases from 43 to 63 % from the policy options illustrated in Fig. 10.9. This is the best policy although the self-sufficiency in rice food security in Malaysia still remains a challenge, and Malaysia needs imports of rice for domestic food security. BERNAS is solely authorised to deal with the imports of rice in Malaysia. A recent study on rice imports in Malaysia suggests that rice import largely from Vietnam and Thailand is a weak strategy, and an effective multi-sourcing strategy for rice could be adopted to avoid such a weak strategy in securing supply from the international market rather than importing largely from Vietnam and Thailand (Tey and Radam 2010). However, it can be concluded that there is a need of further increase in rice productivity which can be realised through R&D developing hybrid variety of much more higher productivity, reducing productivity gaps in the fields and vertical expansion of the cropped area through increase in the cropping intensity to meet this challenge.

10.7 Conclusions

In the recent years, there have been increasing interests in policy issues of food security since the food crisis in 2007–2008. Malaysia's main food security concern is to achieve a certain level of self-sufficiency in rice. In this paper a system dynamics model has been developed to simulate rice food security and also to assess the policy options of the food security. Simulated results show that subsidy liberalisation or withdrawn would worsen SSL. However, gradual transition to bio-fertilisers improves SSL, and R&D with training and extension services and increased cropping intensity improve the rice food security considerably. Simulated results suggest that the best policy to achieve the targeted SSL is through R&D, training and extension services using farmer field schools and vertical expansion of cropped area increasing the cropping intensity. Food self-sufficiency in rice in Malaysia still remains a challenge. The trade-off could be to increase the real income, thereby increasing the purchasing power of the farmers with subsidised cereal production for the land suitable for irrigated rice production and highly taxed high-value crop production in the rest of the land suitable for crop production under bold national food security policy. In conclusion this model can provide better understanding and greater insights of the food security issues, and also the results indicate that the proposed model can be efficiently used for developing policy suggestions and for improving decision support to policy makers in food security.

Exercises

Exercise 10.1 Define food security. Discuss the issues of food security in a broad perspective.

Exercise 10.2 Draw the causal loop diagrams of food security based on the FAO definition of food security.

Exercise 10.3 Draw stock–flow diagram of food security based on the FAO definition of food security and simulate the dynamics of food security.

Exercise 10.4 Simulate the model in Exercise 10.3 for different policy options, and make a comparative analysis of the simulated results to search for the best policy option to ensure the optimal food security.

Exercise 10.5 Discuss the policy implication of food security. How the policy of improving food security can be implemented using participatory approach of farmer field schools?

References

Anderson K, Jha S, Nelgen S, Strutt A (2013) Re-examining policies for food security in Asia. Food Secur 5:195–215

Babatunde RO, Omotesho OA, Sholotan OS (2007) Factors influencing food security status of rural farming households in North Central Nigerian. Agric J 2(3):351–357

Bala BK (2012) Agricultural development strategy for Hill Tracts of Chittagong. Final report UNDP CHTDF, Rangamati, Bangladesh

Bala BK, Hossain MA (2010) Modeling of food security and ecological footprint of the coastal zone of Bangladesh. Environ Dev Sustain 12:511–529

Bala BK, Hossain MA (2013) Modeling of ecological footprint and climate change impacts on food security of the Hill Tracts of Chittagong in Bangladesh. Environ Model Assess 18 (1):39–55

Bala BK, Alias EF, Fatimah MA, Noh KM, Hadi AHA (2014) Modeling of food security in Malaysia. Simul Model Pract Theory 47:152–164

Davis K, Nkonya E, Kato E, Mekonnen DA, Odendo M, Mhro R, Nkuba J (2012) Impact of farmer field schools on agricultural productivity and poverty in East Africa. World Dev 40(2):402–413

Duwayri M, Tran DV (2013) Reflections on yield gaps in rice production: how to narrow the gaps. FAO Corporate document repository. Internet. 26.03.2013, 2013

FAO (Food and Agricultural Organization) (1996a) Implications of economic policy for food security: a training manual, Rome.

FAO (Food and Agricultural Organization) (1996b) Technical background document prepared for the World Food Summit, Rome.

FAO (Food and Agricultural Organization) (2002) The state of food insecurity in the world: eradicating world hunger-taking stock ten years after the world food summit, Rome.

Faridi R, Wadood SN (2010) An econometric assessment of household food security in Bangladesh. Bangladesh Dev Stud 33(3):97–111

Fatimah MA, Emmy FA, Tasrif M (2011) Food security: self-sufficiency of rice in Malaysia. Int J Manag Sci 18(2):83–100

Giraldo DP, Betancur MJ, Arango S (2008) Food security in developing countries: a systemic perspective. Paper presented at the international conference of the System Dynamics Society held on July 20–24 2008 at Athens, Greece

Haile HK, Alemu ZG, Kudhlande G (2005) Causes of household food insecurity in Koredegaga Peasant Association, Oromiya Zone, Ethiopia. Working paper, University of the Free State, Department of Agricultural Economics

Lakshmi VG, Lakshmanan A, Sankar A, Latha P, Sekher NU (2012) Biofertilizers in minimizing climate change impacts in rice farming CLIMARICE: sustaining the rice production in a changing climate uncertainties and validating selected adaptation techniques on farmer's field. CLIMARICE technical Brief 9, August, 2012

Larsen AF, Lilleor HB (2012) The impact of farmer field schools on food security and poverty: evidence from a quasi-experiment in Tanzania. Online in the internet, 2012.

Mahajan K, Joshi N (2011) Determinants of household food security in Nepal: a binary logistic regression. J Mt Sci 8:403–413

Majumder S, Bala BK, Hossain MA (2012) Food security of the Hill Tracts of Chittagong in Bangladesh. J Nat Resour Policy Res 4(1):43–60
Mwagi GO, Onyango CA, Mureithi JG, Mungai PC (2003) Effectiveness of FFS approach on technology adoption and empowerment of farmers: A case of farmer groups in Kisii District, Kenya. The Soil Science Society of East Africa: proceedings of the 21st annual conference, 1–5 December 2003, Eldoret, Kenya, 2003
Panagariya A (2002) Trade and food security: conceptualizing the linkages. In: Proceedings of the conference on trade, agricultural development and food security: the impacts recent economic and trade policy reform. 11–12 July 2002, Rome, 2002
Pinstrup-Anderson P (2009) Food security: definition and measurement. Food Secur 1:5–7
Praneetvatakul S, Waibel H (2003) Farm-level economic analysis of farmer fieldschools in Thailand. Poster and background paper presented at 25th International Conference of Agricultural Economists, Durban, South Africa. Italy, FAO, 2003
Ramli NN, Shamsudin MN, Mohammed Z, Radam A (2012) Impacts of fertilizer subsidy on Malaysian paddy/rice industry using system dynamics approach. Int J Soc Sci Hum 2 (2):213–219
Riely F, Mock N, Cogill B, Bailey L, Kenefick E (1999) Food security indicators and framework for use in the monitoring and evaluation of food aid programs, Food and Nutrition Technical Assistance, Washington, DC, 1999
Rola AC, Jamias SB, Quizon JB (2002) Do farmer field school graduates retain what they learn? J Int Agric Ext Educ 9(1):65–75
Sikwela MM (2008) Determinants of household food security in the semi-arid areas of Zimbabwe: a case study of irrigation and non-irrigation farmers in Lupane and Hwange districts, MS thesis, University of Fort Hare, South Africa, 2008
Sterman JD (2000) Business dynamics: systems thinking and modelling for a complex world. Irwin/McGraw Hill, New York
Tey YS, Radam A (2010) Demand patterns of rice imports in Malaysia: implications for food security. Food Secur 3(2010):253–261
USDA (1999) Food Security Assessment. USDA economic research service situation and outlook series GFA-11, Washington, DC, 1999

Bibliography

Bala BK (1999) Principles of system dynamics. Agrotech Publishing Academy, Udaipur
Barlas Y (1996) Formal aspects of model validity and validation in system dynamics. Syst Dyn Rev 12:183–210
Forrester JW (1968) Principles of systems. Wright-Allen Press, Cambridge, MA

11 Modelling of Supply Chain of Rice Milling Systems in Bangladesh

In the previous chapters, case studies on system dynamics modelling of boom and bust of cocoa production systems in Malaysia, modelling of hilsa fish population in Bangladesh and modelling of food security in Malaysia have been presented. This chapter presents the application of system dynamics modelling to supply chain of rice milling systems to demonstrate how to construct system dynamics models and simulate these models for policy planning and design. The model presented is an illustration of modelling and simulation of practical problem to provide management scenarios and policy options, and such experiences are essential to face the challenge of modelling and simulation of dynamic systems in practice. To achieve these goals, the model of this case study is organised as follows: (1) introduction, (2) dynamic hypothesis, (3) causal loop diagram, (4) stock–flow diagram, (5) model validation, (6) simulation and policy analysis and (7) conclusion. The model is simulated to address supply chain management scenarios and design the policy options for efficient and sustainable supply chain management of rice milling systems in Bangladesh to ensure the availability of rice to the consumers in an economic manner under uncertainty. The model has the potential to identify and study the critical components of the overall supply chain, allowing for the creation of an efficient and sustainable network.

11.1 Introduction

Rice is the staple food grain in Bangladesh, and it covers almost three quarters of the total cropped area in Bangladesh. Per capita consumption of rice in Bangladesh is one of the highest in the world and it is 188 kg (BBS 2011). The production of rice in Bangladesh has increased significantly over last 40 years, and this success has been strongly related to high yielding varieties triggered by the liberalisation of key input markets such as irrigation and fertilisers (Ahmed 2000; Hossain 1988; Hossain et al. 2006). Although rice production has increased in the recent years with a self-sufficiency in rice food security, the availability of rice to consumers has

B.K. Bala et al., *System Dynamics*, Springer Texts in Business and Economics,
DOI 10.1007/978-981-10-2045-2_11

not been stabilised, and hence the changes in the price of rice by days and even by minutes have been common phenomenon. This indicates that it is necessary to identify the appropriate demand and meet them properly in a profitable way, and it is the primary concern of rice supply chain. A proper supply chain management framework is very essential for efficient sourcing, processing, distribution and retailing and hence meeting the customer demands without facing any crisis. Production and business of rice have been one of the most traditional and major concerns of milling industries in Bangladesh. But still no proper supply chain framework has been developed. Of course, dealing with inventory is the major issue in the supply chain management of rice, and it requires that milling, wholesale and retail inventories are properly managed and maintained to ensure the availability of rice to the consumers.

Rice supply chain in Bangladesh is a focus for considerations of food security and climate change impacts. The rice supply chain in Bangladesh is demand driven. Although Bangladesh is self-sufficient in rice, the supply chain does not meet the requirement at the right time at the right place resulting artificial crisis and price volatility. Furthermore, the supply of paddy is also affected by natural calamities such as drought and flood. This requires the supply chain to be more demand driven and all the actors involved in the supply chain to be responsive to the demand information of the ultimate consumers in a timely and cost-effective manner.

Realisation of a demand-driven supply chain is a complex task (Selen and Soliman 2002). Any consumer requirement may trigger execution of different activities by different contributors (Prahalad and Ramswamy 2000), and hence demand-driven supply chains are highly dynamic having different modes of cooperation, control and coordination. This requires that demand-driven supply chains must be modelled and simulated before implementation. Reference process models can be valuable means to support these challenges in the design and implementation of demand-driven supply chains. The reference process models represent specific ordering work activities across time and place (Verdouw et al. 2010; Davenport 1993). Reference process models are predefined models used for the construction of other models (Verdouwa et al. 2010; Thomas 2006). Also the reference process models can be used to construct system dynamics computer models using systems approach developed by Forrester (1968).

Rice milling systems in Bangladesh mainly consist of milling sector, wholesale sector and retail sector. The milling sector procures paddy from the farmers, paddy traders and wholesalers and then prepares the milled rice. The wholesale sector procures the milled rice from this sector and sells the milled rice to the retail sector, while the retail sector sells the milled to the consumer. The product turnover of the retailers is about 22 kg per day. This compares with 10 tons per day for the wholesalers and up to 50 tons per day per mill. In other words, an urban retailer would need on average the sales of 1.5 farmers to assure his rice supplies for the year, whereas a wholesaler would require the sales of 400 farmers and a mill about 2000 farmers (Minten et al. 2013). The supply of paddy to milling sector is seasonal, and there are three seasons of rice production. The main harvesting season is from November to January followed by April to June and July to August. The demand of milled rice to the retail sector depends on the per capita rice consumption and population level. The critical components are supply of paddy in the milling sectors and inventory in the milling, wholesale and retail sector as well as the supply chain costs. Effective management requires timely delivery of rice at right quantity at the right time and right place. The current supply chain of rice milling systems in Bangladesh is lacking efficiency, and this needs reforms. This necessitates a computer model to analyse and support the rice milling systems in Bangladesh. However, for sustainable development of rice supply system in Bangladesh, rice supply chain management must be modelled and simulated for designing policy options before implementation (Bala 1999).

Carter and Easton (2011) reported a systematic review of sustainable supply chain management and demonstrated that it is necessary to develop models. Riddals et al. (2000) reported a critical review of the various mathematical methods used to model and analyse supply chains and concluded that operations research techniques are useful in providing solutions to local tactical problems, but the impact of these solutions on the global behaviour of the whole supply chain can only be assessed using dynamic simulation. System dynamics techniques, a methodology of constructing dynamic simulation model based on feedback concepts incorporating non-linearities and time lags developed by Forrester can be used to model such a dynamic system and analyse performance dynamics of the system (Bala 1999; Forrester, 1968).

Several studies have been reported on system dynamics modelling of supply chain of food industry (Apaiah et al. 2005; Apaiah and Hendrix 2005; Georgiadis et al. 2005). Minegishi and Thiel (2000) developed a system dynamics supply chain model to study the complex logistic behaviour of an integrated food industry and applied it to poultry production and processing to study the influence of different policies like the poultry breeding program and to show the phenomena of instabilities and system controls in supply chains confronted with serious hazards. Vo and Thiel (2008; 2011) reported dynamic behaviour of the entire chicken meat supply chain under bird flu crisis during the period from October 2005 to March 2006 in France, and this model is helpful to decision-makers for other fresh food supply chains when they are facing such crises. Kumar and Nigmatullin (2011) reported a system dynamics model to study the non-perishable product food supply chain performance under a monopolistic environment, and the model was used to study the behaviour and relationships within a supply chain for a non-perishable product and to determine the impact of demand variability and lead time on supply chain performance. The proposed model can be used to analyse 'what if scenarios' of different inventory policies and strategic food supply chain management issues. Also Sachan et al. (2005) reported a system dynamics model of Indian grain supply chain cost to device policies to reduce total supply chain cost and suggested action plans to reduce total supply chain costs.

Several studies have been reported on modelling of food supply chain management (Bosona and Gebresenbet 2013; Dabbene et al. 2008a, b). Dabbene et al. (2008a, b) reported a hybrid model consisting of event-driven dynamics and time-driven dynamics to describe beef meat supply chain, and Bosona and Gebresenbet (2013) conducted route analysis to determine optimal product centres of local food chain. Vo and Thiel (2011) reported an economic simulation of poultry supply chain using system dynamics. More recently detailed frameworks of food supply chain assessment and reference process models for sustainable food supply chain have been reported (Manzini and Accorsi 2013; Verdouw et al. 2010). Although Sachan et al. (2005) reported a system dynamics grain supply chain cost model, it does not include grain supply chain management. Effective and sustainable supply chain management of rice milling systems should not only consider total supply chain cost but also should consider inventory decision and policy development for effective and sustainable management of rice milling systems. To meet this research and literature gap as well the research gap, Bala et al. (2015) and Bhuiyan (2015) motivated us to develop a system dynamics model of rice milling systems that focuses on inventory and policy development. The modelling of the supply chain of the rice milling systems presented here is adopted from Bala et al. (2015) and Bhuiyan (2014).

11.2 Dynamic Hypothesis

The dynamic hypothesis seeks to define the critical feedback loops and stock–flow diagrams that drive the system's behaviour in the reference mode. When the model based on feedback concept is simulated, the endogenous structure of the stock–flow diagram should generate the reference mode behaviour of the system, and thus the endogenous structure causes the changes in dynamic behaviour of the system. In system dynamics modelling, causal loop diagrams and stock–flow diagrams are hypothesised to generate the reference mode of the behaviour over time (Sterman 2000). Supply chain of rice milling systems can be represented by causal loop diagrams and stock–flow diagrams, and the simulation model based on the causal loop diagram and stock–flow diagram would generate dynamic behaviour of the supply chain of the rice milling systems. The supply chain of rice milling systems in the form of causal loop diagram and stock–flow diagram is hypothesised to generate the observed inventory and price in the reference mode. In essence the bullwhip effect in the inventory levels results from the endogenous consequences of the feedback structure of the system.

11.3 Causal Loop Diagram

The causal loop diagram of the supply chain of rice milling systems in Bangladesh is shown in Fig. 11.1. The daily retailer demand is computed from the per capita consumption and population. When the inventory of the retailer reaches the reordering point, the retailer places order to the wholesaler which is equal to economic order quantity. The wholesaler also places order at the reordering point, and the quantity ordered is the economic order quantity. The miller procures the paddy from the farmers, the paddy traders and wholesalers. The rice production is seasonal. The supply chain of the rice milling systems shown in Fig. 11.1 is a multi-loop system having some non-linear cause–effect relationships. For example, three major loops of the milling sector are milling–order for milling–rice inventory–rice for milling–milling; milling–reordering point for milling–milling; and milling–rice for wholesale–wholesale–order for wholesale–milling. Also

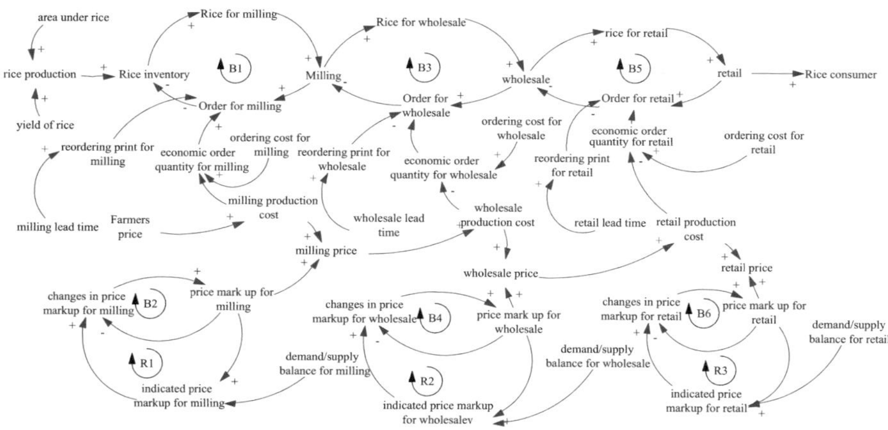

Fig. 11.1 Causal loop diagram of supply chain of the rice milling systems in Bangladesh

demand/supply balance is non-linearly related to price markup. Wholesale and retail also have similar multi-loops and non-linearities.

An economic order quantity could assist in deciding what would be the best optimal order quantity at the lowest cost. Similar to economic order quantity, the reordering point will advise when to place an order for specific products based on the historical demand. The reordering point also allows sufficient stock at hand to satisfy demand while the next order arrives due to the lead time.

The economic order quantity is computed using the following equation (Riggs 1970):

$$Q = \sqrt{\frac{2DS}{H}} \tag{11.1}$$

where

Q = economic order quantity of rice, tons
D = annual demand of rice, tons
S = ordering cost, taka/tons
H = holding and storage cost, taka/ton

and the reordering point for the economic order quantity is computed from the following equation (Riggs 1970):

$$R = d \times L + z \times \sigma_L \times \sqrt{L} \tag{11.2}$$

where

R = reordering point, tons
d = average demand of rice, tons per day

L = lead time, day
z = safety factor
σ_L = standard deviation of average demand, ton

The economic order quantity model gives an optimal solution in closed form, and this model has been effectively employed in automobiles, pharmaceutical industries and retail sectors (Muckstadt and Sapra 2010). Another important technique used along with economic order quantity is the reordering point and safety stock, and this reflects the level of inventory that triggers the placement of an order for another units, whereas the quantity associated with safety stock protects from stock-out (Chen 1998).

The milling price, wholesale price and retail price are determined in this study by change in price markup anchored by demand/supply balance (Kpmaier and Voigt 2013; Teimoury, et al. 2013).

11.4 Stock–Flow Diagram

The detailed STELLA flow diagram of the rice milling systems in Bangladesh is shown in Fig. 11.2. In this study, we consider the simulation of inventories of rice in the milling systems for economic order quantities and different policy options for sustainable optimum operation of the rice milling systems in Bangladesh. The fundamental equations used in this study are as follows.

11.4.1 Rice Milling Sector

Rice inventory is increased by rice production and decreased by sales for milling, and it is computed as

$$\begin{aligned}&\text{rice inventory }(t) = \text{rice inventory }(t-1)\ + \text{rice production rate} \times \Delta t\\&-\text{rice sales for milling} \times \Delta t\end{aligned} \tag{11.3}$$

Rice production depends on area under rice cultivation and yield of rice, and it is calculated as

$$\text{rice production rate} = \text{area under rice} \times \text{yield of rice} \tag{11.4}$$

The rice sales for milling is equal to milling order rate for rice for milling on order and milling received for milling inventory. It is the actual milling placed. When milling on order plus milling inventory is less than milling reordering point, then milling placed is milling reordering point; otherwise, it is zero. This is expressed as

$$\text{milling placed} = \text{IF}(\text{milling on order} + \text{milling inventory}) < (\text{milling reordering point})\text{THEN}(\text{milling to order})\text{ELSE}(0) \tag{11.5}$$

Milling reordering point is defined as the milling demand average over milling lead time plus milling safety stock, and it is computed as

$$\text{milling reordering point} = (\text{milling demand aver} \times \text{milling lead time}) + \text{milling safety stock} \tag{11.6}$$

The safety stock for milling is computes as

$$\text{milling safety stock} = \text{milling safety factor} \times \text{standard deviation of milling demand aver} \times \text{SQRT}(\text{milling lead time}) \tag{11.7}$$

Milling to order, i.e. milling reordering point, is initiated when milling inventory is less than milling reordering point; otherwise, it is zero. This is expressed as

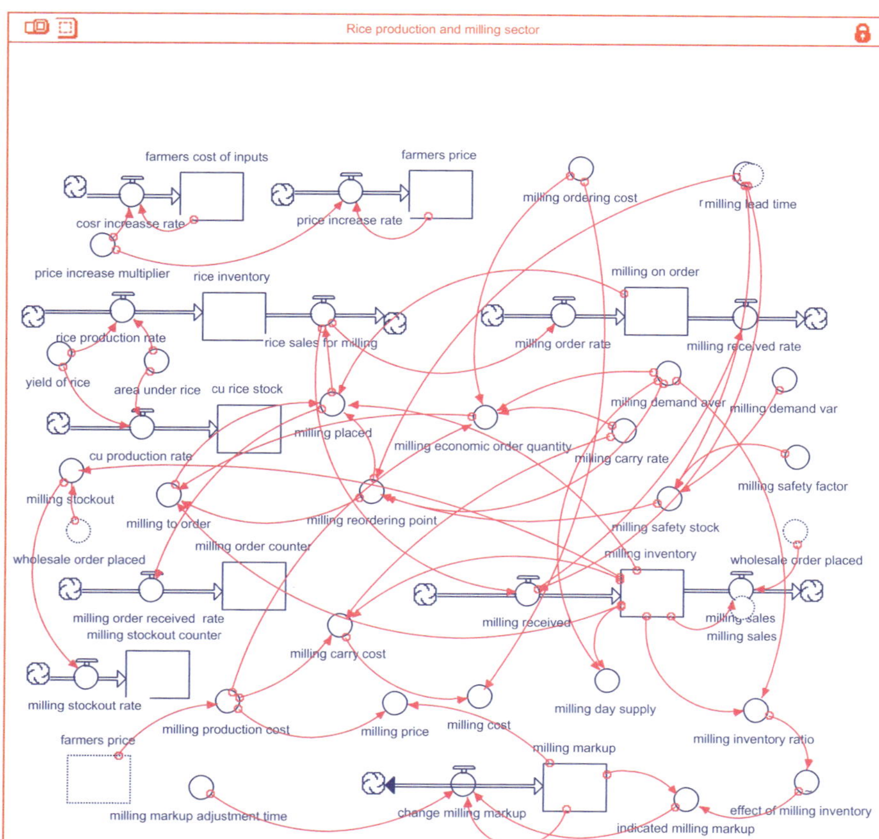

Fig. 11.2 STELLA flow diagram of the rice milling systems in Bangladesh

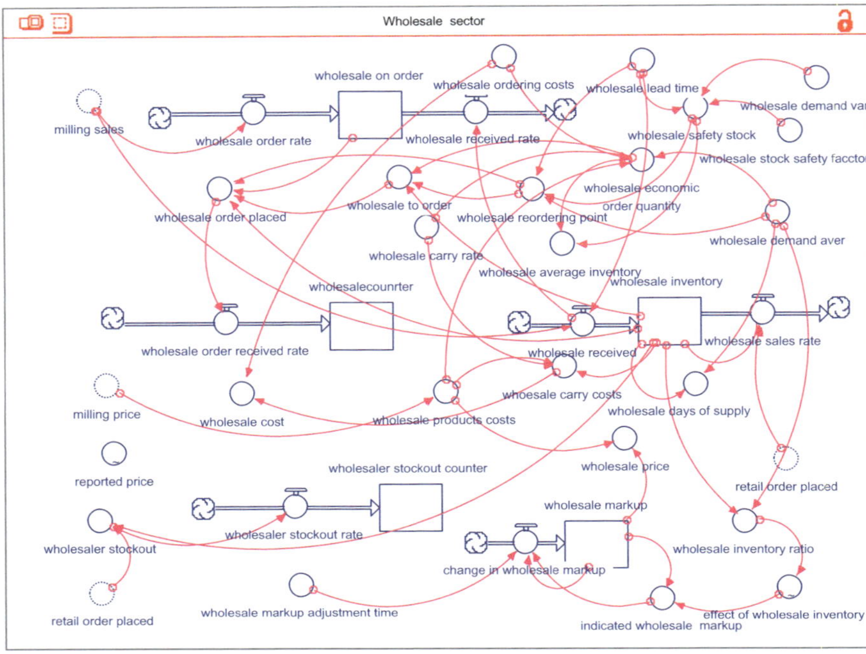

Fig. 11.2 (continued)

$$\text{milling to order} = \text{IF}(\text{milling inventory}) < (\text{milling reordering point})\text{THEN}(\text{milling economic order quantity})\text{ELSE}(0) \tag{11.8}$$

Milling economic order quantity is computed from the following relation:

$$\begin{aligned}&\text{milling economic order quantity}\\&= \text{SQRT}\big((2 \times \text{milling demand aver} \times \text{milling ordering cost})\\&/((\text{milling carryrate}/365) \times \text{milling production cost})\big)\end{aligned} \tag{11.9}$$

The state variable milling on order is increased by milling order rate and decreased by milling received rate by milling inventory, and it is computed as

$$\begin{aligned}&\text{milling on order}(t) = \text{ milling on order } (t-1) \ + \text{milling order rate} \times \Delta t\\&-\text{milling received rate} \times \Delta t\end{aligned} \tag{11.10}$$

Rice milling inventory is increased by milling received and decreased by milling sales of rice, and it is expressed as

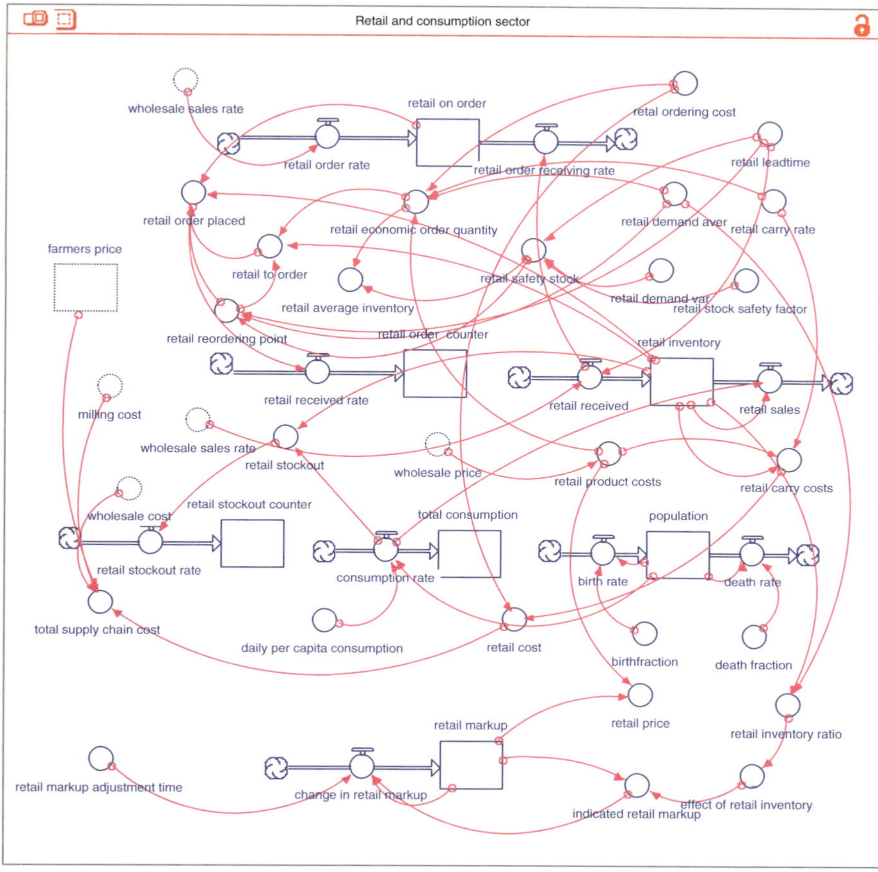

Fig. 11.2 (continued)

$$\text{milling inventory } (t) = \text{milling inventory } (t-1) + \text{milling received} \times \Delta t - \text{milling sales} \times \Delta t \tag{11.11}$$

Milling received is the rice received for milling in the milling inventory with a time delay of milling lead time, and it is expressed as

$$\text{milling received} = \text{DELAY}(\text{rice sales for milling}, \text{milling lead time})) \tag{11.12}$$

Milling sales depends on milling inventory and wholesale order placed. When wholesale order placed is greater than milling inventory, then milling sales is equal to milling inventory; otherwise, it is wholesale order placed, and it is expressed as

$$\begin{aligned}\text{milling sales} = \ & \text{IF}(\text{wholesale order placed}) > (\text{milling inventory}) \\ & \times \text{THEN}(\text{milling inventory})\text{ELSE}(\text{wholesale order placed})\end{aligned} \tag{11.13}$$

The milling price of rice is computed from cost of milled rice and price markup for milling as

$$\text{milling price} = \text{MAX}\big(\text{milling production cost}\ ,\ (1 + \text{milling markup}) \times \text{milling production cost}\big) \tag{11.14}$$

Price markup for milling of rice depends on change in milling markup, and it is computed as

$$\text{milling markup}(t) = \text{milling markup}(t-1) + \text{change in milling markup} \times \Delta t \tag{11.15}$$

Change in milling markup depends on milling markup itself, indicated milling markup and milling markup adjustment time, and it is computed as

$$\text{change in milling markup} = (\text{indicated milling markup} - \text{milling markup}) / \text{milling markup adjustment time} \tag{11.16}$$

Indicated price markup for milling of rice is computed from price markup for milling and effect of relative inventory coverage of the milled rice (Sterman 2000; Teimoury et al. 2013), and it is computed as

$$\text{indicated milling markup} = \text{milling markup} \times \text{effect of milling inventory} \tag{11.17}$$

Effect of milling inventory is expressed as a non-linear graphical function of relative inventory coverage of the milled rice. This formulation of price settling essentially consists of expectation adjustment and price adjustment (Sterman 2000). The concept of expectation adjustment was introduced by Nerlove (1958), and subsequently it was used by Meadows (1970) and Bala (1975, 1990) for formulation price setting of agricultural commodities with a reasonably good agreement between the predicted prices and reported values.

11.4.2 Wholesale Sector

Wholesale on order is increased by wholesale order rate and decreased by wholesale received rate, and it is computed as

$$\text{wholesale on order}(t) = \text{wholesale on order}\ (t-1) + \text{wholesale order rate} \times \Delta t - \text{wholesale received rate} \times \Delta t \tag{11.18}$$

Rice wholesale inventory is increased by wholesale received and decreased by wholesale sales rate of rice, and it is expressed as

$$\text{wholesale inventory } (t) = \text{ wholesale inventory } (t-1) + \text{wholesale received} \times \Delta t \\ -\text{wholesale sales rate} \times \Delta t \tag{11.19}$$

Wholesale received is the rice received for wholesale inventory with a time delay of wholesale lead time, and it is expressed as

$$\text{wholesale received} = \text{DELAY(milling sales, wholesale lead time)} \tag{11.20}$$

Wholesale sales rate depends on wholesale inventory and retail order placed. When retail order placed is greater than wholesale inventory, then wholesale sales rate is wholesale inventory; otherwise, it is retail order placed, and it is expressed as

$$\text{wholesale sales rate} = \text{IF(retail order placed)} \\ > \text{(wholesale inventory)THEN(wholesale inventory)ELSE(retail order placed)} \tag{11.21}$$

When wholesale on order plus wholesale inventory is less than wholesale reordering point, then wholesale order placed is wholesale to order; otherwise, it is zero. This is expressed as

$$\text{wholesale order placed} = \text{IF(wholesale on order } + \text{wholesale inventory)} \\ < \text{(wholesale reordering point)THEN(wholesale to order)ELSE(0)} \tag{11.22}$$

Wholesale reordering point is defined as the sum of wholesale demand average over wholesale lead time and wholesale safety stock, and it is computed as

$$\text{wholesale reordering point} = \text{(wholesale demand aver} \times \text{wholesale lead time)} \\ +\text{wholesale safety stock} \tag{11.23}$$

The safety stock for wholesale is computes as:

$$\text{wholesale safety stock} = \text{wholesale safety factor} \\ \times\text{standard deviation of wholesale demand aver} \times \text{SQRT(wholesale lead time)} \tag{11.24}$$

Wholesale to order, i.e. wholesale reordering point, is initiated as wholesale economic order quantity. When wholesale inventory is less than wholesale reordering point, then wholesale to order is wholesale economic order quantity; otherwise, it is zero. This is expressed as

$$\text{wholesale to order} = \text{IF}(\text{wholesale inventory}) < (\text{wholesale reorderingpoint})\text{THEN}(\text{wholesale economic order quantity})\text{ELSE}(0) \tag{11.25}$$

Wholesale economic order quantity is computed from the following relation:

$$\text{wholesale economic order quantity} = \text{SQRT}\big((2 \times \text{wholesale demand aver} \times \text{wholesale ordering cost})/((\text{wholesale carry rate}/365) \times \text{wholesale product cost})\big) \tag{11.26}$$

The wholesale price of rice is computed from cost of wholesale rice and price markup for wholesale as

$$\text{wholesale price} = \text{MAX}\big(\text{wholesale productscosts}\ ,\ (1 + \text{wholesale markup}) \times \text{wholesale products cost s}\big) \tag{11.27}$$

Price markup for wholesale of rice depends on change in wholesale markup, and it is computed as

$$\text{wholesale markup}(t) = \text{wholesale markup}(t-1) + \text{change in wholesale markup} \times \Delta t \tag{11.28}$$

Change in wholesale markup depends on wholesale markup itself, indicated wholesale markup and wholesale markup adjustment time and it is computed as

$$\text{change in wholesale markup} = (\text{indicated wholesale markup} - \text{wholesale markup}) / \text{wholesale markup adjustment time} \tag{11.29}$$

Indicated price markup for wholesale of rice is computed from price markup for wholesale and effect of relative inventory coverage of the wholesale rice (Sterman 2000; Teimoury et al. 2013), and it is computed as

$$\text{indicated wholesale markup} = \text{wholesale markup} \times \text{effect of wholesale inventory} \tag{11.30}$$

Effect of wholesale inventory is expressed as a non-linear graphical function of relative inventory coverage of the wholesale rice.

11.4.3 Retail Sector

Retail on order is increased by retail order rate and decreased by retail received rate, and it is computed as

$$\text{retail on order}(t) = \text{retail on order } (t-1) + \text{retail order rate} \times \Delta t \\ - \text{retail received rate} \times \Delta t \tag{11.31}$$

Rice retail inventory is increased by retail received and decreased by retail sales of rice, and it is expressed as

$$\text{retail inventory } (t) = \text{retail inventory } (t-1) + \text{retail received} \times \Delta t \\ - \text{retail sales} \times \Delta t \tag{11.32}$$

Retail received is the rice received for retail inventory with a time delay of retail lead time, and it is expressed as

$$\text{retail received} = \text{DELAY}(\text{wholesale sales, retail lead time}) \tag{11.33}$$

Retail sales rate depends on retail inventory and consumption rate. When consumption rate is greater than or equal to retain inventory, then it is retail inventory; otherwise, it is consumption rate, and it is expressed as

$$\text{retail sales} = \text{IF}\big(\text{consumption rate} \\ >= \text{retail inventory}\big)\text{THEN}(\text{retailinventory})\text{ELSE}(\text{consumption rate}) \tag{11.34}$$

When retail on order plus retail inventory is less than retail reordering point, then retail placed is retail to order; otherwise, it is zero. This is expressed as

$$\text{retail placed} = \text{IF}(\text{retail on order } + \text{retail inventory}) \\ < (\text{retail reordering point})\text{THEN}(\text{retail to order})\text{ELSE}(0) \tag{11.35}$$

Retail reordering point is defined as the sum of retail demand average over retail lead time and retail safety stock, and it is computed as

$$\text{retail reordering point} = (\text{retail demand aver} \times \text{retail lead time }) \\ + \text{retail safety stock} \tag{11.36}$$

The safety stock for retail is computed as

$$\text{retail safety stock} = \text{retail safety factor} \\ \times \text{standard deviation of retail demand aver} \times \text{SQRT}(\text{retail lead time}) \tag{11.37}$$

Retail to order, i.e. retail reordering point, is initiated as retail economic order quantity. When retail inventory is less than retail reordering point, then retail to order is retail reordering quantity; otherwise, it is zero. This is expressed as

$$\text{retail to order} = \text{IF}(\text{retail inventory}) \\ < (\text{retail reorderingpoint})\text{THEN}(\text{retail economic order quantity})\text{ELSE}(0) \tag{11.38}$$

Retail economic order quantity is computed from the following relation:

$$\begin{aligned}&\text{retail economic order quantity} = \text{SQRT}\big((2 \times \text{retail demand aver}\\&\times \text{retail ordering cost})/((\text{retail carry rate}/365) \times \text{retail product cost})\big)\end{aligned} \tag{11.39}$$

The retail price of rice is computed from cost of retail rice and price markup for retail as

$$\text{retail price} = \text{MAX}(\text{retail productcosts}\ ,\ (1 + \text{retail markup}) \times \text{retail product cost s}) \tag{11.40}$$

Price markup for retail of rice depends on change in retail markup, and it is computed as

$$\text{retail markup}(t) = \text{retail markup}(t-1) + \text{change in retail markup} \times \Delta t \tag{11.41}$$

Change in retail markup depends on retail markup itself, indicated retail markup and retail markup adjustment time, and it is computed as

$$\begin{aligned}&\text{change in retail markup} = (\text{indicated retail markup} - \text{retail markup})\\&/\text{retail markup adjustment time}\end{aligned} \tag{11.42}$$

Indicated price markup for retail of rice is computed from price markup for retail and effect of relative inventory coverage of the retail rice (Sterman 2000; Teimoury et al. 2013), and it is computed as

$$\text{indicated retail markup} = \text{retail markup} \times \text{effect of retail inventory} \tag{11.43}$$

Effect of retail inventory is expressed as a non-linear graphical function of relative inventory coverage of the retail rice.

These equations were solved numerically using Runge–Kutta fourth-order method, and the solution interval dt was taken to be 0.25 which is less than half of the shortest first-order delay.

11.5 Model Validation

Initial values and the parameters were estimated from the primary and secondary data (BBS 2011; Bhuiyan 2014). The data on milling inventory, wholesale inventory and retail inventory used for simulation of supply chain of the rice milling systems in Bangladesh are shown in Table 11.1.

To build up confidence in the predictions of the model, various ways of validating a model such as checking the structure of the model, comparing the model predictions with historic data, checking whether the model generates plausible behaviour and checking the quality of the parameter values were considered. Time series data of wholesale prices of rice in Bangladesh in 2011 were compared with the simulated wholesale rice prices to build up confidence in the model, and

Table 11.1 Data used for the simulation of rice supply chain in Bangladesh

Item	Initial value, tones	Lead time, days	Demand average, tonnes	Ordering cost, taka[a]	Carry rate, fraction of product cost (1/year)
Milling inventory	2,500,000	7	950,000	500,000	0.15
Wholesale inventory	1,500,000	7	950,000	500,000	0.15
Retail inventory	2,500,00	7	95,000	50,000,000	0.15

Sources: BBS (2011) and Bhuiyan (2014)
[a]1 US dollar = 78.30 Bangladeshi taka

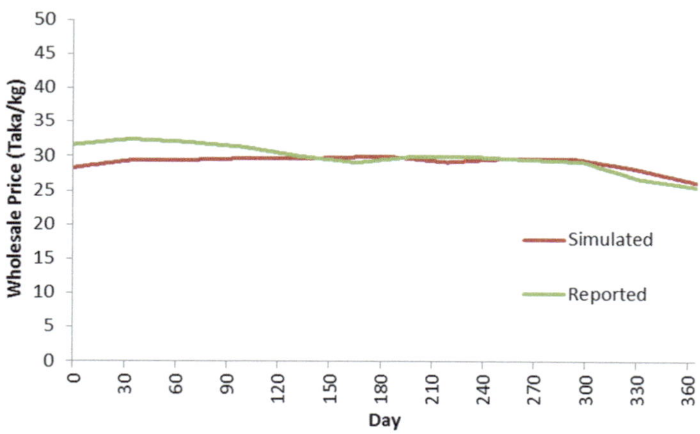

Fig. 11.3 Comparisons of the predicted and reported values of wholesale price in Bangladesh in 2011

this is shown in Fig. 11.3. The model-simulated prediction agrees reasonably well with historical behaviour. Also the model was tested for extreme conditions to detect the defect in the model structure and enhance the usefulness of the model for policy analysis. Figure 11.4 shows simulated milling inventory, wholesale inventory and retail inventory under extreme condition of crop failure, i.e. zero crop production. Under this condition, the milling inventory and then wholesale inventory and retail inventory are reduced to zero since the rice production is zero. The model results confirmed to the expected patterns of results and realities. This model complied with the basic principles of supply chain management and was consistent with supply chain theory and research results. The model is able to provide qualitative and quantitative understanding of the supply chain performances of rice milling systems. Hence, the model is reliable, and the validated model was used for policy analysis.

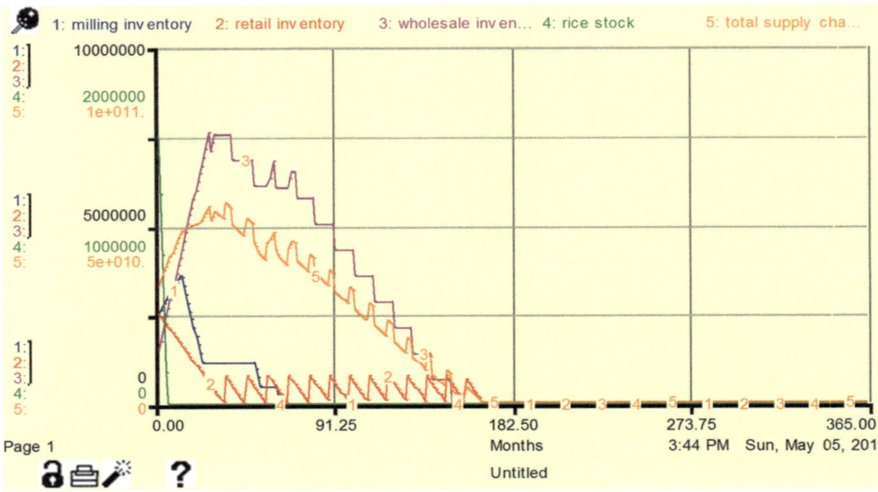

Fig. 11.4 Simulated rice stock, total supply chain cost, milling inventory, wholesale inventory and retail inventory under extreme condition of total crop failure

11.6 Simulation and Policy Analysis

Figure 11.5 shows the simulated milling inventory, wholesale inventory and retail inventory for economic order quantity and reordering point operation of the order rate of milling, wholesale and retail. Economic order quantity and reordering point operation of the order rate of milling, wholesale and retail of the supply chain of rice milling systems stabilises the retail inventory and retail supply of the rice for consumption is ensured. However, there is a bullwhip or Forrester effect due to economic order quantity and reordering point operation of the milling, wholesale and retailing. Four major causes of bullwhip effect are demand updating, order batching, price fluctuation and rationing and shortage gaming. In this study, mainly the demand updating and order batching in the form of economic order quantity and reordering point operation, respectively, caused this bullwhip effect. Ideally inventory level should be as low as possible, but increased inventory is justified to meet the increased demand if the cost is low, and stock-out means failure to meet consumer demand. The milling and wholesale inventories fluctuate due to seasonal supply of rice, and this causes the rice milling industries to starve for rice for milling during the off-season of rice production. The milling inventory reduces to zero during the off-season of rice production, while the wholesale inventory accumulates rice from the harvesting season and reaches the peak at early part of the year to maintain the supply to the retailer for sustainable availability of rice to the consumers during the off-season, and then it decreases until rice from new

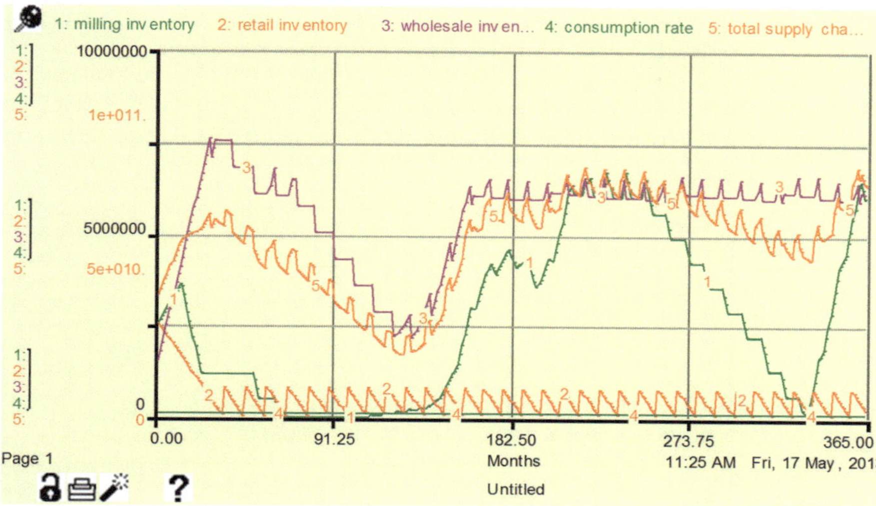

Fig. 11.5 Simulated milling, wholesale and retail inventory, daily rice requirement and total supply chain costs for economic order quantity and reordering point operation of the order rate of milling, wholesale and retailing

harvest is available. Then it increases and remains stable for rest of the period of the year.

To assess how the supply chain of the rice milling systems in Bangladesh would behave for any other ordering policies, the model was simulated for lead time average demand (demand average × lead time) policy. Although lead time average demand (demand average × lead time) order quantity and reordering point operation policy stabilises the retail inventory in a pattern similar to the economic order quantity policy, it increases the bullwhip effect and total supply chain cost. The annual total supply chain cost increases by 2.64 times the annual total supply chain cost of economic order quantity and reordering point operation of the milling, wholesale and retailing. This demonstrated the potentiality of using economic order quantity operation policy for supply chain management of rice milling systems.

The model was also simulated to address the impacts of rice productivity on the supply chain performances. Here the rice productivity is the yield of rice per ha. Rice productivity may be reduced from crop damage due to floods or pest infestation, and also it can be increased by development of higher yield hybrid rice through research and development. Rice productivity for this policy is defined as

$$\text{rice production rate} = \text{area under rice} \times \text{yield of rice} \tag{11.44}$$

$$\text{yield of rice} = 1.81, 2.81 \text{ and} 3.81 \text{ tons/ha} \tag{11.45}$$

Simulated milling inventory, wholesale inventory, retail inventory and total supply chain cost for rice productivity of 1.81 tons per ha, 2.81 tons per ha (present average rice productivity) and 3.81 tons per ha are shown in Figs. 11.6a, 11.6b, 11.6c and

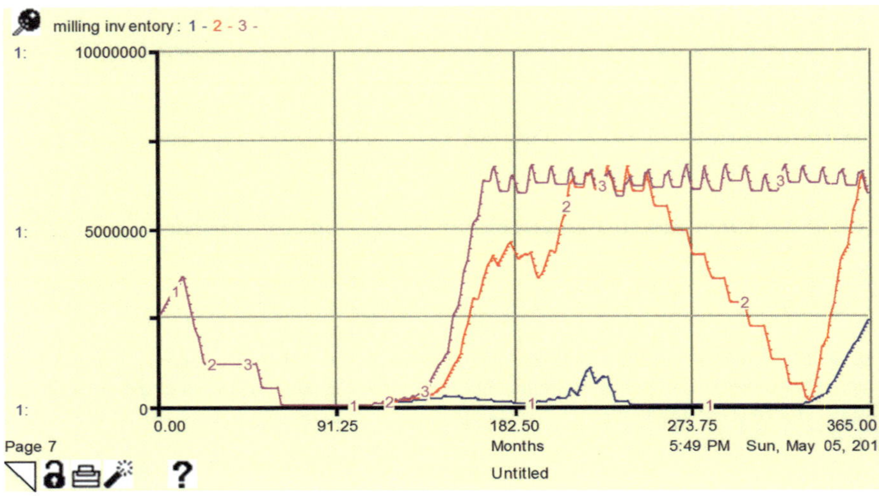

Fig. 11.6a Simulated milling inventory for rice productivity of 1.81 tons per ha, 2.81 tons per ha and 3.81 tons per ha

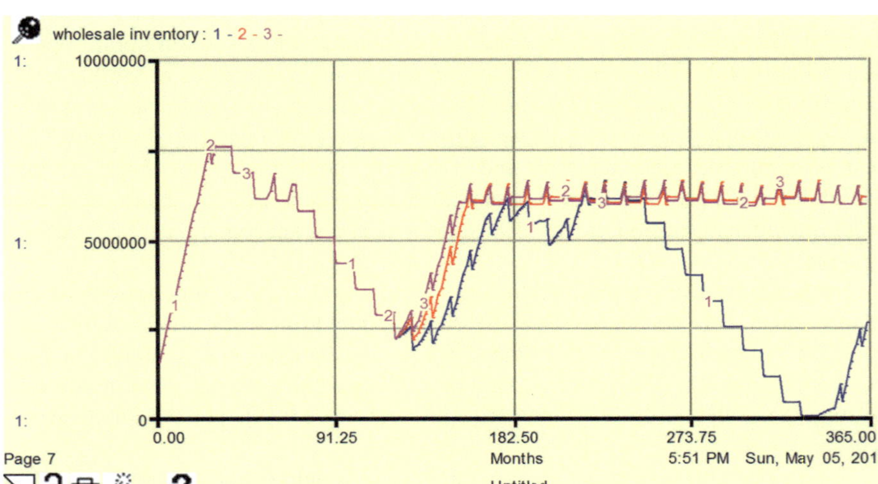

Fig. 11.6b Simulated wholesale inventory for rice productivity of 1.81 tons per ha, 2.81 tons per ha and 3.81 tons per ha

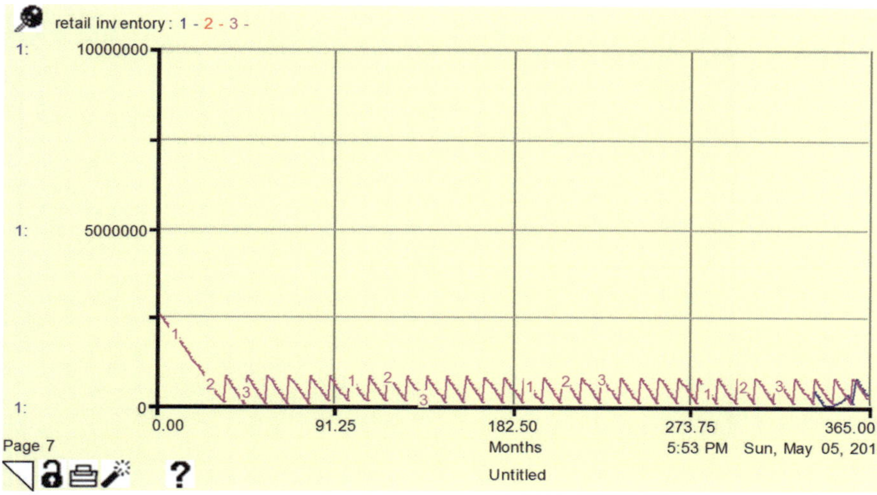

Fig. 11.6c Simulated retail inventory for rice productivity of 1.81 tons per ha, 2.81 tons per ha and 3.81 tons per ha

11.6d, respectively. Milling inventory is reduced to zero for most of the period of the reduced productivity of rice (1.81 tons per ha), while it is high for bumper production of rice (3.81 tons per ha) (Fig. 11.6a). Wholesale inventory is reduced significantly in the fourth quarter of the year for reduced rice productivity (1.81 tons per ha) (Fig. 11.6b). Total supply chain cost is also reduced in the third and fourth quarter of the year for reduced rice productivity (Fig. 11.6d). However, in all the cases, the retail inventory is almost the same except towards the end of the year when both milling and wholesale inventories are empty for reduced productivity (Fig. 11.6c). Thus, increased wholesale inventory is a possible solution for the retail inventory to face the shortage of rice during the off-peak harvesting season of rice production. This implies that as long as wholesale inventory is available, the retail inventory is stabilised based on economic order quantity and reordering point operation of milling, wholesale and retail inventory. This demonstrates that the policy based on economic order quality and reordering point can ensure the availability of rice even under reduced production of rice, i.e. during crop damage/failure unless both wholesale and milling inventories are empty.

The model was also simulated to assess the impacts of lead times on the supply chain performances. Lead time can be defined as the time it takes from when first a need for a product is determined until it arrives on the doorstep and the reordering point and safety stock depend the lead time. Reordering point and safety stock for milling, wholesale and retailing for this policy are defined as

$$\text{reordering point} = (\text{ demand aver} \times \text{lead time }) + \text{safety stock} \tag{11.46}$$

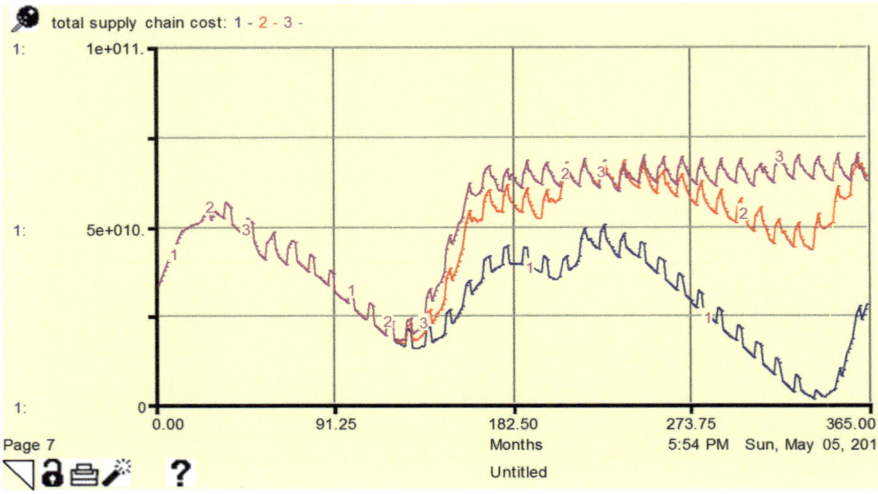

Fig. 11.6d Simulated total supply chain cost for rice productivity of 1.81 tons per ha, 2.81 tons per ha and 3.81 tons per ha

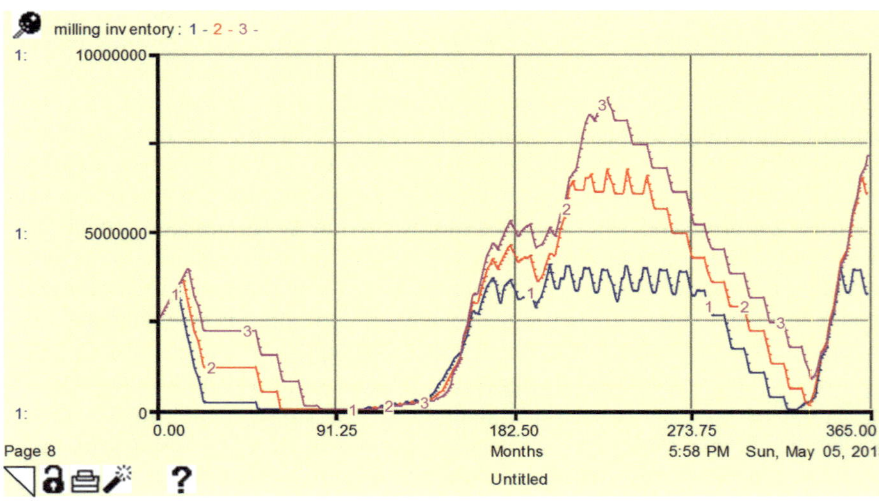

Fig. 11.7a Simulated milling inventory for milling lead time of 4 days, 7 days and 10 days

$$\begin{aligned}\text{safety stock} &= \text{safety factor} \times \text{standard deviation of} \\ &\text{demand aver} \times \text{SQRT(lead time)}\end{aligned} \tag{11.47}$$

$$\text{lead time} = 4,\ 7 \text{ and } 10 \text{ days} \tag{11.48}$$

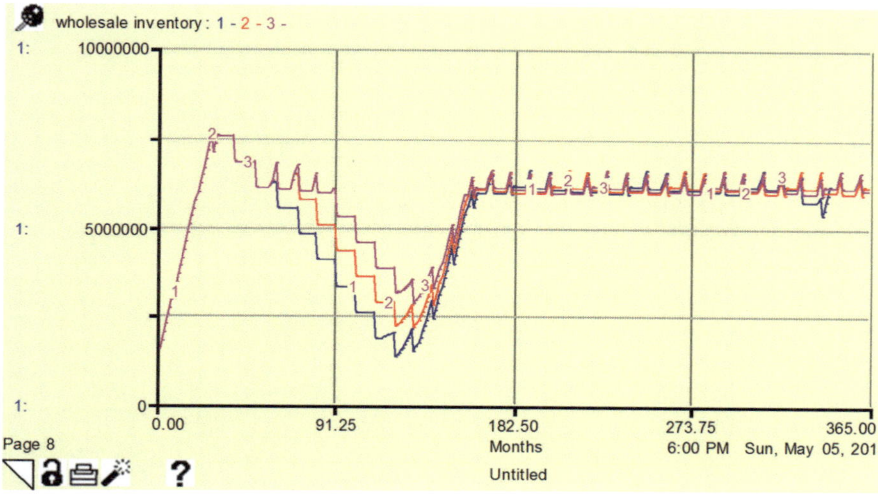

Fig. 11.7b Simulated wholesale inventory for milling lead time of 4 days, 7 days and 10 days

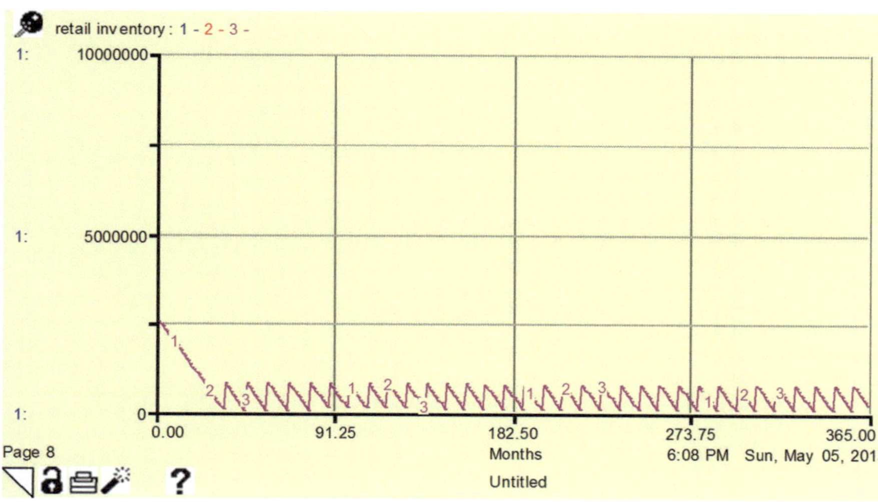

Fig. 11.7c Simulated retail inventory for milling lead time of 4 days, 7 days and 10 days

Simulated milling inventory, wholesale inventory, retail inventory and total supply chain cost for milling lead time (4, 7 and 10 days), wholesale lead time (4, 7 and 10 days) and retail lead time (4, 7 and 10 days) are shown in Figs. 11.7a, 11.7b, 11.7c, 11.7d, 11.8a, 11.8b, 11.8c, 11.8d, 11.9a, 11.9b, 11.9c and 11.9d, respectively.

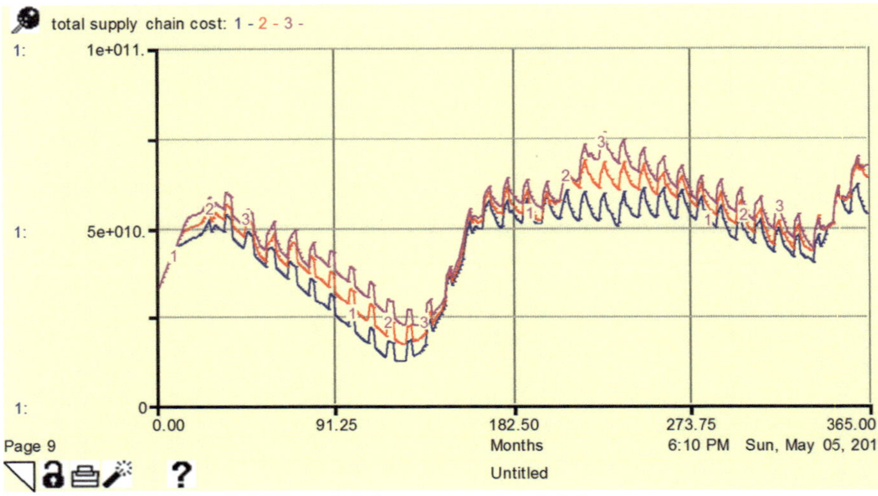

Fig. 11.7d Simulated total supply chain cost for milling lead time of 4 days, 7 days and 10 days

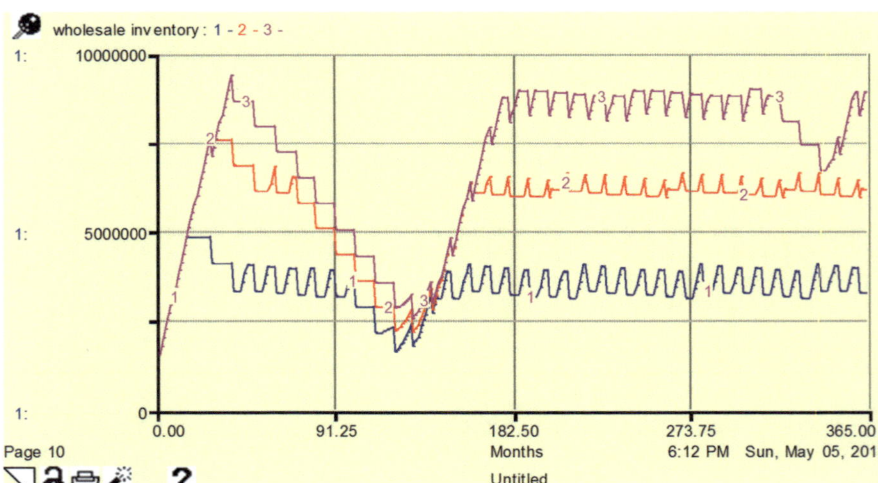

Fig. 11.8a Simulated wholesale inventory for wholesale lead time of 4 days, 7 days and 10 days

Firstly let us consider the impacts of milling lead time on supply chain performances for milling lead time of 4 days, 7 days and 10 days. Milling inventory increases considerably with the increase in the milling lead time (Fig. 11.7a). Wholesale inventory and total supply chain cost increase little with the increase of milling lead time (Figs. 11.7b and 11.7d), but the retail inventory remains almost

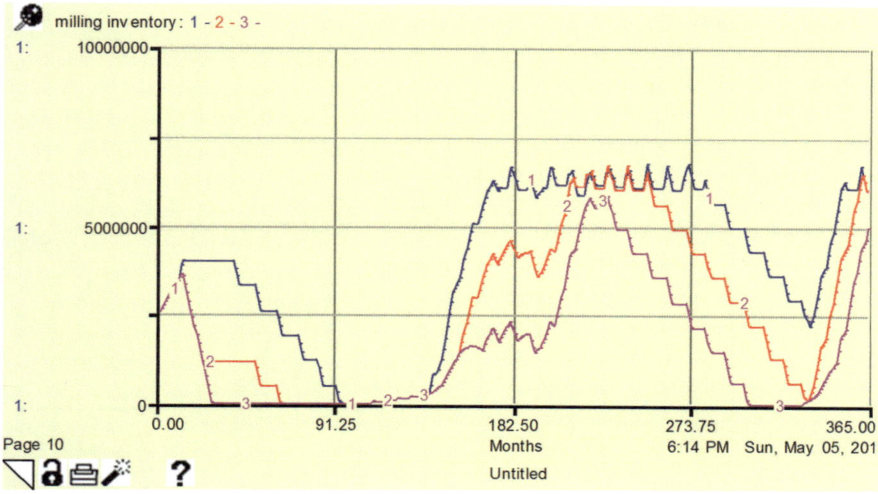

Fig. 11.8b Simulated milling inventory for wholesale lead time of 4 days, 7 days and 10 days

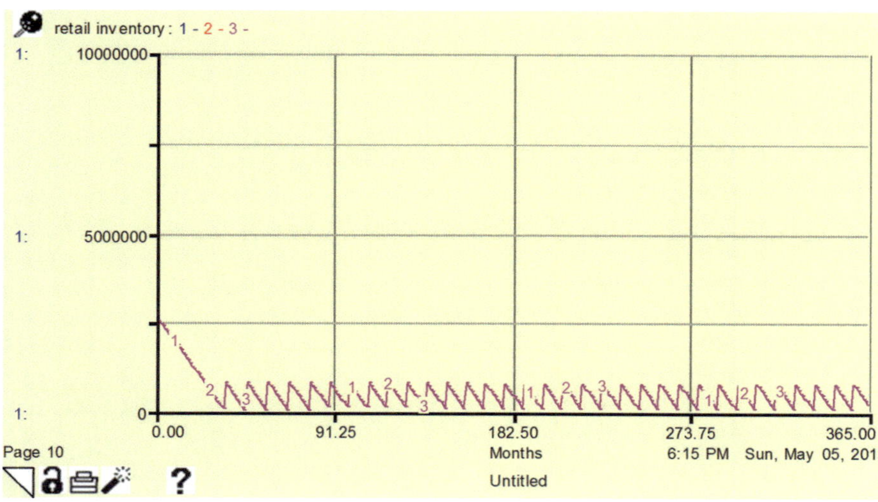

Fig. 11.8c Simulated retail inventory for wholesale lead time of 4 days, 7 days and 10 days

unchanged (Fig. 11.7c). In essence the major impact of the changes in milling lead time is reflected on the milling inventory, but the impact on the retail inventory is a minimum.

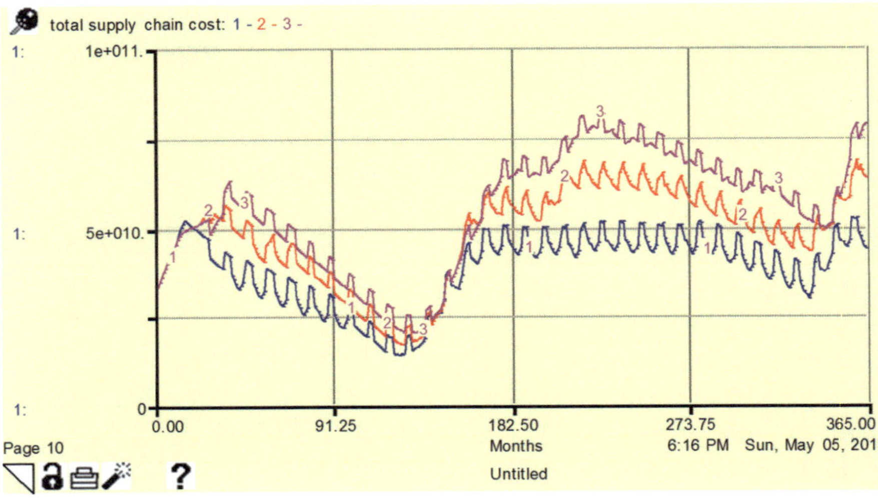

Fig. 11.8d Simulated total supply chain cost for wholesale lead time of 4 days, 7 days and 10 days

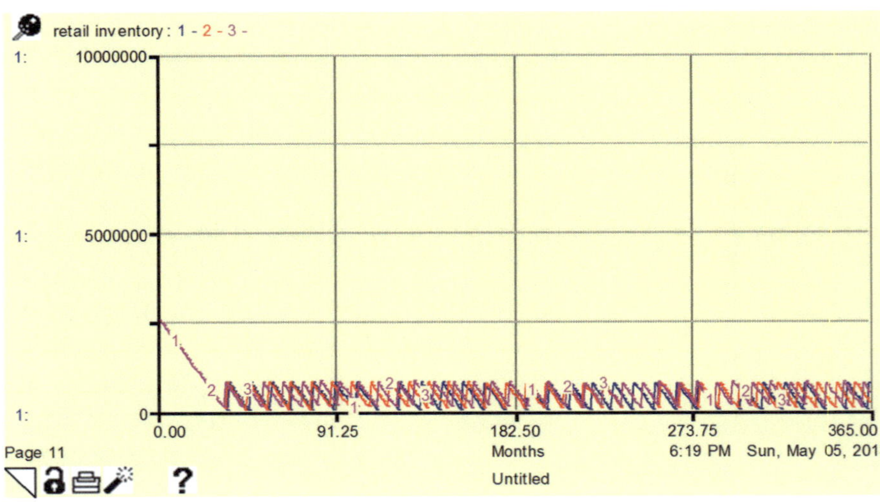

Fig. 11.9a Simulated retail inventory for retail lead time of 4 days, 7 days and 10 days

Secondly let us consider the impacts of wholesale lead time on supply chain performances for wholesale lead time of 4 days, 7 days and 10 days. Simulated wholesale inventory, milling inventory, retail inventory and total supply chain cost for wholesale lead time of 4 days, 7 days and 10 days are shown in Figs. 11.8a,

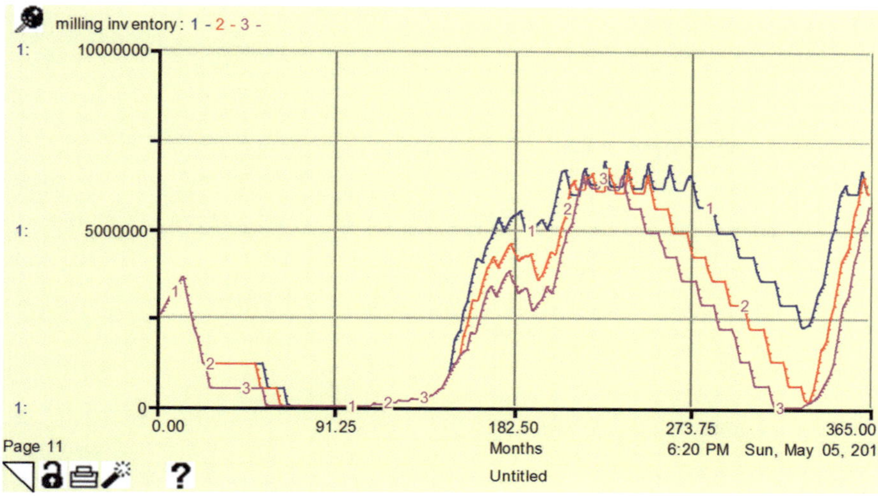

Fig. 11.9b Simulated milling inventory for retail lead time of 4 days, 7 days and 10 days

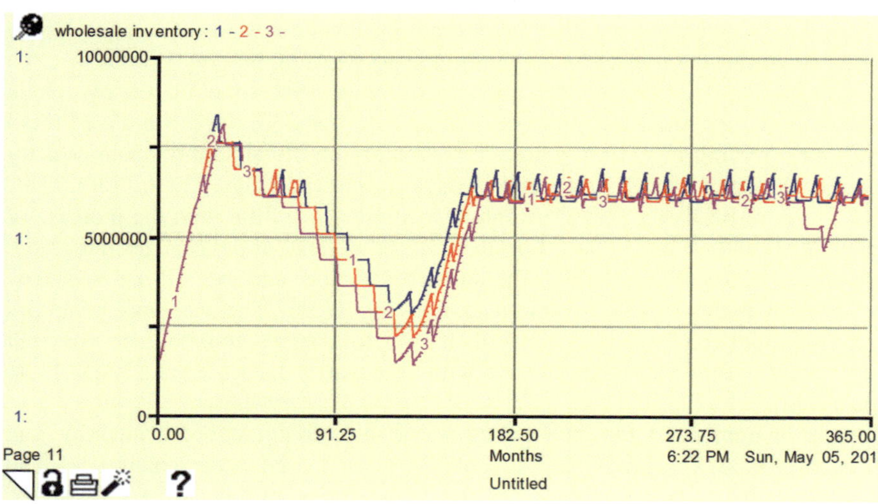

Fig. 11.9c Simulated wholesale inventory for retail lead time of 4 days, 7 days and 10 days

11.8b, 11.8c, and 11.8d, respectively. Wholesale inventory and total supply chain cost increase with the increase of wholesale lead time (Figs. 11.8a and 11.8d), while milling inventory decreases (Fig. 11.8b), and the changes are relatively large, but the retail inventory remains almost unchanged (Fig. 11.8c). In essence the major impacts of the changes in wholesale lead time are reflected not only on the changes

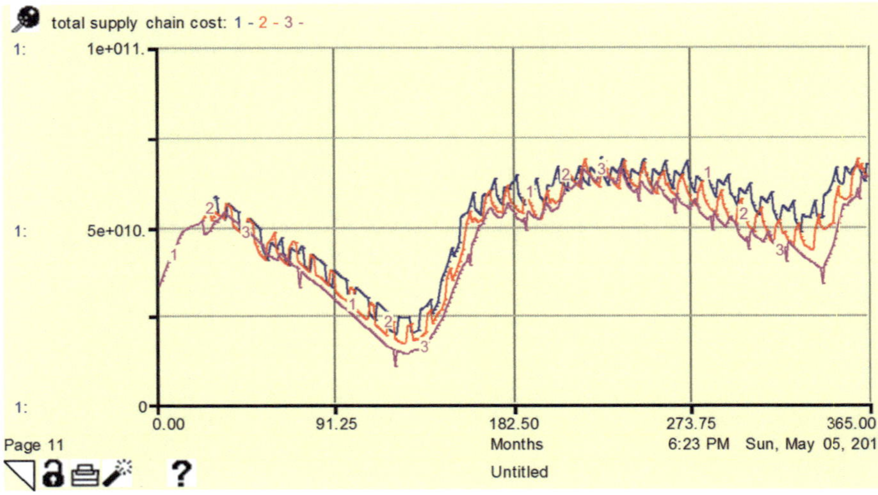

Fig. 11.9d Simulated total supply chain cost for retail lead time of 4 days, 7 days and 10 days

of wholesale inventory but also on the milling inventory and total supply chain costs. However, the impact on the retail inventory is a minimum.

Thirdly let us consider the impacts of retail lead time on supply chain performances for retail lead time of 4 days, 7 days and 10 days. Simulated retail inventory, milling inventory, wholesale inventory and total supply chain cost for retail lead time of 4 days, 7 days and 10 days are shown in Figs. 11.9a, 11.9b, 11.9c and 11.9d, respectively. Retail inventory increases very little with the increase of retail lead time (Fig. 11.9a). But the effects of lead time changes are more prominent in the second half of the simulated period, and the milling inventory decreases with the increase in retail lead time (Fig. 11.9b). The wholesale inventory and total supply chain cost show small changes for changes in retail lead time, but each category follows similar patterns with time within a narrow band (Figs. 11.9c and 11.9d).

It is demonstrated that the policy based on economic order quantity and reordering point can ensure the availability of rice even under changes in lead time and demand average in the supply chain of the rice milling systems. However, the bullwhip or Forrester effect remains present in the system. Forrester (1961) and Sterman (1989) ascribe behavioural causes to the bullwhip effect, where Lee et al. (1997a, b) suggest that the bullwhip effect also occurs due to operational causes. The lack of supply chain coordination results in bullwhip effect which distorts the demand information in the supply chain and increases the supply chain cost. The most obvious remedy to counter demand signal processing is collaboration, but it generally does not completely eliminate the problem (Moll 2013). The largest increase in total supply chain cost occurs for the increase in the wholesale

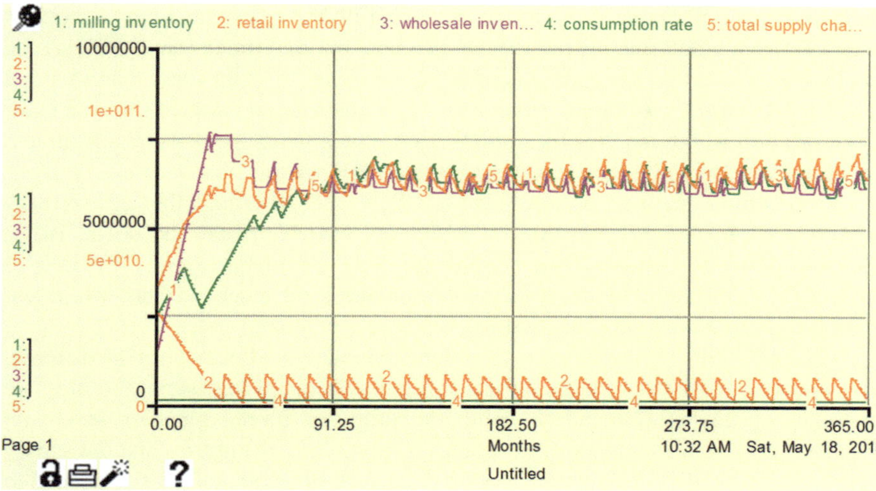

Fig. 11.10 The impacts of the round-the-year supply of rice to the milling industries on wholesale inventory, retail inventory and total supply chain costs

demand average, and this indicates that supply chain cost increases at the wholesale stage of the supply chain. In essence the major influence of the retail lead time and demand average is on the milling inventory.

The model was also simulated to assess the impacts of the round-the-year supply of rice to the milling industries on wholesale inventory, retail inventory and total supply chain costs. Round-the-year equal amount of supply of rice is ensured by the round-the-year equal amount of release of rice from the storage for sales for milling. The rice supply to milling industries is defined as

$$\begin{aligned} \text{rice inventory } (t) = {} & \text{rice inventory } (t-1) + \text{rice supply rate} \times \Delta t \\ & -\text{rice sales for milling} \times \Delta t \end{aligned} \tag{11.49}$$

$$\text{rice suply rate} = 97955 \text{ tons/day} \tag{11.50}$$

Figure 11.10 shows the impacts of the round-the-year supply of rice to the milling industries on wholesale inventory and retail inventory. The simulated results show that round-the-year supply and the use of rice to the milling industries not only stabilise the milling inventory but also the whole system. This implies that if rice is dried initially for safe storage and then used round the year, the whole system can be stabilised. However, the bullwhip or Forrester effect remains present in the stabilised system.

The simulated results of the model demonstrate that the economic order quantity and reordering point for milling, wholesale and retail inventories not only ensure the availability of rice to the consumers during poor harvest (as long as stock is available from wholesale inventory) but also near optimal order quantities. Annual total supply chain cost for economic order quantity is 2.64 times less than the total supply chain cost for lead time demand average, and a major share of this cost is for transportation which requires fossil fuels. Hence economic operation results in less fuel consumption and hence less contribution to global warming.

Sustainability in three dimensions is widely accepted, and this allows easy comprehension of the integration of economic, environmental and social issues (Seuring 2013). Economic order quantity and sustainable supply of rice to consumers in this study address the economic and food security issues while less contribution to global warming addresses the environmental issue.

This study brings systems approach to understanding and managing the different activities needed to coordinate the rice milling systems to best serve the ultimate customer of rice in Bangladesh. With the proposed model planners and plant managers can visualise the milling, wholesale and retail inventories and the supply chain costs as well as even can be used to determine the storage facilities needed to ensure the availability of rice to the consumers. The actors in the rice supply chain make decisions individually and collectively such as milling and ordering activities. Some of these decisions influence the capabilities and effectiveness of the supply chain. Also Seuring and Müller (2008) emphasise the need of increasing cooperation along the supply chain, if sustainability goals are to be reached. Participatory system dynamics modelling uses system dynamics perspective in which stakeholders or clients participate to some degree in different stages of the modelling process, including problem definition, system description, identification of policy levers, model development and/or policy analysis. Participatory system dynamics/multi-agent system modelling of the rice milling supply chain involving all the stakeholders is recommended for future study. This model can be used for participatory modelling of rice milling systems. Overall, the proposed model can be used to study and analyse 'what if' scenarios of the supply chain of the rice milling systems.

11.7 Conclusion

A system dynamics model of supply chain of rice milling systems in Bangladesh is developed for policy analysis, and the retail inventory is fully stabilised for economic order quality and reordering point resulting in an efficient and sustainable supply chain network. Productivity reduction of seasonal production of rice causes changes in the milling inventory, wholesale inventory and total supply chain cost, but the retail inventory remains constant as long as the wholesale inventory is not empty. Lead time changes have positive influences on the directly related inventory level, and the changes are more pronounced in the milling and wholesale inventories. Also the influence of the demand average is dominant on the milling inventory. Round-the-year operation of the milling industries increases the stability of the whole system. Also the less total supply chain cost from economic operation results in less fuel consumption and hence less contribution to global warming. This model facilitates identification and study of the critical components of the overall supply chain, allowing for the creation of an efficient and sustainable supply chain network. This model also provides greater insight and better understanding of the supply chain of rice milling systems, and it can be used as a computer laboratory for developing scenarios for policy analysis

Exercises

Exercise 11.1 Why a computer model is a necessity to analyse and support rice milling systems. Discuss where to use operations research and system dynamics in supply chain management.

Exercise 11.2 Draw the causal loop diagram of the supply chain of rice milling systems. Draw another causal loop diagram for sustainable supply chain management.

Exercise 11.3 Draw stock–flow diagrams of both the causal loop diagrams in exercise 11.2 and compare the simulated results of retail inventory and retail price.

Exercise 11.4 Draw causal loop and stock–flow diagrams of supply chain of rice milling systems incorporating economic order quantity and reordering point for sustainable development. Simulate the retail inventory, retail price and contribution to global warming.

Exercise 11.5 Simulate both the models in exercise 11.2 to assess the impacts of round-the-year supply of rice to the milling industries and also assess the contribution to global warming.

References

Ahmed R (2000) Liberalization of agricultural input markets in Bangladesh. In: Ahmed R, Haggblade S, Choudhury T (eds) Out of the shadow of famine: evolving food markets and food policy in Bangladesh, pp 49–72

Apaiah KA, Hendrix EMT (2005) Design of a supply chain network for pea-based novel protein foods. J Food Eng 70:383–391

Apaiah KA, Hendrix EMT, Meerdink G, Linnermann R (2005) Qualitative methodology for efficient food chain design. Trends Food Sci Technol 16:204–214

Bala BK (1975) Applications of control theory to the dynamic behaviour of agricultural commodities. M. Eng. thesis, Asian Institute of Technology, Bangkok, Thailand

Bala BK (1990) Applications of control theory to the dynamic behaviour of agricultural commodities. Syst Dyn Int J Policy Model 3(1&2):87–100

Bala BK (1999) Principles of system dynamics. Agrotech Publishing Academy, Udaipur

Bala BK, Bhuiyan MGK, Alam M, Fatimah MA, Alias EF, Sidique SF (2015) Modeling of supply chain of rice milling systems in Bangladesh. Institute of Agricultural and Food Policy Studies, Universiti Putra Malaysia

BBS (Bangladesh Bureau of Statistics) (2011) Statistical Pocket Bangladesh – 2011. Statistics and Information Division, Ministry of Planning, Government of People's Republic of Bangladesh

Bhuiyan MGK (2014) Milled rice value chain in Bangladesh. Unpublished PhD thesis, Bangladesh Agricultural University, Mymensingh, Bangladesh

Bosona TG, Gebresenbet G (2013) Cluster building and logistics network integration of local food supply chain. Biosyst Eng 108:299–302
Carter CR, Easton PL (2011) Sustainable supply chain management: evolution and future directions. Int J Phys Distrib Logist Manag 141:46–62
Chen F (1998) Echelon reorder points, installation reorder points, and the value of centralized demand information. Manag Sci 44(12):S221–S234
Dabbene F, Gay P, Sacco N (2008a) Optimisation of fresh – food supply chains in uncertain environments, Part I: background and methodology. Biosyst Eng 99:348–359
Dabbene F, Gay P, Sacco N (2008b) Optimisation of fresh – food supply chains in uncertain environments, Part II: a case study. Biosyst Eng 99:360–371
Davenport TH (1993) Process innovation: reengineering work through information technology. Harvard Business School Press, Boston, MA
Forrester JW (1961) Industrial dynamics. MIT Press, Cambridge, MA
Forrester JW (1968) Principles of systems. MIT Press, Cambridge, MA
Georgiadis P, Vlachos D, Iakavou E (2005) A system dynamics modeling framework for the strategic supply chain management of food supply chains. J Food Eng 70:351–3364
Hossain M (1988) Nature and impact of the green revolution in Bangladesh. J Dev Econ 19 (2):49–166
Hossain M, Bose ML, Mustafi BAA (2006) Adoption and productivity impact of modern rice varieties in Bangladesh. IFPRI Research Report 57
Kpmaier F, Voigt A (2013) Pushing the limits – Using system dynamics to forecast short-term commodity price movements. Internet.15.08.2013
Kumar S, Nigmatullin A (2011) A system dynamics analysis of food supply chains – case study with non-perishable products. Simul Model Pract Theory 19:2151–2168
Lee HL, Padmanabhan V, Whang S (1997a) The bullwhip effect in supply chains. Sloan Manage Rev 38(3):93–102
Lee HL, Padmanabhan V, Whang S (1997b) Information distortion in a supply chain: the bullwhip effect. Manag Sci 43(4):546–558
Manzini R, Accorsi R (2013) The new conceptual framework for food supply chain assessment. J Food Eng 115:251–263
Meadows DL (1970) Dynamics of commodity production cycles. Wright Allen Press Inc., Cambridge, MA
Minegishi S, Thiel D (2000) System dynamics modeling and simulation of a particular food supply chain. Simul Model Pract Theory 8:321–339
Minten B, Murshid KAS, Reardon T (2013) Food quality changes and implications: evidence from the rice value chain of Bangladesh. World Dev 42:100–113
Moll J (2013) The bullwhip effect: analysis of causes and remedies. MS thesis, VU University Amsterdam, The Netherlands
Muckstadt JA, Sapra A (2010) Principles of inventory management. Springer, Heidelberg
Nerlove N (1958) Adaptive expectations and cobweb phenomena. Q J Econ 72:227–240
Prahalad CK, Ramswamy V (2000) Co-opting customer competence. Harv Bus Rev 78(1):79
Riddals CE, Bennett S, Tipi NS (2000) Modeling the dynamics of supply chains. Int J Syst Sci 31 (8):969–976
Riggs JL (1970) Production systems: planning, analysis and control. Wiley, New York
Sachan A, Sahay BS, Sharma D (2005) Developing Indian grain supply chain cost model: a system dynamics approach. Int J Product Perform Manag 54(3):187–205
Selen W, Soliman F (2002) Operations in today's demand chain framework. J Oper Manage 20 (6):667
Seuring S (2013) A review of modeling approaches for sustainable supply chain management. Decis Support Syst 54:1513–1520
Seuring S, Müller M (2008) From a literature review to a conceptual framework for sustainable supply chain management. J Clean Prod 16(15):1699–1710

Sterman JD (1989) Modeling managerial behavior: misperceptions of feedback in a dynamic decision making experiment. Manag Sci 35(3):321–339

Sterman JD (2000) Business dynamics: systems thinking and modeling for a complex world. Irwin McGraw Hill, New York

Teimoury E, Nedaei H, Ansari S, Sabbaghi M (2013) A multi-objective analysis for import quota policy making in a perishable fruit and vegetable supply chain: a system dynamics approach. Comput Electron Agric 93:37–45

Thomas O (2006) Understanding the term reference model in information systems research: history, literature analysis and explanation. In: AI CBE (ed) Satellite workshop of BPM 2005. Springer-Verlag, Nancy, France, pp 484–496

Verdouw CN, Beulensb AJM, Trienekensc JH, Wolferta J (2010) Process modelling in demand-driven supply chains: a reference model for the fruit industry. Comput Electron Agric 73:174–187

Vo TLH, Thiel D (2008) A system dynamics model of the chicken meat supply chain faced with bird flu. Proceedings of system dynamics conferences, Athens, 20–24 July 2008, VO153

Vo TLH, Thiel D (2011) Economic simulation of poultry supply chain facing a sanitary crisis. Br Food J 113(8):1011–1030

Bibliography

Barlas Y (1996) Formal aspects of model validity and validation in system dynamics. Syst Dyn Rev 12:183–210

Modelling of Solid Waste Management Systems of Dhaka City in Bangladesh

12

In the previous chapters of this part, case studies on system dynamics modelling and simulation of some practical problems have been presented. This chapter presents a case study of environmental management of solid waste management system of Dhaka City, Bangladesh. The model of this case study is organised as (1) introduction, (2) dynamic hypothesis, (3) causal loop diagram, (4) stock–flow diagram, (5) model validation, (6) simulation and policy analysis and (7) conclusion to illustrate the system dynamics applications in environmental management based on systems thinking. Population, uncleared waste, untreated waste, composite index and public concern are projected to increase with time for Dhaka City. Simulated composts, CO_2 total, CH_4 and BOD from energy, leachate production and BOD from leachate over a period of 30 years are also presented. Simulated results also show that increasing the budget for collection capacity alone does not improve environmental quality; rather an increased budget is required for both collection and treatment of solid wastes of Dhaka City. This model can be used as a tool to assess and design policy options of solid waste management.

12.1 Introduction

Solid waste consists of the highly heterogeneous mass of discarded materials from the urban community, as well as the more homogeneous accumulation of agricultural, industrial and mining wastes. The principal sources of solid wastes are residences, commercial establishments, institutions and industrial and agricultural activities. Domestic, commercial and light industrial wastes are considered together as urban wastes. The main constituents of urban solid wastes are similar throughout the world, but the quantity generated, the density and the proportion of constituents vary widely from country to country and from town to town within a country according to the level of economic development, geographic location, weather and social conditions. In general, it has been found that as the personal income

B.K. Bala et al., *System Dynamics*, Springer Texts in Business and Economics,
DOI 10.1007/978-981-10-2045-2_12

rises, kitchen wastes decline but the paper, metal and glass wastes increase; the total weight generated increases but the density of the wastes declines (Rao 1992).

Several disposal methods are being used in various parts of the world, and the most prominent of these are open dumping, sanitary landfilling, incineration and composting. Sanitary landfilling is the main method used in industrialised countries, and open dumping is very common in developing countries like Bangladesh and India.

Open dumping of solid wastes is practised extensively in Bangladesh because it is cheap and requires no planning. Generally, the low-lying areas and outskirts of the towns and cities are used for this purpose.

Sanitary landfilling is a controlled engineered operation, designed and operated according to acceptable standards. It may be defined as a controlled method of disposing of refuse onto or into land while minimising nuisances or hazards to public health or safety. The operation is carried out without environmental damage and in areas already spoiled or in need of restoration.

Incineration involves the burning of solid wastes at high temperatures. If incineration is to become an economical method for solid waste disposal, useful materials and energy must be recovered by the process. Heat can be recovered by putting a waste heat boiler or some other recovery devices on an existing solid waste incinerator. The heat so recovered can be utilised for generating electricity or for space heating purposes. In general, solid waste has about one-third the heating value of coal, but unlike coal it has a very low sulphur content. All types of incinerators produce air pollution. The contributions to global warming by incineration are much less than those of landfill but comparable to those by composting. Yongfeng (2005) reported that municipal solid waste incineration has recently emerged as the method of choice for the primary treatment of municipal solid wastes in many cities in China. Alam and Bole (2001) and Sufian and Bala (2006) reported that there exists high electrical energy recovery potential from urban solid waste of Dhaka City.

In contrast to a sanitary landfill, composting of refuse is an aerobic method of decomposing solid waste. Many types of microorganisms already present in the waste biostabilise the organic matter in the waste and produce a soil conditioner as a result of the process. The organisms include bacteria, which predominate at all stages; fungi, which often appear after the first week; and actinomycetes, which assist during the final stages. Zurbrugg et al. (2002) reported a success story of decentralised collection and composting scheme in Mirpur, Dhaka, Bangladesh. The case of Mirpur shows that composting can be a good alternative to conventional solid waste management options, reducing the amount of waste to be transported and dumped by producing a valuable raw material for fertilisers.

Solid wastes contain significant amounts of valuable materials like steel, aluminium, copper and other metals which, if they are recovered and reused, would reduce the volume of the wastes to be collected and at the same time would yield significant salvage and resale income. In addition, better reclamation techniques will help to save valuable natural resources and turn wastes, which could be dangerous, into useful products. Some important solid wastes that have been successfully reclaimed are paper, plastics, glass and metals.

Solid waste generation of the Dhaka metropolitan area in 1998 was 3944 tons/day and 4750 tons/day in 2000. Solid waste generation in Dhaka City is increasing rapidly with rapidly growing population and GDP, and the Japan International Cooperation Agency (JICA) (2005) predicted solid waste generation to increase from 3200 tons/day in 2004 to 4634 tons/day in 2015. Of the total waste produced, nearly 20 % is used for recovery and recycling and about 37 % remains scattered laying around on roadsides, open spaces or drains. The remaining 43 % of the wastes are deposited together in some primary depots from where about 45 % is finally disposed of either by the DCC or community-based organisations (CBOs) in the open landfill sites.

Many studies have been reported on strategies to achieve municipal solid waste management (Pawan et al. 1997; Salvato 1992; Kum et al. 2005). Linear programming, input–output analysis, expert system (a methodology that uses expert knowledge to solve problems of a complex system) and system dynamics have been applied to aid decision-makers in planning and management of solid waste management systems (Everett and Modak 1996; Clayton and McCarl 1979; Barsi 2000; Ming et al. 2000; Heikki 2000; Mashayekhi 1992; Sudhir et al. 1997). Several studies have been reported on life cycle assessment (LCA) of municipal solid waste management systems (Consonni et al. 2005a, b; Velumani and Meenakshi 2007). Consonni et al. (2005a, b) examined environmental impacts and economics of the various strategies using life cycle assessment starting from heat and mass balances. Velumani and Meenakshi (2007) reported life cycle inventory analysis for three different municipal solid waste management scenarios: (i) open landfill; (ii) generation, collection, transportation and sanitary land filling; and (iii) generation, collection, transportation, segregation, incineration, recycling and sanitary land filling for Coimbatore City, India, to aid the decision-makers and planners for integrated management of solid waste management for sustainable development.

Dyson and Chang (2005) emphasised the capability of system dynamics for prediction of solid waste generation, and Sufian and Bala (2006, 2007) and Bala and Sufian (2006) successfully applied system dynamics to model and analyse the policy options of solid waste management system.

Solid waste crisis is emerging in Bangladesh and the majority of the cities are severely affected. The disposal problem is more serious for Dhaka City, the capital of Bangladesh. There is a large gap between the waste generation and management system which results in environmental pollution. To meet this gap, there is a need of a model for solid waste management systems of Dhaka City, Bangladesh. Here the system dynamics model of solid waste management adopted from Sufian and Bala (2006, 2007) is presented, and it has been designed to predict solid waste generation, scientific landfill, incineration, composting and life cycle assessment.

12.2 Dynamic Hypothesis

The dynamic hypothesis is a conceptual model which seeks to define the critical feedback loops and system structure that drive the system's behaviour in the reference mode. When the model based on feedback concept is simulated, the endogenous structure of the model should generate the reference mode behaviour of the system, and thus, the endogenous structure causes the changes in the dynamic behaviour of the system. In system dynamics modelling, causal loop diagrams identify the principal feedback loops of a system, and the causal loop diagram and stock–flow diagram are used to describe the basic cause–effect relationships hypothesised to generate the reference mode of the behaviour over time (Sterman 2000). Solid waste management systems can be represented by causal loop diagram and stock–flow diagram, and the simulation model based on the causal loop diagram and stock–flow diagram would generate dynamic behaviour of the solid generation, uncleared wastes and untreated wastes and composting in the reference mode. The solid wastes and its present management systems in the form of causal loop diagram and stock–flow diagram are hypothesised to generate the observed solid waste generation, uncleared wastes and untreated wastes in the reference mode. In essence there is piling up of uncleared wastes and untreated wastes from the lack of management strategies and fund; this dynamics results from the endogenous consequences of the feedback structure of the solid waste management systems.

12.3 Causal Loop Diagram

Causal loop diagram of solid waste management systems of Dhaka City, Bangladesh, is shown in Fig. 12.1. Municipal solid wastes increase with population (R1) and solid generation creates public concern and pressure to mitigate the waste (B1). The solid wastes are separated into recyclable wastes and nonrecyclable wastes. Both the uncollected waste and unhygienic disposal of waste create environmental pollution, which gives rise to increase public annoyance and anger, and hence public concern develops to reduce waste generation (B2 and B3). Also, the untreated wastes create public concern and pressure (B4). A higher composite index increases management perception which increases fund allocation for solid waste management (B5). The composite index is defined as

$$\text{Composite Index} = (w_1 \times \text{UNCL} + w_2 \times \text{UNTR}) \times \text{POPR} \tag{12.1}$$

where

w_1, w_2 = weighting factor ($w_1 = 0.5$ and $w_2 = 0.5$)
UNCL = the ratio of the uncleared waste at any point of time to the base value
UNTR = the ratio of the untreated waste at any point of time to the base value
POPR = the ratio of the population at any point of time to the base value

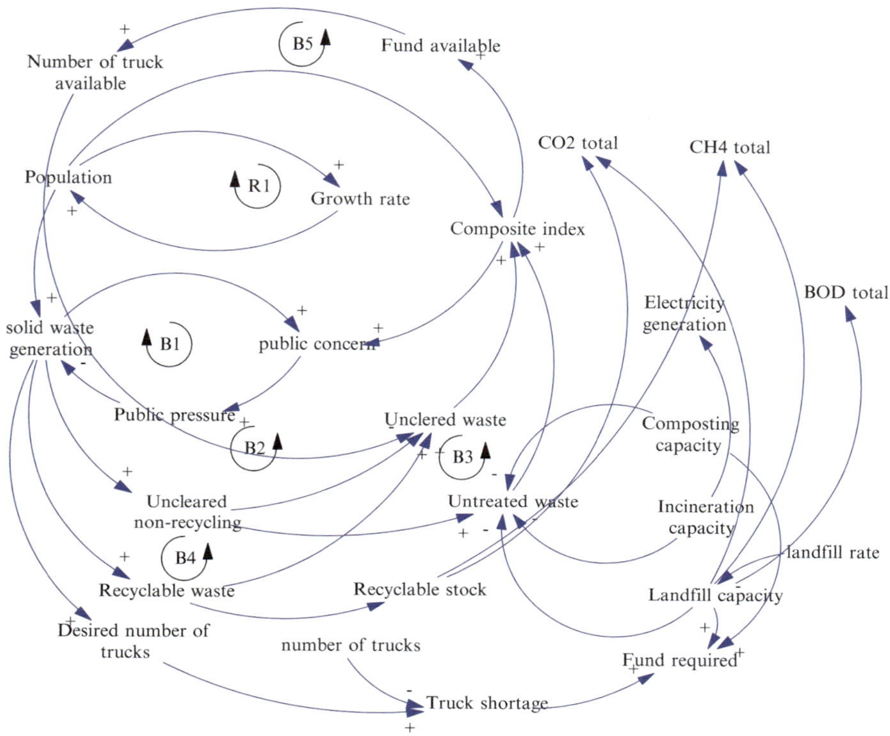

Fig. 12.1 Causal loop diagram of solid waste management systems of Dhaka in Bangladesh

A higher value of composite index indicates a progressive deterioration in health and environmental quality. A fraction of organic solid wastes from the municipal solid wastes is used for composting, and the remaining portion of the nonrecyclable wastes is used for sanitary landfill and incineration.

Life cycle assessment (LCA) of municipal solid waste management of the Dhaka City is also included in this model. The scope of life cycle assessment is to account for all possible environmental impacts 'from cradle to grave', i.e. along the whole sequence of actions required to bring about and later terminate a given product or an activity. Life cycle assessment of municipal solid wastes starts just at the moment when it is put into the waste bin and ends up when it is disposed. Life cycle assessment assesses the use of resources and releases of emissions to air, water and land and the generation of useful products from the wastes. This is done here by compiling an inventory of relevant inputs and outputs of the system and evaluating the potential impacts of those inputs and outputs in relation to the objectives of the study. To evaluate properly the global warming potential, CO_2 emissions are broken down into a fraction originated by fossil carbon (plastics, rubber or fossil fuel) and a renewable fraction originated from biogenic materials (paper, wood or organic fractions).

12.4 Stock Flow Diagram

The flow diagram of the model was constructed using icon-operated stock and flow symbols of STELLA software. The STELLA flow diagram of the urban solid waste management system is shown in Figs. 12.2a, 12.2b, 12.2c and 12.2d. Fundamental equations that correspond to the major state variable are as follows:

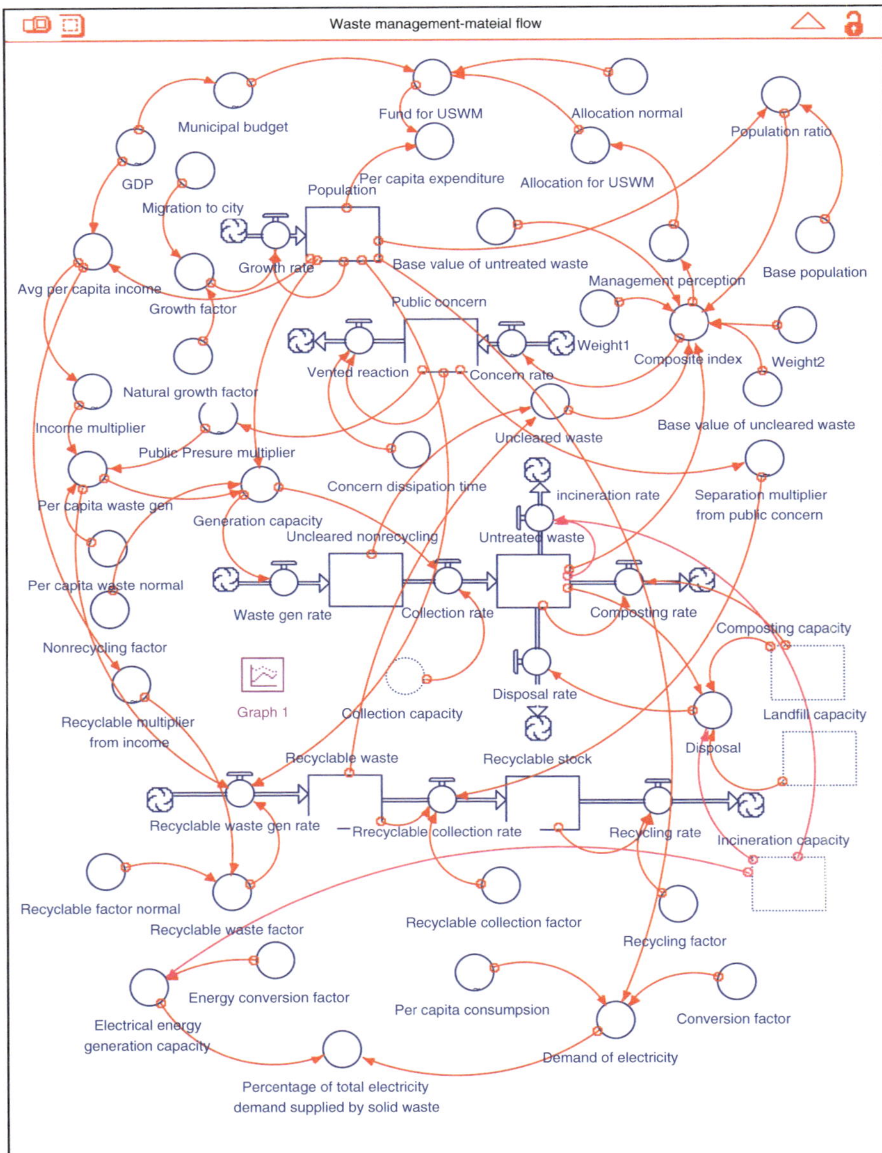

Fig. 12.2a STELLA flow diagram of solid waste management system dynamics—material flow

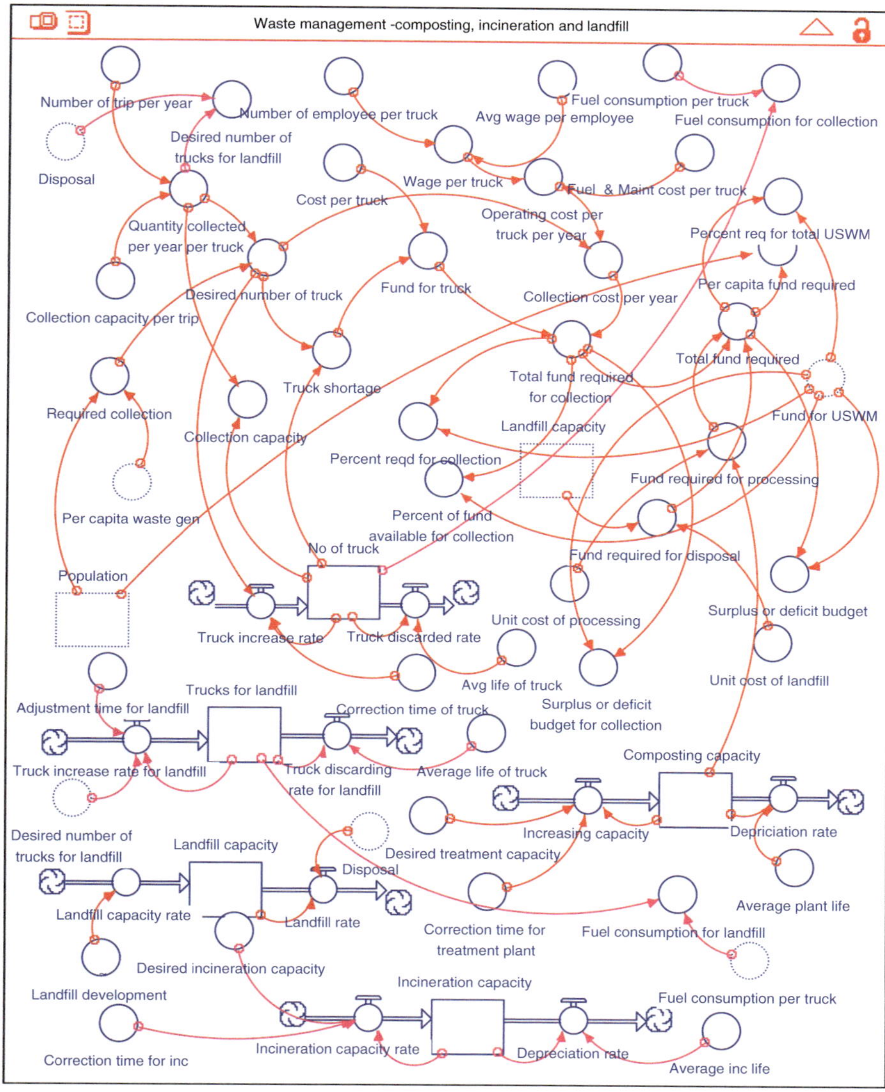

Fig. 12.2b STELLA flow diagram of solid waste management system dynamics—composting, incineration and landfill

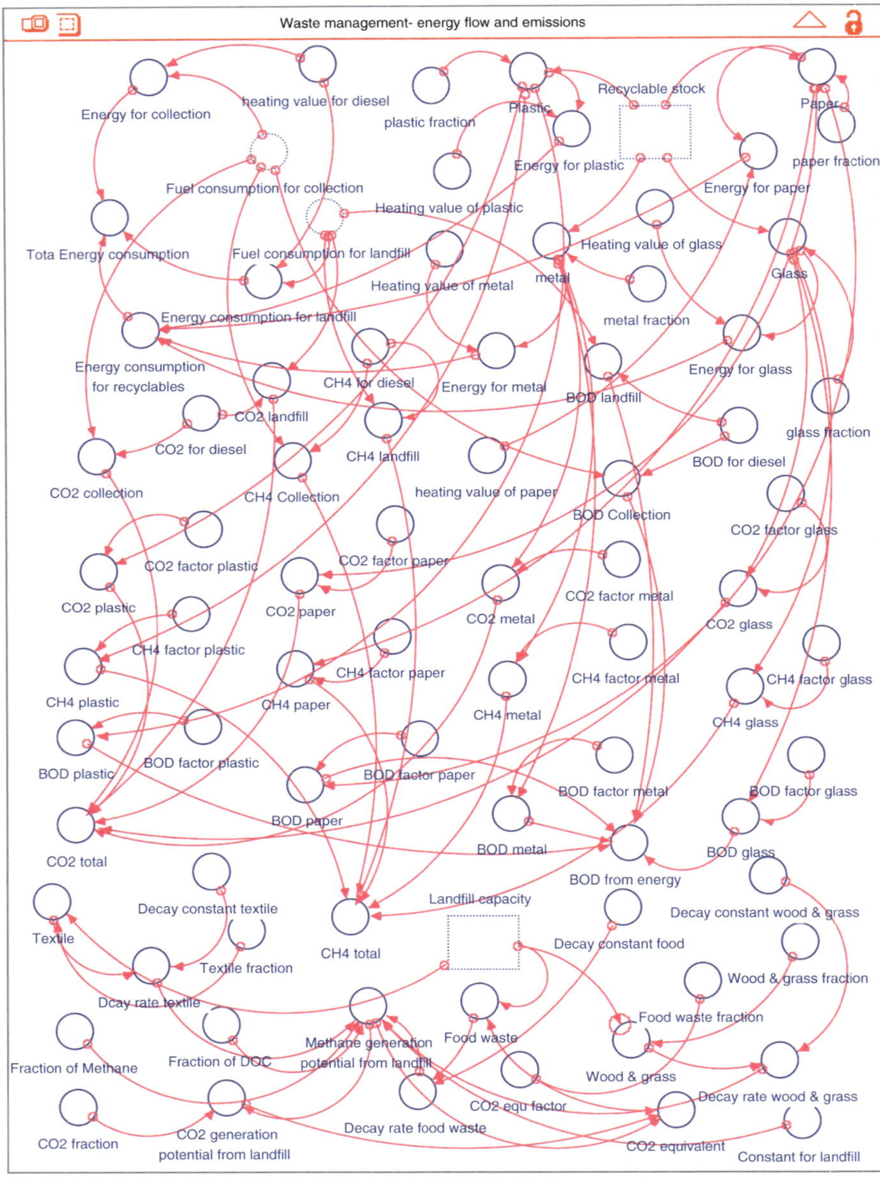

Fig. 12.2c STELLA flow diagram of solid waste management system dynamics—energy flow and emissions

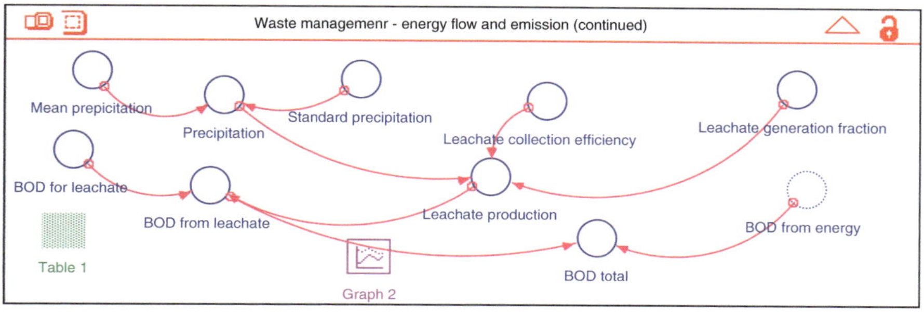

Fig. 12.2d STELLA flow diagram of solid waste management system dynamics—energy flow and emissions

12.4.1 Waste Management–Material Flow

The generated solid wastes consist of uncleared nonrecycling wastes and recyclable wastes.

Uncleared nonrecycling waste is increased by waste generation rate and decreased by collection rate, and it is expressed as

$$\begin{aligned} \text{Uncleared\_nonrecycling}(t) &= \text{Uncleared\_nonrecycling}(t-dt) \\ &+ (\text{Waste\_gen\_rate} - \text{Collection\_rate}) * dt \end{aligned} \quad (12.2)$$

Nonrecyclable waste generation depends on the population, per capita income and GDP. The nonrecyclable waste generation rate, i.e. waste capacity, is computed from population, per capita waste generation and nonrecycling factor as

$$\text{Generation\_capacity} = \text{Population*Nonrecycling\_factor*Per\_capita\_waste\_gen} \quad (12.3)$$

The waste collection rate depends on collection capacity and generation capacity. IF generation capacity is greater than collection capacity THEN collection rate is collection capacity ELSE it is generation capacity, and it is expressed as

$$\begin{aligned} \text{Collection\_rate} = &\ \text{IF(Generation\_capacity} > \text{Collection\_capacity)} \\ &\ \text{THEN(Collection\_capacity)ELSE(Generation\_capacity)} \end{aligned} \quad (12.4)$$

Recyclable waste is increased by recyclable waste generation rate and decreased by recyclable collection rate, and it is expressed as

$$\begin{aligned} \text{Recyclable\_waste}(t) = &\ \text{Recyclable\_waste}(t-dt) \\ &+ (\text{Recyclable\_waste\_gen\_rate} \\ &- \text{Recyclable\_collection\_rate}) * dt \end{aligned} \quad (12.5)$$

The recyclable waste generation rate is computed from population, average per capita income and recyclable waste factor as

$$\begin{aligned}&\text{Recyclable\_waste\_gen\_rate}\\&= \text{Population*Per\_capita\_waste\_gen*Recyclable\_waste\_factor}\end{aligned} \tag{12.6}$$

The recyclable collection rate depends on recyclable waste, recyclable collection factor and separation multiplier from public concern, and it is computed as

$$\begin{aligned}&\text{Recyclable\_collection\_rate}\\&= \text{Recyclable\_waste*Recyclable\_collection\_factor*Separation\_multiplier\_from\_public\_concern}\end{aligned} \tag{12.7}$$

The portion of the uncleared nonrecycling wastes and recyclable wastes that are not collected is termed as uncleared wastes. Uncleaned and untreated wastes create nuisance which causes public concern. This concern is reduced by vented reaction. The public concern is expressed as

$$\begin{aligned}\text{Public\_concern}(t) &= \text{Public\_concern}(t - dt)\\&\quad + (\text{Concern\_rate} - \text{Vented\_reaction}) * dt\end{aligned} \tag{12.8}$$

Concern rate increases with composite index and it is non-linearly related with composite index. This relationship is expressed by a graphical function:

Concern_rate = GRAPH(Composite_index)
(0.00, 1.00), (25.0, 1.11), (50.0, 1.25), (75.0, 1.35), (100, 1.42), (125, 1.47), (150, 1.50)

Concern decreases with time and it is expressed as

$$\text{Vented\_reaction} = \text{Public\_concern/Concern\_dissipation\_time} \tag{12.9}$$

Public concern in turn creates public pressure to reduce the waste generation, and the effect of public pressure resulting from public concern is expressed as a multiplier. The public pressure multiplier is non-linearly related to public concern, and it is expressed as

Public_pressure_multiplier = GRAPH(Public_concern)
(1.00, 1.01), (2.50, 1.04), (4.00, 1.08), (5.50, 1.15), (7.00, 1.25), (8.50, 1.35), (10.0, 1.50)

Untreated waste is increased by collection rate and decreased by composting rate, disposal rate and incineration rate, and it is expressed as

$$\text{Untreated\_waste}(t) = \text{Untreated\_waste}(t - dt) + (\text{Collection\_rate} - \text{Composting\_rate} - \text{Disposal\_rate} - \text{incineration\_rate}) * dt \tag{12.10}$$

IF untreated waste is greater than composting capacity THEN composting rate is composting capacity ELSE it is untreated waste, and it is expressed as

$$\begin{aligned}\text{Composting\_rate} = {} & \text{IF}(\text{Untreated\_waste} > \text{Composting\_capacity}) \\ & \times\text{THEN}(\text{Composting\_capacity})\text{ELSE}(\text{Untreated\_waste})\end{aligned} \tag{12.11}$$

IF untreated waste plus incineration capacity is greater than composting capacity THEN disposal rate is the MIN of (i) (untreated waste plus incineration capacity >composting capacity) and (ii) (landfill capacity) ELSE it is zero, and it is expressed as

$$\begin{aligned}&\text{Disposal\_rate} \\ &= \text{IF}((\text{Untreated\_waste} + \text{Incineration\_capacity}) > \text{Composting\_capacity}) \\ &\text{THEN}(\text{MIN}((\text{Untreated\_waste} + \text{Incineration\_capacity} - \text{Composting\_capacity}), \\ &\qquad \text{Landfill\_capacity})) \times \text{ELSE}(0)\end{aligned} \tag{12.12}$$

IF untreated waste is greater than incineration capacity THEN incineration rate is incineration capacity ELSE it is untreated waste, and it is expressed as

$$\begin{aligned}\text{incineration\_rate} = {} & \text{IF}(\text{Untreated\_waste} > \text{Incineration\_capacity}) \\ & \text{THEN}(\text{Incineration\_capacity})\text{ELSE}(\text{Untreated\_waste})\end{aligned} \tag{12.13}$$

Both uncleared wastes and untreated wastes create nuisance and increase environmental concern. Composting index is an environmental indicator and it is defined by the following expression:

$$\begin{aligned}&\text{Composite\_index} \\ &= (\text{Weight1}*\text{Uncleared\_waste}/\text{Base\_value\_of\_uncleared\_waste} \\ &+ \text{Weight2}*\text{Untreated\_waste}/\text{Base\_value\_of\_untreated\_waste})*\text{Population\_ratio}\end{aligned} \tag{12.14}$$

Recyclable stock increases by recyclable collection rate and decreases by recycling rate. It is expressed as

$$\text{Recyclable\_stock}(t) = \text{Recyclable\_stock}(t - dt) + (\text{Recyclable\_collection\_rate} - \text{Recycling\_rate}) * dt \quad (12.15)$$

Recycling rate is a fraction of recyclable stock and it is expressed as

$$\text{Recycling\_rate} = \text{Recyclable\_stock*Recycling\_factor} \quad (12.16)$$

12.4.2 Waste Management: Composting, Incineration and Landfill

Composing capacity is increased by increasing capacity and decreased by depreciation and it is expressed as

$$\text{Composting\_capacity}(t) = \text{Composting\_capacity}(t - dt) + (\text{Increasing\_capacity} - \text{Depriciation\_rate}) * dt \quad (12.17)$$

Increasing capacity is computed from desired treatment capacity, existing composting and correction time, and it is expressed as

$$\text{Increasing\_capacity} = (\text{Desired\_treatment\_capacity} - \text{Composting\_capacity}) / \text{Correction\_time\_for\_treatment\_plant} \quad (12.18)$$

Depreciation rate is computed from composting capacity and its average life, and it is computed as

$$\text{Depriciation\_rate} = \text{Composting\_capacity}/\text{Average\_plant\_life} \quad (12.19)$$

Incineration capacity is increased by incineration capacity rate and decreased by depreciation, and it is expressed as

$$\text{Incineration\_capacity}(t) = \text{Incineration\_capacity}(t - dt) + (\text{Incineration\_capacity\_rate} - \text{Depreciation\_rate}) * dt \quad (12.20)$$

Incineration capacity rate is computed from desired incineration capacity, existing incineration capacity and correction time, and it is expressed as

$$\text{Incineration\_capacity\_rate} = (\text{Desired\_incineration\_capacity} - \text{Incineration\_capacity}) / \text{Correction\_time\_for\_inc} \quad (12.21)$$

Depreciation rate is computed from incineration capacity and its average life, and it is computed as

$$\text{Depreciation\_rate} = \text{Incineration\_capacity}/\text{Average\_inc\_life} \quad (12.22)$$

Landfill capacity is increased by landfill capacity rate and decreased by landfill rate, and it is expressed as

$$\begin{aligned}\text{Landfill\_capacity}(t) &= \text{Landfill\_capacity}(t-dt) \\ &+ (\text{Landfill\_capacity\_rate} - \text{Landfill\_rate}) * dt \quad (12.23)\end{aligned}$$

Landfill capacity rate is the landfill development and it is expressed as

$$\text{Landfill\_capacity\_rate} = \text{Landfill\_development} \quad (12.24)$$

IF Disposal is less than Landfill capacity THEN Landfill rate is Disposal ELSE it is Landfill capacity, and it is expressed as

$$\begin{aligned}\text{Landfill\_rate} = \; & \text{IF}(\text{Disposal} < \text{Landfill\_capacity})\text{THEN}(\text{Disposal}) \\ & \text{ELSE}(\text{Landfill\_capacity})\end{aligned} \quad (12.25)$$

The landfill development is non-linearly related to time and it is expressed as

Landfill_development = GRAPH(TIME)
(0.00, 175500), (3.00, 252000), (6.00, 366750), (9.00, 460250), (12.0, 558000), (15.0, 647250), (18.0, 732250), (21.0, 791750), (24.0, 855500), (27.0, 893750), (30.0, 915000)

The number of trucks for collection of wastes increases with increase rate and decreases by discard rate, and it is expressed as

$$\begin{aligned}\text{No\_of\_truck}(t) &= \text{No\_of\_truck}(t-dt) \\ &+ (\text{Truck\_increase\_rate} - \text{Truck\_discarded\_rate}) * dt \quad (12.26)\end{aligned}$$

The truck increase rate is computed from the desired number of trucks, number of trucks and correction time as

$$\begin{aligned}\text{Truck\_increase\_rate} = \; & (\text{Desired\_number\_of\_truck} - \text{No\_of\_truck}) \\ & /\text{Correction\_time\_of\_truck}\end{aligned} \quad (12.27)$$

The truck discard rate is calculated from number of trucks and average life of a truck:

$$\text{Truck\_discarded\_rate} = \text{No\_of\_truck}/\text{Avg\_life\_of\_truck} \quad (12.28)$$

The number of trucks for landfill increases with increase rate for landfill and decreases by discard rate for landfill, and it is expressed as

$$\begin{aligned}&\text{Trucks\_for\_landfill}(t) = \text{Trucks\_for\_landfill}(t - dt)\\&+ (\text{Truck\_increase\_rate\_for\_landfill} - \text{Truck\_discarding\_rate\_for\_landfill}) * dt\end{aligned} \tag{12.29}$$

The truck increase rate for landfill is computed from the desired number of trucks for landfill, number of trucks for landfill and adjustment time for landfill as

$$\begin{aligned}&\text{Truck\_increase\_rate\_for\_landfill}\\&= (\text{Desired\_number\_of\_trucks\_for\_landfill} - \text{Trucks\_for\_landfill})\\&/\text{Adjustment\_time\_for\_landfill}\end{aligned} \tag{12.30}$$

The truck discard rate for landfill is calculated from the number of trucks for landfill and average life of a truck:

$$\text{Truck\_discarding\_rate\_for\_landfill} = \text{Trucks\_for\_landfill}/\text{Average\_life\_of\_truck} \tag{12.31}$$

The total fund required is computed from the fund required for disposal, processing and collection as

$$\begin{aligned}\text{Total\_fund\_required} &= \text{Fund\_required\_for\_disposal}\\&+ \text{Fund\_required\_for\_processing}\\&+ \text{Total\_fund\_required\_for\_collection}\end{aligned} \tag{12.32}$$

The total fund required for collection is the sum of collection cost and cost of the truck, and it is expressed as

$$\begin{aligned}\text{Total\_fund\_required\_for\_collection} &= \text{Collection\_cost\_per\_year}\\&+ \text{Fund\_for\_truck}\end{aligned} \tag{12.33}$$

12.4.3 Waste Management: Energy Flow and Emissions

BOD total is the sum of BOD from energy and BOD from leachate, and it is expressed as

$$\text{BOD\_total} = \text{BOD\_from\_energy} + \text{BOD\_from\_leachate} \tag{12.34}$$

BOD from energy is the sum of BOD collection, BOD from the glass, BOD from the landfill, BOD from the metal, BOD from the paper and BOD from the plastic, and it is expressed as

$$\begin{aligned}\text{BOD\_from\_energy} = {} & \text{BOD\_Collection} + \text{BOD\_glass} + \text{BOD\_landfill} \\ & + \text{BOD\_metal} + \text{BOD\_paper} + \text{BOD\_plastic}\end{aligned} \tag{12.35}$$

where

$$\text{BOD\_Collection} = \text{Fuel\_consumption\_for\_collection*BOD\_for\_diesel} \tag{12.36}$$

$$\text{BOD\_glass} = \text{Glass*BOD\_factor\_glass} \tag{12.37}$$

$$\text{BOD\_landfill} = \text{Fuel\_consumption\_for\_landfill*BOD\_for\_diesel} \tag{12.38}$$

$$\text{BOD\_metal} = \text{metal*BOD\_factor\_metal} \tag{12.39}$$

$$\text{BOD\_paper} = \text{Paper*BOD\_factor\_paper} \tag{12.40}$$

$$\text{BOD\_plastic} = \text{Plastic*BOD\_factor\_plastic} \tag{12.41}$$

BOD from leachate is computed from leachate and BOD for leachate as

$$\text{BOD\_from\_leachate} = \text{Leachate\_production*BOD\_for\_leachate} \tag{12.42}$$

where

$$\begin{aligned}&\text{Leachate\_production} \\ &= \text{Leachate\_collection\_efficiency*Leachate\_generation\_fraction*Precipitation}\end{aligned} \tag{12.43}$$

CH_4 total is the sum of CH_4 from collection, CH_4 from the glass, CH_4 from the metal, CH_4 from the paper, CH_4 from the plastic and CH_4 from the landfill, and it is computed as

$$\begin{aligned}\text{CH}_4\text{\_total} = {} & \text{CH}_4\text{\_Collection} + \text{CH}_4\text{\_glass} + \text{CH}_4\text{\_metal} + \text{CH}_4\text{\_paper} \\ & + \text{CH}_4\text{\_plastic} + \text{CH}_4\text{\_landfill}\end{aligned} \tag{12.44}$$

where

$$\text{CH}_4\text{\_Collection} = \text{Fuel\_consumption\_for\_collection*CH}_4\text{\_for\_diesel} \tag{12.45}$$

$$\text{CH}_4\text{\_glass} = \text{Glass*CH}_4\text{\_factor\_glass} \tag{12.46}$$

$$\text{CH}_4\text{\_landfill} = \text{Fuel\_consumption\_for\_landfill*CH}_4\text{\_for\_diesel} \tag{12.47}$$

$$\text{CH}_4\text{\_metal} = \text{metal*CH}_4\text{\_factor\_metal} \tag{12.48}$$

$$\text{CH}_4\text{\_paper} = \text{Paper*CH}_4\text{\_factor\_paper} \tag{12.49}$$

$$CH_4\_plastic = Plastic{*}CH_4\_factor\_plastic \quad (12.50)$$

CO_2 total is the sum of CO_2 from collection, CO_2 from the glass, CO_2 from the landfill, CO_2 from the metal, CO_2 from the paper and CO_2 from the plastic, and it is expressed as

$$CO_2\_total = CO_2\_collection + CO_2\_glass + CO_2\_landfill + CO_2\_metal + CO_2\_paper + CO_2\_plastic \quad (12.51)$$

where

$$CO_2\_collection = Fuel\_consumption\_for\_collection{*}CO_2\_for\_diesel \quad (12.52)$$

$$CO_2\_glass = Glass{*}CO_2\_factor\_glass \quad (12.53)$$

$$CO_2\_landfill = Fuel\_consumption\_for\_landfill{*}CO_2\_for\_diesel \quad (12.54)$$

$$CO_2\_metal = metal{*}CO_2\_factor\_metal \quad (12.55)$$

$$CO_2\_paper = Paper{*}CO_2\_factor\_paper \quad (12.56)$$

$$CO_2\_plastic = Plastic{*}CO_2\_factor\_plastic \quad (12.57)$$

12.5 Model Validation

The various ways of validating a system dynamics model, such as comparing the model results with historical data, checking whether the model generates plausible behaviour and checking the quality of parameter values, were considered. Some of the parameters have been derived from studies in other areas and some were the results of expert guesswork. To judge the plausibility of the model, the behaviour of the key variables in the base run was examined.

12.6 Simulation and Policy Analysis

Computer projections of population, solid waste generation, uncleared waste and untreated waste for Dhaka City are shown in Fig. 12.3. Dhaka City had a population of 4.375 million in 1995 and approaching 12.082 million by 2025. The population growth rate of the city is higher than the average value of the whole country. This might be due to the fact that for job opportunities or other attractive factors, there is a rapid population inflow into the city. More population means more waste, and more waste means more resources needed for waste management. The waste generation increases from 1.027 million tons in 1995 to 4.257 million tons in 2025. An estimate of waste generation is crucially important to collection services and disposal facilities. It is clear from the figure that uncleared waste increases from

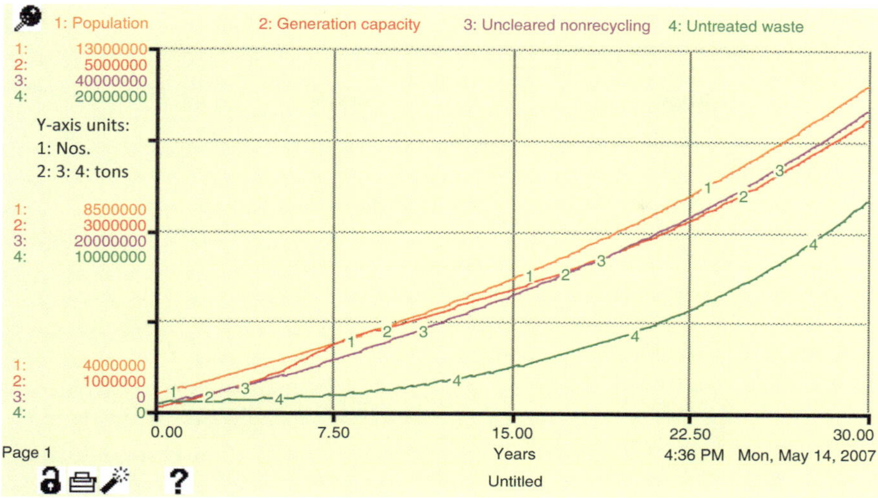

Fig. 12.3 Population, solid waste generation, uncleared waste and untreated waste of Dhaka City for a time horizon of 30 years

519,681 tons in 1995 to 33.183e + 006 tons in 2025, and untreated waste increases from 376,321 tons in 1995 to 11.65e + 006 tons in 2025. The uncleared waste of Dhaka City is increasing with time because of an inadequate collection capacity to transport the wastes to the dumpsites. Untreated waste is also increasing with time due to the lack of treatment facilities.

Figure 12.4 shows simulated incineration capacity, composting capacity, landfill capacity, electrical energy generation potential and composite index for a time horizon of 30 years. For a policy of introduction of incineration plant, composting and sanitary landfill, the landfill capacity increases from 150,000 tons in 1995 to 907,729 tons in 2025, incineration capacity increases from 150,000 tons in 1995 to 0.190e + 006 tons in 2025, corresponding electrical energy generation capacity increases from 66,750 MWh in 1995 to 84,556 MWh in 2025 and composting plant capacity increases up to 15 years, and then it becomes almost constant. The electrical energy recovery from urban solid waste generation of Dhaka City can supply a significant portion of the consumption requirement of electrical energy of the city. Hence, adoption of the policy for electricity from urban solid waste of Dhaka City should be dictated by the economy of adoption of the technology of electricity generation from the solid waste and environmental implications. The composite index increases from 0.86 in 1995 to 101.47 in 2025. The rapid increase in composite index with time means that the quality of the environment is deteriorating rapidly with time. In the early period (0–7.5 years), the composite index is very low with a very small growth rate, while in the later period, it increases very rapidly. The rapid increase in the composite index with time means that the quality of the environment is deteriorating rapidly with time.

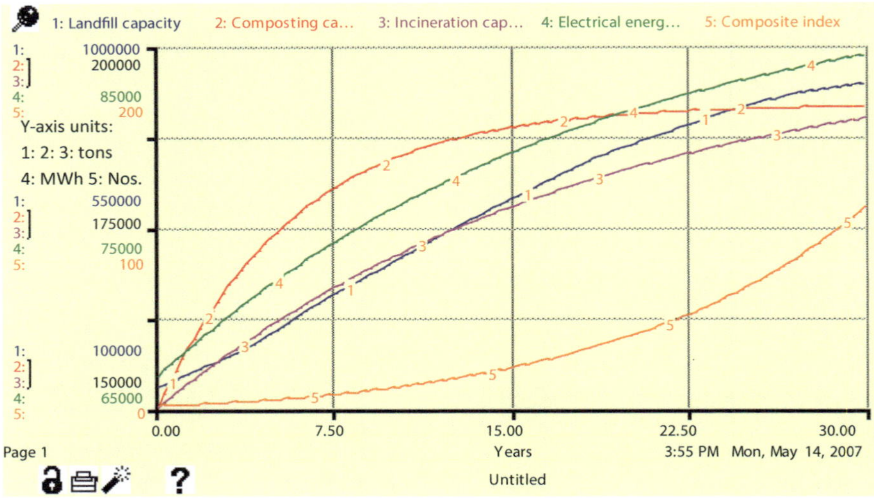

Fig. 12.4 Simulated landfill capacity, composting capacity, incineration capacity and electricity generation potential from solid waste and composite index for a time horizon of 30 years

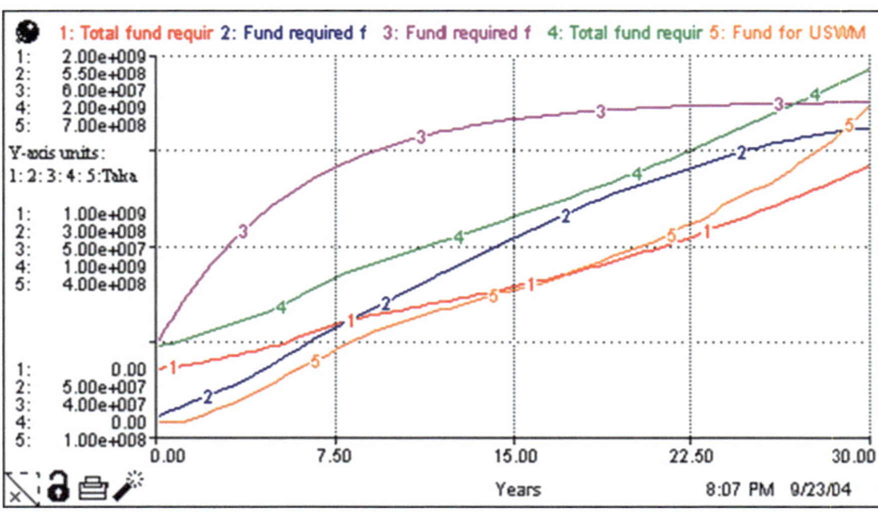

Fig. 12.5 Simulated total fund required for collection, fund required for disposal, fund required for processing, total fund required and fund for USWM for a time horizon of 30 years (one US $ = Taka 70.00)

Thus, solid waste management system is sustainable in the short run and becomes unsustainable in the long run.

Figure 12.5 shows the simulated total funds required for collection, landfill and treatment, total fund required and fund for USWM for a time horizon of 30 years.

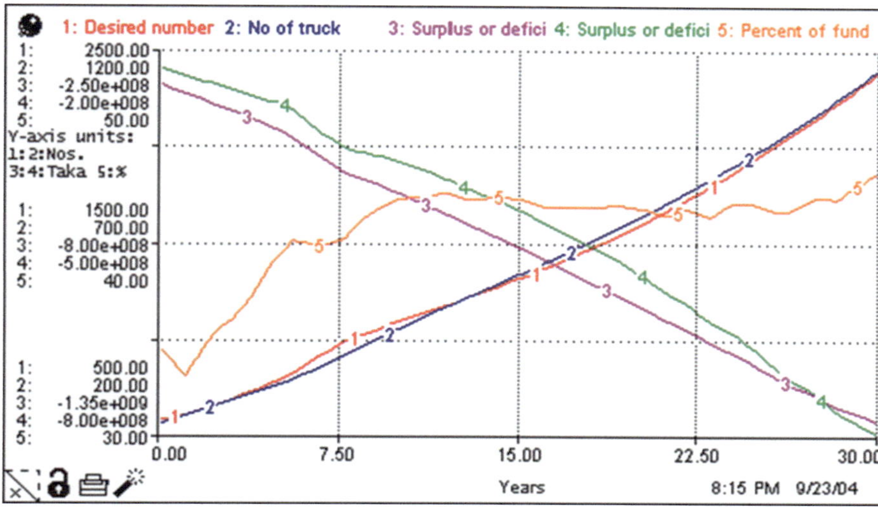

Fig. 12.6 Simulated desired number of trucks, number of trucks used, surplus or deficit budget, surplus or deficit budget for collection and percent of fund available for collection for a time horizon of 30 years

The simulated total fund required for collection, total fund required and fund for USWM are increasing with time. The fund required for collection increases from Tk 3.46e + 008 in 1995 to Tk 14.15e + 008 in 2025 since the generation of solid waste increases. The fund required for disposal also increases from Tk 7.50e + 007 in 1995 to Tk 45.38e + 007 in 2025. The fund required for waste treatment is increasing up to 15 years. After 15 years, the fund required for treatment is more or less constant at a value of Tk 5.70e + 007. This is because of the fact that the capacity utilisation of the treatment plant is fulfilled. The fund required for landfill is also increasing until towards the end of the simulated period. These are due to the facts that treatment capacity and landfill capacity are approaching the proposed desired treatment capacity and proposed desired landfill capacity, respectively. Logically, the total fund required increases from Tk 4.660e + 008 in 1995 to Tk 19.27e + 008 in 2005, and the total fund required for solid waste management increases from Tk 1.18e + 008 in 1995 to Tk 6.16e + 008 in 2005. Thus, increased funds are needed for both collection and solid waste management.

Figure 12.6 shows the simulated desired number of trucks, number of trucks used, surplus or deficit total budget, surplus or deficit budget for collection and percent of fund available for collection for a time horizon of 30 years. The desired number of trucks increases from 575 in 1995 to 2384 in 2025, whereas the number of trucks used increases from 230 in 1995 to 1148 in 2025. There is always a gap between the desired number of trucks and number of trucks used. Thus, the collection service at Dhaka City is deteriorating rather than improving. Moreover, the transportation of the waste to the dumpsite is not properly managed. Wastes are seen flying from the trucks during transport. Since the population and wastes generated are increasing with time, the desired number of trucks is also increasing. The number of trucks never equals the desired number of trucks, since the policy was to reduce the shortage in number of trucks which is dynamic. This indicates that more funds are required to mitigate the shortage of trucks and to meet the collection cost. The patterns of the change of deficit of the budget and the budget for collection decrease with time, but the percent of fund available for collection increases from almost 34 % in 1995 to 39 % in 2000 and then gradually increases to 43.5 % in 2005 and then almost remains constant. Thus, there is high shortage of fund for collection particularly for trucks for collection.

Figure 12.7 shows the simulated percent of fund increase required for only total waste collection and total USWM for a time horizon of 30 years. Initially the required budget for collection is 291 %. The budget requirement for collection decreases from 291 to 254 % sharply within 5 years; then it gradually decreases to a constant value of 238 % within 10 years and it continues up to 25 years. After 25 years, it decreases gradually. But for total urban solid waste management, the required budget for total waste management is 415 %. Then, the budget requirement decreases gradually from 415 % to an almost constant value of 340 % within 10 years. The initial jumps of the budgets for total waste collection and total urban solid waste management are due to the introduction of treatment plant and landfills for solid waste disposal.

Figure 12.8 shows simulated composts, i.e. nitrogen, potassium and phosphorus, and composite index. Composts available from solid waste increase from 0.150e+006 tons in 1995 to 0.191e+006 tons in 2025. Nitrogen, potassium and phosphorus increase from 3600 tons in 1995 to 4602 tons in 2025, from 2550 tons in 1995 to 3260 tons in 2025 and from 1800 tons in 1995 to 2301 tons in 2025, respectively,

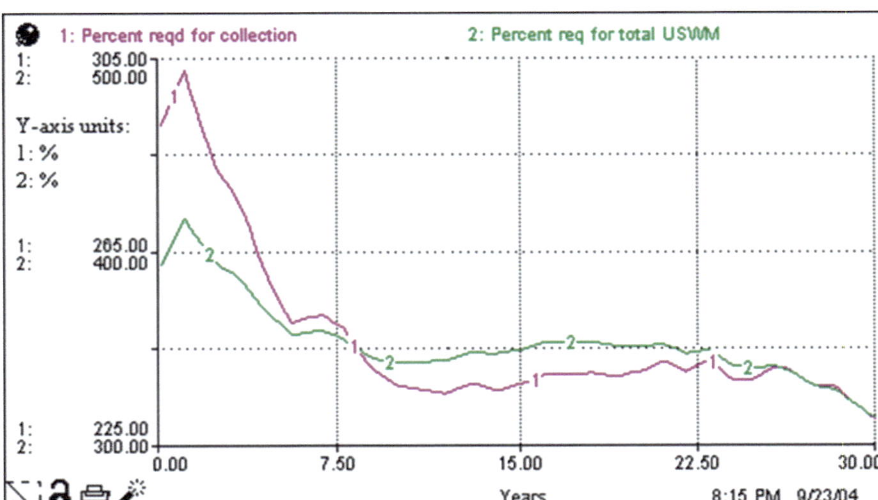

Fig. 12.7 Simulated percent of fund increased required for only total waste collection and total USWM for a time horizon of 30 years

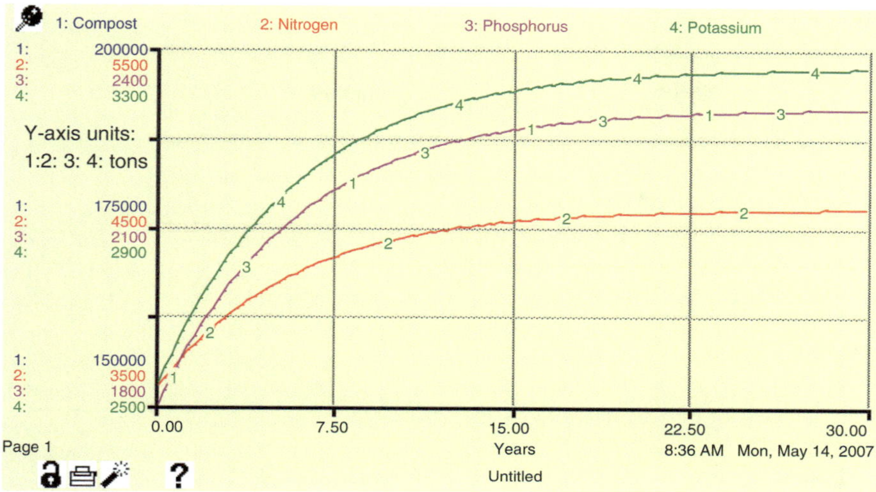

Fig. 12.8 Simulated compost, nitrogen, phosphorus and potassium for a time horizon of 30 years

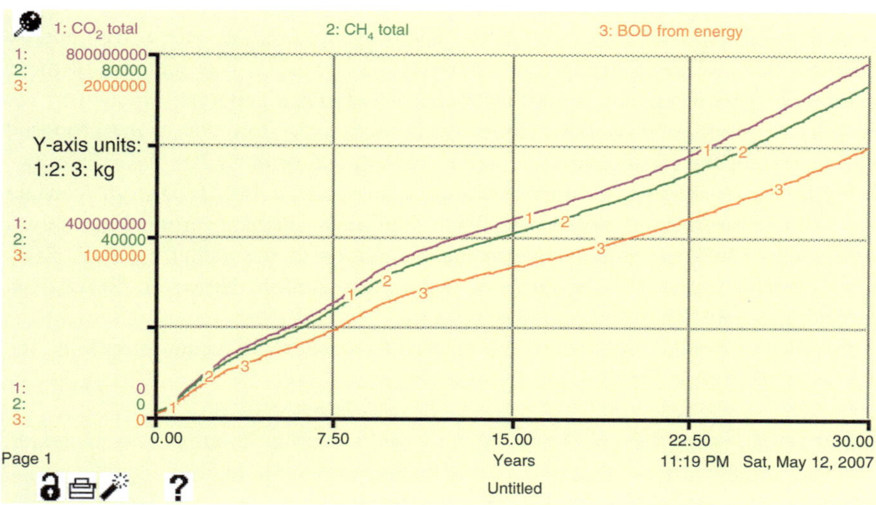

Fig. 12.9 Simulated CO_2 total, CH_4 total and BOD from energy for a time horizon of 30 years

and follow the production pattern of the composts. There is a significant contribution of organic fertilisers from the compost from composting of solid waste. This compost is a good-quality organic fertiliser for garden vegetables and other crops, and it is approved by the Bangladesh Agricultural Research Council (BARC).

Figure 12.9 shows simulated CO_2 total, CH_4 and BOD from energy. Simulated CO_2 total increases from 5.153e + 006 kg in 1995 to 78.112e + 007 kg in 2025. CH_4

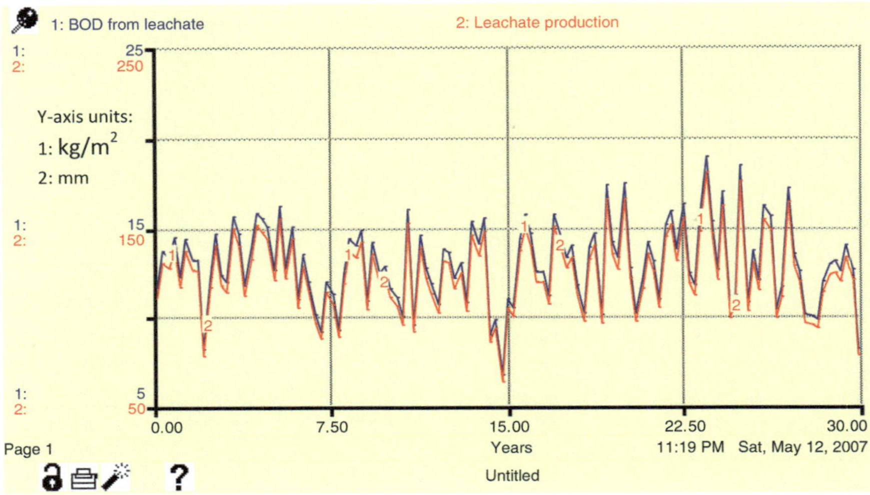

Fig. 12.10 Simulated leachate production (mm) and BOD (kg/m^2) from leachate for a time horizon of 30 years

total increases from 412 kg in 1995 to 72,884 kg in 2005 and BOD from energy increases from 8136 kg in 1995 to 1.48e + 006 kg in 2025. The increase of these emissions with time is due to the increase in solid waste generated. However, the intensity of production of these emissions is moderately low due to the proposed introduction of the composting, incineration plant and sanitary landfill.

Figure 12.10 shows the simulated leachate production and BOD from leachate for a time horizon of 30 years. The simulated leachate production shows fluctuations which are mainly due to the fluctuation in the annual rainfall. BOD from leachate follows the pattern of leachate production. However, BOD from leachate lies between 100 mm and 150 mm.

In order to obtain insight into the effect of the alternative policy options, the following two policy options are considered:

Policy 1: Increasing the collection capacity and assessing its impact on uncleared waste, untreated waste, number of trucks and composite index.

Policy 2: Increasing the collection capacity, treatment capacity and landfill capacity and assessing its impact on uncleared waste, untreated waste, number of trucks and composite index.

Policy 1: Figure 12.11 shows the simulated uncleared waste, untreated waste, number of truck and composite index for increase in collection capacity for a time horizon of 30 years. From the figure it is observed that if we increase the collection capacity by doubling the truck increase rate, the uncleared waste decreases and untreated waste increases, but the composite index remains unchanged as compared with the base scenarios discussed earlier. This means

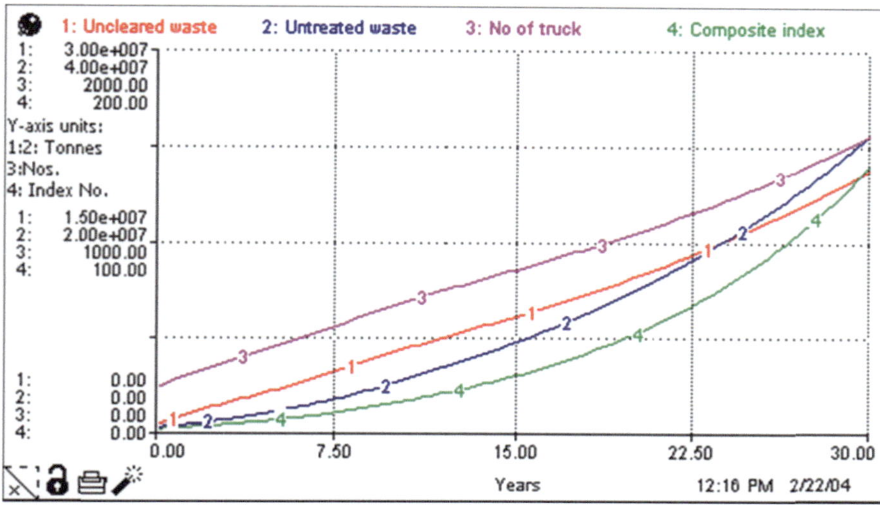

Fig. 12.11 Simulated uncleared waste, untreated waste, number of trucks and composite index with increase in collection capacity for a time horizon of 30 years

that increasing collection capacity alone does not improve the environmental quality because composite index is the indicator of environmental quality.

Policy 2: Figure 12.12 shows the simulated uncleared waste, untreated waste, number of trucks used and composite index with increase in collection capacity, treatment capacity and landfill capacity for a time horizon of 30 years. From Fig. 12.12, it is observed that if we increase collection capacity, treatment capacity and landfill capacity by doubling the truck increase rate, treatment capacity increase rate and landfill capacity increase rate, respectively, the uncleared waste decreases in a similar fashion as in policy 1, but the untreated waste and the composite index also decrease as compared to policy 1. Increased composite index is the sign of environmental quality deterioration, and decreased composite index is the sign of environmental quality improvement. This implies that the increased budget allocation for both clearing and treating the wastes is essential for improving the environmental quality of Dhaka City.

Energy from the waste as well as from the incineration reduces greenhouse gas emission. An analysis of CO_2 equivalent emission per kWh of electricity produced by energy from waste showed that global warming potential of emission from waste is less than coal, fuel and even natural gas.

Finally, in the short run the proposed policy is for sustainable development. But in the long run, it appears to be a dream unless population control within the carrying capacity of the city is achieved, and the treatment plants with energy and material recovery are realised.

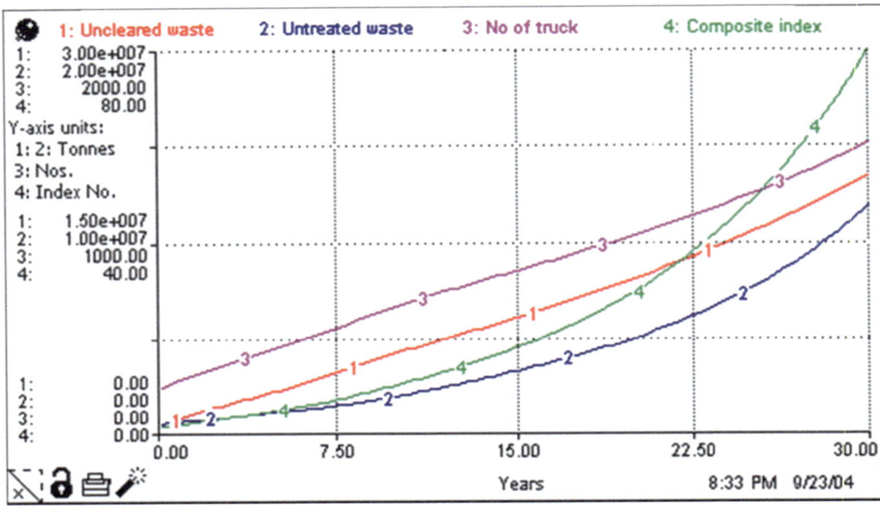

Fig. 12.12 Simulated uncleared waste, untreated waste, number of trucks and composite index with increased collection capacity, treatment capacity and landfill capacity for a time horizon of 30 years

12.7 Conclusions

Population, solid waste generation and electrical energy generation possibility from the solid waste for Dhaka City are increasing with time. The electrical energy recovery from urban solid waste generation of Dhaka City can supply a significant portion of the consumption requirement of electrical energy of the city. Adoption of the policy for electricity from urban solid waste should be dictated by the economy of adoption of the technology of electricity generation from the waste and environmental implications. There exists a potential for organic fertilisers for composting of solid waste, and this compost is a good-quality organic fertiliser for garden vegetables. A policy of introduction of incineration, composting and sanitary landfill is assessed. The intensity of production of the emissions is moderately low due to the proposed introduction of the composting, incineration plant and sanitary landfill. The proposed system is sustainable in the short run but becomes unsustainable in the long run. This model can be used as a tool to assess and design policy options.

Exercises

Exercise 12.1 What is meant by solid waste? Why modelling and simulation of solid waste management systems is important?

Exercise 12.2 Draw the causal loop diagrams of solid waste management systems consisting of solid waste generation and disposal in the form of limited landfill, composting and open dumping.

Exercise 12.3 Draw stock–flow diagram of solid waste management systems consisting of solid waste generation and disposal in the form of limited landfill, incineration, composting and open dumping and simulate the model for solid waste generation, landfill, composting and open dumping.

Exercise 12.4 Simulate the model in Exercise 12.3 for increasing collection and treatment capacity and also analyse the simulated results.

Exercise 12.5 Simulate the model including incineration in the model in Exercise 12.3 and assess the contribution of CO_2, CH_4 and BOD from the solid waste management system.

References

Alam MJ, Bole B (2001) Energy recovery form municipal solid waste in Dhaka city. In: Proceedings of the international conference on Mechanical Engineering, 26–28 December 2001, Dhaka, pp I 125–130

Bala BK, Sufian MA (2006) A. Modeling of urban solid waste management: the case of Dhaka city. J Agric Mach Bioresour Eng 4(1&2):17–25

Barsi HB (2000) An expert system for landfill leachate management. Environ Technol 21 (2):157–166

Clayton KC, McCarl BA (1979) Management of solid waste systems including non-metropolitan areas, with emphasis on resource recovery. North Cent J Agric Econ 1(1):61–72

Consonni S, Guigliano M, Grosso M (2005a) Alternative strategies for energy recovery from municipal solid waste Part A: mass and energy balances. Waste Manag 25:123–135

Consonni S, Guigliano M, Grosso M (2005b) Alternative strategies for energy recovery from municipal solid waste Part B: emission and cost estimates. Waste Manag 25(2005b):137–148

Dyson B, Chang NB (2005) Forecasting municipal solid waste generation in a fast growing urban region with system dynamics modeling. Waste Manag 25(7):669–679

Everett JW, Modak AR (1996) Optimal regional scheduling of solid waste systems. J Environ Eng 122(9):785–792

Heikki TJ (2000) Strategic planning of municipal solid waste management. Resour Conserv Recycl 30(2):111–133

JICA (2005) The study on solid waste management in Dhaka city, clean Dhaka master plan, final report, Japan International Agency, Pacific Consultants International, Yachiyo Engineering Co., Ltd.

Kum V, Sharp A, Harnpornchai N (2005) Improving the solid waste management in Phnom Penh city: a strategic approach. Waste Manag 25:101–109

Mashayekhi AN (1992) Transition in the New York state solid waste system: a dynamic analysis. Syst Dyn Rev 9(1):23–47

Ming ZG, Zhong YX, Yue ZP, Cheg GH, He HGG, Hemelaar L (2000) Environmental input – output model and its analysis with a focus on the waste management sectors. J Environ Sci 12 (2):178–183

Pawan S, Sikka P, Maheshwari RC, Chaturvedi P (1997) Management of municipal solid waste. Bio energy for rural energization. Department of Science and Technology. IIT, New Delhi, India, pp 205–209

Rao CS (1992) Environmental pollution control engineering and sanitation. Willey Eastern Limited, New Delhi, pp 396–414

Salvato JA (1992) Solid waste management. Environmental engineering and sanitation, 4th edn. Willey, Hoboken, pp 662–766
Sterman JD (2000) Business dynamics: systems thinking and modeling for a complex world. Irwin McGraw Hill, New York
Sudhir V, Srinivasan G, Muraleedharan VR (1997) Planning for sustainable solid waste management in urban India. Syst Dyn Rev 13(3):223–246
Sufian MA, Bala BK (2006) Modeling of electrical energy recovery from urban solid waste system: the case of Dhaka city. Renew Energy 31(10):1573–1580
Sufian MA, Bala BK (2007) Modeling of urban solid waste management: the case of Dhaka city. Waste Manag 27:858–868
Velumani A, Meenakshi P (2007) Life cycle inventory analysis of emissions from existing municipal solid wastes management scenarios in coimbatore city. In: Proceedings of the international conference on Sustainable Solid Waste Management held on 5–7 September 2007, Chennai, India, pp 75–82
Yongfeng NIE (2005) Development and prospects of municipal solid waste (MSW) incineration in China. Front Environ Sci Enfin China 2(1):1–7
Zurbrugg C, Drescher S, Sinha M, Enaetullah I (2002) Decentralized composting in Dhaka, Bangladesh: production of compost and its marketing. In: Proceedings of ISWA 2002 Annual Congress, Istanbul, Turkey, 8–12 July 2002

Bibliography

Bala BK (1998) Energy and environment: modeling and simulation. Nova Science Publisher, New York
Bala BK (1999) Principles of system dynamics. Agrotech Publishing Academy, Udaipur
Cavana RY, Maani KE (2000) A methodological framework for systems thinking and modelling (ST&M) interventions. Ist international conference on systems thinking in management. Internet
Diaz R, Warith M (2005) Life cycle assessment of municipal solid wastes: development of the WASTED model. Waste Manag 25(7):669–679
Forrester JW (1968) Principles of systems. Wright-Allen Press/MIT, Cambridge, MA

Index

B.K. Bala et al., *System Dynamics*, Springer Texts in Business and Economics,
DOI 10.1007/978-981-10-2045-2

MIX
Papier aus verantwortungsvollen Quellen
Paper from responsible sources
FSC® C105338

If you have any concerns about our products,
you can contact us on
ProductSafety@springernature.com

In case Publisher is established outside the EU,
the EU authorized representative is:
Springer Nature Customer Service Center GmbH
Europaplatz 3, 69115 Heidelberg, Germany

Printed by Libri Plureos GmbH
in Hamburg, Germany

Lecture Notes in Civil Engineering

Volume 15

Lecture Notes in Civil Engineering (LNCE) publishes the latest developments in Civil Engineering - quickly, informally and in top quality. Though original research reported in proceedings and post-proceedings represents the core of LNCE, edited volumes of exceptionally high quality and interest may also be considered for publication. Volumes published in LNCE embrace all aspects and subfields of, as well as new challenges in, Civil Engineering. Topics in the series include:

- Construction and Structural Mechanics
- Building Materials
- Concrete, Steel and Timber Structures
- Geotechnical Engineering
- Earthquake Engineering
- Coastal Engineering
- Hydraulics, Hydrology and Water Resources Engineering
- Environmental Engineering and Sustainability
- Structural Health and Monitoring
- Surveying and Geographical Information Systems
- Heating, Ventilation and Air Conditioning (HVAC)
- Transportation and Traffic
- Risk Analysis
- Safety and Security

To submit a proposal or request further information, please contact the appropriate Springer Editor:

- Mr. Pierpaolo Riva at pierpaolo.riva@springer.com (Europe and Americas);
- Ms. Swati Meherishi at swati.meherishi@springer.com (India);
- Ms. Li Shen at li.shen@springer.com (China);
- Dr. Loyola D'Silva at loyola.dsilva@springer.com (Southeast Asia and Australia/NZ).

More information about this series at http://www.springer.com/series/15087

Boominathan Adimoolam
Subhadeep Banerjee
Editors

Soil Dynamics and Earthquake Geotechnical Engineering

IGC 2016 Volume 3

Springer

Editors
Boominathan Adimoolam
Department of Civil Engineering
Indian Institute of Technology Madras
Chennai, Tamil Nadu
India

Subhadeep Banerjee
Department of Civil Engineering
Indian Institute of Technology Madras
Chennai, Tamil Nadu
India

ISSN 2366-2557 ISSN 2366-2565 (electronic)
Lecture Notes in Civil Engineering
ISBN 978-981-13-0561-0 ISBN 978-981-13-0562-7 (eBook)
https://doi.org/10.1007/978-981-13-0562-7

Library of Congress Control Number: 2018941216

Printed on acid-free paper

This Springer imprint is published by the registered company Springer Nature Singapore Pte Ltd.
part of Springer Nature
The registered company address is: 152 Beach Road, #21-01/04 Gateway East, Singapore 189721, Singapore

Foreword

The Indian Geotechnical Society (IGS) was started as the Indian National Society of Soil Mechanics and Foundation Engineering in the year 1948, soon after the Second International Conference on Soil Mechanics and Foundation Engineering held at Rotterdam. The Society was affiliated to the International Society in the same year, and since then, it has strived to fulfil and promote the objectives of the International Society. In December 1970, the name 'Indian Geotechnical Society (IGS)' was adopted.

The Society conducted several symposia and workshops in different parts of India since its inception in 1948. In the year of 1983, the Indian Geotechnical Society organized its first annual conference IGC 1983 in Indian Institute of Technology Madras.

Several local chapters of the Society were established over the years, and gradually, the annual conferences were held in different cities under the leadership of the respective local chapters. The conferences were well utilized as the venue for showcasing the research works and the case studies on geotechnical engineering and geoenvironmental engineering, and the papers presented during the deliberations were being published as conference proceedings volume.

The annual conferences IGC 1996 and IGC 2006 were organized by IGS Chennai Chapter, and the papers were published as printed volumes. The practice of printing the volumes was later discontinued in view of large expenditure involved and also because of the easy access to the electronic storage media. The papers then became not accessible to the rest of the geotechnical community.

The responsibility of organizing the annual conference of 2016 was taken up by IGS Chennai Chapter, and the conference was held during 15–17 December 2016. It was felt necessary to publish selected papers for the benefit of the entire geotechnical engineering community as well as of the contributing authors. The Chapter approached Springer with a request to take up this responsibility, and with the great help of Springer, about 150 selected papers are being published in four volumes under the series 'Lecture Notes in Civil Engineering'. The selected papers are distributed among these volumes depending upon the topic of discussion.

Volume 1: Geotechnical Characterisation and Geoenvironmental Engineering
Volume 2: Ground Improvement Techniques and Geosynthetics
Volume 3: Soil Dynamics and Earthquake Geotechnical Engineering
Volume 4: Geotechnical Applications

I am thankful to all the general editors and the reviewers of these papers who provided valuable review comments for the betterment of the papers published in these volumes. The support provided by the former President of the Indian Geotechnical Society, the late Prof. Sreerama Rao, and the former Honorary Secretary of IGS, Shri Jai Bhagwan, is gratefully acknowledged here.

Bringing out these volumes under the banner of prestigious publishers, Springer will immensely improve the visibility of these conference proceedings volumes. The Chennai Chapter of the Indian Geotechnical Society places on record its acknowledgement of the efforts put forth by Springer for bringing out these volumes for the benefit of the geotechnical engineering community.

Chennai, India

Prof. Boominathan Adimoolam
Chairman, Indian Geotechnical
Conference 2016
Department of Civil Engineering IIT Madras

Preface to Volume 3

There are about 29 papers in this volume covering various aspects of *Soil Dynamics and Earthquake Geotechnical Engineering*. India has a wide variety of natural features and is under a constant risk of being hit by disasters involving dynamic loadings such as earthquakes, winds, man-made blasting, waves, traffic loadings. Recent events like the Sikkim earthquake in 2011 and the Nepal earthquake in 2015 had catastrophic effects on the region, resulting in a high death toll and loss of public property. India has seen a booming development in infrastructure in the recent years. Though measures have been taken to strike a balance between nature and development, unavoidably some disasters have been accelerated by these developments. Moreover, in India, there are numerous types of geological and geotechnical conditions encountered. For example, most of the major cities in India are concentrated near or along the coastline. The coastal plains of India consist of predominantly marine clays ranging in thickness of 2–10 m. As a result, many important inland and offshore structures such as bridges, port and harbours, tall structures like liquid storage tanks and chimneys in those areas are built overlying soft to medium stiff marine clay. In such situations, the behaviour of those structural systems under earthquake loading is an important factor affecting the integrity of infrastructures. On the other hand, the entire Indo-Gangetic plain comprises of fine silts and sands which are extremely vulnerable against liquefaction. All these reasons warrant the relevance of the theme of the present volume which includes a wide range of studies on seismic response of dams, foundation–soil systems, natural and man-made slopes, reinforced earth walls, base isolation systems. Moreover, there are papers reporting some of the allied issues such as landslide risk assessments, liquefaction mitigation, dynamic analysis of mechanized tunnelling, advanced seismic soil–structure interaction analysis. In summary, the present volume covers length and breadth of the area of *Soil Dynamics and Earthquake Geotechnical Engineering*.

We thank the technical committee of IGC 2016 for giving us the responsibility to be editors for this volume of the Springer Proceedings.

Chennai, India

Dr. Boominathan Adimoolam
Dr. Subhadeep Banerjee
General Editors

Contents

About the Editors

Prof. Boominathan Adimoolam received his Ph.D. in soil dynamics from Moscow Civil Engineering Institute in 1986. He worked as a CSIR Scientific Pool Officer at the Ocean Engineering Centre, Indian Institute of Technology (IIT) Madras, in 1987. Since 1988, he has been a faculty member of the Geotechnical Engineering Division at the Indian Institute of Technology Madras. His fields of specialization include soil dynamics and earthquake engineering, foundation engineering and ground improvement. He has long since been actively involved in the area of site-specific seismic hazard assessment, which includes seismological, geophysical and geotechnical considerations. He has completed nine sponsored research projects worth roughly Rs. 1.5 crores and is currently coordinating major research projects funded by the Ministry of Earth Sciences of Government of India and the Nuclear Power Corporation of India (NPCIL), Mumbai, in the area of seismic soil–structure interaction and base isolation using geoisolators. He has authored and co-authored over 130 scientific papers and serves as an editorial board member of Journal of Earthquake and Tsunami. He was part of a working group of experts in the Geohazards programme of the National Disaster Management Authority (NDMA), Government of India. He received the IGS-Shri M. S. Jain Biennial Prize in 2004 and IGS-Prof. C. S. Desai Biannual Prize in 2012.

Dr. Subhadeep Banerjee is an associate professor of geotechnical engineering in the Department of Civil Engineering, Indian Institute of Technology Madras. He received his Ph.D. in civil engineering from the National University of Singapore. His research interests include cyclic behaviour of geomaterials, seismic soil–structure interactions and finite element analysis of complex dynamic problems. In addition, he is an expert for constitutive modelling of soils, with a special focus on transient dynamic behaviour. At the IIT Madras, he has formed three research groups to study the following attributes: (i) cyclic characterization of various types of natural soils, improved ground, landfills, etc., (ii) development of constitutive models of soils, with an emphasis on their performance under seismic loadings and (iii) application of fundamentals of dynamics to various complex soil–structure problems such as tunnelling, base isolation techniques, plate anchors, retaining walls.

Part I
Soil Dynamics and Earthquake Geotechnical Engineering

Effect of Dynamic Loading on Properties of Saturated Sand

D. Devi Priyanka, Deepak Rathi and G. K. Kumar

Abstract Liquefaction has become one of the most important and interesting topics in the field of geotechnical earthquake engineering. Also, its study became a part and parcel of seismic microzonation considering its severity. The devastation brought by it due to earthquakes of Niigata (Japan), Bhuj (India), Nepal has turned the spotlight on the seriousness with which this hazard has to be considered. So, in the present study, a comprehensive experimental program was undertaken in which a series of strain-controlled cyclic triaxial tests were performed to evaluate the liquefaction resistance of local sand in Warangal. Tests were carried out at relative densities, i.e., 35, 50, and 65%, and reduction in pore pressure with increase in relative density was observed. Similarly, confining pressures of 100 and 150 kPa for amplitudes ranging from 0.08 to 1% were also considered. From the results, it was also observed that higher confining pressures resulted in decrease in pore pressure whereas higher amplitude of loading resulted in a rise in pore pressure. This proved that densification or increasing confining pressure improves resistance against liquefaction which can be considered as the cheapest soil improvement technique to mitigate liquefaction hazards in unconstrained areas.

Keywords Liquefaction · Cyclic triaxial test · Strain-controlled
Pore pressure ratio

D. Devi Priyanka (✉)
Department of Civil Engineering, Indian Institute of Technology Hyderabad, Hyderabad, India
e-mail: devipriyanka422@gmail.com

D. Rathi · G. K. Kumar
Department of Civil Engineering, National Institute of Technology, Warangal, India
e-mail: deepakrathi.mmec@gmail.com

G. K. Kumar
e-mail: kalyan@nitw.ac.in

B. Adimoolam and S. Banerjee (eds.), *Soil Dynamics and Earthquake Geotechnical Engineering*, Lecture Notes in Civil Engineering 15,
https://doi.org/10.1007/978-981-13-0562-7_1

1 Introduction

Construction of structures like buildings, foundations on loose granular soils (like sands) has become a tough task in earthquake-prone areas. This is due to the fact that behavior of soil under static loading is different from dynamic loading. Several areas in India such as Kashmir, Himalayas, Bihar, Northeast region are frequently prone to earthquakes. Moreover, the recent earthquakes like Bhuj, 2001 and Nepal earthquake (2015) which were the most destructive earthquakes in India in the last 50 years had caused huge damage. The reconnaissance carried by many people showed that those earthquakes were characterized by widespread liquefaction that caused sand volcanoes, ground cracking, lateral spreading, etc. Not only these, there are many other cases also in the past where severe havoc was seen due to induced liquefaction during earthquakes (Seed and Idriss 1967; Sitharam et al. 2004; Boulanger et al. 1997). This resulted in evaluation of liquefaction potential by several researchers either by stress-based (Seed and Idriss 1971; Seed 1979) or strain-based (Sitharam et al. 2012) approaches. The advantage in adopting a strain-based approach is the total or partial replacement of parameters influencing soil properties like method of sample preparation, lateral earth pressure coefficient by the shear modulus in calculation of shear strains (Dobry et al. 1982). A field measurement of shear modulus (G_{max}) automatically incorporates many of the soil characteristics which are necessary for pore pressure buildup during earthquakes. Thus, this approach decreases the need for a detailed knowledge of the site which is essential in the stress-based approach. There are many ways to mitigate liquefaction of which densification is the cheapest technique (Menard and Broise 1975; Lyman 1942). So, in the present work, an attempt is made in the laboratory to study the liquefaction potential of locally available sand in Warangal and determine the effect of parameters like amplitude of loading, confining pressure, and relative density on dynamic soil behavior.

2 Experimental Program

2.1 Soil Sampling

The sand used in this study was collected from the surroundings of Warangal, Telangana, India. Tests were performed as per IS code to obtain the basic index properties as listed in Table 1. The gradation curve of the soil is shown in Fig. 1. It is clear from the properties that sand under consideration is poorly graded.

Table 1 Index properties of sand

Property	Value
Specific gravity	2.63
Maximum void ratio	0.98
Minimum void ratio	0.62
D10 (mm)	0.2
D30 (mm)	0.32
D60 (mm)	0.43
Coefficient of uniformity (Cu)	2.15
Coefficient of curvature (Cc)	1.19
Soil classification	SP

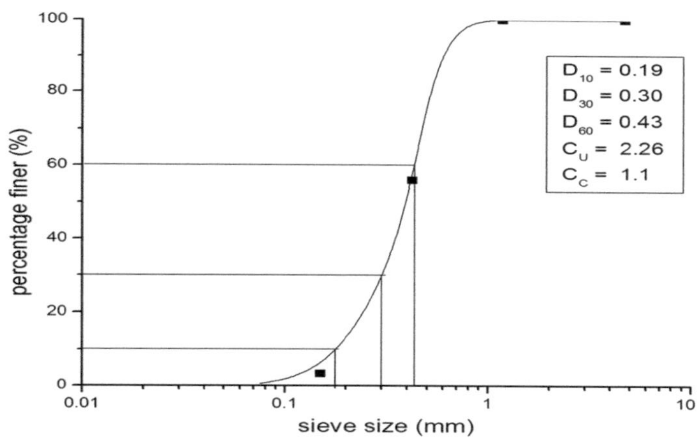

Fig. 1 Grain size distribution of sand used

2.2 *Test Methodology*

Specimens of 7.5 cm diameter and 15 cm height were reconstituted in the laboratory as explained below, and a series of strain-controlled tests were carried as listed in Table 2 for evaluation of liquefaction potential and also to determine effect of various parameters. Basically, each test comprises of three segments, i.e., saturation followed by sample preparation, consolidation, and testing stage.

2.2.1 Saturation

A calculated amount of clean dry soil was taken based on considered relative density for the preparation of sample using water sedimentation technique (Ishihara 1996).

Table 2 List of tests performed

Test no.	Dr (%)	Strain (%)	B	Eff. confining pressure (kPa)
1	50	0.02	0.95	100
2	50	0.06	0.96	100
3	50	0.20	0.95	100
4	50	0.50	0.95	100
5	50	1.00	0.96	100
6	50	0.06	0.96	150
7	50	0.20	0.94	150
8	50	0.50	0.95	150
9	50	1.00	0.95	150
10	35	0.50	0.95	100
11	65	0.50	0.96	100

Later, drainage lines are connected and saturation was carried with the aid of back pressure by maintaining initial net effective stress on the specimen. Full saturation was assumed to be achieved for B values (= $\Delta u/\Delta\sigma_c$) of 0.96 or higher. After attaining it, consolidation stage was initiated. Time taken for this saturation depends on degree of denseness of sample which increases with density of sample.

2.2.2 Consolidation

During consolidation, initially both confining and back pressure valves were closed and test pressures were given maintaining required net effective stress. Later, volume change was allowed which initiates the process of consolidation and is recorded with time. A constant line in volume change versus time graph after some time indicates the completion of consolidation stage.

2.2.3 Cyclic Loading (ASTM D5311)

After consolidation, test was carried at certain frequency (f), amplitude for a given number of cycles (N). All the tests were carried at a frequency 1 Hz as suggested by ASTM D5311. As test is being carried under strain approach, amplitude of loading is given in mm. Test was stopped after noticing a pore pressure almost equal to confining pressure as it indicates the stage of liquefaction. Then, results were analyzed and compared with other tests for conclusions.

3 Results and Discussion

A series of undrained triaxial tests were performed as listed in Table 2. Result of test performed on a specimen subjected to 0.5% strain (ε) for 100 cycles is as shown in Fig. 2. It indicates that with increase in number of cycles, pore pressure builds up and results in decrease of shear stress and stiffness of the specimen which can be noticed by flattening of hysteresis loop. Complete flattening of loop indicates that soil cannot take load anymore which shows that soil is liquefied.

3.1 *Effect of Shear Strain and Number of Cycles*

With an assumption that soil under consideration possess a Poisson's ratio (v) of 0.3 at a 50% relative density (D_r), tests were conducted at strains of 0.06, 0.2, 0.5, and 1% which corresponds to cyclic shear strains (γ_c) of 0.07, 0.26, 0.65, and 1.3% and results were analyzed. To better evaluate the effect of strain, excess pore pressure ratio (r_u) versus shear strain is considered for different loading cycles (N). These curves illustrate generation of pore pressure by each loading cycle and are called pore pressure generation curves. They also showed that r_u is a function of γ_c and N which can be seen in Figs. 3 and 4 at different consolidation stresses. The difference between the r_u values of first and last cycles shows the influence of loading cycles on pore pressure and indicates potential of pore pressure that could develop at a given shear strain. Once liquefaction triggers, the curve flattens and this difference decreases. At lower strains ($\gamma_c < 0.03\%$), pore pressure difference

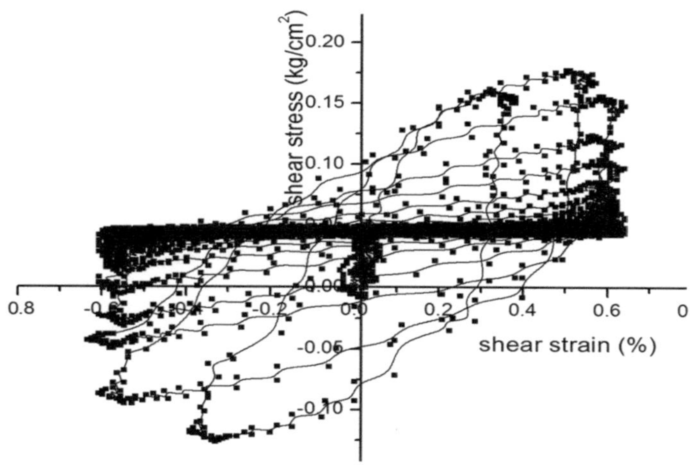

Fig. 2 Variation shear stress versus shear strain for a test at $\varepsilon = 0.5\%$, $f = 1$ Hz, $D_r = 50\%$, $\sigma_c' = 100$ kPa

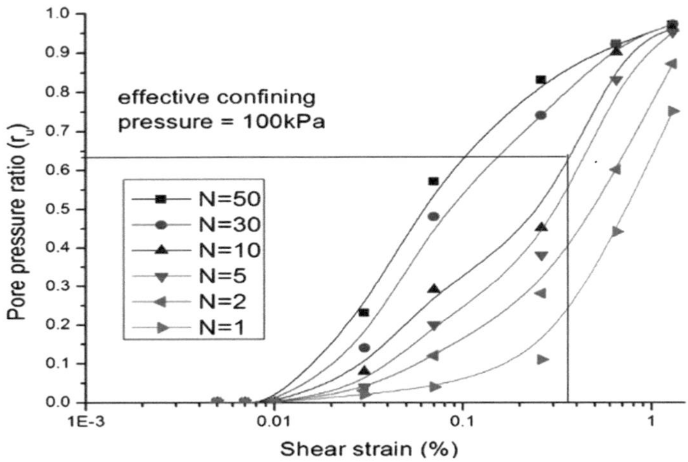

Fig. 3 Variation of r_u with γ_c at $\boldsymbol{\sigma}_c' = 100$ kPa, relative density of 50%, and frequency 1 Hz

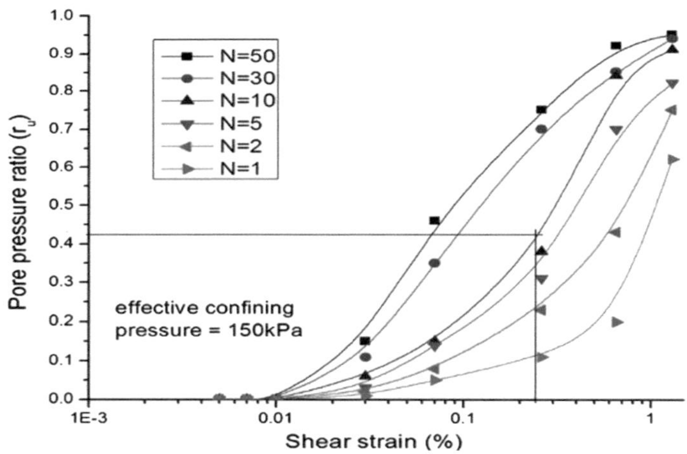

Fig. 4 Variation of pore pressure ratio with cyclic shear strain at $\boldsymbol{\sigma}_c' = 150$ kPa, D_r of 50%, and frequency 1 Hz

between first and last cycles is smaller. With increase in $\gamma_{c,}$ there is significant rise in pore pressure generation for the specimen consolidated to 100 kPa (Fig. 3), and the largest difference was observed at γ_c of 0.1%, whereas at 150 kPa, the influence region shifted to higher strains, as shown in Fig. 4 implying that the number of cycles is more influential for pore pressure buildup at high strains.

3.2 Effect of Confining Pressure

In the field, each depth corresponds to different consolidation stresses. So, it is essential to know the effect of confining pressure in the laboratory which accounts for depth effect in the field. For present work, two different consolidation stresses (100, 150 kPa) were considered. Dobry, 1985 performed similar tests on various sands using different specimen preparation techniques, relative densities, confining stresses and provided upper and lower bounds as shown in Fig. 5. From our present work, at a confining pressure of 100 kPa and 10 cycles, it is noticed that there is a reasonable agreement between the results obtained with the lower bound curve. Results also showed that significant pore pressure development for $\gamma_c < 0.01\%$, suggesting a threshold shear strain of 0.01%, which agrees with Dobry's results for clean sands. Similarly, tests were performed at 150 kPa and results are presented (Fig. 6), which shows 10% deviation in r_u for a smaller increment of $\boldsymbol{\sigma}_c'$, i.e., 50 kPa. At higher increments, more decrease in r_u can be noticed.

Moreover, the influence of the consolidation stress on pore pressure ratio is not very significant at lower ($\gamma_c < 0.03\%$) and higher ($\gamma_c > 1\%$) strain levels. More reduction in pore pressure was seen in the middle range strains. It is important to note that specimens consolidated to 150 kPa also did not produce any excess pore pressure for $\gamma_c < 0.01\%$ showing that threshold shear strain is not influenced by $\boldsymbol{\sigma}_c'$ between 100 and 150 kPa.

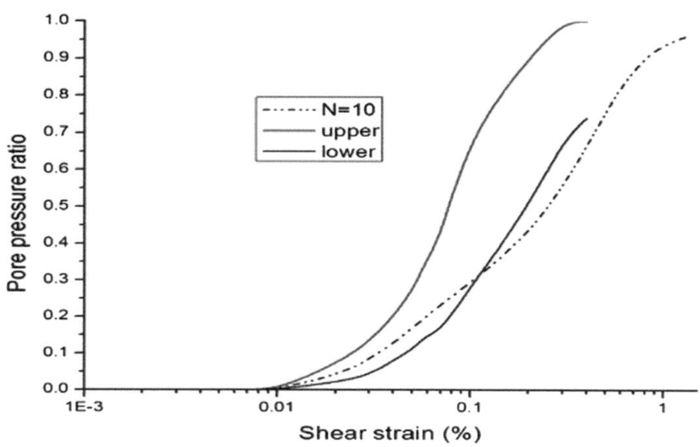

Fig. 5 Variation of pore pressure with shear strain at $\boldsymbol{\sigma}_c' = 100$ kPa, D_r of 50%, and frequency 1 Hz for ten cycles

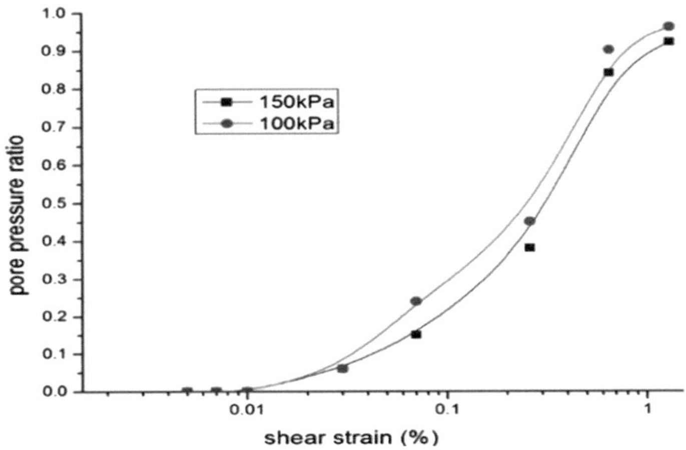

Fig. 6 Variation of pore pressure ratio with shear strain at different consolidation stresses for ten cycles

3.3 *Effect of Relative Density*

Tests were also carried out on samples at relative densities of 35, 50, and 65% by varying strains. Each test was terminated after reaching initial liquefaction and analyzed for results. The comparison of pore pressure generated for loose and medium dense specimens at a γ_c of 0.65% with number of cycles is as shown in Fig. 7. This showed that specimens prepared at lower densities generate higher pore

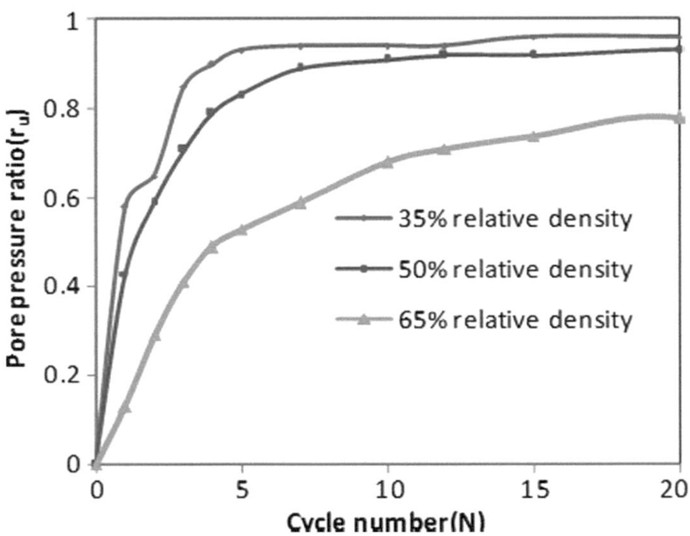

Fig. 7 Variation of pore pressure ratio with cycles at different relative densities

pressure compared to medium dense specimens indicating that liquefaction potential for looser sand deposits is high. This is due to the fact that loose specimens possess more voids filled with water which under load application in undrained conditions results in high pore pressure generation.

4 Conclusions

The following conclusions were drawn based on the investigations carried on saturated sands.

- In cases where site-specific data is incomplete, strain-based approach can be used comfortably.
- Increase in confining pressure results in reduction in excess pore pressure development during cyclic loading. But for strains less than 0.03%, no clear development of pore pressures was observed over the confining pressure range of 100–150 kPa.
- As amplitude of loading or number of cycles increases, the excess pore pressure also raises resulting in the loss in shear strength and stiffness of specimen and leading to failure by liquefaction.
- Threshold shear strain is not influenced by effective confining pressure in a range of 100–150 kPa.
- The higher the relative density more will be liquefaction resistance suggesting densification as one of the methods to mitigation liquefaction hazards.

References

ASTMD5311 M-13. (2013). Standard test method for load controlled cyclic triaxial strength of soil, pp. 1–11.

Boulanger, R., Meija, L., & Idriss, I. (1997). Liquefaction at moss landing during Loma Prieta earthquake. *Journal of Geotechnical and Geoenvironmental Engineering, ASCE, 123,* 453–467.

Dobry, R., Ladd, R. S., Yokel, F. Y., Chung, R. M., & Powell, D. (1982), Prediction of pore water pressure buildup and liquefaction of sands during earthquakes by the cyclic strain method. *National Bureau of Standards Building Science Series 138*, pp. 150.

Ishihara, K. (1996). *Soil behavior in earthquake geotechnics* (1st ed., p. 350). Oxford: Clarendon Press.

Lyman, A. K. B. (1942). Compaction of cohesion less foundation soils by explosives. *Transactions ASCE*, pp. 1330–1348.

Menard, L., & Broise, Y. (1975). Theoretical and practical aspects of dynamic consolidation. *Géotechnique, 15,* 3–18.

Seed, H. B., & Idriss, I. M. (1967). Analysis of soil liquefaction: Niigata earthquake. *Journal of the Soil Mechanics and Foundation Division ASCE, 93,* 83–108.

Seed, H. B., & Idriss, I. M. (1971). Simplified procedure for evaluating soil liquefaction potential. *Journal of the Soil Mech and Foundations Division, ASCE, 97*(9), 1249–1273.

Seed, H. B. (1979). Soil liquefaction and cyclic mobility evaluation for level ground during earthquakes. *Journal of the Geotechnical Engineering Division, ASCE, 105*(GT2), 201–255.
Sitharam, T. G., Govinda Raju, L., & Srinivasa Murthy, B. R. (2004). Cyclic and monotonic undrained shear response of silty sand from Bhuj region in India. *ISET Journal of Earthquake Technology, 41*(2–4), 249–260, paper 450.
Sitharam, T. G., Ravishankar, B. V., & Patil, S. M. (2012). Liquefaction and pore water pressure generation in sand: Cyclic strain controlled triaxial tests. *IJGEE, 3*(1), 57–85.

A Study on Mitigation of Tilting Problems in Liquefiable Soil

Puja Dutta and Ashim Kanti Dey

Abstract Tilting of foundations during liquefaction of soil is one of the primary concerns for stability of structures constructed in saturated sand in seismically active areas. Niigata earthquake resulted in tilting of a number of buildings. Once tilted, the structure is to be dismantled. Tilting also results in loss of life and property. The main aim of this study is to restrict tilting and allow the structure to float on liquefied soil. During a high-magnitude earthquake, the structure may not float for a long time, but may sink instead of tilting. Archimedes principle has been applied in the present study. The shape of conventional rectangular footing is changed to spherical and trapezoidal section. Shake table tests were performed at different frequencies like 0.5, 1 and 1.5 Hz. Comparisons of tilting and sinking of all the three types of footings were performed. Accelerations imposed on the footings were also measured. Results showed a positive response for a footing with spherical cross sections.

Keywords Liquefaction · Tilting of foundation · Shake table test

1 Introduction

A lot of studies have been carried out on seismic performance of shallow foundations resting on saturated sand (Liu and Dobry 1997; Knappett et al. 2004; Dunga et al. 2006, and so on). Post-liquefaction deformations have also been reported (Shamoto et al. 1996). Most of these deformations result in tilting of buildings. Once tilted, the structure cannot be reverted back to its original position. However, no study has been reported to withstand the tilting deformation during liquefaction

P. Dutta (✉) · A. K. Dey
Civil Engineering Department, National Institute of Technology,
Silchar 788010, India
e-mail: puja.dutta11121990@gmail.com

A. K. Dey
e-mail: ashim_kanti@yahoo.co.in

B. Adimoolam and S. Banerjee (eds.), *Soil Dynamics and Earthquake Geotechnical Engineering*, Lecture Notes in Civil Engineering 15,
https://doi.org/10.1007/978-981-13-0562-7_2

of soil. The present study consists of laboratory model tests on specially designed footings to withstand the deformations. One-dimensional shake table tests were conducted to observe the deformations of the footings at different frequencies. It is observed that sinking of footings could not be avoided at higher frequency but tilting deformation could be avoided. Uniform settlement can be accepted like most of the buildings in Mexico City.

2 Methodology

The basic concept consists of modifying the geometrical shape of footing so that during liquefaction of soil the modified footing displaces enough liquefied soil for floating stability. This is achieved by designing the bottom surface of the footing in the form of spherical or trapezoidal. In order to reduce the self-weight of the footing, the midsection of the footing is kept void. Three footings of identical plan area were considered for the present study—conventional rectangular cross section, top flat and bottom spherical cross section, top flat and bottom trapezoidal cross section. Although all the tests were carried out in the laboratory, for designing the shape of the models, a prototype plan of length = 5.5 m and width = 2.1 m was considered initially.

For designing a spherical section in both directions, first a spherical surface was generated along transverse direction. The surface equation was considered to be parabolic, i.e.

$$f(x) = ax^2 + c \tag{1}$$

The surface equation was obtained considering Archimedes principle for floating stability. Theoretically, for any submerged body, the Metacentric height is calculated as,

$$\mathrm{GM} = \mathrm{BM} - \mathrm{BG} \tag{2}$$

where G is the centre of gravity of the submerged body, M is the Metacentric position, and B is the centre of buoyancy.

BM = Ratio of moment of inertia of the cross-sectional area of the submerged body at the liquid surface about its longitudinal axis and volume of liquid displaced by the submerged body.

$$\mathrm{BM} = \mathrm{I/V} \tag{3}$$

where I (Moment of inertia of the body at liquid surface of spherical section) = $1/12 \times 5.5 \times 2.1^3 = 4.244\ \mathrm{m}^4$.

For a structure to be in stable equilibrium, GM > 0.

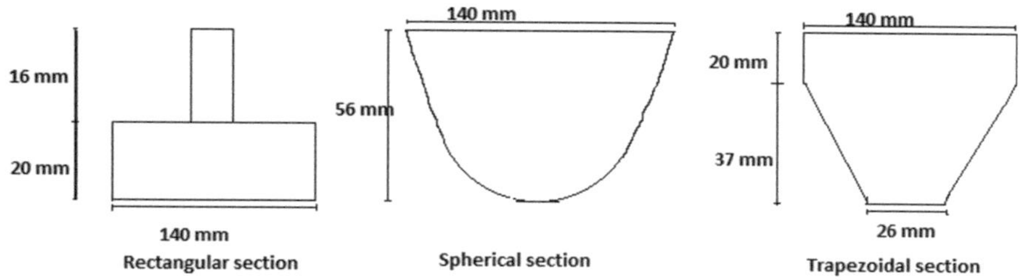

Fig. 1 Cross sections of model footings

And moreover, it was assumed that weight of the fluid displaced by spherical section is more than the rectangular section.

Weight of fluid displaced = Unit weight of fluid × volume of fluid displaced. The unit weight of fluid considered in this work is 14 kN/m^3. The unit weight of fluid is higher than water, because during liquefaction the density of the liquid is more than that of water.

It was proposed to maintain volume of fluid displaced by modified footing > volume of fluid displaced by the conventional rectangular footing.

Satisfying the above equations, a parabolic surface has been achieved for spherical footing: $f(x) = 0.581x^2 - 0.845$ (x in metres), along the transverse direction. Along the longitudinal direction, the depth of footing remains constant till 2.328 m from the centre on both sides and then depth starts decreasing with rate given by the equation $f(x) = 0.024x^2 - 0.460$. Similarly by taking the assumption of GM (for trapezoidal) > GM (for rectangular), a section was obtained for trapezoidal footing with height 0.85 m and width 2.1 m.

For the present model test, a linear scaling factor was adopted as suggested by Harris and Sabnis in the second edition of 'Structural Modeling and Experimental Techniques' (CRC Press, Taylor and Francis Group, 2nd edition, 1999). Linear scaling of length, width and height for model was taken in the ratio of 0.067 of the prototype. Models were prepared with concrete and string rods as reinforcement for footings. A general outline of the footings is shown in Fig. 1. The outline of the cross section was obtained by making two transverse beams and one longitudinal beam passing through the columns. A combined footing with two columns was adopted for the study. Figure 2 shows the reinforcement cage made of string rods.

The spherical-shaped footing with internal void space is shown in Fig. 3. One additional concrete cover was placed on the top of the footing.

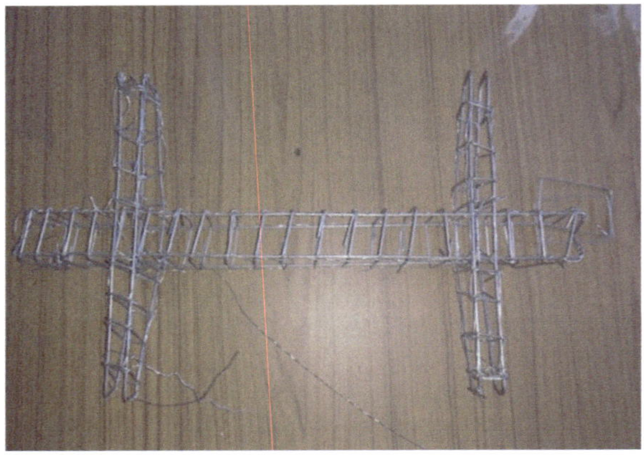

Fig. 2 Reinforcement cage

Fig. 3 Modified footing with voids inside

3 Experimental Set-up

A steel tank of dimension $1 \times 1 \times 1$ m was filled up to 0.9 m with a required quantity of dry sand for achieving a relative density around 50%. Water was slowly poured in the tank so that the sand became fully saturated. The tank was connected with the shake table, and harmonic loading was imparted at a frequency of 0.5, 1 and 1.5 Hz. Footing was placed on the sand with zero depth of foundation. Four accelerometers were connected as shown in Fig. 4, two on the footing and two on the tank in vertical and horizontal directions to record the acceleration time history.

Figures 5, 6 and 7 show the peak vertical acceleration imposed on the footing against time at 0.5, 1.0 and 1.5 Hz frequency, respectively. In all the cases, it is observed that the peak vertical acceleration of the rectangular conventional footing is the highest and that of the spherical footing is the lowest. Figures 8, 9 and 10 show the peak horizontal acceleration imposed on the footing against time at 0.5, 1.0 and 1.5 Hz frequency, respectively. In all the cases, it is observed that the peak

Fig. 4 Accelerometers placed both vertically and horizontally at column face

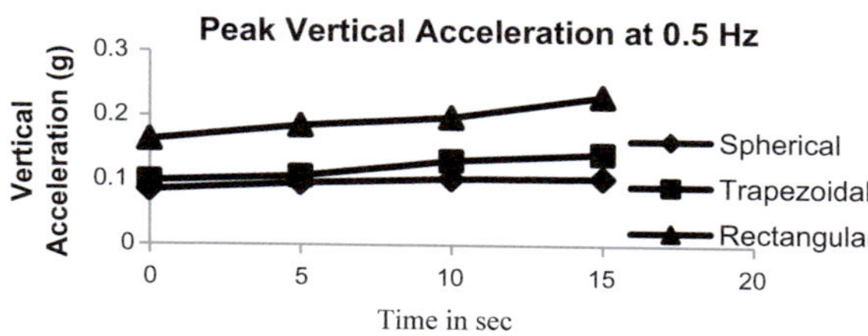

Fig. 5 Peak vertical acceleration versus time at 0.5 Hz

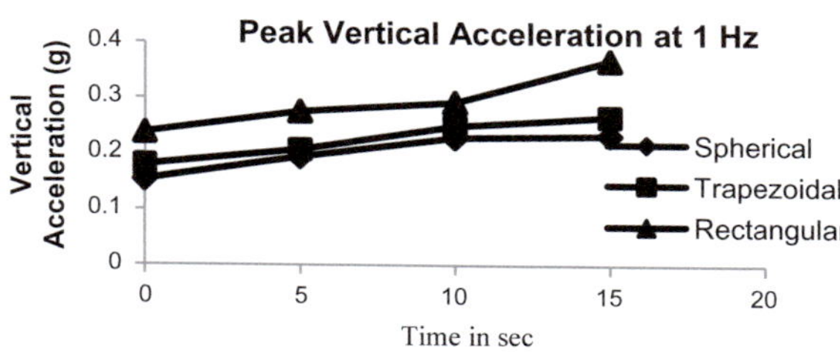

Fig. 6 Peak vertical acceleration versus time at 1.0 Hz

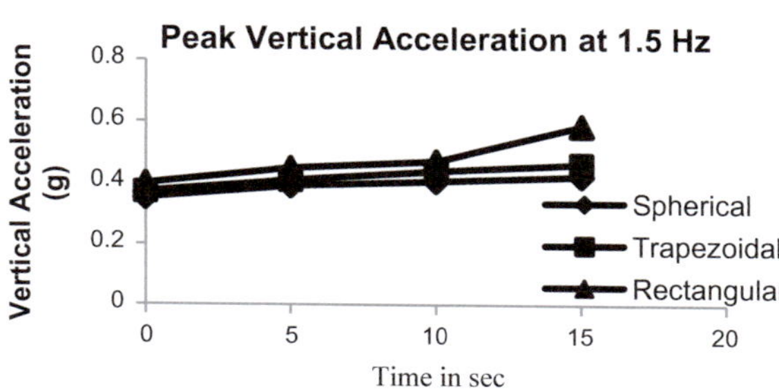

Fig. 7 Peak vertical acceleration versus time at 1.5 Hz

horizontal acceleration of the rectangular conventional footing is the least and that of the spherical footing is the highest. This acceleration response indicates that the spherical footing oscillates more along horizontal direction than the rectangular footing does. These oscillations make the spherical footing float more in the liquefied soil.

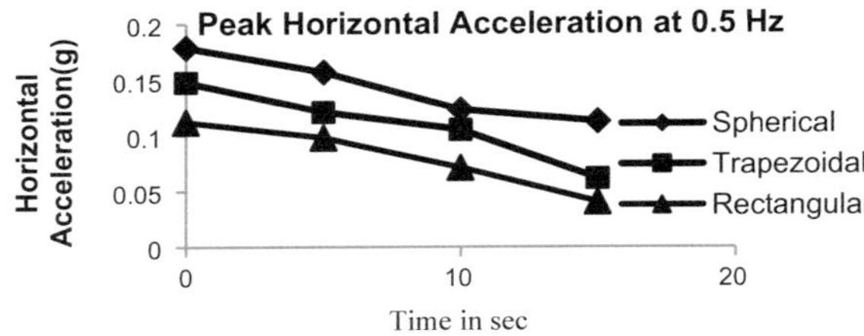

Fig. 8 Peak horizontal acceleration versus time at 0.5 Hz

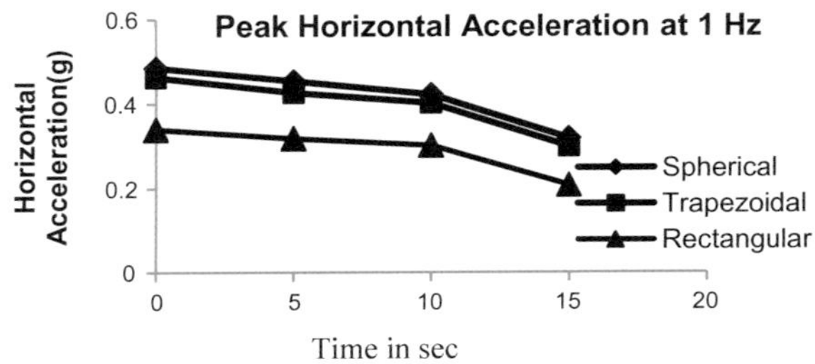

Fig. 9 Peak horizontal acceleration versus time at 1 Hz

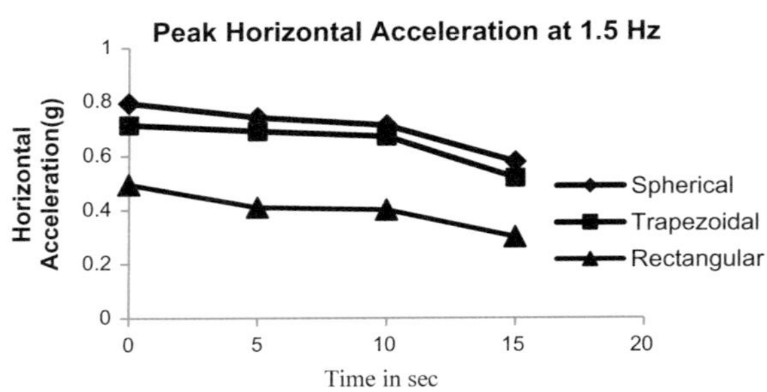

Fig. 10 Peak horizontal acceleration versus time at 1.5 Hz

In order to measure the deformation of the footing during vibration, a special arrangement was made after finding it difficult to use the LVDTs. One high-resolution camera was fixed on the tank, and on the other side, one graph sheet was fixed on a hardboard. The photographic image of the footing on the graph sheet was used to note the tilting and settlement of the footings. Thus during the dynamic tests, videography was taken from the start to the end of the test.

Tests were repeated with different dead loads placed over the footings. A comparison of vertical settlement of different footings carrying different dead weights under different frequencies like 0.5, 1.0 and 1.5 Hz is shown through Tables 1, 2 and 3. It is observed that the vertical settlement is more with increase in dead load. Moreover, the rectangular footing suffered the maximum settlement; spherical footing shows the lease settlement. Vertical settlement also increases with increase in frequency.

Table 1 Vertical settlement at 0.5 Hz

Vertical settlement (mm) at 0.5 Hz				
Shape of section	Without any vertical load	With 5 kg load	With 10 kg load	With 15 kg load
Spherical	15	22	28	35
Trapezoidal	20	28	35	42
Rectangular	30	40	47	55

Table 2 Vertical settlement at 1 Hz

Vertical settlement (mm) at 1.0 Hz				
Shape of section	Without any vertical load	With 5 kg load	With 10 kg load	With 15 kg load
Spherical	32	45	53	61
Trapezoidal	40	55	66	72
Rectangular	55	68	80	87

Table 3 Vertical settlement at 1.5 Hz

Vertical settlement (mm) at 1.5 Hz				
Shape of section	Without any vertical load	With 5 kg load	With 10 kg load	With 15 kg load
Spherical	50	67	82	93
Trapezoidal	65	84	102	114
Rectangular	77	101	115	128

The results obtained above show that the vertical settlement is greatly reduced in terms of spherical and trapezoidal section than conventional rectangular footing.

4 Analysis of Data Obtained from Static Load Tests

To check the suitability of the new designs in terms of load-carrying capacity, static load tests were performed on them and the load versus settlement curves were drawn. The test bed was made of the same foundation sand compacted at 50% relative density and saturated completely by water. Fig. 11 shows the load versus settlement curves obtained from the static load test on the footings. The ultimate load-bearing capacity was found by applying double tangent method in the figure and is shown in Table 4.

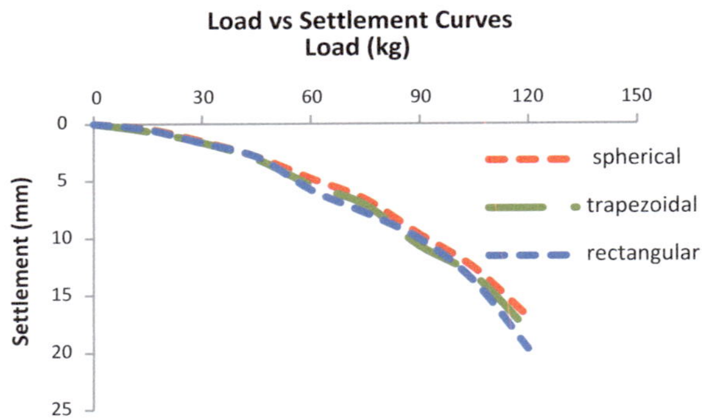

Fig. 11 Load versus settlement curves

Table 4 Ultimate load-carrying capacity of three footings under static vertical loadings

Shape of the footing	Ultimate load-bearing capacity (kg) experimental
Rectangular	97
Spherical	118
Trapezoidal	107

5 Tilting of Footing

To measure the tilting of footing, if any, a bamboo piece was attached with one of the columns and photographs were taken to observe the initial, during the test and final position of the footing. Figures 12 and 13 show the initial and final positions of spherical footing and rectangular footing with 15 kg surcharge load. It can be observed that the spherical footing did not tilt but the rectangular footing tilted after the shake table test. The amount of the tilt was so high that proper measurement could not be taken.

Fig. 12 Images of spherical footing before (left) and after (right) soil liquefaction

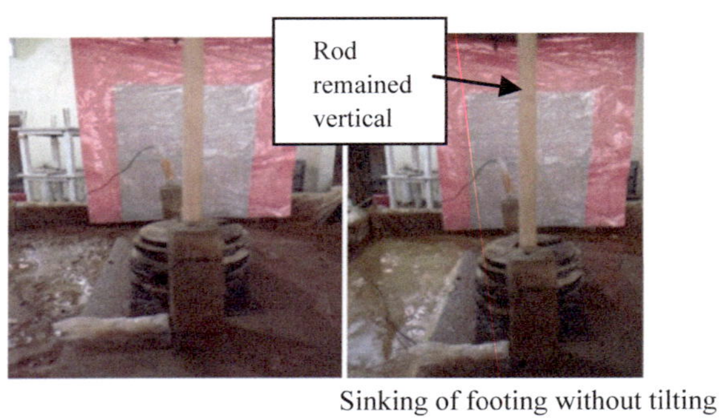

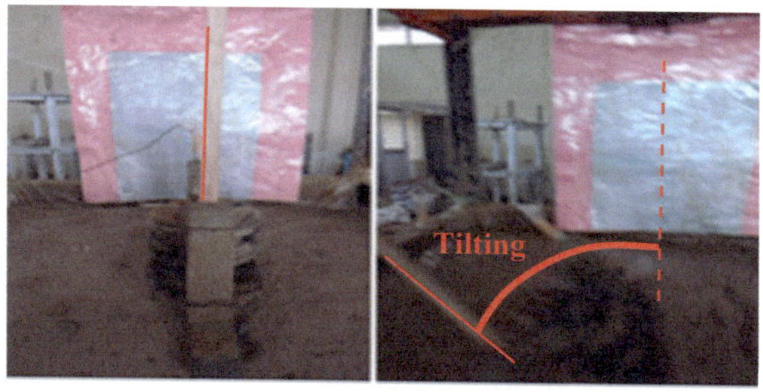

Fig. 13 Images of rectangular footing before (left) and after (right) soil liquefaction

6 Conclusions

The main conclusions derived from the experimental investigations of the conventional rectangular footing and the modified footings are that under static load; all the three footings behaved similarly with little difference in ultimate load-carrying capacity, whereas under dynamic loading, the spherical-shaped footing shows more stable behaviour than trapezoidal and rectangular footing. The spherical footing floats for a longer time during liquefaction and does not tilt, whereas the rectangular footing tilts very quickly.

References

Dungca, J. R., Kuwano, J., Takashashi, A., Saruwatari, T., & Tokimatsu, K. (2006). Shaking table tests on the lateral response of a pile buried in liquefied sand. *Soil Dynamics and Earthquake Engineering, 26*(2006), 287–295.

Knappett, J. A., Haigh, S. K., & Madabushi, S. P. J. (2004). Mechannism of shallow foundations under earthquake loading. *Soil Dynamics and Earthquake Engineering, 26*(2006), 91–102.

Liu, L., & Dobry, R. (1997). Seismic Response of Shallow Foundation on liquefiable sand. *Journal of Geotechnical Geoenvironmental Engineering, 123*(6), 557–567.

Shamoto Y., Zhang, J.-M., Gotosigera, & Tokimatsu, K. (1996). *A new approach to evaluate post-liquefaction permanent deformation in saturated sand*. Paper no. 1097, Eleventh WCEE.

Effect of Soil Heterogeneity on Soil–Structure Interaction

Ravi Kant Mittal and Sanket Rawat

Abstract Soil–structure interaction due to its significant effect on the response of the structure had already gained importance and therefore been incorporated in the Indian seismic code IS 1893-Part 4:2005. Since the advent of soil–structure interaction, continuous researches have been done on this topic and hence the concepts kept on modifying for more realistic results. The existing literature has already shown some of the critical parameters that are missing in Indian seismic code. Still, there is research gap in this area as some of the other important parameters too need to be included. Due to lack of proper geotechnical reports before the start of construction, thereby the misinterpretation of soil properties by the designer increases the susceptibility of the structure to failure in spite of including the code-specified effects such as soil–structure interaction. Critical situation occurs when the soft layer is underlying a hard stratum which the designers may misinterpret as the hard layer to the full depth due to improper geotechnical reports. Also, there is not any guideline in Indian seismic code to direct the designers to deal with the heterogeneous soil condition. In this research, a parametric study is carried out on a 150 m tall RCC chimney considering the effect on structural response of it by incorporating the effect of different conditions of layered soil mass. The results obtained show that soil heterogeneity patterns affect the design forces considerably and hence there is a critical need for inclusion of these factors in Indian seismic code. The paper subsequently concludes that, if appropriate provisions from some international literature are properly adopted, a substantial variation in the seismic response can be achieved and hence in many cases may result in structural-work cost saving. Overall with this study, the paper attempts to address the legitimate concern of design engineers regarding the potential variation in design values, especially for the sites when the soft soil underlies a hard stratum.

R. K. Mittal · S. Rawat (✉)
Department of Civil Engineering, Birla Institute of Technology and Science, 333031 Pilani, India
e-mail: sanketrawat@gmail.com

R. K. Mittal
e-mail: ravimittal@pilani.bits-pilani.ac.in

B. Adimoolam and S. Banerjee (eds.), *Soil Dynamics and Earthquake Geotechnical Engineering*, Lecture Notes in Civil Engineering 15,
https://doi.org/10.1007/978-981-13-0562-7_3

Keywords Heterogeneous soil · Soil–structure interaction · Indian seismic code Layered stratum

1 Introduction

Earthquake force contemplation in design has always been given prominence due to its unpredictable nature. Earlier soil characteristics were not considered to be of significant role in the behavior of structure due to ground motion. However, certain devastating collapses in the history (e.g., Kobe Earthquake, 1995) made it clear that soil characteristics have significant role in the behavior of structure as much as other factors corresponding to the earthquake, i.e., source mechanism and the geology of the seismic-wave path possess. Soil–structure interaction (SSI) due to its extensive impact on the design parameters has been included in Indian seismic code IS 1893 (Part 4):2005. This consideration during design makes a structure more flexible and hence increases the natural period of the structure (Chowdhury and Dasgupta 2008). Furthermore, SSI effect also intensifies the effective damping ratio of the system. This assumption of smooth idealization of design spectra leads to the consent that SSI can be conveniently neglected for conservative design of the structure though it is only valid for specific classes of structures and soil conditions, such as light structures in relatively stiff soil. However, the population rise and limitation of land have forced the user to construct a tall and stiff structure without concerning about the type of soil. Moreover, the increase in natural period of the structure due to consideration of SSI is not always advantageous as recommended by the simplified design spectrums. Soft soil sediments can considerably lengthen the period of seismic waves, and the increase in natural period of structure may lead to resonance with long period ground vibrations (Chowdhury and Dasgupta 2008). The existing literature (Mittal and Gajinkar 2014) has suggested that the formulae specified in IS 1893-Part 4 should be modified to take damping, effective shear modulus, and embedment effect into account for more realistic results. The major factor which still has not been addressed in the code and the related literature is the effect of non-homogenous strata. The stabilization of top soil generally tends to neglect the effect on othersoil strata present beneath the top soil. This neglection proves to be conservative in several cases; however, its effect may not be ignored completely. It has been observed that on radiation damping and energy dissipation are adversely affected by layering. Layer boundaries cause the outgoing waves to reflect back. These layers act as waveguides, entrapping, and reinforcing wave energy (Hadijan and Ellsion 1985). Thus, energy disperses more gradually than it would have been in an analogous homogeneous medium, and damping is effectively reduced. Results derived from several existing literature (Gazetas 1983; Hadijan and Ellsion 1985) have been taken as reference, and the above-mentioned factor has been incorporated in the calculation of effective horizontal seismic coefficient. The results obtained show the significant influence of the effect of the layered stratum on the response of the structure and hence highlight the necessity of inclusion of this factor in the Indian seismic code for a better earthquake resistant and cost-effective design.

2 Existing Provisions in Indian Seismic Code

Soil–structure interaction criteria are specifically included in IS 1893-Part 4:2005, criteria for earthquake resistant design for structures such as industrial structures including stack-like structures. Different clauses are present in which formulae and methods are mentioned. Clause 14 of code mainly deals with the time period of vibration of stack-like structure. Furthermore, methods for calculating design base shear and bending moment have been specified in clause 17 of the code. Formulae to get horizontal and rocking stiffness have been mentioned in Table 12 of the code.

3 SSI Provisions in the Existing Literature

Various parameters such as effective damping ratio, embedment correction factor, and effective shear modulus obtained from the existing literature (Richart et al. 1970; Arya et al. 1984; Kramer 2004) and international codes (EN 1998-5:2004) have been suggested for inclusion in Indian seismic code (Mittal and Gajinkar 2014).

4 Modified Provisions for Present Study

The stiffness (both horizontal and rocking) is modified in the present study due to the presence of layered strata. The modified formulae and provisions have been adopted from Gazetas (1983), and Hadijan and Ellsion (1985).

4.1 Static Stiffness of Circular Foundations on a Stratum-Over-Half Space

Horizontal stiffness ($1 \leq \frac{H}{R} < 4$)

$$K_{\mathrm{H}} = \frac{8G_1R}{(2-\upsilon_1)}\left[\frac{1+\frac{R}{2H}}{1+\frac{RG_1}{2HG_2}}\right] \tag{1}$$

Rocking stiffness $\left(0.75 \leq \frac{H}{R} < 2\right)$

$$K_{\mathrm{R}} = \frac{8G_1R^3}{3(1-\upsilon_1)}\left[\frac{1+\frac{R}{6H}}{1+\frac{RG_1}{6HG_2}}\right] \tag{2}$$

4.2 Elastic Stratum on a Rigid Base

Horizontal stiffness $\left(\frac{H}{R} > 1\right)$

$$K_{\mathrm{H,surface}} = \frac{8GR}{2-\upsilon}\left(1+\frac{R}{2H}\right) \tag{3}$$

Rocking stiffness $\left(4 \geq \frac{H}{R} > 1\right)$

$$K_{\mathrm{R,surface}} = \frac{8GR^3}{3(1-\upsilon)}\left(1+\frac{R}{6H}\right) \tag{4}$$

4.3 Elastic Stratum on a Rigid Base, Embedded Circular Foundation $\left(\frac{D}{R} < 2\right)$

Horizontal stiffness

$$K_{\mathrm{H,embedded}} = \frac{8GR}{2-\upsilon}\left(1+\frac{R}{2H}\right)\left(1+\frac{2D}{3R}\right)\left(1+\frac{5D}{4H}\right) \tag{5}$$

Rocking stiffness

$$K_{\mathrm{R,embedded}} = \frac{8GR^3}{1-\upsilon}\left(1+\frac{R}{6H}\right)\left(1+\frac{2D}{R}\right)\left(1+\frac{0.7D}{H}\right) \tag{6}$$

Here, G_1 is the shear modulus of upper layer and G_2 is the shear modulus of lower stratum. H is the depth of the top stratum. Suffixes 1 and 2 represent the corresponding property of top and bottom layers, respectively. R is equivalent radius of footing. D is the depth of embedment.

5 Problem Statement

To check for the response of RCC chimney subjected to earthquake force resting on two-layered soil strata, SSI with inclusion of modified stiffness has been included in the analysis.

The chimney height is taken as 150 m with raft diameter of 18 m and area of shell 8.5 m^2. The time period of vibration for the structure fixed at base and the flexible base period are computed separately. The damping ratio for the flexible base is also calculated, and the horizontal acceleration coefficient is calculated using the modified time period and damping ratio. The stiffness is first calculated on the basis of provisions available in code, and then for comparison, the modified stiffness is computed based on formulae as mentioned above. The equivalent lateral force static method (clause 17.1, IS 1893-Part 4) of analysis is used to compute the design base shear and design bending moment. The parametric problem data has been obtained from the text by Chowdhury and Dasgupta (2008). The properties of different types of soils that are used in study are mentioned in Table 1 which is extracted from the paper presented by Mehta and Gandhi (2008).

6 Results and Discussions

MS Excel spreadsheet is developed to analyze for response variation (shear at the base of chimney) using the code-specified equivalent static lateral force method for tall chimneys. The below-mentioned graphs describe the results of analysis for different responses and hence illustrate the importance of layered strata effect on seismic response of tall chimneys when it is struck by a long or a short duration earthquake.

Table 1 Properties of soil used in the study (Mehta and Gandhi 2008)

Velocity of shear waves (m/s)	Soil type	Unit weight, γ (kN/m^3)	Poisson's ratio ν	Shear modulus, G (kN/m^2)	Elastic modulus (kN/m^2)
150	Soft soil	16	0.49	36,700	14.95×10^4
300	Stiff soil	20	0.45	183,500	25.84×10^5
600	Dense soil	22.4	0.35	822,000	50.53×10^7
1200	Rock	25.6	0.30	3,758,900	30.42×10^7

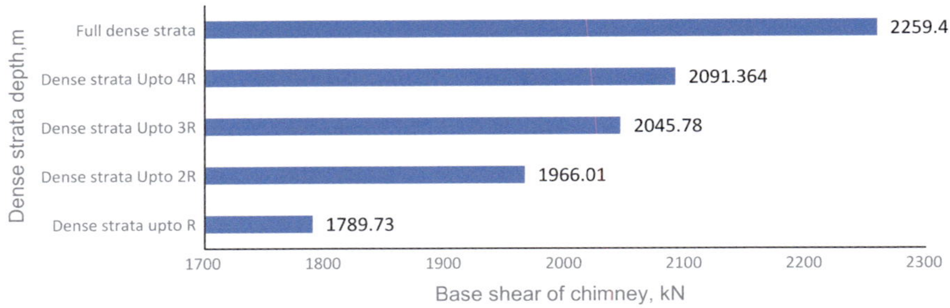

Fig. 1 Variation in base shear on increase in depth of top strata

Fig. 2 Percentage change in base shear at the base on increase in depth of top strata

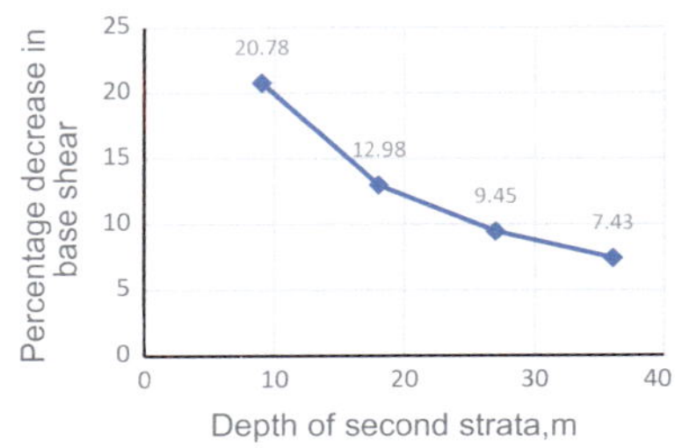

Figures 1 and 2 depict the variation in shear at the base of chimney in the two cases: The first one being in which the soil was assumed to be consisting of a dense stratum to the full depth or top layer was stabilized using natural or mechanical methods making it dense. In the second case, the less dense stratum is considered to be underlying the denser one. The depth of the top stratum is varied, and the results derived in the literature have been incorporated to make the analysis more realistic. In this study, soft stratum is considered below the dense stratum. However, the user can input the data of the soil type in the Excel sheet as per his requirements and consequently results can be obtained.

It can be clearly observed that if the structure is analyzed using the conventional process rather than following the realistic method, the percentage base shear variation in the two cases comes out to be 20.78% when the depth of top (dense) stratum is equal to the radius of circular foundation. Furthermore, this percentage difference decreases as the depth of the top dense stratum is increased. The results of the study are in compliance with the fact that the strata role becomes almost negligible after depth equal to two times the width of foundation.

7 Conclusions

The results of the present study primarily highlight the necessity of inclusion of the layered strata effect in soil–structure interaction analysis. Effect of two-layered soil strata is considered in the analysis along with the effect of increase in depth of the top stratum. To accentuate its significance, effect of the presence of soft stratum underneath dense soil is studied. Noteworthy decrease in shear at the base of chimney was observed when the effect of soil underneath the top layer is considered which further emphasized on the consideration of layered strata effect in Indian seismic code. The paper subsequently concludes that, if appropriate provisions related to the modification in stiffness in case of layered stratum, from some international literature, are properly implemented, a substantial variation in the seismic response can be achieved and hence in many cases may result in structural-work cost saving. On the whole, with this study, the paper attempts to address the genuine concern of design engineers regarding the potential variation in design values in both embedded foundation and foundation at surface, especially for the sites when the soft soil underlies a hard stratum.

References

Arya, S. C., O'Neill, M., & Pincus G. (1984). *Design of structures and foundations for vibrating machines*. Houston, Texas: Gulf Publishing Company.

Chowdhury, I., & Dasgupta, S. P. (2008). *Dynamics of structure and foundation—A unified approach*. Boca Raton: CRC Press.

EN, 1998-5. (2004). *Design of structures for earthquake Resistance-Part 5: Foundations, retaining structures and geotechnical aspects*. UK: British Standard.

Gazetas, G. (1983). Analysis of machine foundation vibrations: State of the art. *Soil Dynamics and Earthquake Engineering, 4*(4) (1985).

Hadjian, A. H., & Ellison, B. (1985). Equivalent properties for layered media. *Soil Dynamics and Earthquake Engineering, 2*(1) (1985).

Indian Standard Code. (2002). IS: 1893 (Part 4), *Criteria for earthquake resistant design of structures (Part 4) Industrial stack-like structures*. New Delhi: Bureau of Indian Standards, India

Kramer, S. L. (2004). *International series in civil engineering and engineering mechanics: Geotechnical earthquake engineering*. Upper Saddle River, New Jersey: Prentice Hall.

Mehta, D., & Gandhi, N. J. (2008). Time response study of tall chimneys, under the effect of Soil Structure Interaction and long period earthquake impulse. In *Proceedings of the 14th World Conference on Earthquake Engineering*, Beijing, China.

Mittal, R. K., & Gajinkar, V. (2014). Evaluation of soil-structure interaction guidelines in Indian seismic codes. *IJCIET, 5*(5), 97–104.

Richart, F. E., Jr., Hall, J. R., Jr., & Woods, R. D. (1970). *Vibrations of soils and foundations*. Englewood Cliffs, N.J.: Prentice Hall.

Nonlinear 1D Ground Response Analysis of Soil Profile Using Different Procedures

Devdeep Basu and Arindam Dey

Abstract Earthquakes are natural phenomena that have caused the loss of millions of lives and property across the globe, and they pose a severe threat to mankind. The earthquake-induced damage at any particular site will be governed by the response of the soil deposit present there. In this paper, ground response analysis (GRA) studies are done for a typical soil stratigraphic profile of IIT Guwahati region. Nonlinear 1D GRA is performed for the soil profile using six different procedures. Both total stress and effective stress analyses are done. The Sikkim earthquake motion recorded at IIT Guwahati with peak bedrock level acceleration (PBRA) of 0.02 g, and three of its scaled-up components with PBRA of 0.06, 0.18, and 0.36 g are used as input motions. Based on some of the computed response parameters, a comparison is made among these different analysis procedures. The suitability of each of these procedures under different circumstances is also scrutinized.

Keywords 1D · Nonlinear GRA · Masing · Non-masing · Ground response parameters

1 Introduction

Numerous moderate to large magnitude earthquakes have occurred in Northeast India in the past, and as a result, this entire region has been designated as zone *V* as per IS:1893 Part 1 (2002), which is the zone of highest seismicity in the country. Severe damages as a result of both structural failures as well as settlement and liquefaction-related failures have been reported during the occurrence of some of

D. Basu · A. Dey (✉)
Department of Civil Engineering, Indian Institute of Technology Guwahati, Guwahati 781039, India
e-mail: arindamdeyiitk@gmail.com

D. Basu
e-mail: basudevdeep@gmail.com

B. Adimoolam and S. Banerjee (eds.), *Soil Dynamics and Earthquake Geotechnical Engineering*, Lecture Notes in Civil Engineering 15,
https://doi.org/10.1007/978-981-13-0562-7_4

the most notable seismic events for this region, like the 1897 Shillong Plateau earthquake ($M_w = 8.3$), the 1950 Assam–Tibet earthquake ($M_w = 8.6$).

In this paper, a 1D GRA study for one typical soil profile of IIT Guwahati campus, situated on the banks of river Brahmaputra, has been presented. The various procedures of nonlinear 1D GRA have been compared based on some of the computed ground response parameters like peak horizontal acceleration (PHA), maximum shear strain, excess pore water pressure (PWP) ratio (which is the ratio of excess PWP to effective vertical stress). All the wave propagation problems presented herein have been solved with the aid of site response analysis code DEEPSOIL v6.0 (2015).

2 Fundamentals of Nonlinear 1D Ground Response Analysis

In nonlinear GRA, any soil profile is discretized into a number of layers using multiple degree of freedom lumped parameter model. Subsequently, the dynamic equation of motion is solved numerically in small time steps using Newmark β method.

Any nonlinear stress–strain model following Masing or non-masing rules can be used during the integration process. The stress–strain models developed by Kondner and Zelasko (1963), Hashash and Park (2001) are commonly used for performing nonlinear GRA incorporating Masing load/unload/reload criteria. The nonlinear stress–strain model developed by Phillips and Hashash (2009) is implemented in the site response code DEEPSOIL 6.0 (2015), for incorporating non-masing criteria. This model uses a hysteretic damping reduction factor (MRDF procedure) which modifies the Masing rules and thus incorporates the non-masing load/unload/reload criteria.

The different procedures of nonlinear GRA employed in DEEPSOIL 6.0 (2015) are listed in Table 1.

Table 1 Summary of different procedures of nonlinear ground response analysis

Abbreviated notation	Description of procedure
NLM1	Masing criteria, Total stress analysis (no PWP is considered)
NLM2	Masing criteria, Effective stress analysis with PWP dissipation
NLM3	Masing criteria, Effective stress analysis without PWP dissipation
NLNM1	Non-Masing criteria, Total stress analysis (no PWP is considered)
NLNM2	Non-Masing criteria, Effective stress analysis with PWP dissipation
NLNM3	Non-Masing criteria, Effective stress analysis without PWP dissipation

3 Methodology

3.1 Modeling of Soil Profile

The soil properties of different layers in the profile are taken from an existing borehole log report (Fig. 1). Shear wave velocities (V_s) of the soil layers are computed from the specified SPT-N values (measured at intervals of 1.5 m) using the correlation suggested by Imai and Tonouchi (1982). The modulus reduction and damping curves for the soil layers are obtained based on the formulations of Ishibashi and Zhang (1993), and they are subsequently fitted using MRDF procedure, which defines the parameters for the nonlinear stress–strain model. For all the analysis presented herein, elastic bedrock having a shear wave velocity of 5000 m/s is considered underlying the soil profile.

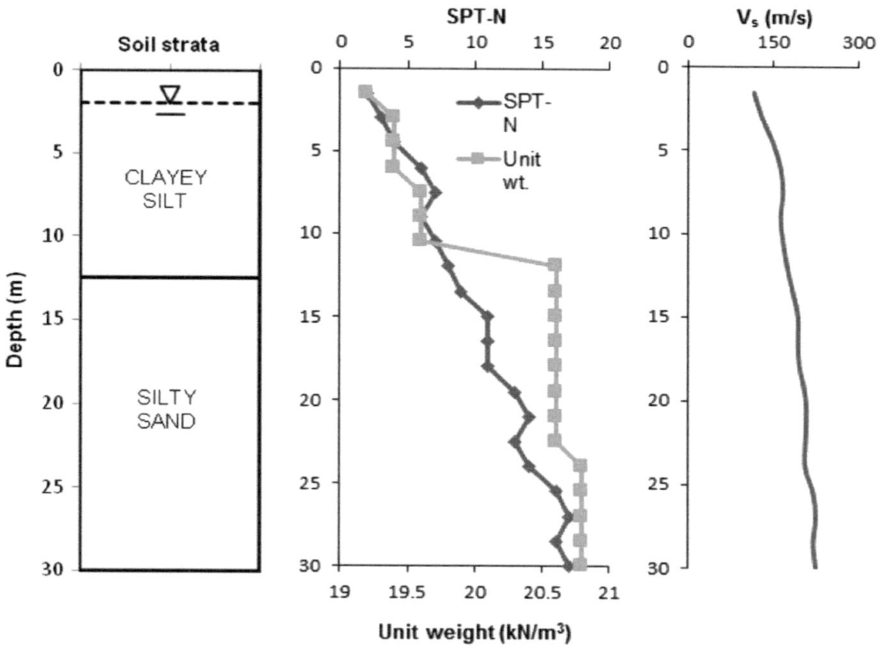

Fig. 1 Soil deposit of IIT Guwahati along with plots of SPT-N, unit weight, and V_s variation with depth

3.2 Description of Seismic Loading

The 2011 Sikkim earthquake motion recorded at IIT Guwahati with a peak bedrock level acceleration (PBRA) of 0.02 g (Fig. 2) and three of its scaled-up components having PBRAs of 0.06, 0.18, and 0.36 g are applied at the soil profile base for all the analyses.

4 Results and Discussions

4.1 Comparison Between NLM2 and NLM3/NLNM2 and NLNM3

No significant difference is observed between the peak horizontal acceleration (PHA) profiles (Fig. 3) and maximum strain profiles (Fig. 4) obtained using NLM2

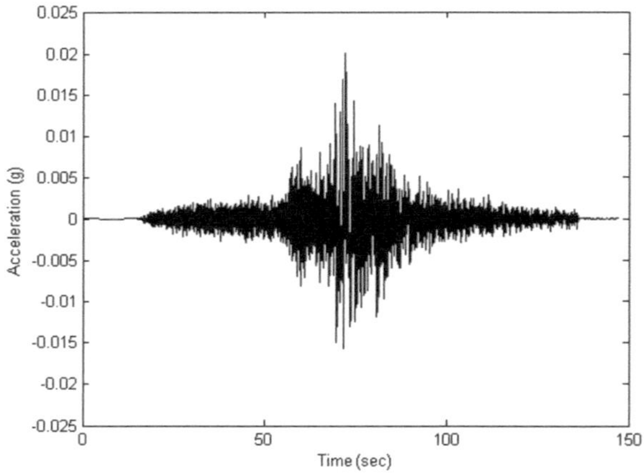

Fig. 2 Acceleration time history of the 2011 Sikkim earthquake motion recorded at IIT Guwahati

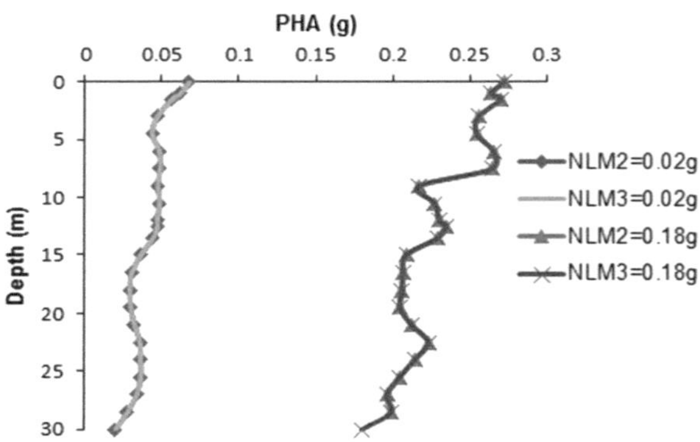

Fig. 3 PHA variation with depth for two motion components having PBRA of 0.02 and 0.18 g

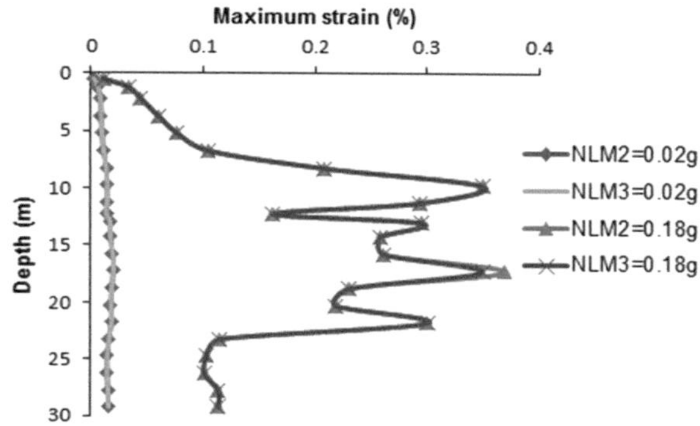

Fig. 4 Maximum strain versus depth for two motion components having PBRA of 0.02 and 0.18 g

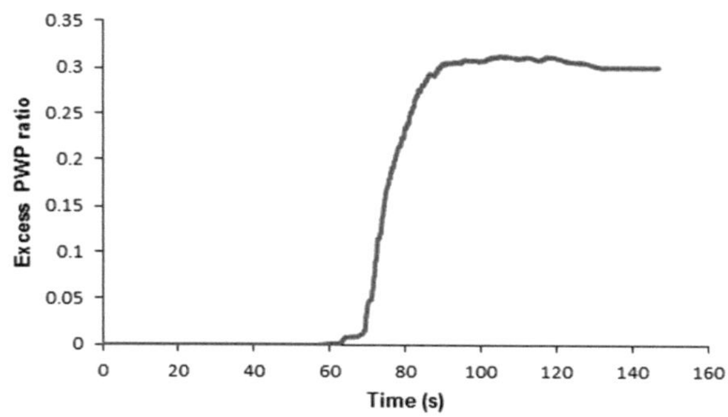

Fig. 5 Excess PWP dissipation with time for 0.18 g PBRA seismic motion in the silty sand soil strata (NLM2 procedure)

and NLM3 procedures of GRA. This implies that the excess PWP dissipation (Fig. 5) that occurs during the time span in which the seismic loading acts is negligible enough to have any role in altering the response parameters computed using these two procedures. The same logic holds good for NLNM2 and NLNM3 GRA procedures which shows similar observations (Figs. 6 and 7).

4.2 *Comparison Between NLM2 and NLNM2*

For the seismic motions with PBRA of 0.02 and 0.06 g, the obtained PHA profiles (Fig. 8) and maximum strain profiles (Fig. 9) are almost similar for both NLM2 and NLNM2 procedures. However, for seismic motions having higher PBRA (greater than 0.06 g), the difference in obtained PHA profiles (Fig. 10) and maximum strain profiles (Fig. 11) for these two procedures is more prominent.

The basic difference between Masing and non-masing procedures lies in the way in which the target modulus reduction and damping curves are fitted to obtain the

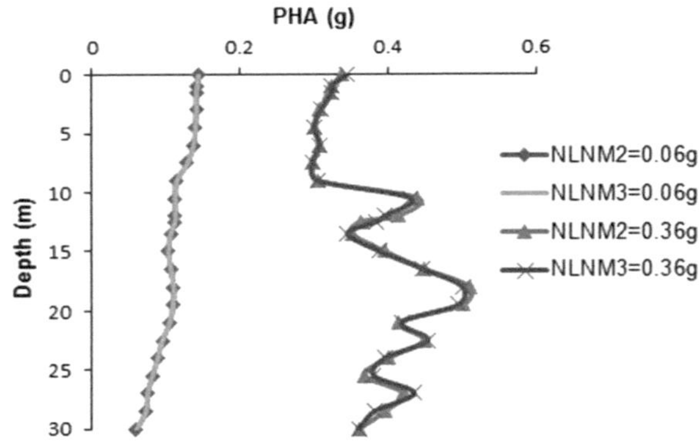

Fig. 6 PHA variation with depth variation for two motion components having PBRA of 0.06 and 0.36 g

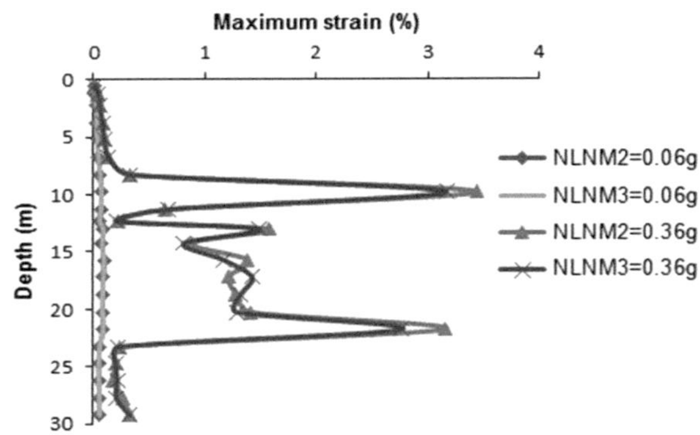

Fig. 7 Maximum strain versus depth for two motion components having PBRA of 0.06 and 0.36 g

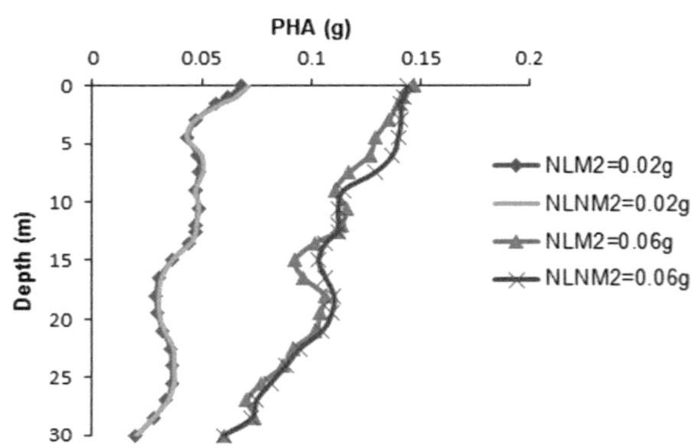

Fig. 8 PHA profiles for seismic motions having PBRA of 0.02 and 0.06 g

parameters of nonlinear stress–strain model. At low strains, Masing procedure gives a good fit with the target modulus reduction and damping curves, whereas at high strains, it overestimates the hysteretic damping significantly. In non-masing MRDF curve fitting procedure, even at high strains the target damping curve is fitted

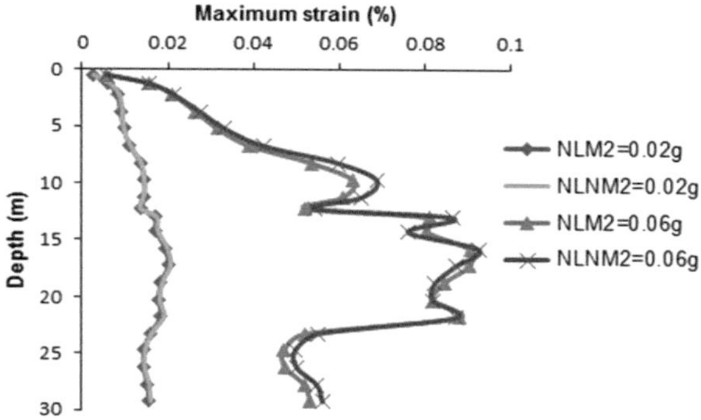

Fig. 9 Maximum strain profiles for seismic motions having PBRA of 0.02 and 0.06 g

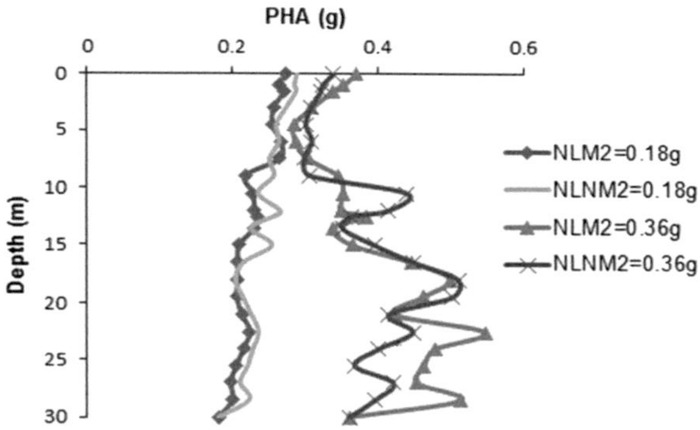

Fig. 10 PHA profiles for seismic motions having PBRA of 0.18 and 0.36 g

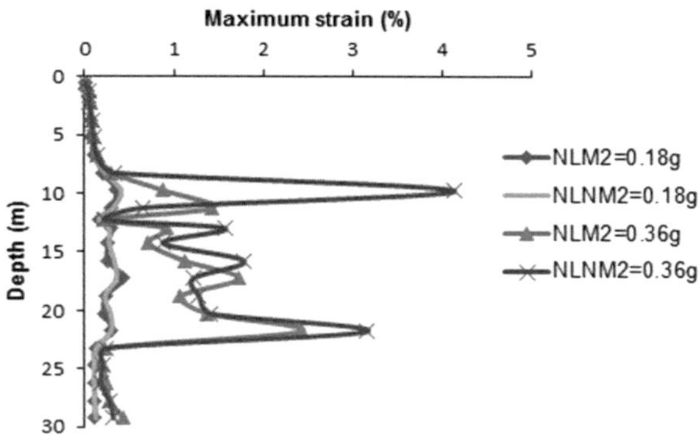

Fig. 11 Maximum strain profiles for seismic motions having PBRA of 0.18 and 0.36 g

accurately. For lower PBRA input motions, the maximum strains developed in the soil layers are less. Thus, at low strains, results obtained for both Masing and non-masing procedures are similar and accurate. However, at high strains, usually developed for motions with high PBRA values, the difference between Masing and non-masing procedures becomes significant.

Fig. 12 PHA profiles obtained incorporating masing criteria for motions with PBRA of 0.02 and 0.36 g

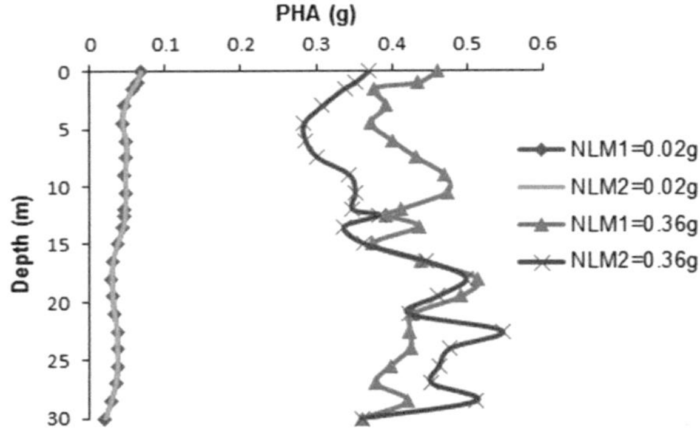

4.3 Comparison Between NLM1 and NLM2/NLNM1 and NLNM2

It is observed that total stress analysis (NLM1 or NLNM1) without PWP generation produces higher values of peak ground acceleration (PGA, which is the peak horizontal acceleration at ground surface) than effective stress analysis. This is because of the fact that in effective stress analysis, the PGA values are calculated based on effective stresses developed in soil, whereas total stresses are used to compute the PGA in total stress analysis procedure. For larger amplitude motions with high PBRA values (greater than 0.06 g), significant PWP develops in a soil profile which brings down the effective stresses. As a result, a marked difference in results is observed between total stress and effective stress analysis procedures. For motions having low PBRA values, PWP developed in a soil profile is minimal (almost zero), and thus the two analysis procedures produce similar results. Figure 12 shows the PHA profiles developed for total stress and effective stress analyses incorporating Masing criteria. PHA profiles obtained using non-Masing criteria are shown in Fig. 13 for the two procedures.

Fig. 13 PHA profiles obtained incorporating non-masing criteria for motions with PBRA of 0.06 and 0.36 g

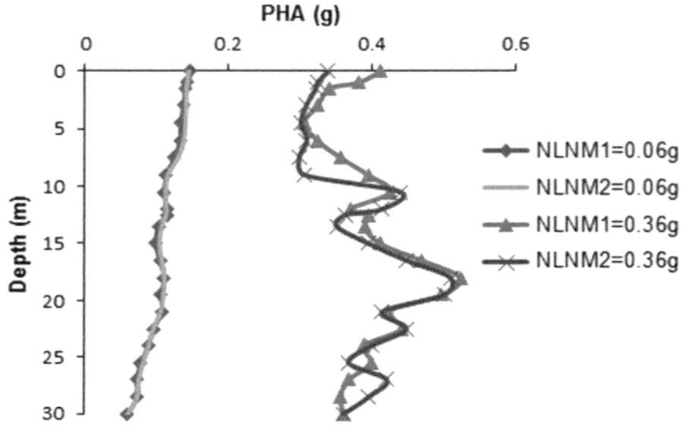

However, NLM2 procedure produced higher PHA values than NLM1 procedure (for 0.36 g PBRA motion) at depths greater than 20 m. This is possibly on account of higher strains that might have developed in the soil profile for NLM2 procedure in the deeper layers. Higher strains induce greater hysteretic damping and, thus, lowers the stress developed and subsequently the PHA.

5 Conclusions

In this paper, a comparison among different procedures of nonlinear 1D GRA and their suitability under different circumstances has been presented. Results have been presented in terms of parameters like PHA, PGA, excess PWP ratio, and maximum shear strain. For both Masing and non-Masing analyses, the PHA and maximum strain profiles obtained when PWP dissipation is allowed have been observed to be similar to the profiles of the same when PWP dissipation is not allowed. A comparison has been made between Masing and non-Masing analysis procedures using motions having different PBRA values. It has been observed that for motions having high PBRA values (greater than 0.06 g), the difference between Masing and non-Masing analysis procedures becomes significant. Finally, total stress and effective stress analysis procedures (both Masing and non-Masing) have been compared. Higher PGA values are obtained for total stress analysis when seismic motions selected have high PBRA values (greater than 0.06 g). Thus, an idea about the seismic site response for IIT Guwahati region, quantified with respect to some response parameters, has been developed in this study.

References

Hashash, Y. M. A., Musgrove, M. I., Harmon, J. A., Groholski, D. R., Phillips, C. A., & Park, D. (2015). DEEPSOIL 6.0, User Manual, pp. 1–116.

Hashash, Y. M. A., & Park, D. (2001). Non-linear one-dimensional seismic ground motion propagation in the Mississippi embayment. *Engineering Geology, 62*(1–2), 185–206.

Imai, T., & Tonouchi, K. (1982). Correlation of N value with S-wave velocity and shear modulus. In *Proceedings of 2nd European Symposium on Penetration Testing,* Amsterdam, pp. 67–72.

IS: 1893 Part 1. (2002). *Criteria for earthquake resistant design of structures*. India: Bureau of Indian Standards.

Ishibashi, I., & Zhang, X. (1993). Unified dynamic shear moduli and damping ratios of sand and clay. *Soils and Foundations, 33*(1), 182–191.

Kondner, R. L., & Zelasko, J. S. (1963). A hyperbolic stress-strain formulation of sands. In *Proceedings of the 2nd Pan American Conference on Soil Mechanics and Foundation Engineering*, Sao Paulo, Brasil, pp. 289–324.

Phillips, C., & Hashash, Y. M. A. (2009). Damping formulation for non-linear 1D site response analyses. *Soil Dynamics and Earthquake Engineering, 29,* 1143–1158.

Fuzzy Probability Approach in Seismic Hazard Analysis

Prasad Hiremath, Mohammad Muzzaffar Khan and G. Kalyan Kumar

Abstract Seismic hazard assessment like many other problems in seismology is a complicated problem, owing to the variety of parameters affecting the occurrence of the earthquake. Assessment of seismic hazard based on the theory of probability becomes hard particularly under highly uncertain conditions, i.e., where neither the statistical data nor the physical knowledge required for a purely probabilistic risk analysis is sufficient. Insufficiency of information will afflict the calculated risk probabilities with imprecision and will to the underestimation of the risk. This uncertainty, which is a result of vagueness and incompleteness of the data, should be considered in a rationale way. Essentials of seismic hazard analysis are identified and the concept of fuzzy logic is applied to seismic hazard analysis, generalizing the conventional probabilistic seismic hazard analysis (PSHA) to fuzzy probabilistic seismic hazard analysis (FPSHA). The β and λ values which are the input parameters for the PSHA are evaluated using earthquake data for the Warangal region. The uncertainty associated with these values will be too high for a site with moderate earthquake activity, and the previous earthquake data is scarce. Fuzzy logic will help in containing these uncertainties by its property of considering the possibility of a certain event occurring. This study presents an approach for seismic hazard analysis based on fuzzy set theory. Seismic activity is based on the seismic activity rate, λ, which is equal to the number of events with magnitudes equal to or greater than a defined magnitude level, say M_0, during a specified time period, T; the parameter b or β ($\beta = b$ ln10). The variables β and λ are first converted into Gaussian fuzzy sets using α-cut method. The ranges of β and λ are chosen based on previous studies. The fuzzified variables are used in seismic hazard analysis. The outputs are defuzzified using the center of area method, and fuzzy hazard curve is

P. Hiremath (✉) · M. M. Khan · G. K. Kumar
Department of Civil Engineering, National Institute of Technology Warangal, Warangal, India
e-mail: prasadbvbcet@gmail.com

M. M. Khan
e-mail: mmkmatrix@gmail.com

G. K. Kumar
e-mail: kalyanu.g@gmail.com

B. Adimoolam and S. Banerjee (eds.), *Soil Dynamics and Earthquake Geotechnical Engineering*, Lecture Notes in Civil Engineering 15,
https://doi.org/10.1007/978-981-13-0562-7_5

developed for the study region. The output is compared with PSHA results. The horizontal PGA expected in Warangal on stiff ground, with a 10% probability of exceedance in 50 years (which corresponds to a return period of 475 years) is 0.0836 g, whereas, that with a 2% probability of exceedance in 50 years (return period = 2475 years) is 0.153 g.

Keywords Fuzzy logic · PSHA · Gaussian fuzzy sets · Peak ground acceleration

1 Introduction

The study of earthquakes dates back many centuries. Written records of earthquakes in China date back as far as 3000 years. Still, compared with millions of years where earthquakes have been occurring, humankind's experience with earthquake is very brief. Today, hundreds of millions of people throughout the world live with a significant risk to their lives and property from earthquakes. The health of many local, regional, and even national economies is also at risk from earthquakes. It is impossible to prevent earthquakes from occurring, but it is possible to mitigate their adverse effects: to reduce life losses, injuries, and damage. In order to realize this, the starting point is being able to estimate the adverse outcomes of the earthquake by means of risk assessment methods in a reliable and realistic way.

Seismic hazard such as ground shaking, liquefaction, landslides, fault movement, which are associated with an earthquake, may produce adverse effects on human activities. Seismic hazard analyses involve the quantitative estimation of ground shaking hazards at a particular site. Seismic hazards may be analyzed deterministically, as when a particular earthquake scenario is assumed, or probabilistically, in which uncertainties in earthquake size, location, and time of occurrence are explicitly considered.

The accuracy of PSHA depends on the accuracy with which uncertainty in earthquake size, location, recurrence, and effects can be characterized. Therefore, it is quite sensible to employ a concept which can model different forms of uncertainty present in natural hazards. A reasonable option is complementing probability theory with an extra dimension of uncertainty provided by fuzzy set theory. This concept has received the generic name of fuzzy probability after Zadeh. However, this generic term has been interpreted and mathematically formalized in various ways. One of the most attractive interpretations of fuzzy probability is where probability of a crisp event, due to sparsity of data sample, is expressed in terms of a fuzzy number. It will be shown that this concept is especially appropriate for situations for which our physical knowledge and available statistical data are insufficient to characterize their likelihood of occurrence with a probability distribution.

The term fuzzy logic was first coined for the proposal of fuzzy set theory by Zadeh (1965). Brown and Yao (1983) in their pioneering research on the application of fuzzy set theory in structural engineering have tried to make use of fuzzy set theory for two of the structural engineering-related problems. Initially, a justification for the theory is provided, and then, simple fuzzy operations are developed and contrasted with those of conventional set theory. Frangopol et al. (1987) proposed a model which is able to reproduce both the randomness and the imprecision in conjunction with earthquake occurrences. Elham et al. (2015) in their study presented a new model for probabilistic seismic hazard assessment based on fuzzy set theory.

In the current study, fuzzy sets have been developed which can be used in the probabilistic seismic hazard assessment of Warangal, Telangana. For this, all the variables to be used in seismic hazard assessment must be defined in terms of fuzzy sets. Then, using α-cut method, the seismic hazard equations are evaluated. The outputs will be in fuzzy sets which will be defuzzified using the center method. At last, the expected peak ground acceleration values from the seismic sources are calculated. The results obtained are compared with the results of PSHA.

2 Methodology

In the present study, four input parameters, i.e., distance (R), magnitude (M), λ, and β, are defined as fuzzy sets by the discrete membership functions $\mu(R)$, $\mu(M)$, $\mu(\lambda)$, and $\mu(\beta)$, respectively. The input parameters, λ and β, for single seismic zone were taken from the previous study of the Warangal region by Deviprasad (2014). Due to the uncertainty of the four mentioned parameters, maximum and minimum values were determined by a Gaussian membership function. It may be noted that λ and β are parts of *G–R* relationship and R and M are uncertain inputs of an attenuation relationship. These uncertain parameters used in *G–R* relation and attenuation relationship are assessed using fuzzy logic approach. The results obtained with these fuzzified inputs will be defuzzified using centroid method. The defuzzified values are used for the seismic hazard assessment by Cornell–McGuire approach.

2.1 Gutenberg–Richter Frequency–Magnitude Recurrence Relationship

Gutenberg–Richter parameters '*a*' and '*b*', which are the basic input for the seismic hazard analysis, have been carefully evaluated through regression analyses using the established catalogue. Table 1 presents the completeness intervals and exceedance rates computed by the CUVI method (Mulargia and Tinti 1985) and Stepp's (1972) method.

The prime input values for the probabilistic seismic hazard analysis, β and λ, will be further processed with the fuzzy approach. To minimize the vagueness, a suitable range for β value and λ value was chosen. For the chosen ranges, using

Table 1 Seismicity parameters from CUVI and Stepp's method

Single source zone	N	a	b	$M_{\min}$	$\lambda = 10^{(a-bM)}$	$\beta = 2.303b$
Stepp's method	248	3.43	0.80	3	10.69	1.85
CUVI method	248	3.33	0.78	3	9.75	1.80

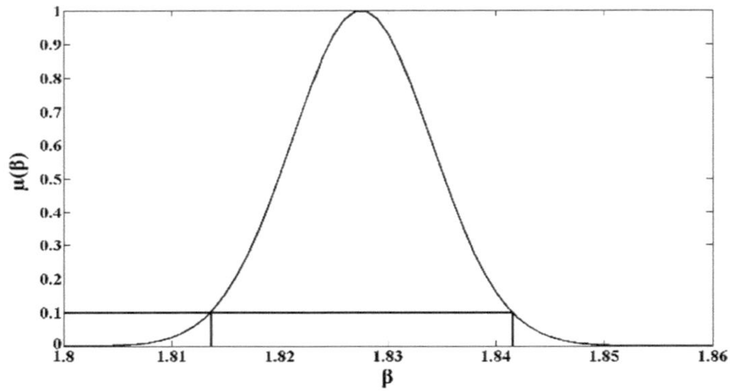

Fig. 1 Gaussian membership curve for β values

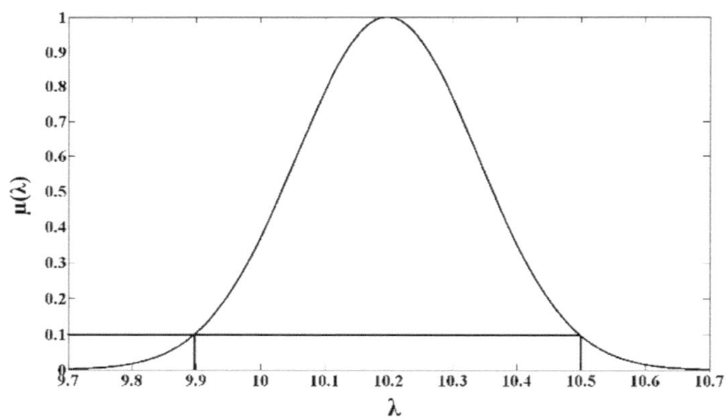

Fig. 2 Gaussian membership curve for λ values

their standard deviation and mean Gaussian membership curve was drawn as shown in Figs. 1 and 2.

As an example, in the first trial minimum value of β from the α-cut = 0.1 and minimum value of λ from the α-cut = 0.1 can be used as inputs. In the second trial, minimum value of β and maximum value of λ can be given as inputs. In that case, we will be having four possible combinations of β and λ values.

2.2 *Fuzzy Attenuation Relation*

Attenuation relationship given by Abrahamson and Silva has been used for calculating PGA values. Fuzzy logic approach is used for different ranges of magnitude (M) and distance (R). Gaussian curves were developed for both M and R as it was done in the case of G–R relations.

2.2.1 Fuzzification of the Magnitude

The magnitude range for a PSHA analysis that was done earlier was 3–6.5. The interval chosen was 0.4375. For each interval, using standard deviation and mean, Gaussian membership curves were drawn. α-cut of 0.1 was applied to get minimum and maximum values of magnitude. One interesting observation that can be made here is that the magnitude range is 6.2812–6.7187 which is greater than the highest value chosen in the conventional PSHA. 6.2812 is the median value of the last but one magnitude range. The fact that the magnitude value higher than the maximum magnitude value for the region being covered under the fuzzy universe makes this method more credible.

Gaussian membership curve for magnitude range 3–3.4375 is shown in Fig. 3. The thick horizontal line at $\mu(M) = 0.1$ cutting the Gaussian membership curve at two points is the α-cut at 0.1.

2.2.2 Fuzzification of Hypocentral Distance

The distance is also divided into ranges for the conventional PSHA. The least distance is 11.46 km, and 350 km is the maximum distance. As it is done in the

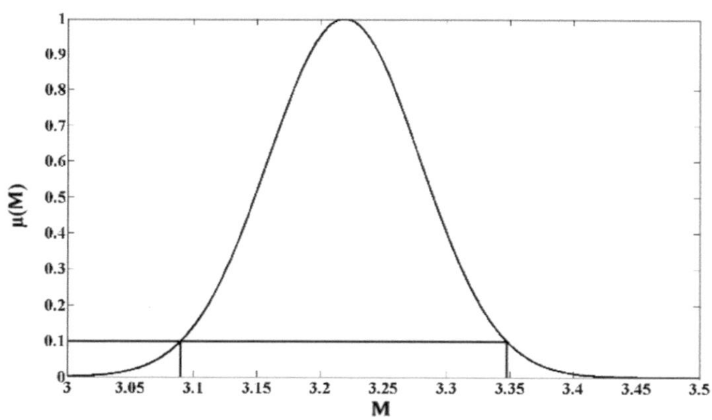

Fig. 3 Gaussian membership curve for magnitude range of 3–3.4375

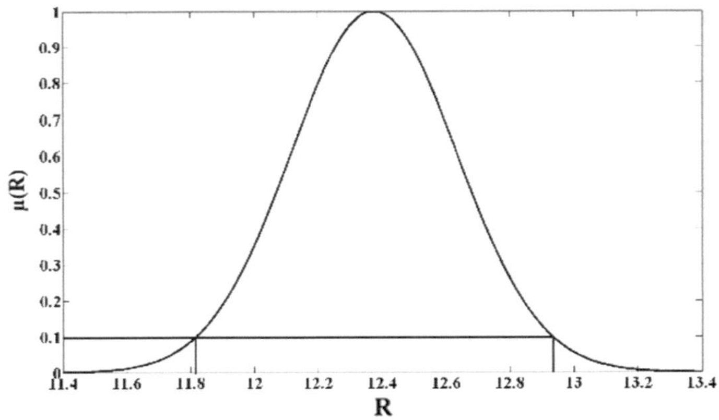

Fig. 4 Gaussian membership curve for the distance range 11.46–13.29

earlier cases of fuzzification, here also for each distance range, a Gaussian membership curve is drawn and from α-cut method. Minimum and maximum values of distance were noted for further calculations. Gaussian membership curve drawn for the distance of range 11.46–13.29 km is shown in Fig. 4.

Distance and magnitude are to be given as inputs in the attenuation relationship of Abrahamson and Silva (1997). We are having two values for magnitude and two values for distance. Similar to the case of *G–R* relationship, here also four combinations of magnitude and distance are possible. Spectral acceleration values are calculated for all these four combinations of *M* and *R*. As there are nine ranges of magnitude chosen, for each of these nine ranges four values of spectral acceleration are obtained. Out of these four values, minimum and maximum values are chosen.

2.3 *Probabilistic Seismic Hazard Analysis*

In the present study, Cornell–McGuire approach is used to estimate the seismic hazard. The computations needed in the PSHA are performed using the program CRISIS 2007 (Ordaz et al. 2007).

In the analysis, we have four possible combinations of λ and β values and two sets of spectral acceleration values, i.e., minimum and maximum spectral acceleration values sets. It makes a total of eight possible combinations. Hence, eight trials have been run in CRISIS 2007 software. One of the best ways to represent the values obtained using fuzzy logic is to present the results in the form of ranges. From the values obtained from all the eight trials, minimum and maximum values of annual rate of exceedance are extracted and are shown in Fig. 5.

2.3.1 Seismic Hazard Curves

The effect of all the earthquakes at different locations, sizes, sources with different probability of exceedance is integrated into one curve. It shows the probability of different ground motions at a site, exceeding during a specified period of time in the form of a range.

The expected value of the PGA ranges corresponding to a return period of typical engineering interest, e.g., for 475 years (10% probability of exceedance in 50 years exposure time), and the range of peak horizontal acceleration can be read from Fig. 5 as 0.038–0.064 g. Different return periods can also be selected for reporting PGA values depending on the type and importance of the structure. All the hazard curves are developed for the rock sites. For any particular site, the local soil profile (thickness of the layers and type of soil) will transform the predicted ground motions.

2.3.2 Uniform Hazard Spectrum

Uniform hazard spectrum (UHS) is the spectral amplitudes of acceleration which can be evaluated at all the natural periods for a constant probability of exceedance at a site. Such a response spectrum is commonly known as the UHS. Fig. 6 shows the UHS corresponding to the upper bounds of 10% probability of exceedance in 50 years (475 years return period).

Fig. 5 Hazard curve for single seismic source zone

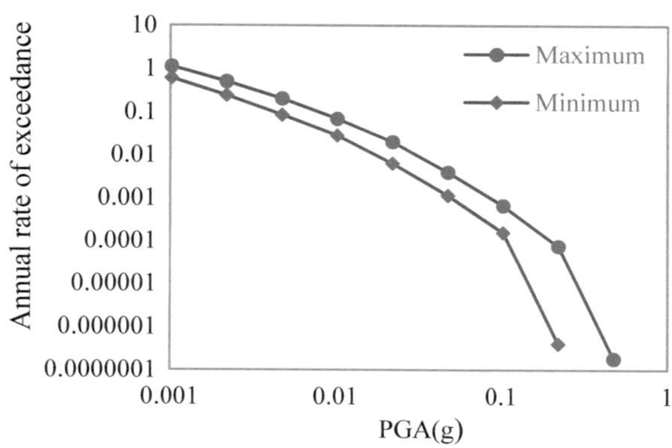

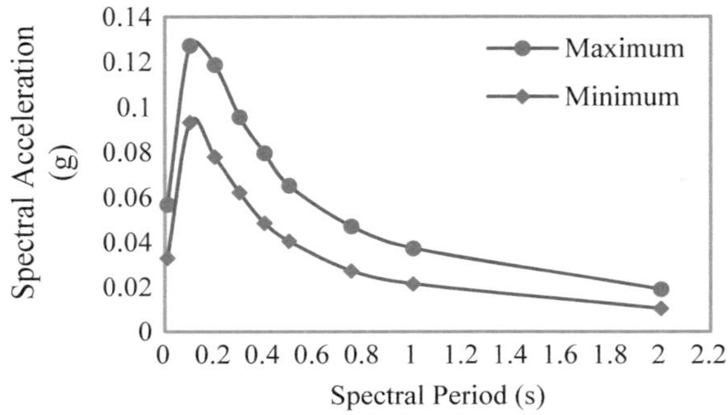

Fig. 6 UHS corresponding to 10% probability of exceedance in 50 years

3 Conclusion

In this study, a modular framework for seismic hazard assessment was presented. The quintessential strategy in developing the new framework is to improve and enhance each component of the conventional probabilistic seismic hazard analysis.

- The horizontal PGA expected in Warangal, with a 10% probability of exceedance in 50 years (which corresponds to a return period of 475 years), is varying from 0.038 to 0.064 g, whereas that with a 2% probability of exceedance in 50 years (return period = 2475 years) is in range of 0.08–0.113 g.
- The annual rates of exceedance and expected PGA calculated using PSHA method are observed to be lying in the ranges that are given by the FPSHA.
- The current study revealed that fuzzy sets could help to incorporate all the uncertainties in PSHA. However, a complete comparison could not be pulled off with previous studies, since not only the applied methods are different, as we have used fuzzy set theory, but also many of the input parameters and attenuation relationships are different.

References

Abrahamson, N. A., & Silva, W. J. (1997). Empirical response spectral attenuation relations for shallow crustal earthquakes. *Seismological Research Letters, 68,* 94–127.

Brown, C. B., & Yao, J. T. (1983). Fuzzy sets and structural engineering. *Journal of Structural Engineering, 109*(5), 1211–1225.

Deviprasad, B. S. (2014). *Probabilistic seismic hazard analysis of Warangal district*' (M. Tech dissertation thesis). NIT Warangal.

Elham, B., Tahernia, N., & Shafiee, A. (2015). Fuzzy—probabilistic seismic hazard assessment, casestudy: Tehran region, Iran. *Natural Hazards, 77,* 525–541.

Frangopol, D. M., Ikejima, K., & Hong, K. (1987). Seismic hazard prediction using a probabilistic-fuzzy approach. *Structural Safety, 5*(2), 109–117.

Mulargia, E., & Tinti, S. (1985). Seismic sample areas defined from incomplete catalogues: An application to the Italian territory. *Physics of the Earth and Planetary Interiors, 40,* 273–300.

Ordaz, M., Aguilar, A., & Arboleda, J. (2007). *CRISIS2007—Ver. 1.1: Program for computing seismic hazard*. UNAM, Mexico: Instituto de Ingenieria.

Stepp, J. C. (1972). Analysis of completeness in the earthquake sample in the puget sound area and its effect on statistical estimates of Earthquake Hazard. In *Proceedings of the International Conference on Microzonation*, Seattle, Washington.

Zadeh, L. A. (1965). Fuzzy probabilities. *Information Processing and Management, 8,* 363–372.

Effect of Saturation on Dynamic Properties of Solani Sand

Priyanka Sharma and B. K. Maheshwari

Abstract Degree of saturation influences significantly the properties of soil. Therefore, dynamic soil properties will also have effect of water content. Hence, it is important to evaluate these dynamic properties of soil, namely the shear modulus and the damping ratio for different degree of saturation. In the present study, a series of cyclic triaxial tests have been carried out on samples of Solani Sand (collected from the bed of Solani River near Roorkee) at three different degrees of saturation, i.e., (S_r = 0, 50, and 98%) at the same effective confining pressure of 100 kPa and at a frequency of 1 Hz in medium-to-large shear strain values (between 0.01 and 1%). The objective of the analysis is to examine the effect of degree of saturation on the behavior of dynamic properties of sands. Thus, results of dry, partially saturated, and fully saturated sand samples are presented and compared.

Keywords Shear modulus · Damping ratio · Degree of saturation
Solani sand · Cyclic triaxial test

1 Introduction

Earthquake damage distribution is highly influenced by the dynamic properties of the soil. For any geotechnical earthquake engineering problem such as ground response analysis, soil–structure interaction, and liquefaction, dynamic soil properties are very essential input parameters. The most important properties are the shear modulus and the damping ratio. The major factors that influence the dynamic properties of soil are the void ratio, effective confining pressure, and degree of saturation. Since 1980, many efforts have been reported on the study of the effects

P. Sharma (✉) · B. K. Maheshwari
Department of Earthquake Engineering, Indian Institute of Technology Roorkee, Roorkee 247667, India
e-mail: pri88.deq2015@iitr.ac.in

B. K. Maheshwari
e-mail: bkmahfeq@iitr.ac.in

B. Adimoolam and S. Banerjee (eds.), *Soil Dynamics and Earthquake Geotechnical Engineering*, Lecture Notes in Civil Engineering 15,
https://doi.org/10.1007/978-981-13-0562-7_6

of saturation on small strain stiffness of unsaturated soils obtained from resonant column (RC) or bender element (BE) testing. A series of resonant column tests were conducted in the torsional mode of vibration to assess the effect of saturation on damping of sands (Madhusudhan 2011). Strain-controlled cyclic triaxial element test has been carried out on dry and saturated soil samples in medium-to-large shear strain levels of Ahmedabad sands (RaviShankar et al. 2005). Very few researches are done to analyze the behavior of dynamic properties with the saturation at high strain. Moreover, the nonlinear dynamic responses of grounds caused by strong earthquake motions demand the evaluation of dynamic soil properties corresponding to large strain levels (Maheshwari et al. 2012).

Hence, in the present study, a series of cyclic triaxial tests have been carried out on sand samples of Solani sand (collected from the bed of Solani River near Roorkee) at three different degrees of saturation level (S_r = 0, 50, and 98%) at the same effective confining pressure in medium-to-large shear strain values. The objective of the analysis is to examine the behavior of dynamic properties of sands based on the degree of saturation of the sand samples such as dry, partially saturated, and fully saturated condition.

2 Testing Procedure Adopted

Cyclic triaxial test was carried out on Solani sand samples of 50 mm diameter and 100 mm height with 50% relative density and at confining pressure of 100 kPa, using a modern system (Wykeham 2008) as per ASTM standard D3999-91 (2003). For a particular test, harmonic excitation of a given frequency of 1 Hz was applied as a sine wave in axial direction (ASTM: D5311/D5311M 2013). The samples were tested at different shear strain values, i.e., at 0.03, 0.08, 0.32, 0.90, 1.005, 1.5, and 2.01%. When cyclic triaxial test is performed, a hysteresis loop is formed (Kramer 1996). The slope of the line connecting the extreme points on the hysteresis loop is the secant dynamic modulus E. The shear modulus G is given as,

$$G = E/2(1 + v) \tag{1}$$

where v is the Poisson ratio. The damping ratio is obtained as

$$D = \frac{1}{4\pi}\frac{\Delta W}{W} \tag{2}$$

where ΔW is the energy loss per cycle and W is the maximum elastic energy that can be stored in a unit volume of viscoelastic body.

3 Results and Discussions

In this paper, a comparative study of the dynamic properties of Solani sand for varying degree of saturation by testing samples under strain controlled on cyclic triaxial system is presented. Figure 1 presents the variation of shear modulus of dry sample (B = 0.01), partially saturated sample (B = 0.50), and fully saturated sample (B = 0.95) of Solani sand at relative density of 50% under an effective confining pressure of 100 kPa and at frequency of 1 Hz with the shear strain. It can be observed from Fig. 1 that the shear modulus decreases with the increase in shear strain, which was expected. Moreover, it is observed that the shear modulus of dry sand is greater than that of partially saturated sand and fully saturated sand. However, there is not much difference noticed in the shear modulus of partially saturated sand and that of fully saturated sand. When the shear strain is higher than 1%, the decrease in shear modulus almost reduces to a very low value for all the three states of soil. However, for dry sand, shear modulus is higher as compared to partially saturated and fully saturated sand. Similar trend for decrease in shear modulus for dry to fully saturated state was observed for Ahmedabad sands at high strain levels by RaviShankar et al. (2005).

In Fig. 2, the variation of the damping ratio with shear strain for different degrees of saturation is shown. As expected and observed, damping ratio increases with the increase in shear strain. However, as degree of saturation increases, an increase in the damping ratio is observed with the increase in strain. It can be observed from Fig. 2 that the damping ratio of saturated sands is greater than that of partially saturated and fully saturated sands at most of the strains. However, the difference in damping ratios of partially saturated and fully saturated sand is smaller. Madhusudan and Kumar (2013) conducted series of resonant column tests to study the effect of saturation on damping ratio and concluded that that minimum damping ratio is always associated with the dry state and reaches its maximum

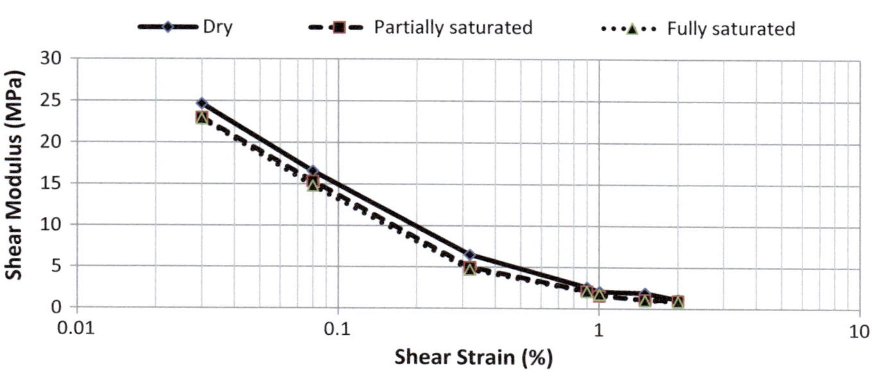

Fig. 1 Variation of shear modulus (G) with shear strain for different degrees of saturation

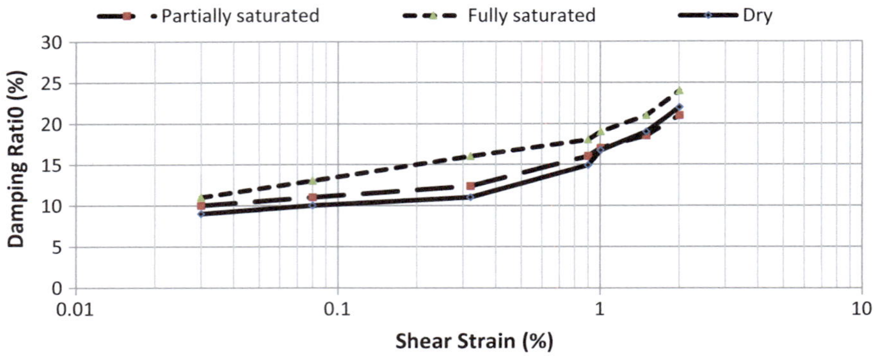

Fig. 2 Variation of damping ratio with shear strain for different degrees of saturation

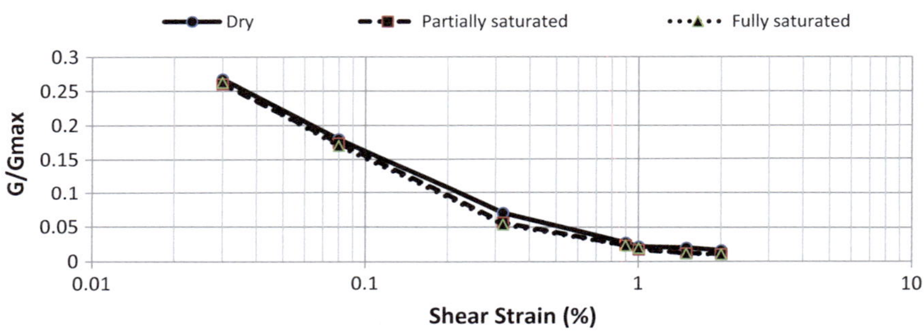

Fig. 3 Variation of normalized shear modulus (G/G_{max}) with shear strain for different degrees of saturation

value for fully saturated sands. Hence, the trend of results in the present study at high strain is similar to that observed by Madhusudan and Kumar (2013) at low strain.

In Fig. 3, the variation of the normalized shear modulus ratio with the increase in shear strain is shown for different degrees of saturation. It can be observed that as the soil is in unsaturated condition, the normalized shear modulus ratio is relatively higher than that of the saturated sand sample. However, smaller difference is noticed for partially saturated and fully saturated samples of Solani sand. Also comparing Figs. 3 with 1, it can be observed that these three curves are closer than before due to normalization with G_{max}.

4 Conclusion

The dynamic properties such as shear modulus and damping ratio on sand samples of Solani sand (collected from the bed of Solani River near Roorkee) have been examined at three different degrees of saturation (S_r = 0, 50, and 98%) using a series of cyclic triaxial tests, at the same effective confining pressure in large shear strain range. Based on the results presented, the following conclusions can be drawn:

(1) With the increase in shear strain, shear modulus decreases while damping ratio increases, as expected.
(2) It was observed that as degree of saturation decreases, the shear modulus increases with the increase in strain. The shear modulus is relatively higher for dry sands as compared to saturated sands but there is a slight difference noticed for partially and fully saturated sands.
(3) The damping ratio is greater for the saturated sand sample than that of the dry and partially saturated sands.
(4) Effect of saturation on normalized shear modulus (G/G_{max}) is smaller than that on shear modulus (G).

The above conclusions are based on the limited test data and to arrive on a quantitative outcome, a more number of samples need to be tested. Nonetheless, the research presented has very much practical significance, as not much study has been reported for this theme.

References

ASTM: D3999-91. (2003). Test method for the determination of the modulus and damping properties of soils using cyclic triaxial apparatus. Annual book of ASTM standards. West Conshohocken, PA: ASTM International.

ASTM: D5311/D5311M. (2013). Standard test method for load controlled cyclic triaxial strength of soil.

Kramer, S. L. (1996). *Geotechnical earthquake engineering*. New Delhi, India: Pearson Education.

Madhusudhan, B. N. (2011). Dynamic properties of dry to fully saturated states using resonant column and bender element tests, Ph.D. thesis, Indian Institute of Science, Bangalore, India.

Madhusudhan, B. N., & Kumar, J. (2013). Damping of sands for varying saturation. *Journal of Geotechnical and Geoenvironmental Engineering, 139*(9), 1625–1630.

Maheshwari, B. K., Kale, S. S., & Kaynia, A. M. (2012). Dynamic properties of Solani sand at large strains: a parametric study. *International Journal of Geotechnical Engineering, 6*(3), 353–358.

RaviShankar, B., Sitharam, T.G., & Govinda Raju, L. (2005). Dynamic Properties of Ahmedabad Sands at Large Strains. In *Proceedings of Indian Geotechnical Conference,* Ahmedabad, pp. 369–372.

Wykeham F. (2008). Manuals of automatic dynamic triaxial system. In *Soil Mechanics Division*. U.K.: Controls Testing Equipment.

Effects of Discontinuities on Rock Slope

Aniruddha Bhaduri and B. K. Maheshwari

Abstract Discontinuities possess a major role in the stability of a rock slope and also influence the behavior and failure patterns of it under different loading environments. In this paper, the effects of different joint patterns on the stability of a 260-m-high slope with a face angle (ψ_f) of 55° are examined under different loading environments. Under gravity loading, when the slope is having out-of-plane joints, it becomes more stable when the dip of the discontinuity (ψ) increases, although failure pattern changes. Plane and rotational failure patterns are observed when ψ becomes less than and greater than ψ_f, respectively. In case of joints dipping into the plane, the shear displacement becomes significantly less than the previous cases, which indicates a toppling failure. Similar observations can be made under rainfall condition and under seismic loading. In the seismic case, shear displacement and block rotations are obtained. The failure mechanisms of the slopes under different joint sets are investigated by obtaining displacements at different time intervals of the seismic loading. Universal distinct element code (UDEC 4.0) is used for all the simulations.

Keywords Discontinuity · Failure pattern · Plane · Rotational
Toppling

1 Introduction

Rockslides are a common phenomenon in hilly areas. If the rock slope consists of discontinuities, then the fractured rock slope becomes vulnerable toward failure. The failure patterns of a rock slope depend on the alignments and the orientations of

---

A. Bhaduri (✉) · B. K. Maheshwari
Department of Earthquake Engineering, Indian Institute of Technology Roorkee,
Roorkee 247 667, India
e-mail: aniruddha241991@gmail.com

B. K. Maheshwari
e-mail: bkmahfeq@iitr.ac.in

B. Adimoolam and S. Banerjee (eds.), *Soil Dynamics and Earthquake Geotechnical Engineering*, Lecture Notes in Civil Engineering 15,
https://doi.org/10.1007/978-981-13-0562-7_7

the discontinuities. The basic terminology which defines the orientation of the discontinuities is termed as dip or dip inclination (ψ). It is defined as the maximum inclination of the discontinuity measured from the horizontal. Different failure patterns can be observed depending upon the joint orientations (i.e., out-of-plane and into-the-plane joints) with respect to the dip of the slope face (ψ_f).

Many researchers have used different numerical techniques to get a proper view about the failure mechanism of a jointed rock slope, of which continuum (finite element) and discontinuum (distinct element) techniques are the most common. But continuum techniques have some major difficulties in analyzing a discontinuous medium consisting of several intersecting interfaces. On the other hand, discontinuum codes (like UDEC) are suitable in these cases due to some advantages like allowing finite displacements and rotations of discrete bodies and also recognizing new contact surfaces with progress. Thus, use of discontinuum codes is preferable over the continuum codes in case of discontinuous rock slopes. Bhasin and Kaynia (2004), Noorzad et al. (2008), Pal et al. (2012), Kainthola et al. (2012), and many other researchers have successfully applied the universal distinct element code UDEC (Itasca 2004) for analyzing discontinuous rock slopes.

This paper presents the effects of discontinuities on a 260-m-high jointed rock slope under different loading conditions like static gravity loading and seismic loading. The main objective of the study is to investigate the effects of joint orientations on the mode of failures like plane, rotational sliding, and toppling failures. Modeling and analysis of the slopes have been carried out by using distinct element code UDEC 4.0. Understanding the failure patterns and mechanisms will be helpful while implementing proper method for stabilizing a rock slope.

2 Failure Patterns

There are three common types of failure patterns observed in a rock slope: plane failure, rotational sliding failure, and toppling failure.

2.1 Plane Failure

This type of failure occurs when the sliding plane becomes parallel or sub-parallel to the slope face. Also here, ψ_f is greater than the dip of the sliding plane (ψ) and ψ is greater than the joint friction angle (ϕ or Jfric) of the discontinuity.

2.2 *Rotational Sliding Failure*

This type of failure occurs when the unstable rock mass slides down along a curved surface. Shear displacement can occur along a circular or non-circular path. When the rock mass becomes densely fractured and fragmented, then this type of failure occurs.

2.3 *Toppling Failure*

In this mode, the unstable blocks of rocks rotate with respect to a fixed base. When the center of gravity of the column slab of block lies outside the base of the column block, then the rotation takes place. In the jointed rock mass, the discontinuities should be dipping into the face and going away from the face. ψ should be steep, and joints should be closely spaced.

3 Problem Statement

Here, the stability of a 260-m-high rock slope has been simulated. The geometry of the slope is taken from Hammah et al. (2007) and also given by Duncan and Mah (2004). Figure 1 shows the basic geometry of the slope.

The material properties taken for the analysis are given in Table 1.

4 Results and Analysis

Mainly two types of analyses are done: gravity analysis and seismic analysis.

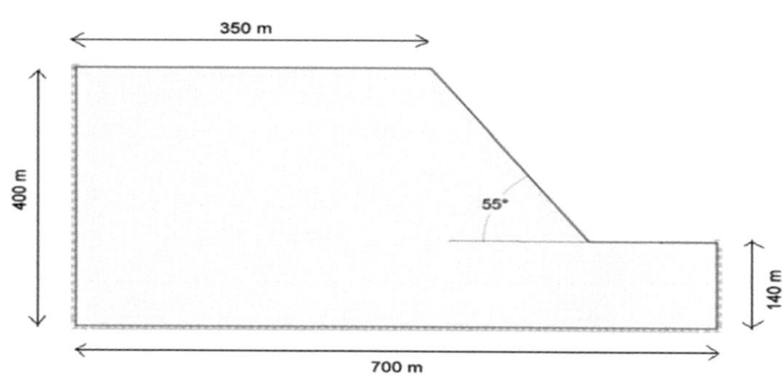

Fig. 1 Basic slope geometry

Table 1 Material properties

Rock mass	Discontinuities
Density = 2610 kg/m^3	Joint normal stiffness (Jk_n) = 100 GPa
Angle of internal friction = 43°	Joint shear stiffness (Jk_s) = 10 GPa
Poisson's ratio = 0.26	Cohesion = 0.1 MPa
Cohesion = 0.675 MPa	Joint friction angle (Jfric) = 40°
Bulk modulus = 9.072 GPa	Joint spacing = 20 m
Shear modulus = 5.184 GPa	–
Tensile strength = 0 MPa	–

4.1 Static Gravity Analysis

Under this analysis, four types of joint alignments are considered. Slope having one joint set with a (1) ψ of 35° out of the plane, (2) ψ of 60° out of the plane, (3) ψ of 75° out of the plane, and (4) ψ of 70° into the plane. Mohr–Coulomb plasticity model has been considered for the material. The factor of safety and the maximum shear displacement values are calculated for all the cases. The results of the analysis are shown in Table 2.

Figure 2 shows the failure patterns for different cases.

From Table 2, we can see that when ψ increases in the out-of-plane direction, the factor of safety also increases and the maximum shear displacement decreases. Thus, the slope becomes more stable with the increase in ψ. But if we compare Fig. 2a–c, we can see that the failure patterns are different. Plane, rotational sliding, and circular sliding failures are observed in the first, second, and third cases, respectively. Plane failure is observed in the first case because here ψ is less than ψ_f. As this postulate does not hold in the second and third cases, rotational sliding failure takes place. In the case when joints are dipping into the plane, we can see that the shear displacement is very small as compared to the other cases. This indicates that it is not a case of shear failure rather a toppling failure. Further parametric studies have been conducted in slopes having two joint orientations (in one case, ψ is 35° out of plane, and in another case ψ is 75° into the plane) by reducing the Jfric in a gradual manner which can correspond to rainfall condition. It has been observed in both the cases that as the joint friction angle reduces, the shear displacement increases and FOS decreases. But a comparative study between the two cases shows that the shear displacement in the case when ψ dips into the plane

Table 2 Results of gravity analysis

Joint alignment	Cases	Dip angle(°)	FOS	Shear displacement (cm)
Out of the plane	First case	35	1.32	5.636
	Second case	60	1.44	4.525
	Third case	75	1.56	3.848
Into the plane	–	70	1.35	0.036

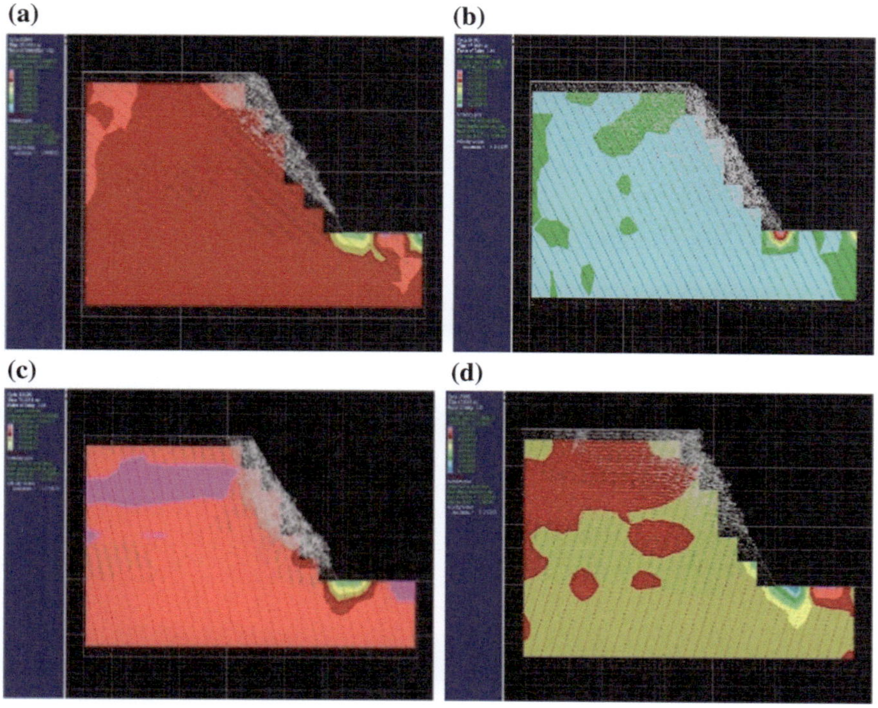

Fig. 2 **a** Plane, **b** rotational sliding, **c** circular sliding, and **d** toppling failure

is very less with respect to the other case when ψ dips out of plane. This observation shows that slope is prone to shear failure when it consists of joints dipping in the out-of-plane direction, while the slope is prone to toppling failure when joint set dips in the into-the-plane direction.

4.2 Slope Under Seismic Loading

Under this condition, again two types of joint alignments are considered. In the first case, the joint set is dipping 70° out of plane, and in the second case, the joint set is dipping 70° into the plane. Here, elastic, isotropic constitutive model has been taken for the material. The seismic loading has been applied as a sinusoidal stress history at the base of the model. The stress history represents the seismic loading of Uttarkashi Earthquake on October 20, 1991. The sinusoidal shear wave which has been applied at the base of the model has a frequency of 3 Hz for a period of 3 s and PGA of 0.3 g. When the sinusoidal loading is applied, the boundary condition has been changed to free field/viscous boundary at the base and in the sides of the model. The viscous boundary conditions are applied to absorb the incoming waves

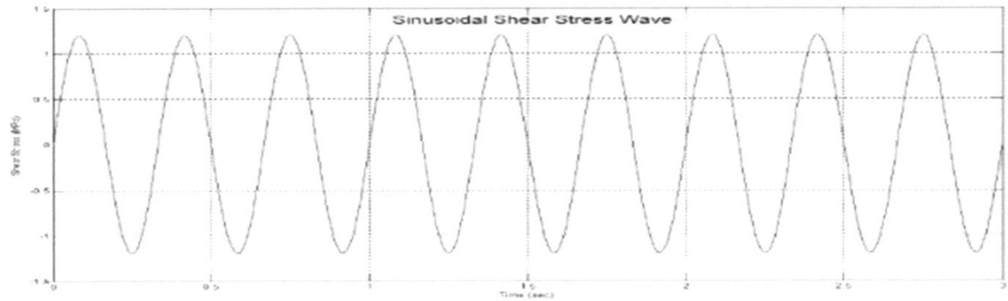

Fig. 3 Input sinusoidal shear stress waveform at the base of the model

Table 3 Results of seismic analyses

Time duration (s)	Max. shear displacement (cm)		Block rotation (°)	
	First case	Second case	First case	Second case
0	2.10	2.20	0.305	0.580
1	2.38	2.55	0.365	0.756
3	3.51	2.68	0.780	1.315
5	6.87	3.68	1.105	1.902
7	10.26	6.71	1.396	2.133
9	14.09	10.35	1.678	2.340
11	16.70	14.94	1.997	2.585
13	–	18.86	–	2.789

hitting the boundaries. It prevents the reflection of the waves back into the model. The sinusoidal shear stress wave which is applied into the model is shown in Fig. 3.

In these analyses, the behavior of the slope has been observed during and after the earthquake event took place. Maximum shear displacement and the block rotations are observed at each time-step. The results of the analyses have been shown in Table 3.

From Table 3, it is observed that in the first case, the slope failed in shear after 11 s due to great contact overlap. In other observations, it can be seen that in the second case, the maximum shear displacement is lesser, while the block rotation is greater than the first case.

Figures 4 and 5 represent the comparison between two cases with respect to their shear displacements and block rotations, respectively. These observations indicate that when joint sets are dipping in the out-of-plane direction, it is prone to shear failure, while when joint sets are dipping into the plane, toppling failure occurs.

The failure mechanisms for both the first and second cases are shown in Figs. 6 and 7, respectively, at 3-s, 5-s, 9-s, and 11-s time interval.

In Figs. 6 and 7, the plots of velocity vectors along with the total displacements are shown. We can see from both the figures that the displacements are increasing

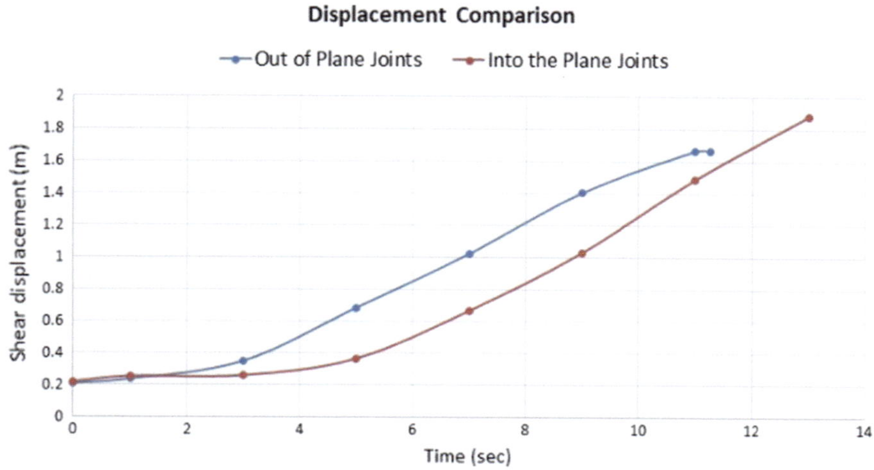

Fig. 4 Comparisons between displacements

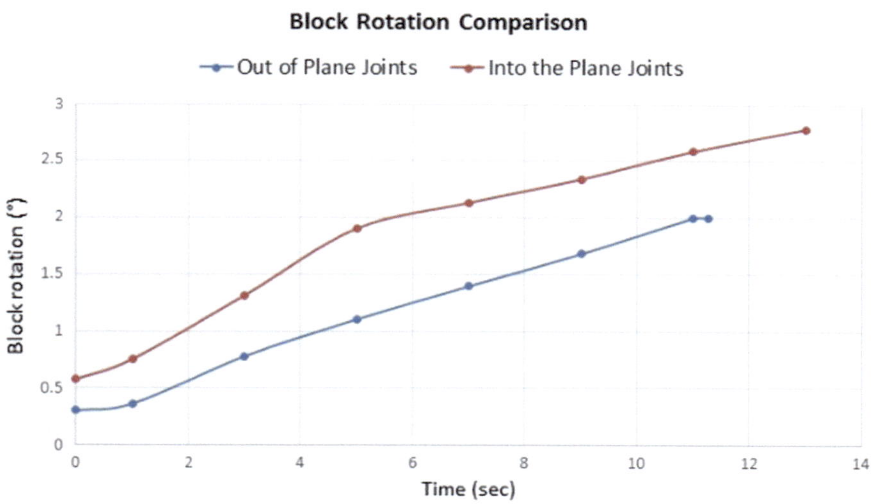

Fig. 5 Comparisons between block rotations

with time. But in Fig. 6, it can also be noted that the lateral and downward shear displacement is more with respect to Fig. 7, while in Fig. 7, in the left portion of slope, the slabs of the blocks are coming out. These observations along with the values obtained from Table 3 indicates that the slope in the first case is prone to shear failure, while in the second case, it is prone to toppling failure.

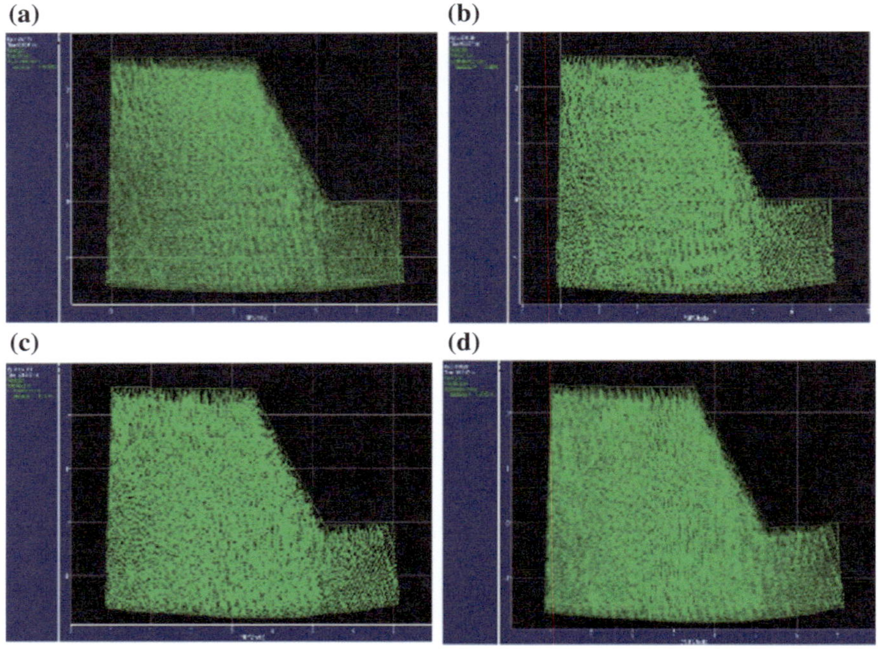

Fig. 6 Failure mechanisms of slopes having out-of-plane joints at **a** 3 s, **b** 5 s, **c** 9 s and **d** 11 s

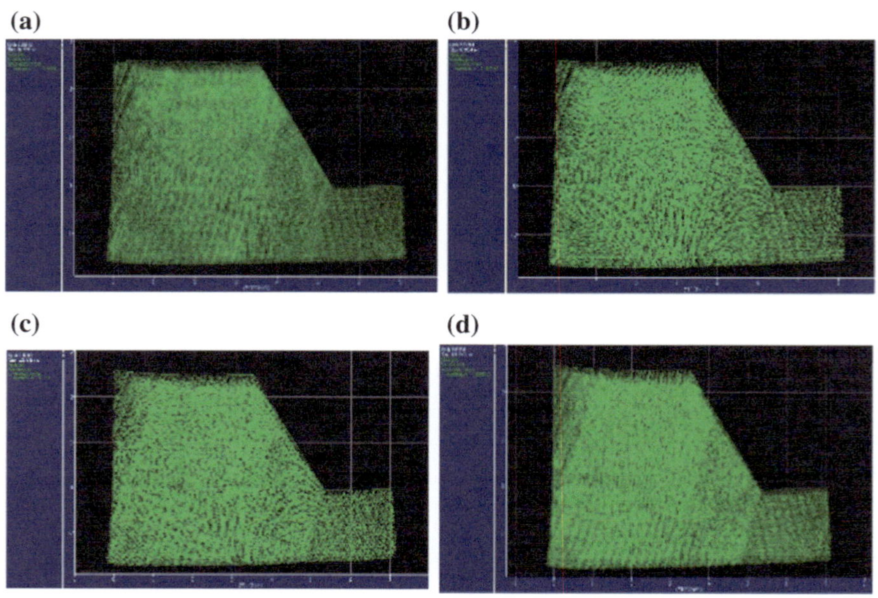

Fig. 7 Failure mechanisms of slopes having into-the-plane joints at **a** 3 s, **b** 5 s, **c** 9 s and **d** 11 s

5 Conclusions

Numerical analysis has been carried out for a 260-m-high rock slope to simulate the effects of joints and their orientations on their respective failure mechanisms. Two types of joint orientations are considered: out-of-plane and into-the-plane joints. The slope has been analyzed under different loading conditions. In the static gravity analysis, it has been observed that when the joint set dips in the out of the plane direction, plane and sliding failures occur. As the dip inclination of the joint set (ψ) gets stepper, the slope becomes more stable against sliding although failure pattern differs. Plane failure takes place when ψ is less than ψ_f, and rotational sliding occurs otherwise. When ψ dips into the plane, then sliding is not the failure pattern, whereas toppling failure occurs. Under rainfall condition, when joint sets dip into the plane, it is more stable against sliding failure in comparison with the case when joint set dips out of plane.

In the seismic analysis, sinusoidal shear stress wave has been applied into the model. Two cases have been investigated. It has been observed that the block rotation is more and the shear displacement is less in the case when slope is having joint sets dipping into the plane in comparison with the case where joint set dips out of the plane. This again indicates a toppling failure for into-the-plane joint alignments and shear failure for the out-of-plane joint sets. These understandings help in suggesting suitable remedial measures for stabilizing a rock slope under different circumstances.

References

Bhasin, R., & Kaynia, A. M. (2004). Static and dynamic simulation of a 700-m high rock slope in western Norway. *Journal of Engineering Geology, 71,* 213–226.

Duncan, C. W., & Mah, C. W. (2004). *Rock slope engineering* (pp. 129–213). Spoon Press.

Hammah, R. E., Yacoub, T. E., Corkum, B., Wibowo, F., & Curran, J. H. (2007). Analysis of blocky rock slopes with finite element shear strength reduction analysis. In *1st Canada–US Rock Mechanics Symposium* (pp. 329–334).

Itasca Consulting Group, Inc. (2004). UDEC theory and background, version 4.0.

Kainthola, A., Singh, P. K., Wasnik, A. B., & Singh, T. N. (2012). Distinct element modelling of Mahabaleshwar road cut hill slope. *Geomaterials, 2*(4), 105–113.

Noorzad, A., Aminpoor, M., & Salari, H. (2008). Seismic slope stability analysis of jointed rock masses on the northern abutment of Gotvand dam, Iran. In *14th World Conference on Earthquake Engineering (14WCEE)*, Bejing, China.

Pal, S., Kaynia, A. M., Bhasin, R. K., & Paul, D. K. (2012). Earthquake stability analysis of rock slopes: A case study. *Journal of Rock Mechanics and Rock Engineering, 45,* 205–215.

Indirect Estimation of Local Soil Response in Light of Past as well as Recent Earthquakes in the Shillong Plateau

Abhishek Kumar and Olympa Baro

Abstract The northeastern region of India is a seismically active zone. The Shillong Plateau (SP) located in the southwestern portion of northeast India surrounded by an intense network of active tectonic faults. The 1897 Assam earthquake (EQ) (M_W 8.1) and other major EQs ($M_W \geq 7.0$) had originated in the faults surrounding the SP, thus highlighting the SP as a zone of high seismicity. During these past major to great EQs, widespread damages in the form of destruction to buildings, excessive ground shaking, uneven settlements, occurrence of ground fissures, and sand vents were reported across the SP and its adjoining regions. Such large-scale catastrophes during an EQ are the combined effect of the ground motions generated during an EQ and its modification by the subsoil at a site. Presence of local soil influences the frequency content, duration, and amplitude of the ground motions generated during an EQ. To understand the ground motion amplification potential at a site due to local soil, recorded ground motions, dynamic soil properties, in situ subsoil characteristics, etc., are required to be known at the site of interest. For majority of location, however, regional dynamic soil properties and in situ subsoil properties are not readily available. In the absence of regional dynamic soil properties, site response studies considering available dynamic soil properties from other regions are followed worldwide. In the present study, response of local soil in the SP is assessed considering the observed ground motion scenario during different EQs at selected sites. Ground motion amplification during each EQ is determined considering the peak horizontal acceleration (PHA) at the bedrock level and the peak ground acceleration (PGA) at the ground surface level. The PHA during each EQ is estimated using regional ground motion prediction equations (GMPEs). The PGA on the other hand is derived based on felt intensities during various EQ at considered sites from isoseismal maps. Thus, the response of in situ soil over a wide range of ground motions is assessed based on actual scenario

A. Kumar (✉) · O. Baro
Department of Civil Engineering, Indian Institute of Technology Guwahati, Guwahati 781039, India
e-mail: abhiak@iitg.ernet.in

O. Baro
e-mail: olympa.baro@iitg.ac.in

B. Adimoolam and S. Banerjee (eds.), *Soil Dynamics and Earthquake Geotechnical Engineering*, Lecture Notes in Civil Engineering 15,
https://doi.org/10.1007/978-981-13-0562-7_8

developed at the surface. This work will be helpful to understand the response of soil during probable future EQs in the absence of regional dynamic soil properties for the SP.

Keywords Shillong Plateau · Local soil effects · Isoseismal maps
Amplification factor

1 Introduction

The northeastern region of the Indian subcontinent consists of a number of tectonic faults. The intraplate tectonic movements along these faults have led to the origin of several earthquakes (EQs) in the past (Baro and Kumar 2015). Along with the occurrences of these EQs, the fault movements had also led to the formation of the Shillong Plateau (SP) within the region (Baro and Kumar 2015). The SP is thus surrounded by active faults which have evidenced numerous EQs in the past. The 1897 Assam EQ (M_W-8.1) is one of the great EQs of India which had originated on a fault lying toward the northern boundary of the SP. The tectonic activity of the faults surrounding the SP makes it essential to understand the seismic hazard potential of the plateau (Baro and Kumar 2016).

Seismic hazard analysis of a region involves the estimation of level of ground shaking that could occur at a site due to the presence of seismic sources surrounding the site. The level of ground shaking at a site is expressed in terms of ground motion parameters such as peak ground acceleration (PGA) or peak horizontal acceleration (PHA). To arrive at a particular ground motion parameter, ground motion prediction equations (GMPEs) are used. GMPEs express above ground motion parameters as functions of magnitude, epicentral or hypocentral distances, fault type, site class, etc. Further, GMPEs are developed by performing regression analysis of the recorded ground motions within a region and hence are best suited for that region. Instrumental recording of EQs is a recent phenomenon, especially in India which began only after 1986. In case of non-availability of recorded ground motion data, multiple GMPEs developed for other regions with similar tectonic characteristics are selected to estimate the ground motion parameter for a particular region. Nath and Thingbaijam (2011) emphasized that even though several GMPEs have been developed for India, however while performing seismic hazard analysis for Indian regions during past studies, due consideration was not given to the selection of GMPEs matching the tectonics of the region. Further, Nath and Thingbaijam (2011) based on the log-likelihood (LLH) method of Scherbaum et al. (2009) considered multiple GMPEs and ranked these GMPEs according to their suitability for the Himalayan, northeastern, and peninsular regions of India. For the northeastern region which is an intraplate region, Nath and Thingbaijam (2011) selected eight GMPEs which were developed for various intraplate regions of the world. Among eight GMPEs selected by Nath and Thingbaijam (2011), the top three are used in this study to estimate the PHA at bedrock level for the SP. These include GMPEs given by Hwang and Huo (1997), Raghukanth and Iyengar (2007) and Nath et al. (2009). In addition, the GMPEs were given by NDMA (2010) and Anbazhagan et al. (2013) are also employed to estimate the PHA for the SP. The GMPE proposed by NDMA (2010) was developed for the entire country with separate set of coefficients for seven different regions within the country including northeast India. Anbazhagan et al. (2013) developed a GMPE for the Himalayan region considering both recorded and synthetically generated ground motion data. The synthetic EQ data was developed for past EQs with no ground motion records, which also included the 1897 Assam EQ originating on the northern edge of the SP. The GMPEs proposed by NDMA (2010) and Anbazhagan et al. (2013) are capable of estimating the PHA for EQs of $M_W \geq 8.0$.

Present work also tests the suitability of the GMPEs proposed by NDMA (2010) and Anbazhagan et al. (2013) for the SP by comparing with the three highly ranked GMPEs for northeast India as per Nath and Thingbaijam (2011).

2 Analysis

The five GMPEs mentioned above are used to estimate the PHA values at selected sites within the SP at the bedrock level. The PHA values thus estimated could be used for hazard estimation of the SP in the future. However, such seismic hazard analysis performed without taking into consideration the effects of the local soil is of limited use. The local soil has the characteristics to modify the EQ ground motion as it travels from the bedrock to the ground surface. Hence, it is essential to consider the local soil effects when estimating the hazard potential of a region. In this study, the effect of local soil is taken into consideration by estimating the ground motion parameter PGA at the surface level for the SP. Thus, the PGA values of some of the past major EQs originating in the faults around the SP are estimated in this study.

To estimate the PGA values, a correlation is developed between Modified Mercalli Intensity (MMI) and PGA in the absence of any correlation existing at present for the SP. Proposed correlation is developed taking into account MMI and PGA values obtained during the 2016 Myanmar EQ (M_W-6.9) and the 2016 Imphal EQ (M_W-6.7) as per United States Geological Survey (USGS) (https://www.usgs.gov/). Even though these two EQs had not occurred on the faults surrounding the SP, the shaking due to the EQs was felt within the plateau. Further, there are no recent major EQs in the vicinity of the SP to develop such correlation. Using the MMI and PGA values reported during the above-mentioned EQs a MMI versus PGA plot is developed as shown in Fig. 1. Figure 1 also shows the correlation developed between MMI and PGA for the SP.

The developed correlation shown in Fig. 1 is used to estimate PGA values during 1885 Bengal EQ (M_L-7.0), 1869 Cachar EQ (M_W-7.5), 1918 Srimangal EQ (M_S-7.6), and 1930 Dhubri EQ (M_S-7.1) based on reported MMI values. It has to be highlighted here that above-mentioned four EQs had originated on faults lying at a

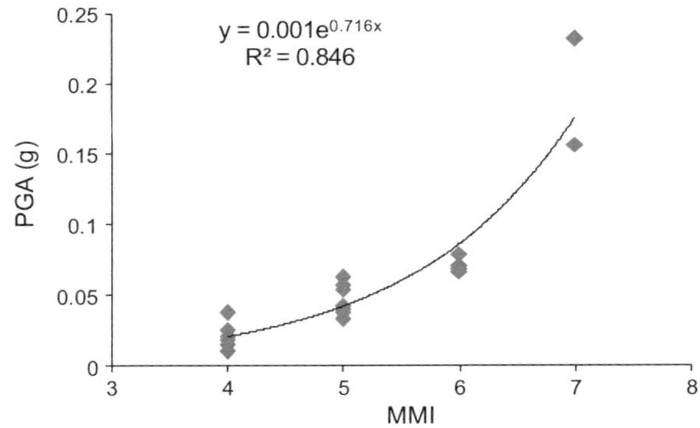

Fig. 1 Correlation between MMI and PGA for the SP

close vicinity of the SP and had caused damages in the SP. The MMI felt within the SP during the above-mentioned past EQs are collected from isoseismal maps. The isoseismal maps were developed by the Geological Survey of India. The Geological Survey of India developed the isoseismal maps for each of these EQs in different intensity scales. Sabri (2002) re-evaluated the intensities for each of the EQs and redrew the isoseismal maps using the European Macroseismic Scale (EMS). In this study, the isoseismal maps developed by Sabri (2002) are used and the EMS scale is converted to MMI scale as per Musson et al. (2010). It was observed from the isoseismal maps that during the various past EQs different sites across the SP experienced different shaking intensities. MMI of VI and V were felt across Shillong city during the 1869 Cachar EQ and 1918 Srimanagal EQ, respectively. Similarly, during the same 1918 Srimanagal EQ, MMI of VI was felt in Cherrapunji. Further, MMI of V and VI during the 1885 Bengal EQ and the 1930 Dhubri EQ, respectively, were reported in Cherrapunji. At Tura as well, MMI of V was reported during the 1885 Bengal EQ. It can be observed from Fig. 1 that the proposed correlation is valid only till a MMI value of 7. For this reason, the PGA value for the 1897 Assam EQ is not estimated using the proposed correlation since a MMI of X was reported due to this EQ within the SP. Thus, employing various MMI values from the isoseismal maps into the proposed correlation, PGA values at the surface level across the SP are estimated.

It has been mentioned earlier that five GMPEs proposed by Hwang and Huo (1997), Raghukanth and Iyengar (2007), Nath et al. (2009), NDMA (2010), and Anbazhagan et al. (2013) are selected in this study to estimate the PHA values at the bedrock level. Using these five GMPEs, PHA values within the SP are estimated as listed in Table 1. It has to be highlighted that soil response which is examined earlier from MMI values is a function of PHA as highlighted by Kumar et al. (2015). In addition, from Table 1 it can be observed that a wide range of PHA from 0.02 to 0.19*g* has been considered in the present work representing a wide range of ground motion scenario. Taking into account, PHA estimated using each of the five selected GMPEs and above-estimated PGA, amplification factor (AF) which is the

Table 1 PHA and AF values estimated for the past EQs in the SP

Site		Shillong city	Shillong city	Tura	Cherrapunji	Cherrapunji	Cherrapunji
GMPEs		EQ					
		1869 Cachar	1918 Srimangal	1885 Bengal	1885 Bengal	1918 Srimangal	1930 Dhubri
Hwang and Huo (1997)	PHA	0.14	0.13	0.07	0.04	0.20	0.06
	AF	0.51	0.28	0.54	1.00	0.37	1.26
Raghukanth and Iyengar (2007)	PHA	0.12	0.10	0.02	0.01	0.16	0.02
	AF	0.59	0.37	1.93	4.31	0.45	3.80
Nath et al. (2009)	PHA	0.19	0.17	0.11	0.06	0.12	0.10
	AF	0.38	0.21	0.34	0.56	0.64	0.72
NDMA (2010)	PHA	0.02	0.11	0.05	0.03	0.14	0.05
	AF	3.86	0.34	0.66	1.04	0.52	1.34
Anbazhagan et al. (2013)	PHA	0.07	0.06	0.03	0.02	0.09	0.03
	AF	1.08	0.56	1.30	2.11	0.78	2.61

Fig. 2 PHA versus AF obtained from present work

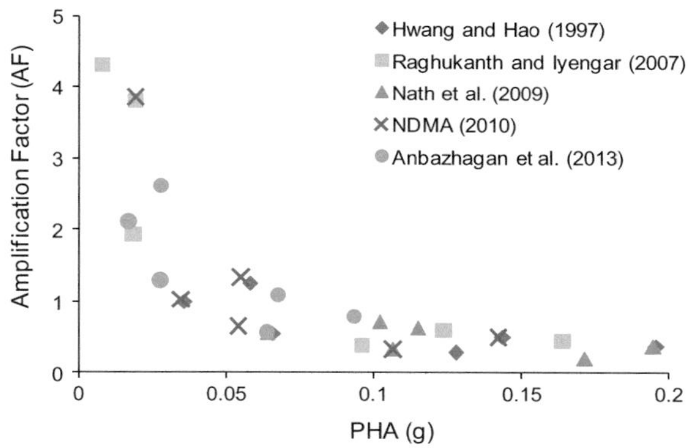

ratio of the PGA to PHA are estimated. Table 1 lists PHA and AF values estimated at various sites within the SP during different EQs.

Kumar et al. (2016b) highlighted that overall soil response can be understood by analyzing collectively the response of soil during different bedrock scenario known as dynamic soil response curve (DSRC). In order to generate an overall understanding of the soil response, DSRC from this work is shown in Fig. 2.

It can be observed from Fig. 2 that for higher PHA values lower values of AF are obtained. This is in accordance to the findings of EPRI (2012), Romero and Rix (2005), and Kumar et al. (2016a, b) where it was found that higher values of PHA give lower values of AF. EPRI (2012) and Kumar et al. (2016a) reported that for PHA higher than 0.5*g* deamplification of AF occurs. From Fig. 2, it can be observed that in this study the deamplification of AF has started to occur at 0.05*g*. It has to be highlighted here that most of the sites across the SP are of site classes of A (firm or hard rocks) or B (soft to firm rocks) as per Mittal et al. (2012). The sites Shillong, Tura, and Cherrapunji chosen in this study are of site classes A, B, and A, respectively. Since the sites are mostly rocks the ground motion amplification is very less within the SP. Hence in Fig. 2, as the PHA value begins to increase the AF starts to decrease.

Another observation which can be made from Fig. 2 is that the GMPEs proposed by NDMA (2010) and Anbazhagan et al. (2013) are closely matching with the other three highly ranked GMPEs.

Nath and Thingbaijam (2011) did not consider GMPEs proposed by NDMA (2010) and Anbazhagan et al. (2013) for northeast India. However, collectively based on PHA observed from Table 1 as well as DSRC from Fig. 2 it can be concluded that the GMPEs proposed by NDMA (2010) and Anbazhagan et al. (2013) are also equally suitable for northeast India both in arriving at seismic hazard values as well as to understand soil response. Similar validation for PHA $> 0.2g$ has not been attempted here and can be attempted during future works.

3 Conclusion

The SP has experienced several damaging EQs in the past. To reduce the risk of damages from similar EQs in the future it is essential to understand the level of ground shaking which occurred within the SP during the past EQs. Hence, in this study an attempt is made to estimate the PHA values at the bedrock level with the help of GMPEs as well as the PGA at the ground surface from isoseismal maps of past EQs. A total of five GMPEs are used to estimate the PHA values. Three of the GMPEs are found to be suitable for northeast India by previous works. Suitability of the remaining two GMPEs for the SP however is tested in this study. Further in situ soil response is studied over a wide range of PHA which is in accordance with existing literature.

References

Anbazhagan, P., Abhishek Kumar, Sitharam, T.G. (2013) Ground motion prediction equation considering combined dataset of recorded and simulated ground motions. *Soil Dynamics and Earthquake Engineering* 53:92–108

Baro, O., & Kumar, A. (2015). A review on the tectonic setting and seismic activity of the Shillong plateau in the light of past studies. *Disaster Advances, 8,* 34–45.

Baro, O., & Kumar, A. (2016). Seismic source characterization for the Shillong plateau in northeast India. *Journal of Seismology* (Under Review).

EPRI. (2012). *Guidelines for determining design basis ground motion*. Palo Alto, CA: Electric Power Research Institute, 1, EPRI TR-102293.

Hwang, H., & Huo, J. R. (1997). Attenuation relations of ground motion for rock and soil sites in eastern United States. *Soil Dynamics and Earthquake Engineering, 16,* 363–372.

Kumar, A., Harinarayan, N. H., & Baro, O. (2015). High amplification factor for low amplitude ground motion: Assessment for Delhi. *Disaster Advances, 8*(12), 1–11.

Kumar, A., Baro, O., & Harinarayan, N. H. (2016a). Obtaining the surface PGA from site response analyses based on globally recorded ground motions and matching with the codal values. *Natural Hazards, 81,* 543–572.

Kumar, A., Harinarayan, N. H., & Baro, O. (2016b). Predicting the surface response spectrum for Nepal based on nonlinear soil response evidenced collectively during recent and past earthquakes. *Natural Hazards* (Under Review).

Mittal, H., Kumar, A., & Ramhmachhuani, R. (2012). Indian national strong motion instrumentation network and site characterization of its stations. *International Journal of Geosciences, 3,* 1151–1167.

Musson, R. M. W., Grünthal, G., & Stucchi, M. (2010). The comparison of macroseismic intensity scales. *Journal of Seismology, 14*(2), 413–428.

Nath, S. K., Raj, A., Thingbaijam, K. K. S., & Kumar, A. (2009). Ground motion synthesis and seismic scenario in Guwahati City–A stochastic approach. *Seismological Research Letters, 80,* 233–242.

Nath, S. K., & Thingbaijam, K. K. S. (2011). Peak ground motion predictions in India: An appraisal for rock sites. *Journal of Seismology, 15,* 295–315.

National Disaster Management Authority (NDMA). (2010). *Development of Probabilistic Seismic Hazard Map of India*. Government of India: National Disaster Management Authority.

Raghukanth, S. T. G., & Iyengar, R. N. (2007). Estimation of seismic spectral acceleration in Peninsular India. *Journal of Earth System Science, 16,* 199–214.

Romero, S. M., & Rix, G. J. (2005). *Ground motion amplification of soils in the upper Mississippi embayment*. Report No. GIT-CEE/GEO-01-1, National Science Foundation Mid America Earthquake Centre.

Sabri, M. S. A. (2002). *Earthquake intensity-attenuation relationship for Bangladesh and its surrounding region* (Master's thesis). Bangladesh University of Engineering and Technology, Dhaka, Bangladesh.

Scherbaum, F., Delavaud, E., & Riggelsen, C. (2009). Model selection in seismic hazard analysis: An information theoretic perspective. *Bulletin of Seismological Society of America, 99,* 3234–3247.

United States Geological Survey (USGS). https://www.usgs.gov/. Assessed July 28, 2016.

Development of Predictive Equation for Vibration Due to DMC Piling

V. Jaya, P. T. Prijil and K. Balan

Abstract Conventional casting of bored cast-in-situ pile which is also known as direct mud circulation (DMC) method was accepted as the most reliable piling system all over the world. Compared to precast pile driving, chiseling action in DMC piling produces low energy. This low energy can produce stress waves in the ground which may cause problems to the adjacent structures. This paper deals with the development of predictive equation for the peak particle velocity (PPV) for different types of soil due to ground vibrations caused during DMC method of piling. Ground vibration analyses were carried out for different soil profiles with different hammer energy impacts. The ground acceleration produced during DMC piling at different distances and depths from the source of vibration were measured on layered soft clay and lateritic soil mediums. The PPV depends on the soil constants k and n. These soil constants for different soil types, such as soft clay and lateritic clay, are compared in this paper. Developed soil constants of DMC piles are then compared with those of driven piles.

Keywords DMC piling · Peak particle velocity · Soil constants
Vibration prediction

V. Jaya
Department of Civil Engineering, Government Engineering College,
Barton Hill, Trivandrum 695035, India
e-mail: jayasraj@gmail.com

P. T. Prijil (✉) · K. Balan
Department of Civil Engineering, College of Engineering Trivandrum,
Trivandrum 695016, India
e-mail: pt.prijil@gmail.com

K. Balan
e-mail: drkbalan@gmail.com

B. Adimoolam and S. Banerjee (eds.), *Soil Dynamics and Earthquake Geotechnical Engineering*, Lecture Notes in Civil Engineering 15,
https://doi.org/10.1007/978-981-13-0562-7_9

1 Introduction

Piling methods are essential for the construction of buildings, transmission towers, and high-rise structures where adequate ground support is not directly available. Conventional bored cast-in-situ piling methods are widely used nowadays over impact-type piling methods. In contrast, vibrations originating from the piling instruments have serious adverse effects on adjacent structures and their foundations, as well as on vibration sensitive installation and occupants of buildings. Thus, the vibration assessment and prediction has become a topic of great interest. Due to the subjective nature, it is difficult to determine the accurate vibration level associated with the piling. In order to analyze the vibration-related problems, the combined effect of several factors such as site characteristics, layering of soil structure, propagation of body and surface waves in the particular soil type are to be considered. Accurate prediction of vibration levels are required to protect the construction companies from fraudulent claims.

Current practices for dealing with the ground vibration during piling operation mainly consider peak particle velocity (PPV) values determined by empirical relations which are easy to predict and compare. The literatures associated with the prediction of PPV values are originally established via empirical correlation determined from actual observation of low-rise residence structural damages (Siskind et al. 1980). Even though there are several empirical methods available for prediction of PPV, most important empirical relations are based on the pioneer generalized form by Wiss (1981). Majority of the studies in ground vibration prediction are related to blasting or mining operations, like predictor equation developed by Aloui et al. (2016). Standards for limiting the ground vibration based on PPV values for blasting and mining operations such as USBM standards, 50.8 mm/s criteria by Svinkin (2003, 2013), can be applied to piling vibrations.

2 Measurement of Ground Vibration

2.1 Site Description

Eight sites with different types of soil strata were selected for the study. Three sites were located at Ernakulam District, three at Alappuzha District, and two at Thiruvananthapuram District in Kerala State.

2.1.1 Type 1: Soft Clay Stratum

Kuttanad region in Alappuzha District is a low-lying area. The soil in this region is highly compressible soft clay overlain by either poorly graded sand or filled laterite with water table near the ground. Typical borehole details of sites with soft clay are shown in Fig. 2 in Balan et al. (2015).

2.1.2 Type 2: Hard Lateritic Clay Stratum

Lateritic soil is present almost all over the state of Kerala except in coastal and hilly regions. Soil strata present in five sites located at Thiruvananthapuram and Ernakulam are generally laterite with high gravel content as top layer followed by hard lateritic clay with silt content with SPT N value in the range of 25–50 in shallow depths and >50 in deeper layers. Typical bore log details are shown in Fig. 3 in Balan et al. (2015).

2.2 Field Setup

Ground vibrations are induced as the piling operations progressed. Four auxiliary boreholes of diameter 10 cm and at a distance of 2, 4, 8, and 16 m were made at radially opposite directions of piling spot as shown in Fig. 1. Accelerometers placed over steel rods are themselves placed in these boreholes at different depths.

2.3 Peak Particle Velocity

Accelerometers were placed over steel rods which themselves were placed in boreholes at different depths. The ground vibration signals obtained by accelerometers were acceleration time histories, which had amplitude in *g*'s. These acceleration signals are, for engineering purposes, usually quantified in terms of peak particle velocity (PPV). Acceleration signals are processed by doing numerical integration to get corresponding PPV. Digital filter (IIR—infinite impulse response filter) was used to eliminate noise. Figures 2 and 3 represent a typical acceleration time history measured using an accelerometer and PPV time history after application of IIR filter, respectively.

Fig. 1 Field layout showing the position of point of impact and accelerometers

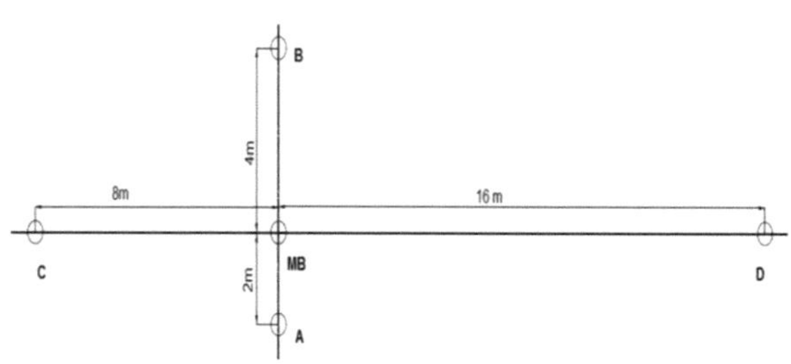

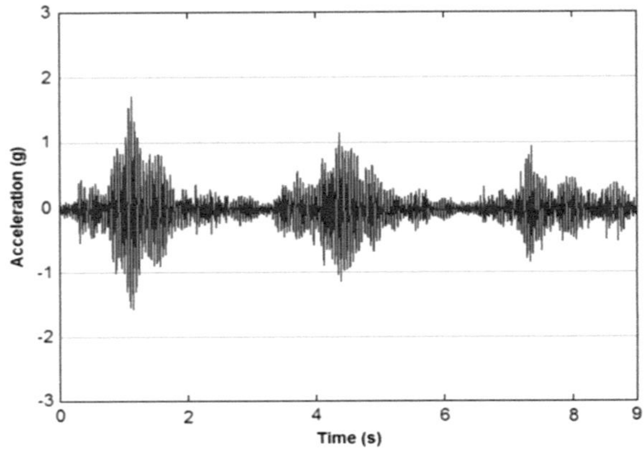

Fig. 2 Acceleration time history of pile driving vibrations

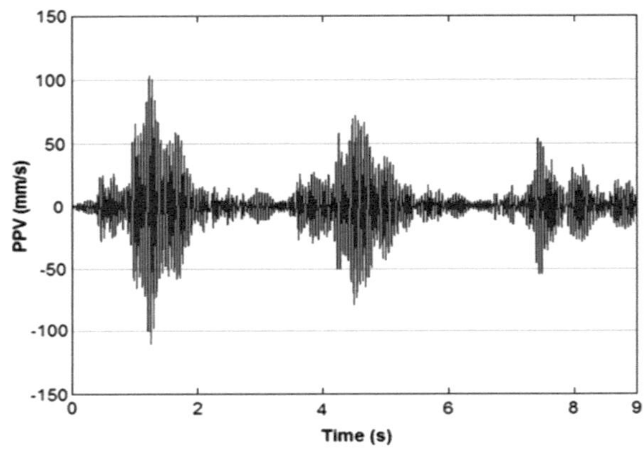

Fig. 3 PPV signal after application of IIR filter

3 Determination of Soil Constants

Vibration predictors generally have two constants such as k and n. These constants are entirely dependent on soil condition and impact source type. The PPV is related to chisel impact energy and distance from the impact by the prediction Eq. (1) (Wiss 1981).

$$V = k\left[\frac{\sqrt{E_n}}{D}\right]^n \quad (1)$$

where

E_n hammer energy in joules
D distance from the source in meters
k and n soil constants and
$\frac{\sqrt{E_n}}{D}$ scaled distance.

Soil constants are determined for each mean depth of impacts and for each site. Table 1 shows the calculated soil constants for various depths and their average values for sites where stiff lateritic clay is present.

Magnitude of soil constant k shows a sudden decrease after a depth of three meters. With the increase in the depth of impact, the magnitude of the surface waves produced will be decreased. Average values are taken accordingly, viz. up to three meters and below three meters. Average values of n are also calculated in a similar manner.

Soil constants are determined for soft clay for four sites having soft clay stratum present below a depth of two meters, and the average values are shown in Table 2.

Similar to the previous case, average values of soil constants are taken up to four meters depth and below four meters depth.

Medium stiff lateritic clay was present in two out of eight study sites. Water table was present near the ground. Average values of soil constants are calculated for these sites, and the results are listed in Table 3. Average soil constants are determined for shallow depths and for depths more than three meters.

Table 1 Average soil constants for stiff lateritic clay

Depth (m)	k	n	Average	
			k	n
1	0.627	1.296	0.610	1.315
2	0.615	1.361		
3	0.589	1.287		
4	0.479	1.194	0.362	1.147
5	0.296	1.144		
7	0.310	1.102		

Table 2 Average values of soil constants for soft clay

Depth (m)	k	n	Average	
			k	n
2	0.621	0.809	0.521	0.898
3	0.432	1.051		
4	0.510	0.834		
5	0.494	0.804	0.411	0.808
7	0.328	0.812		

Table 3 Average soil constants for medium stiff lateritic clay

Depth (m)	k	n	Average	
			k	n
1	0.409	0.937	0.418	0.927
2	0.440	0.929		
3	0.406	0.915		
4	0.386	0.814	0.385	0.812
5	0.399	0.796		
7	0.371	0.826		

Table 4 Average values of soil constants for surface fills

Depth (m)	Sand		Gravelly Laterite	
	k	*n*	*k*	*n*
0	0.391	1.111	0.326	1.194
1	0.609	1.102	0.306	1.079
2	0.304	1.032	0.274	1.041
Average	0.435	1.082	0.302	1.105

Table 5 Predictor equation for different soil types

Soil type	Average soil constants		Predictor equation, (PPV)
	k	*n*	
Hard lateritic clay	0.610	1.315	$0.610 \times \left[\frac{\sqrt{E_n}}{D}\right]^{1.315}$
Soft clay under WT	0.521	0.898	$0.521 \times \left[\frac{\sqrt{E_n}}{D}\right]^{0.898}$
Medium stiff clay under WT	0.418	0.927	$0.418 \times \left[\frac{\sqrt{E_n}}{D}\right]^{0.927}$
Sand	0.435	1.082	$0.435 \times \left[\frac{\sqrt{E_n}}{D}\right]^{1.082}$
Gravelly laterite	0.302	1.105	$0.302 \times \left[\frac{\sqrt{E_n}}{D}\right]^{1.105}$

Surface fills in the study sites were either poorly graded sand or gravelly laterite. Depth of surface fills varied from 0.8 to 2 m. Table 4 shows calculated values of soil constants and their average values.

PPV predictors are developed using the average values of soil constants for the five different types of soil strata. Table 5 shows the developed predictors for different soil strata. Since magnitudes of soil constants at shallow depths are higher, it determines the maximum possible values of PPV; average values at shallow depths are selected for three cases, viz. hard lateritic clay, medium stiff lateritic clay, and soft clay.

Figure 4 shows variation of predicted PPV for different soil types with scaled distances for shallow depths. These variation will have more engineering significance since it will represent the surface waves, which will propagate more distance than the body waves with higher amplitudes.

Figure 5 represents variation of PPV with scaled distance of impact for different soil types using developed predictor equation for a depth of impact greater than three meters. Scaled distance decreased when either energy of the impact decreased or distance from the source increased. For a particular site where impact energy remains constant, distance from the source alone is the indication of scaled distance.

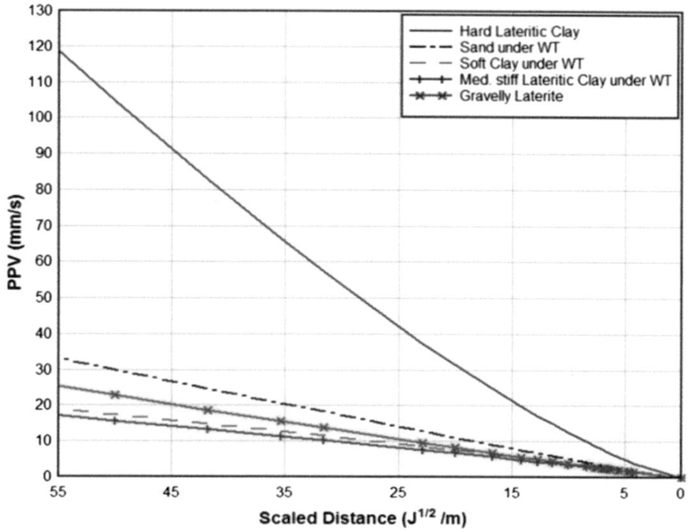

Fig. 4 Variation of predicted PPV values with scaled distances for shallow depths

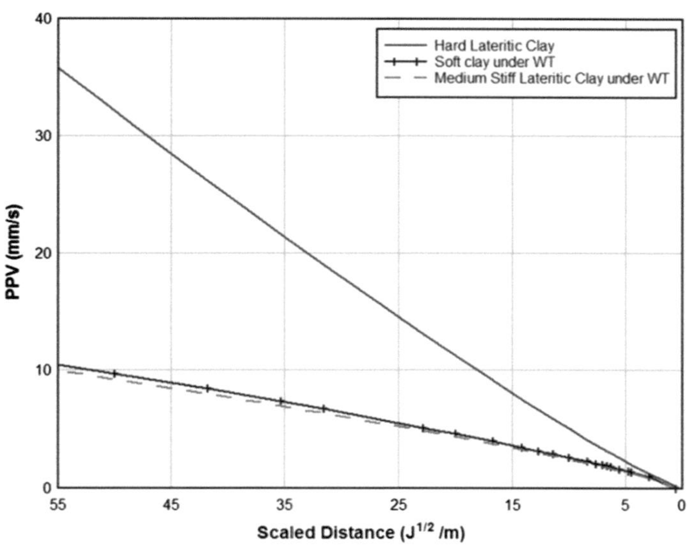

Fig. 5 Variation of predicted PPV values with scaled distances for depth greater than 3 m

Threshold values set by different country's standards by considering safety for low-rise buildings include 50.8 mm/s criteria (Svinkin 2003). A maximum PPV of 2 in./s or 50.8 mm/s is permissible for most of the modern structures if the frequency of excitation is higher.

4 Conclusions

Predictor equations for PPV are developed for each soil type. Efficiency of pile impact was not considered in the study. It is evident that presence of water table decreases the ground vibration amplitude. Medium stiff lateritic clay, highly compressible soft clay, poorly graded sand and gravelly laterite will propagate ground waves similar in terms of magnitude in immersed condition. Structures situated in these types of sites will be safe against damage during piling operations except for damage due to liquefaction of surface sand.

Sites where hard lateritic clay is present will show higher magnitudes of PPV and susceptibility for damage at shallow depths. Maximum possible PPV will exceed the permissible limit of 50.8 mm/s when the scaled distance is more than 30 $J^{1/2}$/m. Piling vibrations at depths more than three meters will induce only 1/3rd of that produced in shallow depths for higher scaled distances.

References

Aloui, M., Bleuzen, Y., Essefi, E., & Abbes, C. (2016). Ground vibrations and air blast effects induced by blasting in open pit mines: Case of Metlaoui Mining Basin, South western Tunisia. *Journal of Geology and Geophysics, 5*, 3.

Balan, K., Jaya, V., & Arun, C. (2015). Prediction of ground vibrations produced in DMC type of piling through soft and hard strata. *Indian Geotechnical Journal*.

Siskind, D. E., Stagg, M. S., Kopp, J. W., & Dowding, C. H. (1980). *Structure response and damage produced by ground vibration from surface mine blasting*. Report of Investigations 8507, Office of Surface Mines, US department of Interiors, USA.

Svinkin, M. R. (2003). *Drawbacks of blast vibration regulations*. www.vulcanhammer.net/svinkin/BLST-CRT.pdf.

Svinkin, M. R. (2013). Forensic engineering of construction vibrations. *Forensic Engineering, 166* (2), 81–93.

Wiss, J. F. (1981). Construction vibrations: State-of-the-art. *Journal of Geotechnical Engineering Division, 107*(No. GT2), 167–181.

Effect of Fines on Pore Pressure Development During Cyclic Loading

Muttana S. Balreddy, S. V. Dinesh and T. G. Sitharam

Abstract Earthquake is one of the natural calamities that occur on large geographical area. During earthquakes, liquefaction of saturated soil deposit is a major cause for failure of various infrastructure as well as lifeline facilities. The cause for liquefaction is the build-up of excess hydrostatic pressure due to the application of cyclic shear stresses induced during ground shaking. From the literature review, it is understood that the mechanism of pore pressure generation of sand and sand–fines mixtures is still not clear and requires further investigation/research. In this paper, a series of stress-controlled cyclic triaxial undrained shear tests were conducted on reconstituted sample of sand–fines mixtures for evaluating and understanding the pore pressure build-up in soils. Sand was procured from Cauvery river bed, Karnataka, and locally available low plasticity clay fines were used to prepare sand–fines mixtures. The results indicate that the presence of fines reduces the pore pressure build-up when compared to clean sand.

Keywords Pore pressure generation · Liquefaction · Sand–fines mixtures
Relative density

M. S. Balreddy (✉) · S. V. Dinesh
Department of Civil Engineering, Siddaganga Institute of Technology,
Tumakuru 5720103, Karnataka, India
e-mail: muttanab@gmail.com

S. V. Dinesh
e-mail: dineshsv2004@gmail.com

T. G. Sitharam
Department of Civil Engineering, Indian Institute of Science,
Bangalore 560012, Karnataka, India
e-mail: sitharam@civil.iisc.ernet.in

B. Adimoolam and S. Banerjee (eds.), *Soil Dynamics and Earthquake Geotechnical Engineering*, Lecture Notes in Civil Engineering 15,
https://doi.org/10.1007/978-981-13-0562-7_10

1 Introduction

The liquefaction phenomenon has received intensive attention since the devastating earthquakes of Alaska and Niigata in Japan. The development of excess pore pressure under undrained loading condition is a hallmark of all liquefaction phenomena. Since then, many investigations have been carried out research in this area. For cyclic loading, two methods are used to study the rate and magnitude of pore pressure generation. The first method is stress-controlled cyclic tests, in which the generation of excess pore pressure is evaluated as a function of the number of loading cycles. The second method is strain-controlled cyclic tests with measurement of excess pore pressure with loading cycles for a given shear strain level. Lee and Albaisa (1974) found a continuous increase in excess pore pressure as the cyclic loading progressed. They suggested that the excess pore pressure falls within a narrowband for a given sand and is insensitive to the soil's initial density and consolidation stress. Dobry et al. (1982) examined pore pressure generation during strain-controlled, undrained cyclic loading as a function of the shear strain. They too observed that the measured excess pore pressure fell within a narrowband and found that it did not develop until a threshold shear strain is reached. Polito and Martin (2003) and Sitharam and Govindaraju (2007) studied the behaviour pore pressure generation, in which the parameters such as amount and the nature of the fines (plastic or non-plastic) play a critical role.

Hazirbaba and Rathje (2009) reported decrease in excess pore water pressure in silty sands and increase in threshold strain. They also reported that limited data are available wrt effect of plastic fines on pore pressure build-up.

Mohtar et al. (2013) examined the pore pressure generation in sand with bentonite. They concluded that the presence of bentonite fines reduces the mean pore pressure generated per loading cycle.

Bayat et al. (2014) showed that pore pressure build-up in sandy soil mixed with plastic fines is slower than pure sand.

Benghalia et al. (2015) conducted tests on Chelf sand with Chelf fines having PI = 13 (low plastic fines). They concluded that the pore pressure decreases up to 10% fines content and then rises again moderately.

Eswara Rao et al. (2015) studied the effect of plastic fines (PI = 33) content on pore pressure generation. They found that the pore pressure increases for 10% plastic fines but there is decrement with increase in fines content up to 15% for 20% fines and 40% for 40% fines.

Therefore in the present paper, an attempt is made to understand the pore pressure generation during cyclic loading in sand–fines mixtures. The mixtures are silty sand and low plasticity clayey sands.

2 Experimental Investigation

2.1 Index Properties of Sand, Fines and Sand–Fines Mixtures

The sand sample used in the present investigation was collected from the Cauvery river bed, Karnataka. In order to study the effect of fines in the sandy soil on pore pressure build-up, clayey sand with low plasticity was collected from local area. Figure 1 shows the grain size distribution curves of both soil samples used in the present investigation. Table 1 gives the summary of the index properties of these soils.

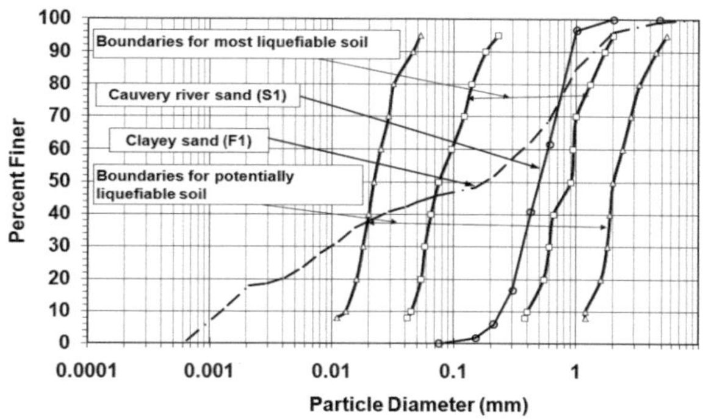

Fig. 1 Grain size distribution curves of sand, clayey sand and those susceptible to liquefaction proposed by Tsuchida (Iwasaki, 1986)

Table 1 Index properties of soils

Properties	Cauvery River sand (S1)	Clayey sand (F1)
Specific gravity (G)	2.62	2.65
Gravel (%)	–	1
Sand (%)	100	53
Silt (%)	–	28
Clay (%)		18
C_u	2.40	–
C_c	0.90	–
e_{max}	0.99	–
e_{min}	0.72	
LL (%)	0	46
PL (%)	0	27
PI	0	19
IS classification	SP	SC

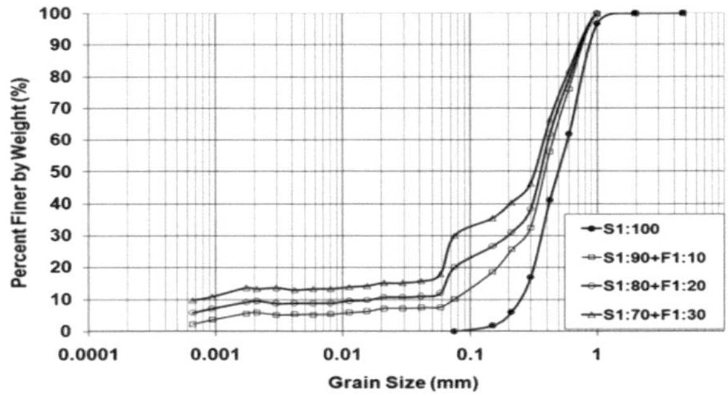

Fig. 2 Grain size distribution curves of soil mixture

Table 2 Index properties of sand and sand–fines mixtures

Soil mixtures	S1:90 + F1:10	S1:80 + F1:20	S1:70 + F1:30
Fines (%)	10	20	30
D_{10}	0.075	0.021	0.0008
e_{max}	0.98	1.06	1.31
e_{min}	0.60	0.61	0.72
LL	22	26	37
PI	0	6	17
IS classification	SW-SM	SC	SC

The fines <75 μm size used in the experimental program were obtained from the clayey sand. These fines were mixed with the clean Cauvery River sand in 90:10, 80:20 and 70:30 (i.e. S1:70 + F1:30) proportions. Figure 2 shows the grain size distribution curves of sand–fines mixtures used in the present investigation. Table 2 shows the index properties of sand–fines mixtures.

2.2 *Sample Preparation*

Triaxial cylindrical soil specimens of 50 mm diameter and height 100 mm were prepared by dry deposition method. The required mass of oven-dried Cauvery River sand (S1) and F1 fines corresponding to gross void ratio were weighed separately, and both mixtures were divided into ten equal parts. The sand and fines were mixed uniformly, and each part was gently poured into the mould by using a funnel having a 12-mm-diameter nozzle in dry state with zero fall height so that a loose soil sample is formed. The required relative density is achieved by giving uniform blows on all four sides of the mould by wooden mallet. In this way, the sample was prepared by filling the mould in ten layers of equal height. Then, the sample was subjected to small vacuum of 5 kPa and carbon dioxide gas was passed for one hour.

2.3 Saturation, Consolidation and Shearing

The specimens were drained by de-aired distilled water through soil sample at a very small head of 5 kPa. The specimens were saturated by applying a back pressure of 100 kPa, and saturation time was maintained till Skempton's pore pressure coefficient '*B*' value exceeds 0.95. Required cell pressure (100 kPa) was applied, and specimens were isotropically consolidated till the end of Casagrande's primary consolidation. After the consolidation process, the specimens were subjected to undrained shearing at constant confining pressure.

2.4 Stress-Controlled Test

The isotropically consolidated cylindrical specimens were subjected to undrained stress-controlled, constant amplitude cyclic loading tests at a frequency of 0.1 Hz. ASTM D 5311 (1996) code was followed for stress-controlled technique. The loading system consists of a submersible load cell of 5 kN capacity, and the deformation was measured by LVDT. Four transducers were used to record the lateral deformation, volume change, pore water pressure and cell pressures. The data was acquired by a computer-controlled data acquisition system (CCDAS). 128 data points were captured in each loading cycle. The cyclic stress, axial strain, pore water pressure and confining pressures are the test parameters and were recorded by CCDAS. Tests were continued till excess pore water pressure is equal to confining pressure or the double amplitude axial strain is in excess of 5%.

3 Results and Discussion

Figure 3 shows the plot of residual pore pressure ratio versus cycle ratio for sand–fines mixtures with 0–30% fines content at relative density 50%. The pore pressure ratio was computed based on residual pore water pressure at zero deviator stress at the end of each loading cycle. The term r_u is a function of the cycle ratio (N/N_L), which is the ratio of number of applied uniform cycles of loading (N) to the number of cycles required to cause liquefaction (N_L). It is observed that all the specimens failed by flow liquefaction. It is clearly observed from Figs. 3 and 4 that the pore pressure development in sand and silty sand with 10% fines shows the same trend.

Sand–fines mixtures containing 20% fines with PI = 6 and 30% fines with PI = 17 also fail by flow liquefaction. However, the rate of pore pressure generation is faster than compared to sand and sand–fines mixtures containing 10% fines. The results indicate that pore pressure generation is faster with increase in PI.

Figure 5 shows the plot of residual pore pressure ratio versus cycle ratio for sand–fines mixtures with 0–30% fines content at relative density 70%. It is

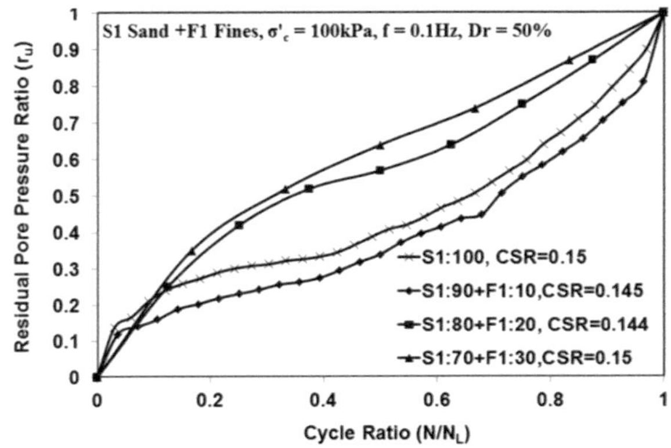

Fig. 3 Residual pore pressure ratio versus cycle ratio of sand–fines mixtures at Dr = 50%

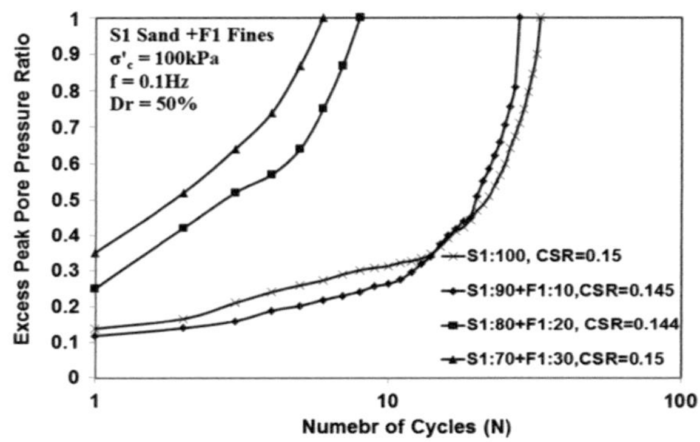

Fig. 4 Excess peak pore pressure generation at Dr = 50% for sand–fines mixtures

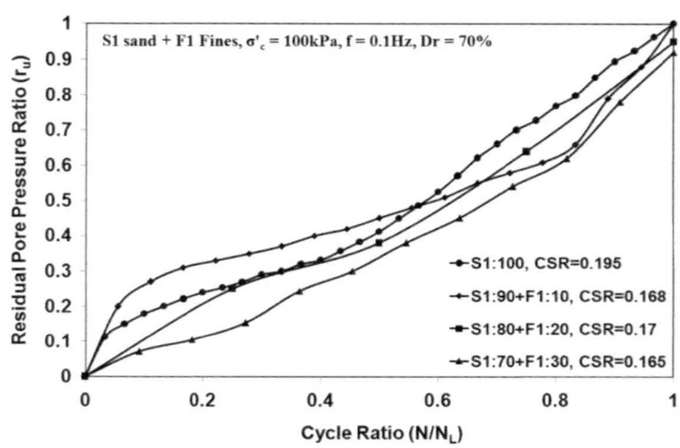

Fig. 5 Residual pore pressure ratio versus cycle ratio of sand–fines mixtures at Dr = 70%

observed that the specimens of sand–fines mixtures containing 20 and 30% fines fail by cyclic mobility due to residual pore pressure not reaching one and effective stress not becoming zero. The residual pore pressure values vary in a narrowband more or less uniformly for all the mixtures.

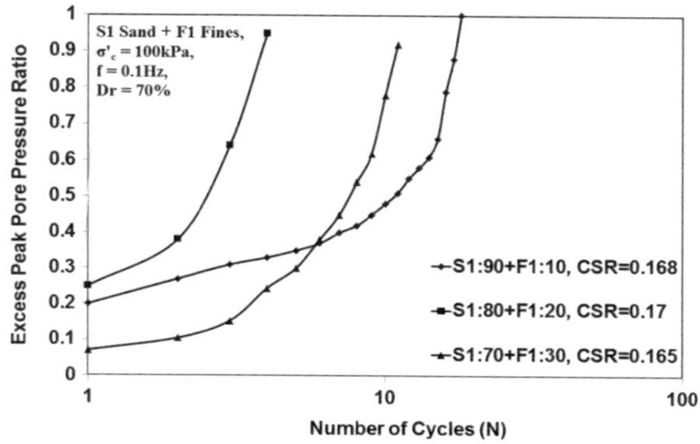

Fig. 6 Excess peak pore pressure generation at Dr = 70% for sand–fines mixtures

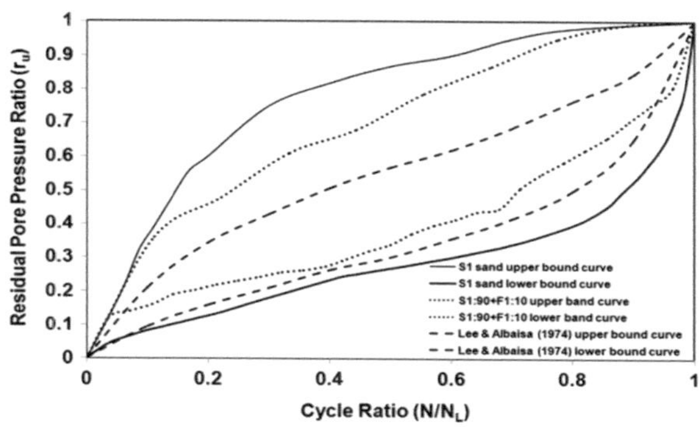

Fig. 7 Residual pore pressure ratio versus cycle ratio of present work with the work of Lee and Albaisa (1974)

Figure 6 shows that the specimens with 20 and 30% fines with PI = 6 and 17 compacted at Dr = 70% fail by cyclic mobility. The results also indicate that the pore pressure build-up is faster with 20% fines than specimen with 30% fines. This indicates that in dense samples the pore pressure build-up decreases with increase in PI > 6 and shows clay-like behaviour, and these results are in agreement with Bayat et al. (2014).

Figure 7 shows the plot of residual pore pressure ratio (r_u) versus cycle ratio (N/N_L) for S1 sand and sand–fines mixtures containing 10% fines at loose and dense state. In addition, the band of pore pressure ratio for sample is shown along with the data of Lee and Albaisa (1974). The Lee and Albaisa (1974) band is generated from large number of tests conducted at 30–100% relative density for Monterey sand. The shape of pore pressure ratio band of S1 sand and sand–fines mixtures is similar to that of Lee and Albaisa (1974). The upper band of present work clearly locates above that of Lee and Albaisa (1974) and lower band of S1 sand locates below that of Lee and Albaisa (1974) but the lower band of S1:90 + F1:10 locates above the lower band of S1 sand and Lee and Albaisa (1974). The pore pressure ratio is much wider at any cycle ratio for S1 sand compared to sand–fines mixtures.

4 Conclusions

The following conclusions are drawn from the experimental evidence.

1. The clean sand and sand–fines mixtures containing 10–30% fines attain pore pressure ratio of 1 and fails by flow liquefaction at Dr = 50%. However, sand–fines mixtures with 20 and 30% fines content at Dr = 70% will have $r_u < 1$ and fails by cyclic mobility.
2. The excess pore water pressure in dense sand–fines specimens (Dr = 70%) decreases with increase in PI (PI > 6).
3. The pore pressure build-up is faster with increase in PI in loosely compacted specimens (Dr = 50%) when compared to dense specimens (Dr = 70%).
4. The shape of pore pressure ratio curves of sand and sand–fines mixtures is similar to published Lee and Albaisa (1974). The S1 sand and sand–fines mixture band are broader than Lee and Albaisa (1974).

References

ASTM D 5311. (1996). *Standard test methods for load controlled cyclic triaxial strength of soil*. American Society for Testing and Materials, USA.

Bayat, M., Bayat, E., Aminpour, H., & Salarpour, A. (2014). Shear strength and pore-water pressure characteristics of sandy soil mixed with plastic fine. *Arabian Journal of Geosciences, 7*(3), 1049–1057.

Benghalia, Y., Bouafia, A., Canou, J., & Claude Dupla, J. (2015). Liquefaction susceptibility study of sandy soils: effect of low plastic fines. *Arabian Journal of Geosciences, 8,* 605–618.

Dobry, R., Ladd, R. S., Chang, R. M., & Powell, D. (1982). *NBS Building Science Series 138: Prediction of pore water pressure build up and liquefaction of sands during earthquakes by the cyclic strain method* (pp. 1–150). Washington DC.

Eswara Rao, S., Varghese, V., & Patel, S. (2015). An experimental study on effect of plastic fines content on liquefaction properties of sand. In *50th Indian Geotechnical Conference*, Pune, India.

Hazirbaba, K., & Rathje, E. M. (2009). Pore pressure of silty sands due to induced cyclic shear strains. *Journal of Geotechnical and Geoenvironmental Engineering, ASCE, 135*(12), 1892–1905.

Lee, K. L., & Albaisa, A. (1974). Earthquake induced settlements in saturated sands. *Journal of Geotechnical Engineering Division, ASCE, 100*(GT4), 387–406.

Mohtar, C. S., Bobet, A., Drnevich, V. P., Johnston, C. T., & Santagata, M. C. (2013). Pore pressure generation in sand with bentonite: from small strains to liquefaction. *Géotechnique, 64* (2), 108–117.

Polito, C. P., & Martin, J. R. (2003). A reconciliation of the effects of non-plastic fines on the liquefaction resistance of sands reported in the literature. *Earthquake Spectra, 19*(3), 635–651.

Sitharam, T. G., & Govindaraju, L. (2007). Pore pressure generation in silty sands during cyclic loading. *Geomechanics and Geoengineering: An International Journal, 2*(4), 295–306.

Comparative Study of Dynamic Response Analysis of Shallow Foundation on Layered Soils

M. V. Ravi Kishore Reddy, Supriya Mohanty and Pradeep Kumar Ramancharla

Abstract Dynamic response analysis of shallow foundation in time-varying motion is very much essential in the present days. As the seismic events are unpredictable, it draws the attention of researchers towards the response of the soil foundation system against dynamic loading. Hence, in the present study, an attempt has been made to investigate the dynamic response of a shallow foundation (continuous footing) resting on different layered soil deposits subjected to seismic loadings. The layered soil deposits considered for the analysis are weak clay underlain by stiff clay (case I) and loose sand underlain by stiff clay (case II). In addition, comparative studies have been made considering the parameters like layered soil types (case I and II), water table effects (dry and partially saturated) and seismic loadings (Bhuj earthquake: M_w—7.7 and Nepal earthquake: M_w—7.8) conditions. The modelling of the soil foundation system has been carried out using two-dimensional finite element software CyclicTP. The response of the dynamic analysis of shallow foundation on layered soils is presented in the form of peak acceleration; horizontal and vertical displacement; shear stress versus shear strain; and excess pore pressure variations at different locations. The results of the analysis indicate that the peak acceleration, horizontal and vertical displacements are more in both the cases (I and II) under the excitation of Nepal earthquake than the Bhuj earthquake motion. Also, it is observed that the response in terms of peak acceleration, horizontal and vertical displacement is more in partially saturated state than in dry state for both the earthquake motions.

Keywords Dynamic response · Seismic loading · Nonlinear analysis
Peak acceleration · Displacement history

M. V. Ravi Kishore Reddy · S. Mohanty (✉) · P. K. Ramancharla
Earthquake Engineering Research Centre, International Institute of Information Technology, Hyderabad 500032, Telangana, India
e-mail: supriya.mohanty@iiit.ac.in

M. V. Ravi Kishore Reddy
e-mail: meegada.reddy@research.iiit.ac.in

P. K. Ramancharla
e-mail: ramancharla@iiit.ac.in

B. Adimoolam and S. Banerjee (eds.), *Soil Dynamics and Earthquake Geotechnical Engineering*, Lecture Notes in Civil Engineering 15,
https://doi.org/10.1007/978-981-13-0562-7_11

1 Introduction

The response analysis of a system to any dynamic load such as blasting, explosions, earthquake is very much essential in the present days. As the earthquake events are unpredictable, it draws the attention of researchers towards the response analysis of soil foundation system against earthquake loading. The response of the soil foundation system depends on the nature of the soil, type of loading and the type of foundation. The response of the soil foundation system during seismic events is of great importance, and it may be directly related to the damage and loss of life and property. Many researchers have performed numerical analysis on the seismic response and settlement of foundation and structures resting on different soils (Asgari et al. 2014; Liu and Dobry 1997; Ragheb 1994; Stokoe et al. 2013; Yoshimi and Tokimatsu 1977). Study on liquefaction of soil, its effects on footings and buildings have been investigated by many researchers (Elgamal et al. 2005; Ishihara et al. 1993; Kishida 1966; Seed and Idriss 1967). Also, studies have been done on dynamic behaviour and liquefaction evaluation of Indian pond ash samples (Mohanty and Patra 2012; Mohanty and Patra 2014; Mohanty and Patra 2016a, b).

As the determination of the response of a soil foundation system is demanding in a layered soil deposit than in homogeneous soil deposit, it is attempted to investigate the dynamic response of a shallow foundation (continuous footing) resting on different layered soil deposits (case I: weak clay underlain by stiff clay, and case II: loose sand underlain by stiff clay) subjected to seismic loadings. Also, comparative studies have been made considering the parameters like layered soil types (case I and II), water table effects (dry and partially saturated) and seismic loadings (Bhuj: M_w—7.7 and Nepal: M_w—7.8 earthquakes) conditions. The results of the analysis were presented in terms of acceleration, displacement and pore pressure time history, shear stress variation with shear strain at different locations.

2 Input Data

The following parameters are considered for the dynamic response analysis of shallow foundation resting on layered soil deposit.

2.1 Foundation Parameters

Foundation width (B): 1.5 m, height above ground surface: 0.4 m, depth below ground surface: 1 m, Young's modulus of concrete: 25E+6 kPa, Poisson's ratio: 0.3, mass density: 2300 kg/m^3.

2.2 Soil Parameters

The soil domain has a width of 15 m and a depth of 15 m, and the water table is considered at a depth of 2 m below the ground surface. Hard rock has been considered at a depth of 15 m below the ground surface. Table 1 represents the details of soil parameters considered in the present study.

2.3 Input Motions

Input motions considered for the analysis are Bhuj earthquake (M_w = 7.7, 2001) and Nepal earthquake (M_w = 7.8, 2015). Peak acceleration values noticed for Bhuj and Nepal earthquake are 0.106 g at 46.94 s and 0.1639 g at 49.31 s, respectively. Typical diagram of input motion of Bhuj and Nepal earthquake used in the present analysis is shown in Fig. 1.

Table 1 Details of soil parameters considered

Case	Layer no.	Thick (m)	Soil type (V_s in m/s)
I	1	2.5	Cohesive soft (V_s = 100)
	2	3	Cohesive stiff (V_s = 300)
	3	3	Cohesive stiff (V_s = 300)
	4	6.5	Cohesive stiff (V_s = 300)
II	1	2.5	Cohesionless loose (V_s = 185)
	2	3	Cohesive stiff (V_s = 300)
	3	3	Cohesive stiff (V_s = 300)
	4	6.5	Cohesive stiff (V_s = 300)

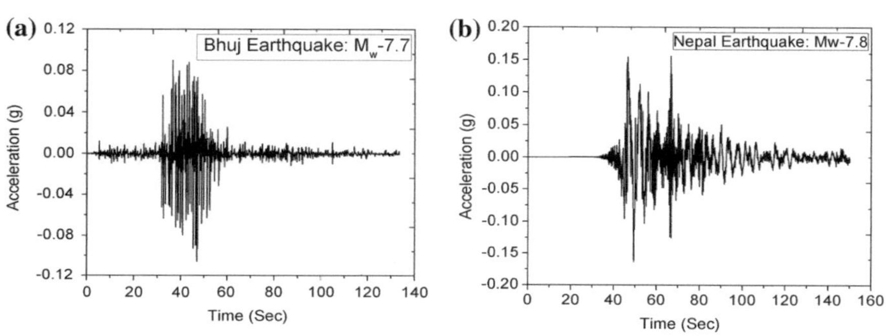

Fig. 1 Acceleration time history of **a** Bhuj and **b** Nepal earthquake

3 Numerical Modelling

The modelling of the soil foundation system has been carried out using two-dimensional finite element software CyclicTP. Cyclic TP is a PC-based graphical pre-processor and post-processor (user-interface) for two-dimensional finite element analysis of shallow foundations subjected to seismic excitations. Both linear analysis (with viscous damping) and nonlinear analysis (plasticity-based nonlinear soil compatibility) can be performed by this program. 9 (solid phase)–4 (fluid phase)-noded quadratic element has been employed for modelling the soil domain. The soil domain has a total number of 1246 elements. A typical representation of generated finite element mesh of the soil domain (case I) is shown in Fig. 2. The boundary conditions considered in the finite element modelling of the foundation soil system are: (i) at any given depth, displacement degrees of freedom of the left and right boundaries were tied together both horizontally and vertically, (ii) the soil surface was traction free, with zero prescribed pore pressure, and (iii) the base and lateral boundaries were impervious.

4 Results and Discussions

Here, the results of 2D dynamic response analysis of shallow foundation are presented in terms of acceleration, displacement, stress–strain and excess pore pressure responses. The results are discussed in the following sections.

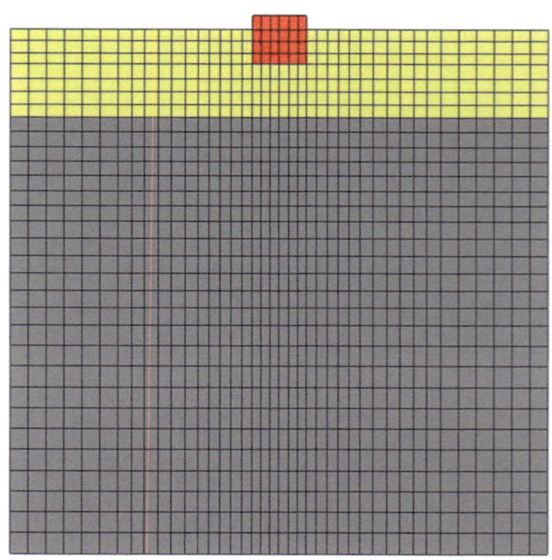

Fig. 2 Generated finite element mesh of the soil domain (case I)

4.1 Acceleration Responses

The acceleration responses have been recorded at different depths of the soil domain. Typical diagrams of acceleration time history and PGA variation with depth for partially saturated soil of case I under the excitation of Bhuj earthquake are presented in Figs. 3 and 4, respectively. The peak ground acceleration (PGA) value of 1.7745 m/s^2 is observed at 47.07 s for case I partially saturated soil under Nepal earthquake as input motion.

It is observed that PGA value is more in case I than in case II soil (Table 2). Also, PGA values are found to be more in partially saturated condition than in dry condition for both the cases under the considered input motions.

4.2 Displacement Responses

The displacement responses have been recorded at different depths of the soil domain. Typical diagram of horizontal displacement variation with depth for partially saturated soil of case I under the excitation of Bhuj earthquake is presented in

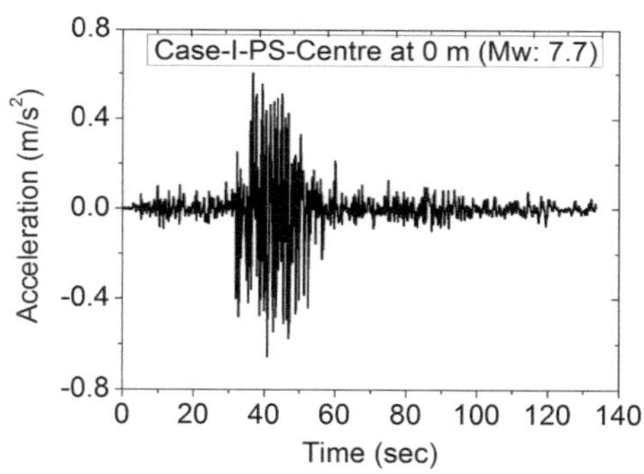

Fig. 3 Acceleration time history for partially saturated soil of case I at 0 m from the centre of footing (M_w: 7.7)

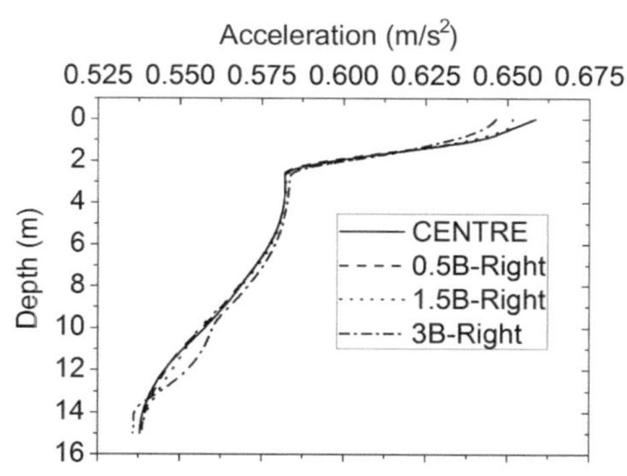

Fig. 4 Variation of acceleration with depth for partially saturated soil of case I (M_w: 7.7)

Table 2 Variation of peak ground acceleration (PGA) at different locations

Case	Condition	M_w	PGA (m/s^2)			
Distance from footing centre →			Centre	0.5B (R)	1.5B (R)	3B (R)
I	Dry	7.7	0.6510	0.6510	0.6429	0.6417
		7.8	1.7688	1.7688	1.7689	1.7684
	Partially saturated (PS)	7.7	0.6585	0.6585	0.6515	0.6466
		7.8	1.7740	1.7740	1.7744	1.7745
II	Dry	7.7	0.5938	0.5938	0.5922	0.5953
		7.8	1.7405	1.7405	1.7412	1.7415
	Partially saturated (PS)	7.7	0.597	0.597	0.596	0.599
		7.8	1.7439	1.7439	1.7447	1.744

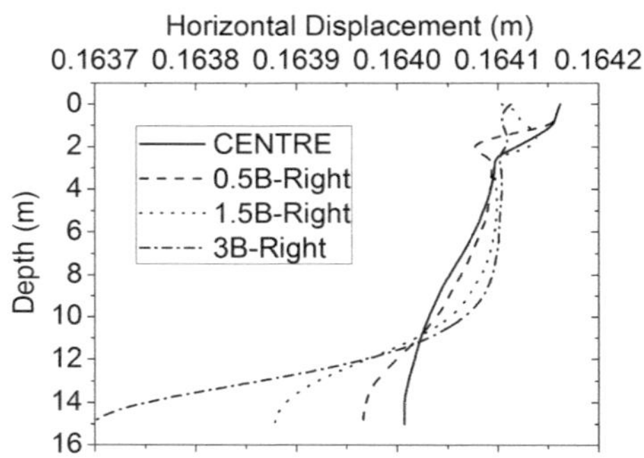

Fig. 5 Variation of horizontal displacement with depth for partially saturated soil of case I (M_w: 7.7)

Table 3 Variation of horizontal displacement at different locations

Case	Condition	M_w	Horizontal displacement (m)			
Distance from footing centre →			Centre	0.5B (R)	1.5B (R)	3B (R)
I	Dry	7.7	0.161	0.161	0.161	0.161
		7.8	1.2495	1.2495	1.2492	1.2493
	Partially saturated (PS)	7.7	0.16416	0.16416	0.16414	0.16411
		7.8	1.3083	1.3083	1.3081	1.3081
II	Dry	7.7	0.1648	0.1648	0.1648	0.1648
		7.8	1.3042	1.3042	1.304	1.3041
	Partially saturated (PS)	7.7	0.1674	0.1674	0.1673	0.1673
		7.8	1.3805	1.3805	1.3803	1.3804

Fig. 5. The peak horizontal and vertical displacements are found to be 1.3805 and 0.0138 m, respectively, under Nepal input motion (Tables 3 and 4). The horizontal displacement is observed to be more at the centre location. The vertical displacement values are found to be very less as compared to horizontal displacement.

Table 4 Variation of vertical displacement at different locations

Case	Condition	M_w	Vertical displacement (m)			
Distance from footing centre →			Centre	0.5B (R)	1.5B (R)	3B (R)
I	Dry	7.7	6e−5	8e−5	0.0001	6e−5
		7.8	0.0135	0.0135	0.0131	0.0137
	Partially saturated (PS)	7.7	0.0002	0.0002	6e-5	0.0002
		7.8	0.0136	0.0136	0.0133	0.0138
II	Dry	7.7	0.0002	0.0002	0.0003	0.0002
		7.8	0.0131	0.0132	0.0132	0.0135
	Partially saturated (PS)	7.7	8e−5	9e−5	9e−5	7e−5
		7.8	0.0132	0.0132	0.0133	0.0135

Hence, the possibility of lateral spreading is more in partially saturated soils under the considered input motions.

4.3 Stress–Strain Responses

Shear stress versus shear strain responses was obtained at different depths and also at different locations from the centre of the footing. Typical shear stress versus shear strain variation recorded at a distance of 3B away from the centre of the footing (right) and at a depth of 11.38 m for dry soil of case II under the excitation of Bhuj earthquake is presented in Fig. 6. From the results, it can be noticed that the shear stress value decreases with the increase in duration of input motions at all the locations. However, the peak value of shear stress increases with increase in depth from the ground surface and distance away from the centre of the footing. The maximum value of shear stress was found to be 85.67 kPa for dry soil of case II under Nepal earthquake at distance of 3B away from the centre of the footing (right).

Fig. 6 Variation of shear stress with shear strain at a distance of 3B away from the centre of the footing and at a depth of 11.38 m for dry soil of case II (M_w: 7.7)

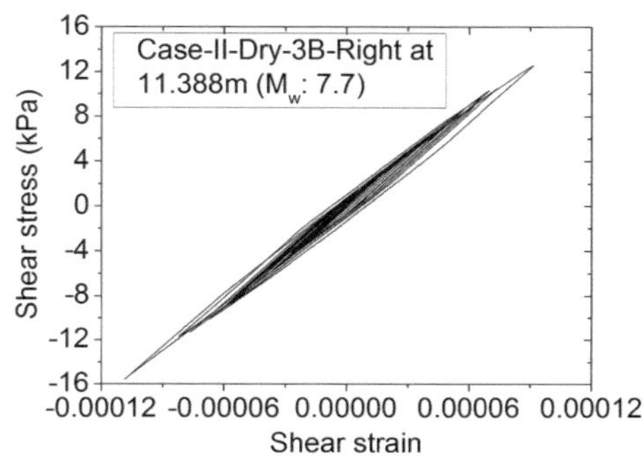

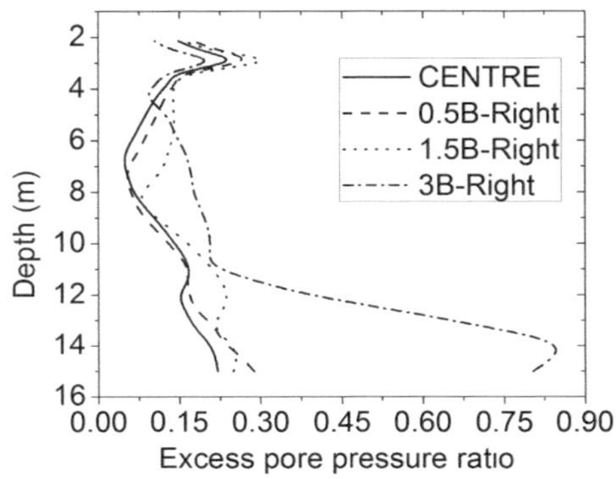

Fig. 7 Variation of excess pore pressure ratio with depth for partially saturated soil of case II (M_w: 7.8)

4.4 *Excess Pore Pressure Responses*

Excess pore pressure time histories were obtained at different depths and different locations from the centre of the footing. It is observed that excess pore pressure increases with increase in depth. The maximum value of excess pore pressure was found to be 114.8 kPa for partially saturated soil of case II under Nepal input motion. The excess pore pressure is observed to be more at distance away from the centre of the footing (3B). The maximum value of excess pore pressure ratio was observed to be 0.85 for partially saturated soil of case II under Nepal input motion. As the excess pore pressure ratio is found to be less than one, the foundation is safe against liquefaction. Typical diagram of excess pore pressure ratio variation with depth for partially saturated soil of case II under the excitation of Nepal earthquake is presented in Fig. 7.

5 Conclusion

In this paper, comparative study of dynamic response of a shallow foundation resting on different layered soil deposits subjected to seismic loadings has been investigated. It can be observed that peak response of acceleration, displacement and excess pore pressure values are found to be more for partially saturated soil under the excitation of Nepal earthquake. This may be due to the more PGA value of Nepal earthquake (0.1639 g) than that of Bhuj earthquake motion (0.106 g). The maximum PGA value (1.7745 m/s^2) was noticed for partially saturated soil of case I under Nepal earthquake. The horizontal displacement (1.3805 m) values are observed to be more as compared to vertical displacement (0.0138 m). Here, the peak excess pore pressure ratio (0.85) is less than one but in extreme saturated situations, the excess pore pressure ratio may exceed one, which may cause foundation failure due to liquefaction. Hence, it is always advisable to perform dynamic response analysis of foundation resting on layered soil deposits subjected to seismic loadings.

References

Asgari, A., Golshani, A., & Bagheri, M. (2014). Numerical evaluation of seismic response of shallow foundation on loose silt and silty sand. *Journal of Earth System Science, 123*(2), 365–379.

Elgamal, A., Lu, J., & Yang, Z. (2005). Liquefaction induced settlement of shallow foundations and remediation: 3D numerical simulation. *Journal of Earthquake Engineering, 9*(1), 17–45.

Ishihara, K., Acacio, A. A., & Towhata, I. (1993). Liquefaction induced ground damage in Dagupan in the July 16, 1990 Luzon earthquake. *Soils and Foundations, 33*(1), 133–154.

Kishida, H. (1966). Damage to reinforced concrete buildings on Nigata city with special reference to foundation engineering. *Soils and Foundations, 6*(1), 71–88.

Liu, L., & Dobry, R. (1997). Seismic response of shallow foundation on liquefiable sand. *Journal of Geotechnical and Geoenvironmental Engineering, ASCE, 123*(6), 557–567.

Mohanty, S., & Patra, N. R. (2012). Assessment of liquefaction potential of pond ash at Panipat in India using SHAKE2000. In *Proceedings of GeoCongress-2012, State of the Art and Practice in Geotechnical Engineering*, GSP: 225 (pp. 1829–1838). California: ASCE.

Mohanty, S., & Patra, N. R. (2014). Cyclic behaviour and liquefaction potential of Indian pond ash located in seismic zone III and IV. *Journal of Materials in Civil Engineering ASCE, 26*(7), 06014012.

Mohanty, S., & Patra, N. R. (2016a). Liquefaction and earthquake response analysis of Panipat pond ash embankment in India. *Journal of Earthquake & Tsunami*. https://doi.org/10.1142/s1793431116500093.

Mohanty, S., & Patra, N. R. (2016b). Dynamic response analysis of Talcher pond ash embankment in India. *Soil Dynamics and Earthquake Engineering, 84,* 238–250.

Ragheb, A. M. (1994). *Numerical analysis of seismically induced deformations in saturated granular soil strata*. Ph.D. thesis, Department of Civil Engineering, Rensselaer Polytechnic Institute, Troy, NY.

Seed, H. B., & Idriss, I. M. (1967). Analysis of soil liquefaction: Niigata earthquake. *Journal of the Soil Mechanics and Foundations Division, ASCE, 93*(3), 83–108.

Stokoe, K. H., Kacar, O., & Van Pelt, J. (2013). Predicting settlements of shallow footings on granular soil using nonlinear dynamic soil properties. In *Proceedings of 18th International Conference on Soil Mechanics and Geotechnical Engineering* (pp. 3467–3470). Paris.

Yoshimi, Y., & Tokimatsu, K. (1977). Settlement of buildings on saturated sand during earthquakes. *Soils and Foundations, 17*(1), 23–38.

Shake Table Studies on Embankments on Liquefiable Soil

G. Unni Kartha, K. S. Beena and C. P. Mohamed Thahir

Abstract Experimental studies using shaking table test on liquefaction have been widely conducted in the countries like Japan, where a number of earthquakes occur every year. Kerala, one of the coastal states, having large area of sand deposit is included in earthquake zone III. Hence, the liquefaction studies are of prime importance as many developmental projects like the coastal highway, high-speed railway tracks are on the anvil. The objective of the present study was to conduct model tests on embankments over liquefiable soil using one-dimensional shaking table. The system behaviour during shaking, effects of liquefaction on the embankment, effect of providing geotextile reinforcement to soil, pore water pressure build-up, failure mode also comes under the scope of the study. The experiments were conducted using an ingeniously developed shake table and data acquisition system. The model behaviour of embankment with and without geotextile reinforcements under shaking was conducted. The acceleration and frequency of shaking were kept as 0.2 g and 1 Hz, respectively. It was observed that geotextile reinforcement layer below the embankment can improve the liquefaction resistance of the soil. Additional reinforcement inside the soil bed further reduced the liquefaction, and the pore water pressure reduced considerably.

Keywords Liquefaction · Shake table · Embankment

G. Unni Kartha (✉) · K. S. Beena · C. P. Mohamed Thahir
Division of Civil Engineering, School of Engineering, Cochin University of Science and Technology, Kochi 682022, India
e-mail: unnikartha@gmail.com

K. S. Beena
e-mail: beenavg@gmail.com

C. P. Mohamed Thahir
e-mail: cpthahir@gmail.com

B. Adimoolam and S. Banerjee (eds.), *Soil Dynamics and Earthquake Geotechnical Engineering*, Lecture Notes in Civil Engineering 15,
https://doi.org/10.1007/978-981-13-0562-7_12

1 Introduction

Earthquakes are the most devastating natural phenomena which results in loss of life and damage of building structures such as residential and industrial buildings, dams roads and railway embankments. "Liquefaction" denotes a condition in which soil undergoes continued deformation at constant low residual stress or with low residual resistance, due to the build-up and maintenance of high pore water pressures that reduce the effective confining pressure to a very low value (Seed 1968).

Experimental studies using shaking table test on liquefaction are widely conducted in the countries like Japan, where a number of earthquakes occur every year. But in countries like India, these types of studies are limited. Earthquake geotechnical engineering in India had received tremendous boost after the Gujarat earthquake of 2001. Earthquakes are not uncommon, but the damage suffered during earthquakes is on the rise because of the population growth, overcrowding of civil engineering facilities in urban areas and improper understanding of ground behaviour among many. Recent earthquakes in Gujarat (2001), Chamoli (1999), Jabalpur (1997), Latur (1993) and Uttarkashi (1991) in India have increased the research activities in the country in the field of earthquake engineering. Kerala, one of the coastal states, having large area of sand deposit, is included in earthquake zone III. The liquefaction studies are of prime importance in Kerala as many developmental projects like the coastal highway are about to start in the state.

This paper presents the results conducted on models of embankments over liquefiable soil using one-dimensional shaking table. The system behaviour during shaking, effects of liquefaction on the embankment, effect of providing geotextile reinforcement to soil, pore water pressure build-up, failure mode also comes under the scope of the study.

2 Previous Studies

Since the understanding of ground behaviour during shaking is still incomplete, different approaches like element tests, model tests, analytical/numerical model tests and field tests are required to understand the phenomenon. Unfortunately in India, lack of facilities for laboratory studies and non-availability of instrumentation in the field can be attributed as the main reasons for lesser number of studies. At the same time, it should be noted that uniaxial shake table tests are sufficient to understand the response and failure mechanism of earth structures like embankment, retaining wall (Prasad et al. 2004). Many researchers have used small-scale shake table with lengths in the range 2.0–1.5 m, width 0.5–1.0 m and depths 0.3–1.0 m and acceleration in the range 0.1–0.5 g and frequency 1–3 Hz (Prasad et al. 2004; Maheshwari et al. 2012; Varghese and Latha 2013; Orense et al. 2003; Ramakrishnan et al. 1998; Beena and Unni Kartha 2014).

The type of soil most susceptible to liquefaction is one in which the resistance to deformation is mobilized by friction between particles. If other factors such as grain shape, uniformity coefficient and relative density are equal, the frictional resistance of cohesion less soil decreases as the grain size of soils becomes smaller. Tsuchida (1970) summarized the results of sieve analyses performed on a number of alluvial and diluvial soils that were known to have liquefied or not to have liquefied during earthquakes. He proposed ranges of grain size curves separating liquefiable and non-liquefiable soils.

The shake table tests conducted by Koga and Matsuo (1990) on reduced scale embankment models founded on saturated sandy ground indicate that the ground beneath the embankment is not prone to liquefaction and it slumps and cracks as a result of softening of underlying ground. They also point that the extent of damage is more extensive with longer duration and lower frequency earthquake input motion. Studies by Prasad et al. (2004) have demonstrated densification as an effective technique for liquefaction countermeasure. Numerical studies have shown that techniques like sheet pile and stone column are effective methods in reducing liquefaction in soil under embankments (Bhatnagar et al. 2003). Reinforcement of soil with geogrid can be used in many field applications. This technique is found to be useful and cost-effective in the construction of roads, embankments, dams, and low-rise residential buildings. Recent researches have proved that inclusion of geogrids as reinforcement in layers of soil is potential solution for reducing liquefaction (Maheshwari et al. 2012).

In this study, the effect of geotextile reinforcement when it is placed as a separator and also within the supporting soil is studied and compared.

3 Properties of Materials Used in the Study

Figure 1 represents the particle size distribution of Aluva River sand, in comparison with the boundaries defined by Tsuchida. It is found that the particle size distribution of the sand lies within the boundaries of potentially liquefiable soil. Table 1 shows the properties of Aluva River sand.

The tests are conducted keeping the density of sand layer as 14.1 kN/m^3. The details of embankment model and soil are as shown in Table 2. Figure 2 shows the schematic diagram of the embankment over the soil bed. The properties of geotextile used as reinforcement are shown in Table 3.

4 Experimental Set-up

The experiments are conducted using an ingeniously developed shake table and data acquisition system. The shake table consists of a rigid box of size 1500 mm × 500 mm × 600 mm (L × B × H) made of Perspex and Steel. It is

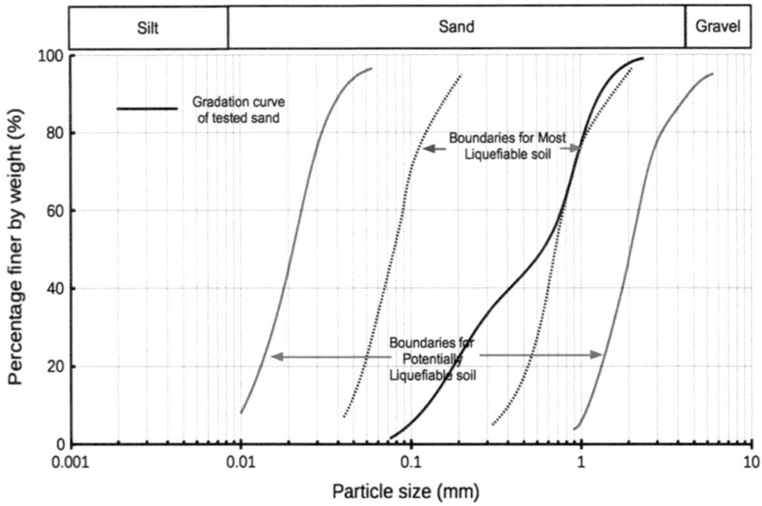

Fig. 1 Particle size range proposed by Tsuchida

Table 1 Properties of Aluva river sand

1.	% Silt size	1.37
2.	% Sand size	98.63
3.	% Gravel	0
4.	D_{10} (mm)	0.18
5.	D_{30} (mm)	0.35
6.	D_{60} (mm)	0.71
7.	Uniformity coefficient (C_u)	3.94
8.	Coefficient of curvature (C_c)	0.96
9.	Specific gravity	2.65
10.	Maximum void ratio (e_{max})	0.88
11.	Minimum void ratio (e_{min})	0.63

Table 2 Properties of embankment model

1.	Soil type	Moist red earth
2.	Height of embankment (mm)	200
3.	Slope of embankment	1 in 1.5
4.	Total weight of embankment (kN)	0.73
5.	Density (kN/m^3)	14.60

mounted on a platform connected to a pulley system that runs by a 5 HP three-phase electric motor. The ratio of pulleys is so designed to create a frequency of 1 Hz. The amplitude of shake is selected to keep the peak acceleration at 0.2 g. The data acquisition system consists of accelerometers and pore water pressure

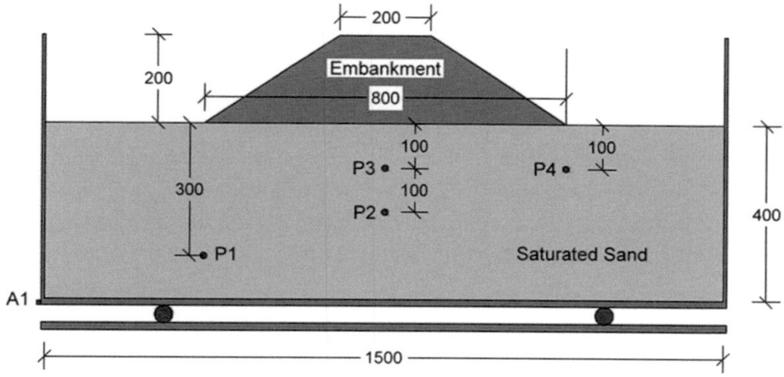

Fig. 2 Schematic diagram of test set-up

Table 3 Properties of geotextiles used

1.	Thickness (mm)	0.31
2.	Opening size (mm)	0.075
3.	Specific gravity	0.880
4.	Tensile strength (breaking strength at 200 mm wide and 100 mm length) Lengthwise (kN)	3.6
5.	Tensile strength (breaking strength at 200 mm wide and 100 mm length) Widthwise (kN)	2.0

sensors. The location of the sensors is shown in Fig. 2. The container box has baffle walls at it bottom to allow water uniformly to saturate in the sand bed.

Four experiments were conducted as part of the study.

(1) Soil without embankment;
(2) Soil with embankment;
(3) Soil with embankment and geotextile as separator at its base;
(4) Soil with embankment and geotextile reinforcement at its base and 100 mm within the soil.

The dry sand bed is formed first at the required density, and the embankment is created on top of it. After forming the model, it is saturated from the bottom very slowly without disturbing the layers. The saturation process took 4 h on an average.

All experiments are conducted in a controlled manner keeping density and input motions as constant. The sensor locations are shown in Fig. 2.

5 Results and Discussions

The excess pore water pressure that builds up during shaking is measured using the sensors kept at various locations. Figure 3 shows the variation of pore water pressure developed in the sensor P2 kept 200 mm below the embankment for all the cases studied. It is observed that the pore water pressure build-up is more when the embankment is present in the model. The maximum value observed is 3 kPa. When geotextile reinforcement is provided at the base, the pore water pressure builds up to

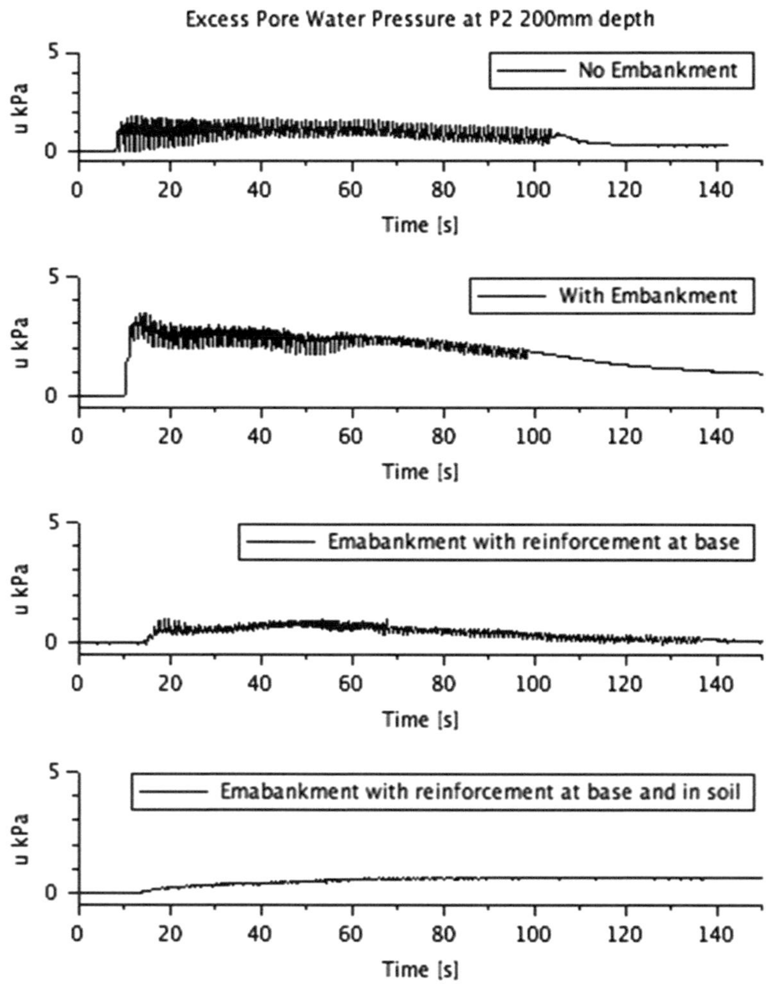

Fig. 3 Excess pore water pressure at P2, 200 mm below the embankment

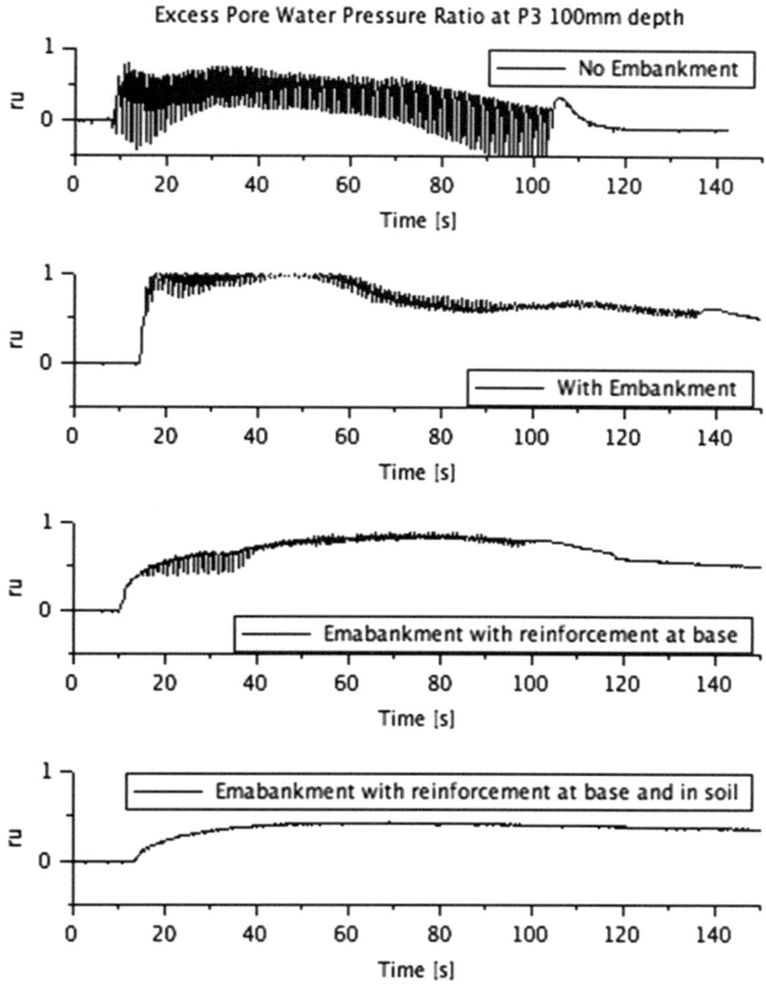

Fig. 4 Excess pore water pressure ratio at P3, 200 mm below the embankment

a lesser value of 1.0 kPa. The pressure build-up is minimum in the case where an additional reinforcement is provided at a depth of 100 mm in the sand layer.

Figure 4 shows the plot of excess pore water pressure ratio r_u, which is the ratio of excess pore water pressure to effective stress, at P3, 200 mm below the ground level. The ground nearly liquefied when the embankment was present, indicated by a value of 1.0 of pore water pressure ratio (r_u). The failure of the embankments is

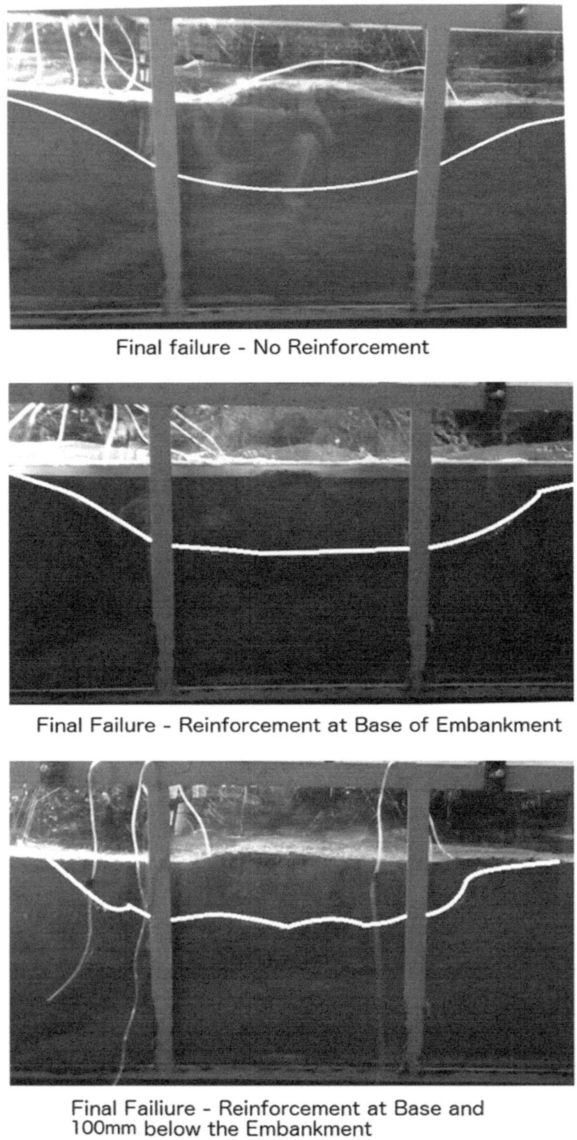

Fig. 5 Failure modes of test cases

shown in Fig. 5. Intense deformation of embankment due to soil liquefaction in foundation ground was observed, which can be classified as type 3 failure (Sasaki and Tamura 2007). This type of failure is induced by the deformation of the subsoil layer during shaking.

6 Conclusion

The experimental investigation provided a better understanding of the response of embankments on liquefiable deposits, the results interpreted qualitatively. The results from the experiments show that the pore pressure build-up will be more in the presence of embankment. This behaviour is due to the fact that the embankment will prevent the dissipation of excess pressure developed through the top surface. Providing the geotextile reinforcement at the base reduces the pore water pressure build-up. Providing an additional layer of geotextile reinforcement also leads to a reduction the pore water pressure build-up. It may be concluded that providing geotextile reinforcements can mitigate the effects of liquefaction.

References

Beena, K. S., & Unni Kartha, G. (2014). Stone columns for liquefaction mitigation: An experimental investigation. In *6th International Geotechnical Symposium on Disaster Mitigation in Special Geoenvironmental Conditions* (pp. 169–172). IIT Chennai.

Bhatnagar, S., Kumari, S., & Sawant, V. A. (2003). Numerical analysis of earth embankment resting on liquefiable soil and remedial measures. *International Journal of Geomechanics,* 1–13.

Koga, Y., & Matsuo, O. (1990). Shaking table tests of embankment resting on liquefiable sandy ground. *Soils and Foundations, 30*(4).

Maheshwari, B. K., Singh, H. P., & Saran, S. (2012). Effects of reinforcement on liquefaction resistance of Solani sand. *Journal of Geotechnical and Geoenvironmental Engineering, 138*(7), 831–840.

Orense, R. P., Morimoto, I., Yamamoto, Y., Yumiyama, T., Yamamoto, H., & Sugawara, K. (2003). Study on wall-type gravel drains as liquefaction countermeasure for underground structures. *Soil Dynamics Earthquake Engineering, 23,* 19–39.

Prasad, S. K., Towhata, I., Chandradhara, G. P., & Nanjundaswamy, P. (2004). Shaking table tests in earthquake geotechnical engineering. *Current Science, 87*(10).

Ramakrishnan, S., Budhu, M., & Britto, A. (1998). Laboratory seismic tests on geotextile wrap-faced and geotextile-reinforced segmental retaining walls. *Geosynthetics Internationals, 5*(1–2).

Sasaki, Y., & Tamura, K. I. (2007). Failure mode of embankments due to recent earthquakes in Japan. In *4th International Conference on Earthquake Geotechnical Engineering* (No. 1479).

Seed, H. B. (1968). Landslides during earthquakes due to soil liquefaction. *Journal of Soil Mechanics and Foundations, Division, 94*(5), 1055–1123.

Tsuchida, H. (1970). Prediction and countermeasure against the liquefaction in sand deposits. *Abstract of the Seminar in the Port and Harbor Research Institute*.

Varghese, R. M., & Latha, G. M. (2013). Initiation of liquefaction in sand under different surcharge and overburden stresses. In *Proceedings of Indian Geotechnical Conference*, Roorkee.

Development of Design Charts for the Dynamic Active Thrust from c–ϕ Soil Backfills

Ashish Gupta, Vinay Kumar Chandaluri, Vishwas A. Sawant and Sanjay Kumar Shukla

Abstract This paper presents design charts to calculate dynamic active thrust on retaining walls resulting from c–ϕ soil backfills. Shukla (J Geotech Geoenviron Eng ASCE 135:690–696, 2015) derived an explicit generalized analytical expression to calculate dynamic active thrust using the pseudo-static approach. This approach evaluates the dynamic active thrust accounting for all possible and the influencing factors. Using the explicit generalized analytical expression, the influence of various key factors such as backfill inclination, dimensionless cohesion, and angle of shearing resistance, horizontal and vertical seismic coefficients are studied. The design charts are presented in non-dimensional form. These charts are straightforward and can be adopted for the design of retaining walls in seismic zone. The active earth pressure coefficient and the critical value of inclination of the failure plane, obtained from the present study, are validated with studies reported in the literature.

Keywords Retaining walls · c–ϕ soil backfills · Seismic loads Surcharge · Dimensionless cohesion · Dynamic active thrust · Active earth pressure coefficient

---

A. Gupta (✉) · V. K. Chandaluri · V. A. Sawant
Department of Civil Engineering, Indian Institute of Technology, Roorkee, Roorkee 247667, Uttarakhand, India
e-mail: shi_g2000@rediffmail.com

V. K. Chandaluri
e-mail: vinay2744@gmail.com

V. A. Sawant
e-mail: sawntfce@iitr.ac.in

S. K. Shukla
SOE, Faculty of Health, Engineering and Science, Edith Cowan University, Joondalup, Australia
e-mail: s.shukla@ecu.edu.au

B. Adimoolam and S. Banerjee (eds.), *Soil Dynamics and Earthquake Geotechnical Engineering*, Lecture Notes in Civil Engineering 15,
https://doi.org/10.1007/978-981-13-0562-7_13

1 Introduction

In practical situations, the dynamic active thrust on retaining walls is calculated using explicit analytical expressions. Mononobe-Okabe expression is typically used for calculating the dynamic active thrust for the cohesionless soils. But this expression cannot be used for analyzing the real field situations where c–ϕ soil backfills are present.

Shukla et al. (2009) gave a simple expression for calculating the dynamic active thrust from the c–ϕ soil backfills. The wall friction and the adhesion between soil backfill and retaining wall face are ignored in this study. The back face of the wall is assumed to be vertical.

Kim et al. (2010) obtained an expression for calculating the total dynamic active thrust in terms of the inclination of the failure plane. But the calculation of inclination of the failure plane was based on the trial and error procedure. Thus, this expression has limited applications in real design practices.

Shukla (2011) studied the effects of tension cracks, wall–backfill adhesion, and surcharge loading on dynamic active thrust. The effect of wall friction and other factors on dynamic active thrust was presented by Shukla and Bathurst (2012). Shukla (2013) presented an expression to calculate the dynamic active thrust for sloping soil backfill.

Shukla (2015) developed a generalized analytical expression for the dynamic active thrust on retaining wall supporting a c–ϕ soil sloping backfill. Effect of wall geometry, different properties of backfill and wall–backfill interface, tension cracks, surcharge, horizontal and the vertical ground acceleration were examined. The expression for the critical value of inclination of the failure plane is also developed in this study.

In the present study, design charts have been developed for dynamic active thrust by using the explicit generalized analytical expression proposed by Shukla (2015). These charts can be adoptable for the design of retaining walls for the seismic stability.

2 Analytical Derivation

Figure 1 shows a retaining wall supporting a c–ϕ soil backfill of height H with an active trial failure wedge $A_1A_2A_3$ of weight W. The backfill face of wall A_1A_2 is inclined at an angle β with the horizontal. Sloping backfill A_1A_3 is inclined at an angle i with the horizontal. The failure plane is assumed along A_2A_3 and passes through the bottom of the wall. A_2A_3 is inclined at an angle α to the horizontal. k_hW and k_vW are the horizontal and vertical seismic inertial forces, where k_h and k_v are the horizontal and the vertical seismic coefficients. A surcharge q per unit surface area is placed at the top of the sloping backfill. k_hqB and k_vqB are the surcharge loads along with the horizontal and vertical seismic inertial forces, where B is the length of sloping backfill A_1A_3.

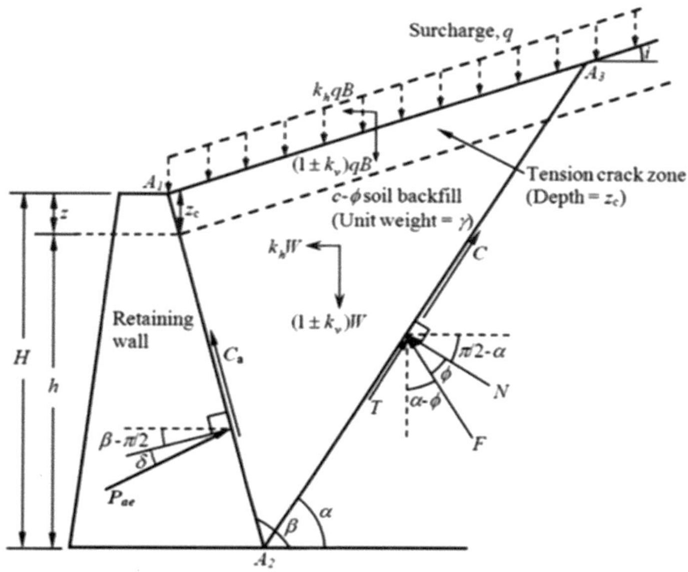

Fig. 1 Forces shown in a trial failure wedge supporting a c–ϕ soil backfill in active state (after Shukla 2015)

z_c is the depth of tension crack from backfill surface. To simplify, the wall height H is taken as the sum of z and h. The force F is the resultant of frictional force T and normal force N acting on the failure plane A_2A_3. C is the total cohesive force on the failure plane. C_a is the total adhesive force mobilized along the wall–backfill interface, and a_f is the adhesion factor. P_{ae} is the dynamic active thrust, inclined at an angle δ.

Taking equilibrium of forces in the vertical and horizontal direction, respectively,

$$P_{ae}\cos(\beta+\delta) + (1 \pm k_v)(W + qB) - F\cos(\alpha-\phi) - \frac{\bar{c}H\sin(\beta-i)\sin\alpha}{\sin\beta\sin(\alpha-i)} - a_f\bar{c}H = 0 \quad (1)$$

$$P_{ae}\sin(\beta+\delta) - k_h(W + qB) - F\sin(\alpha-\phi) + \frac{\bar{c}H\sin(\beta-i)\cos\alpha}{\sin\beta\sin(\alpha-i)} - a_f\bar{c}H\cot\beta = 0 \quad (2)$$

Eliminating F from Eqs. (1) and (2) and further simplifying, P_{ae} can be expressed as,

$$P_{\mathrm{ae}} = \frac{1}{2}\left(\frac{a_1 \tan^2 \alpha - b_1 \tan \alpha + c_1}{a_2 \tan^2 \alpha - b_2 \tan \alpha + c_2}\right)\gamma H^2 \tag{3}$$

In which,

$$\theta = \tan^{-1}\left[\frac{k_{\mathrm{h}}}{1 \pm k_{\mathrm{v}}}\right] \tag{4}$$

$$a_{\mathrm{f}} = \frac{c_{\mathrm{a}}}{c} \tag{5}$$

where c_{a} is the adhesion between the wall back face and the soil backfill.

$$m_1 = \left(\frac{1 \pm k_v}{\cos \theta}\right)\left[\frac{2q}{\gamma H} + \frac{\sin(\beta - i)}{\sin \beta}\right]\mathrm{cosec}\beta \tag{6}$$

$$m_2 = a_f\left(\frac{2c}{\gamma H}\right)\left(1 - \frac{z_{\mathrm{c}}}{2H}\right)\mathrm{cosec}\beta \tag{7}$$

$$m_3 = \left(\frac{2c}{\gamma H}\right)\left(1 - \frac{z_{\mathrm{c}}}{2H}\right)\sin(\beta - i)\cos \phi \mathrm{cosec}\beta \tag{8}$$

$$a_1 = m_1 \cos \beta \cos(\theta - \phi) + m_2 \cos i \ \sin(\beta + \phi) + m_3 \tag{9}$$

$$b_1 = m_1 \sin(\beta - \theta + \phi) - m_2 \cos(i + \beta + \phi) \tag{10}$$

$$c_1 = m_3 - m_1 \sin \beta \sin(\theta - \phi) - m_2 \sin i \cos(\beta + \phi) \tag{11}$$

$$a_2 = \cos i \ \cos(\beta + \delta + \phi) \tag{12}$$

$$b_2 = \sin(\beta + \delta + \phi + i) \tag{13}$$

$$c_2 = \sin i \ \sin(\beta + \delta + \phi) \tag{14}$$

a_1, b_1, c_1 and a_2, b_2, c_2 are the non-dimensional constants.
In Eqs. (7) and (8),

$$z_{\mathrm{c}} = \frac{2c}{\gamma}\tan\left(\frac{\pi}{4} + \frac{\emptyset}{2}\right) \tag{15}$$

For optimizing the value of dynamic active thrust P_{ae}, the following condition should be satisfied.

$$\frac{\partial P_{\mathrm{ae}}}{\partial(\tan \alpha)} = 0 \tag{16}$$

Or,

$$(a_2b_1 - a_1b_2)\tan^2\alpha - 2(a_2c_1 - a_1c_2)\tan\alpha + (b_2c_1 - b_1c_2) = 0 \quad (17)$$

Above quadratic equation (15) is solved for tanα, to get the critical value of inclination of the failure plane, $\alpha = \alpha_c$ as,

$$\alpha_c = \tan^{-1}\left[\frac{(a_2c_1 - a_1c_2) \pm \sqrt{(a_2c_1 - a_1c_2)^2 - (a_2b_1 - a_1b_2)(b_2c_1 - b_1c_2)}}{(a_2b_1 - a_1b_2)}\right] \quad (18)$$

Substituting $\alpha = \alpha_c$ into Eq. (3), P_{ae} is obtained as by Shukla (2015),

$$P_{ae} = \frac{1}{2}\left(\frac{a_1\tan^2\alpha_c - b_1\tan\alpha_c + c_1}{a_2\tan^2\alpha_c - b_2\tan\alpha_c + c_2}\right)\gamma H^2 \quad (19)$$

Or,

$$K_{ae} = P^*_{ae} = \frac{P_{ae}}{0.5\gamma H^2} = \left(\frac{a_1\tan^2\alpha_c - b_1\tan\alpha_c + c_1}{a_2\tan^2\alpha_c - b_2\tan\alpha_c + c_2}\right) \quad (20)$$

where K_{ae} is the active earth pressure coefficient.

3 Results and Discussions

Based on the generalized expression derived (Eq. 20), design charts are presented for calculating the total active thrust on retaining wall for different values of k_h, k_v, and c^* with and without q^*, where c^* and q^* are non-dimensional cohesion and non-dimensional surcharge defined in Eqs. (21) and (22), respectively.

$$c* = c/\gamma H \quad (21)$$

$$q* = q/\gamma H \quad (22)$$

The charts are presented for a vertical wall (β = 90°). In the presented charts, k_h ranges from 0.0 to 0.5 and three different values of k_v = 0.0, 0.25 k_h and 0.5 k_h are considered. The vertical inertial force is considered as positive in upward direction. Design charts obtained from the present study are shown in Figs. 2a–e and 3a–e. Figure 2a–e are design charts for calculating the total active thrust from c–ϕ soil backfill without surcharge loading (q^* = 0.0). In Fig. 3a–e, the effect of surcharge on the active earth pressure coefficient is additionally considered. The non-dimensional surcharge is taken as 0.1 (q^* = 0.1).

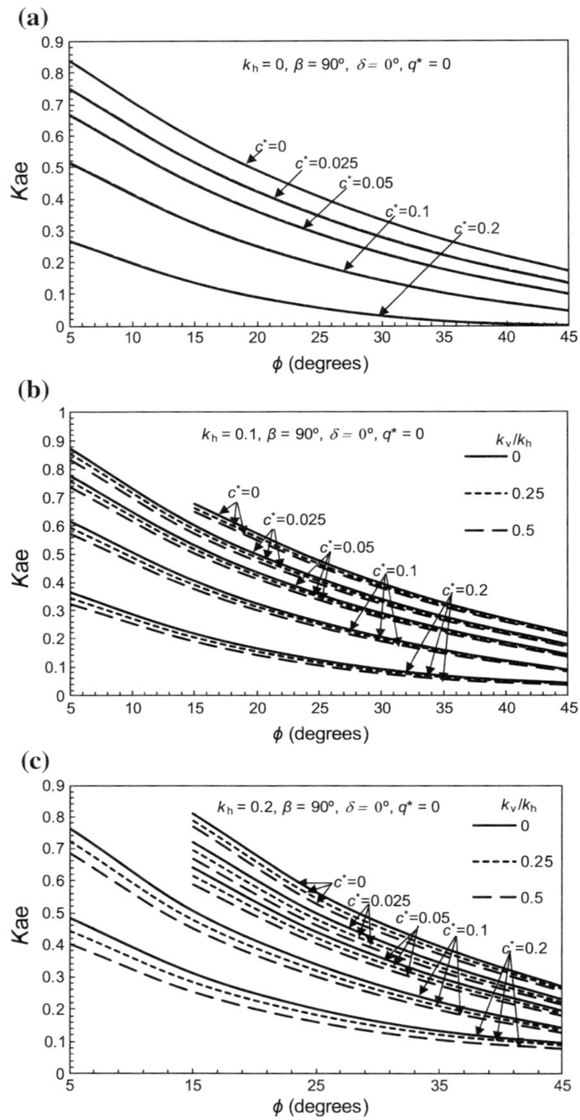

Fig. 2 **a–e** Design charts for calculating total active thrust from c–ϕ soil backfill for different values of k_h, k_v, and $c*$ without backfill surcharge loading

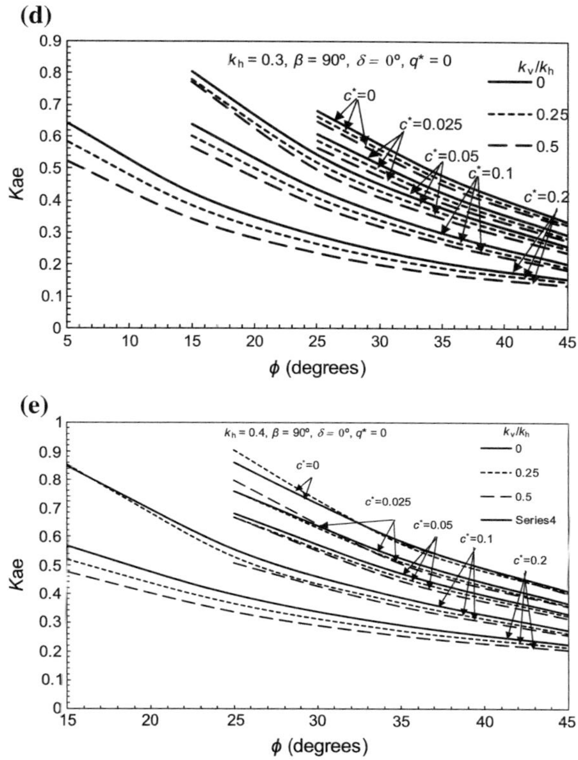

Fig. 2 (continued)

From Figs. 2 and 3, it can be clearly observed that the value of K_{ae} decreases with increase in cohesion as well as increase in angle of shearing resistance of soil. For $\phi = 15°$, $c^* = 0.1$, when q^* increases from 0.0 to 0.1, the value of K_{ae} increases about 30–36%. However, the percentage increase is decreasing with increase in the value of k_h.

For $\phi = 15°$, $c^* = 0.1$, the value of K_{ae} increases by about 25.8, 57.1 and 98.1%, when k_h value increases from 0.0 to 0.1, 0.2, and 0.3, respectively.

The value of α_c and K_{ae} obtained from the present study for different values of k_h and ϕ has been compared with studies reported in the literature as shown in

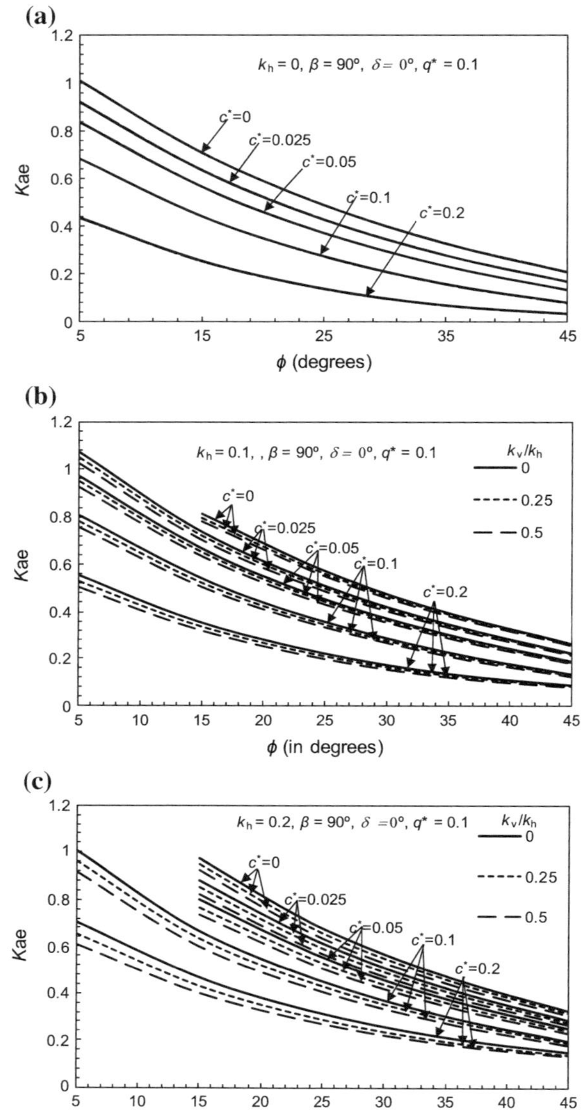

Fig. 3 a–e Design charts for calculating total active thrust from c–ϕ soil backfill for different values of k_h, k_v, and c^* with $q^* = 0.1$

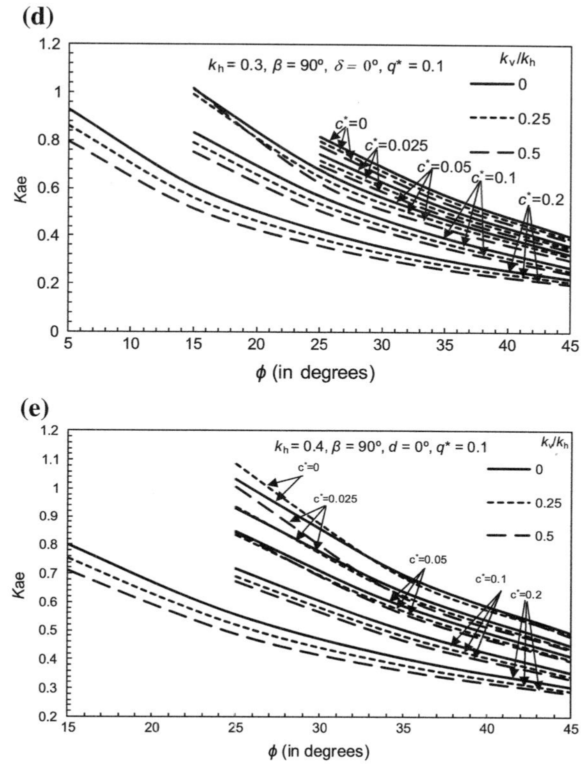

Fig. 3 (continued)

Tables 1 and 2, respectively. From Tables 1 and 2, it can be noted that all the results are in good agreement.

4 Conclusions

The present study provides the design charts for calculating total active thrust from c–ϕ soil backfills on retaining wall considering both the horizontal and vertical seismic coefficients and non-dimensional cohesion with and without non-dimensional surcharge. Based on the design charts from the present study, the following conclusions can be summarized:

1. The value of active earth pressure coefficient decreases with increase in non-dimensional cohesion as well as increase in angle of shearing resistance of soil backfills irrespective of the backfill non-dimensional surcharge loading.

Table 1 Comparison of results for calculation of critical value of inclination of failure plane ($i = 0°$, $k_v = 0$, $c = 0$, $\delta = (2/3)\phi$, and $\beta = 90°$)

ϕ	$k_h = 0.0$					$k_h = 0.05$					$k_h = 0.1$					$k_h = 0.2$				
	LEM	HSM	M-O	Chang	Present analysis	LEM	HSM	M-O	Chang	Present analysis	LEM	HSM	M-O	Chang	Present analysis	LEM	HSM	M-O	Chang	Present analysis
20°	50	49.9	50	50	50.01	47	47.1	46.6	47	46.65	43	43.9	42.8	43	42.83	33	36.3	33.4	33	33.37
25°	53	52.9	53	53	53.01	50	50.5	50.1	50	50.1	47	47.8	46.9	47	46.89	39	41.7	39.4	39	39.38
30°	55	55.9	56	56	55.98	53	53.7	53.4	53	53.37	51	50.6	50.5	51	50.53	44	46.1	44.1	44	44.11

LEM: Ghanbari and Ahmadabadi (2010) (Based on Limit Equilibrium)
HSM: Ghanbari and Ahmadabadi (2010) (Based on Horizontal Slice Method)
M-O: Mononobe and Matsuo (1929)

Table 2 Comparison of results for calculation of active earth pressure coefficient ($i = 0°$, $k_v = 0$, $c = 0$, $\delta = (2/3)\phi$, and $\beta = 90°$)

ϕ	$k_h = 0.0$					$k_h = 0.05$					$k_h = 0.1$					$k_h = 0.2$				
	LEM	HSM	M-O	Chang	Present analysis	LEM	HSM	M-O	Chang	Present analysis	LEM	HSM	M-O	Chang	Present analysis	LEM	HSM	M-O	Chang	Present analysis
20°	0.438	0.440	0.438	0.426	0.438	0.478	0.479	0.479	0.656	0.478	0.526	0.526	0.525	0.511	0.525	0.647	0.645	0.647	0.629	0.647
25°	0.361	0.362	0.361	0.346	0.361	0.397	0.398	0.397	0.380	0.397	0.438	0.438	0.438	0.419	0.438	0.539	0.539	0.539	0.516	0.539
30°	0.297	0.299	0.297	0.279	0.297	0.330	0.330	0.330	0.310	0.330	0.366	0.366	0.366	0.344	0.366	0.454	0.453	0.454	0.426	0.454

LEM: Ghanbari and Ahmadabadi (2010) (Based on Limit Equilibrium)
HSM: Ghanbari and Ahmadabadi (2010) (Based on Horizontal Slice Method)
M-O: Mononobe and Matsuo (1929)

2. The active earth pressure coefficient increases with increase in the value of horizontal seismic coefficient. However, the percentage increase is marginally decreasing with the increase in horizontal seismic coefficient value.
3. The critical value of inclination of the failure plane and the active earth pressure coefficient obtained from the present study for different values of horizontal seismic coefficient and angle of shearing resistance has been compared with studies reported in the literature, and the results found are in good agreement.

References

Ghanbari, A., & Ahmadabadi, M. (2010). Active earth pressure on inclined retaining walls in static and pseudo-static conditions. *International Journal of Civil Engineering, 8*(2), 159–173.

Kim, W. C., Park, D., & Kim, B. (2010). Development of a generalised formula for dynamic active earth thrust. *Geotechnique, 60*(9), 721–727.

Mononobe, N., & Matsuo, H. (1929). On the determination of earth pressure during earthquake. In *Proceedings of the World Engineering Congress* (pp. 177–185), Tokyo, Japan.

Shukla, S. K. (2011). Dynamic active thrust from c-ϕ soil backfills. *Soil Dynamics and Earthquake Engineering, 31*(3), 526–529.

Shukla, S. K. (2013). Seismic active earth pressure from the sloping c-ϕ soil backfills. *Indian Geotechnical Journal, 43*(3), 274–279.

Shukla, S. K. (2015). Generalized analytical expression for dynamic active thrust from c-ϕ soil backfills. *International Journal of Geotechnical Engineering, 9*(4), 416–421.

Shukla, S. K., & Bathurst, R. J. (2012). An analytical expression for the dynamic active thrust from c-ϕ soil backfills on retaining wall with wall friction and adhesion. *Geomechanics and Engineering, 4*(3), 209–218.

Shukla, S. K., Gupta, S. K., & Sivakugan, N. (2009). Active earth pressure on retaining all for c-ϕ soil backfill under seismic loading condition. *Journal of Geotechnical and Geoenvironmental Engineering, ASCE, 135*(5), 690–696.

Pseudo-static Stability Analysis of MSW Landfills in Goa

Tanvi Mardolkar and Purnanand Savoikar

Abstract In the conventional method for determining the seismic stability of geotechnical structures using pseudo-static or pseudo-dynamic analysis, the horizontal seismic acceleration coefficient is assumed arbitrarily or based upon past experience, neglecting effect of seismic wave velocities, seismic displacements, time period, spectral acceleration, faulting pattern, etc. In the present study, pseudo-static limit equilibrium slope stability analysis of landfill (proposed in Goa) is performed by considering the effects of the above parameters using conventional and rational method of selection of horizontal seismic coefficient considering the effect of vertical seismic acceleration. The effect of water table and leachate buildup is also considered. It is observed that due to the loss of shear strength as a result of pore pressure buildup in case of wet condition, there is decrease in F.O.S. up to 30%.

Keywords MSW landfills · Pseudo-static · Leachate buildup

1 Introduction

There is an immense increase in waste generation with rapid increase in industrialization. The unscientific disposal of municipal solid waste (MSW) in the past several years has led to the contamination of environment and natural resources in and around such dumpsites thereby posing serious risk to human and animal health. MSW has therefore emerged as a growing challenge for various states and municipal authorities in the country. MSW is heterogeneous mixtures of wastes consisting of residential and commercial origin. Municipal solid waste landfills are

T. Mardolkar · P. Savoikar (✉)
Department of Civil Engineering, Goa College of Engineering,
Farmagudi 403401, Goa, India
e-mail: psavoikar@gmail.com

T. Mardolkar
e-mail: sweettanu045@gmail.com

B. Adimoolam and S. Banerjee (eds.), *Soil Dynamics and Earthquake Geotechnical Engineering*, Lecture Notes in Civil Engineering 15,
https://doi.org/10.1007/978-981-13-0562-7_14

huge geotechnical structures, and their heights may be as small as 10 m to as high as 100 m or even more. The aim of engineered landfills is to avoid any leachate released into subsurface environment and controls the exposure of the environment and humans to the detrimental effect.

The stability of landfills against these seismic forces can be evaluated by computing factor of safety for different modes of failure, and the permanent safety can also be evaluated by permanent displacement to check the serviceability of landfills after earthquake. Most of the landfill failures occur due to the excessive leachate buildup in the waste mass. Proper management of leachate is very important for any landfill construction.

2 Literature Review

Behavior of MSW landfills during earthquakes, along with the recordings of the past events containing static and dynamic properties of landfills, helps in analyzing the seismic behavior of landfills.

Various researchers such as Kandolkar et al. (2010), Bray and Travassarou (2009), Savoikar and Aldonkar (2011) carried out pseudo-static limit equilibrium slope stability analysis of hill type embankment by considering rational method for selecting horizontal seismic acceleration coefficient. By varying the angle of failure surface, F.O.S. was evaluated. Design charts were developed for F.O.S. for different values of various parameters such as width-to-height ratio, failure plane inclination. The papers clearly illustrated the importance of rational method for selecting horizontal seismic acceleration coefficient. However, the effect of water table or leachate buildup is ignored. Qian (2006) developed calculation method of limit equilibrium analysis for translational failure. Four leachate buildup conditions were considered during development of calculation method for analyzing translational failure to represent various conditions in landfills. Excessive leachate buildup is a triggering mechanism for translational failure. It was concluded that as the leachate level increases the F.O.S. decreases.

In this paper, the effect of water table/leachate buildup is considered for landfill analysis using rational method for selecting the horizontal seismic coefficient suggested by Bray and Travassarou (2009).

2.1 *Landfills in India*

Unfortunately, even today MSW is continued to be disposed of as uncontrolled dump in developing countries such as India. Increasing prices of land and in the amount of waste forces engineers to create higher and steeper dumps and landfills to

Fig. 1 Landfill construction at Saligao

accommodate more waste. In its 2009–10 annual report, the Ministry of New and Renewable Energy (MNRE) estimated that approximately 55 million tons of MSW are generated in urban areas of India annually. Among the four geographical regions in India, Northern India generates the highest amount of MSW.

As per MSW Rules 2000, Government of Goa has taken an initiative to establish a Centralized Municipal Solid Waste Management Facilities at an existing MSW dumpsite located at Calangute/Saligao Villages in North Goa district (Fig. 1). Major portion of the MSW generated in the North Goa district is dumped haphazardly and unscientifically at this site by nearby panchayats for last 20 years. Unscientific dumping of MSW at this site resulted in contamination of environment. Major focus of the project is on the issues pertaining to odor and leachate generation and their scientific management to minimize the same. Plant operation shall be carried out using the electricity produced in-house from the organic fraction of the MSW.

Efforts are also taken for providing an in-house auditorium/resource center for displaying the recycling and waste treatment process, gardens, public conveniences, and café for encouraging public tours. The facility shall not only serve as a tourist attraction, but also as learning center to create awareness about the positive environmental impact of the solid waste treatment facility in the State of Goa. This would also facilitate image building of the state government with respect to environmental restoration.

3 Methodology

A hill type landfill model is used for the present analysis with height H, top width B, side slope of θ (tan − 1(1/S)). Biplanar failure surface makes an angle β with horizontal. Limit equilibrium method is used to analyze the pseudo-static stability of the landfill with and without considering the effect of leachate buildup. Four ideal seepage cases are considered during the development of calculation methods for analyzing translational failures with seepage as shown in Fig. 2.

Case—1 is parallel-to-subgrade and back slope seepage buildup. In this case, the seepage flows parallel to the subgrade. Such type of leachate buildup normally occurs when landfill is in normal operating condition.
Case—2 is parallel-to-subgrade seepage buildup. The seepage flow is parallel to the landfill bottom slope, and the leachate depth exceeds the limited depth of 300 mm.
Case—3 is horizontal seepage buildup with seepage parallel to back slope. Such situation may occur when there is breakage of leachate collection pump or has lost power in an active landfill with a temporary or partial final cover. The incoming liquid can cause a rapid rise of the leachate level in the waste mass.
Case—4 is horizontal seepage buildup. In fourth case, horizontal seepage builds up and either ends at the liner system or extends deeply into the waste mass. Such type of buildup of seepage buildup might occur in partially or fully closed landfills during power shutdown, when pump shutdown problems cause high liquid accumulation in the lower part of the landfill.

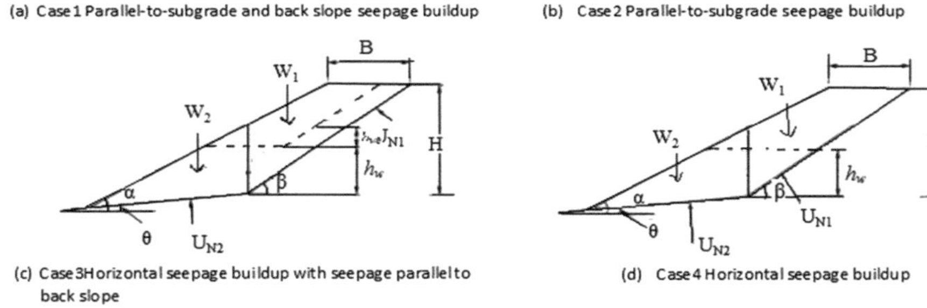

(a) Case 1 Parallel-to-subgrade and back slope seepage buildup

(b) Case 2 Parallel-to-subgrade seepage buildup

(c) Case 3 Horizontal seepage buildup with seepage parallel to back slope

(d) Case 4 Horizontal seepage buildup

Fig. 2 Different cases of leachate buildup

The equation derived for factor of safety without considering the effect of water table is given below:

$$\mathrm{FS} = \frac{\begin{array}{l}(1 \mp k_{\mathrm{v}})\{W_1.\cos^2\theta + W_2.\cos^2\beta\}\tan\phi \\ -k_{\mathrm{h}}.\tan\phi\{W_1.\sin\theta.\cos\theta + W_2.\sin\beta.\cos\beta\} \\ +c(L_1.\cos\theta + L_2.\cos\beta)\end{array}}{\begin{array}{l}(1-k_{\mathrm{v}})\{W_1.\cos\theta.\sin\theta + W_2.\cos\beta.\sin\beta\} \\ +k_{\mathrm{h}}\left(W_1 + W_2 - W_1.\sin^2\theta - W_2.\sin^2\beta\right)\end{array}}$$

The destabilizing forces due to pore pressure generated as the result of leachate buildup acts at the bottom of active and passive wedges. They are defined as U_{N1}, U_{N2} in Fig. 3.

The pressures caused by pore water due to leachate buildup either in active or closed condition can be evaluated using the equations recommended by Qian (2008).

The equation derived for factor of safety considering the effect of water table is given below:

$$\mathrm{F.O.S.} = \frac{\begin{array}{l}(1 \mp k_{\mathrm{v}})\{W_1.\cos^2\theta + W_2.\cos^2\beta\}\tan\phi \\ -k_{\mathrm{h}}.\tan\phi\{W_1.\sin\theta.\cos\theta + W_2.\sin\beta.\sin\beta\} + (U_{\mathrm{N1}}\cos\theta + U_{\mathrm{N2}}\cos\beta)\tan\phi \\ +c(L_1.\cos\theta + L_2.\cos\beta)\end{array}}{\begin{array}{l}(1-k_{\mathrm{v}})\{W_1.\cos\theta.\sin\theta + W_2.\cos\beta.\sin\beta\} + (U_{\mathrm{N1}}\sin\theta + U_{\mathrm{N2}}\sin\beta) \\ +k_{\mathrm{h}}(W_1 + W_2 - W_1.\sin^2\theta - W_2.\sin^2\beta)\end{array}}$$

where W_1 and W_2 are the weight of the active and passive wedges, respectively, and U_{N1} and U_{N2} are the forces due to pore water pressure. k_{h} and k_{v} are horizontal and vertical seismic coefficients, respectively. The vertical seismic coefficient k_{v} is taken as $0.5k_{\mathrm{h}}$. The forces due to pore pressure are found for active and closed conditions. Active condition is when waste is being dumped and the front slope of the fill is not assumed to be covered by the final cover. In a closed condition, the filling is assumed to be complete and the front slope is assumed to be covered with an impermeable or very low permeable.

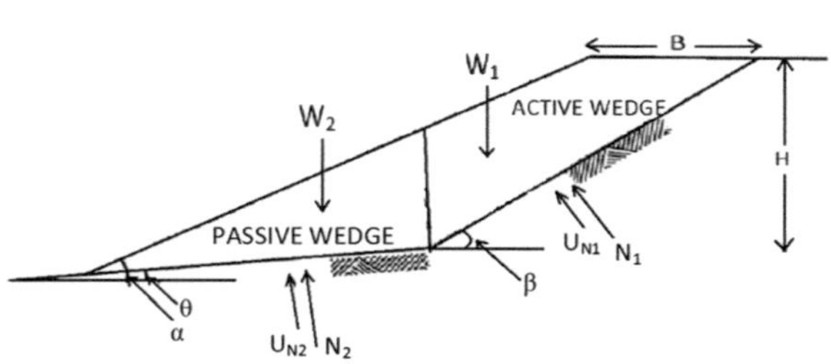

Fig. 3 Leachate buildup condition

4 Results and Discussions

A parametric study is done to analyze the effect of various parameters such as landfill side slope (1:S), failure angle, magnitude, acceleration coefficient on the factor of safety. To obtain the values of k_h (horizontal seismic coefficient), both rational as well as conventional methods are used. The mathematical model suggested by Bray and Travasarou (2009) is used in the analysis. The attenuation relationship suggested by Idriss (2002) is used to obtain spectral acceleration (S_a).

The analyses are performed for both cases, i.e., with or without considering the effect of leachate buildup, and the values are compared. The following parameters are assumed as per the literature survey and considered for the analysis: the unit weight of solid waste, γ_{sw}, of 10.5 kN/m^3 (Qian 2008); a saturated unit weight of solid waste, $(\gamma_{sw})_{sat}$, of 13 kN/m^3 (Qian 2008), Gs, of 1.60; a porosity, n, of 0.50. The height is 20 m, and top width taken into account is 15 m (Kandolkar et al. 2010). The shear strength parameters of solid waste are taken as: c = 15 kPa and ϕ = 20 (Kandolkar et al. 2010). The leachate level is assumed to be present at the depth of 1.5 m, and side slope taken into account is 1:1. The failure angle is of 0.667θ, i.e., 30° (approximately). It must be noted that assumption of $h_{wb} = 0.5h_w$ (Qian 2008) for Case 1 and Case 3 is used in the following analysis, where h_w is the vertical leachate depth in landfill measured from the toe of back slope, and h_{wb} is the vertical leachate depth in landfill measured along back slope for Case 1 and Case 3. Constant earthquake magnitude of 5 is considered. For parametric variation, three different types of earthquakes with moment magnitude M_w = 5, 6.5, and 7.0 are used. The value of failure angle β varies from 0.35 to 0.85 θ.

4.1 *Effect of Failure Plane Angle, β*

Failure angle β is varied as mentioned above for various values of embankment side slopes such as 1:1, 1:2, 1:2.5 (Fig. 4). It can be seen that as failure slope angle becomes steeper, factor of safety increases for both the cases. In case of dry condition, the F.O.S. remains almost constant and then increases slightly, whereas in case of pore pressure buildup conditions, the F.O.S. decreases gradually with increase in failure angle. This behavior is observed in wet condition since shear strength is reduced.

4.2 *Effect of Earthquake Magnitude*

Figure 5 depicts variation of factor of safety with failure plane angle β for various values of moment magnitude and horizontal seismic coefficient. As the value of moment magnitude increases, the value of horizontal seismic coefficient also

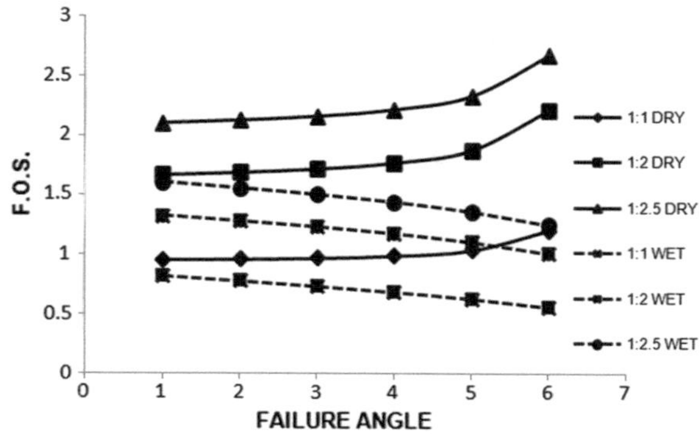

Fig. 4 Effect of side slope in dry and wet conditions

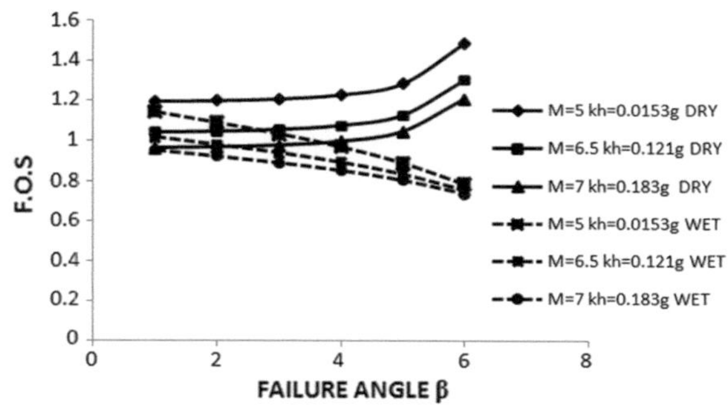

Fig. 5 Effect of magnitude with and without considering leachate buildup condition

increases, and hence, the driving force increases which in turn lowers the factor of safety. The trend of variation of factor of safety in this case (Fig. 5) is similar to that of side slope variation as in the previous case (Fig. 4). The relation between the moment magnitude and k_h is given below:

$$k_h = e^{\frac{-a+\sqrt{b}}{0.665}}$$

where

$$\begin{aligned} a &= 2.83 - 0.55\ln(S_a) \\ b &= (a^2 - 1.33(\ln(D_a) + 1.1 - 3.04\ln(S_a) \\ &\quad + 0.224(\ln(S_a))^2 - 1.5T_s - 0.278(M_w - 7) - \varepsilon) \\ T_s &= 4H/V_s \end{aligned}$$

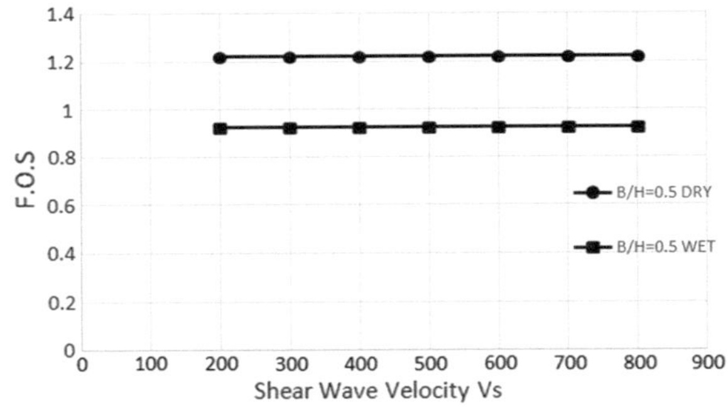

Fig. 6 Effect of stiffness with and without considering leachate buildup condition

4.3 *Effect of Stiffness of Landfill*

Effect of stiffness of the landfill (i.e., shear wave velocity, V_s) is observed in Fig. 6. The velocity is varied from 200 to 800 m/s. There is no effect on the F.O.S. values in both dry as well as wet cases.

4.4 *Effect of Leachate Buildup for All Four Cases*

Figure 7 shows the variation of F.O.S. with leachate buildup depths in all the four typical cases for both active and closed conditions.

The graph (Fig. 7) clearly shows a downward fall with increasing leachate buildup level. Case 1 and Case 2 conditions are on a safer side, whereas Case 3 and Case 4 conditions of F.O.S are unsafe. The fall in the factor of safety is steeper for Case 1 and Case 2, whereas in Case 3 and Case 4 the factor of safety is almost constant.

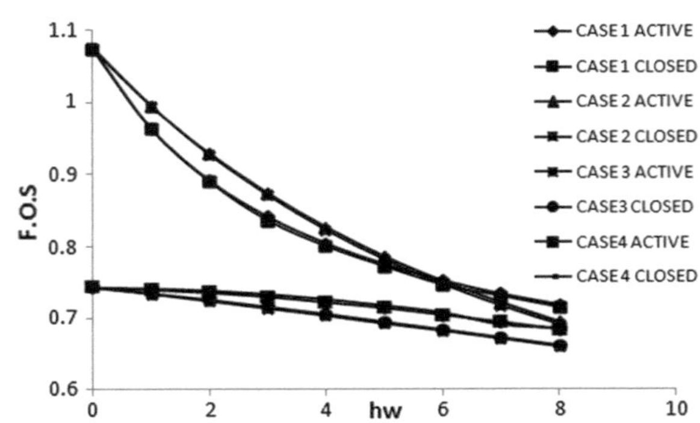

Fig. 7 Relationship between h_w and FS in various leachate buildup cases for active and closed conditions

5 Conclusions

The F.O.S. increases as the slope becomes steeper for both dry and wet conditions. The F.O.S. increases as the failure angle increases in case of dry condition. Due to the loss of shear strength as a result of pore pressure buildup in case of wet condition, there is decrease in F.O.S. up to 30%. The similar trend is followed also when the moment magnitude values are changed. This also illustrates the importance of k_h value by rational method where moment magnitude is one of the factor on which k_h value depends upon. There is no effect of shear wave velocity on F.O.S. Such similar trend is observed by Kandolkar et al. (2010). For all the four cases in active and closed conditions, overall there is decrease in F.O.S. with increase in leachate level. But there is higher fall in the value for Case 1 and Case 2, whereas the values tend to be constant in Case 3 and Case 4. The design charts developed can be helpful in the preliminary design of landfills.

References

Bray, J., & Travasarou, T. (2009). Pseudostatic coefficient for use in simplified seismic slope stability evaluation. *Journal of Geotechnical and Geoenvironmental Engineering, 135*(9), 1336–1340.

Idriss, I. (2002). Attenuation relationship derived by I. M. Idriss in 2002. *Interim Report Issued for USGS Review*. PEER Centre, University of California, Berkeley.

Kandolkar, S., Aldonkar, S., & Savoikar, P. (2010). Rational pseudo-static stability analysis of embankment. In *Indian Geotechnical Conference–2010, GEOtrendz*, December 16–18, 2010.

Qian, X. (2006). Translational failure of geosynthetic lined landfills under different leachate buildup condition. In *Advances in unsaturated soil, seepage and environmental geotechnics, GSP148*.

Qian, X. (2008 March). Limit equilibrium analysis of translational failure of landfills under different leachate buildup conditions. *Water Science and Engineering, 1*(1), 44–62.

Savoikar, P., & Aldonkar, S. S. (2011). Rational pseudo-static seismic stability analysis of MSW landfills. In *International Congress on Geo-Environmental Engineering 2011*, Takamatsu, Japan.

Study of Local Site Effects for Strong Motion Recording Stations of Delhi

Bhavesh Pandey, Ravi S. Jakka and Ashok Kumar

Abstract Effects of local site conditions on strong ground motion characteristics are a well-known phenomenon studied by many researchers. Similar studies were also carried out in past for Delhi to understand the amplification of ground motion in various locations using different methods. In this study, an attempt is made to study the amplifications in eight locations of Delhi where strong motion recording stations are located. For this study, shear wave velocity profiles were used for performing ground response analysis using equivalent linear approach. For input motion, response spectra compatible motion was used, compatible to response spectra for Rock sites as per IS 1893:2002. Transfer functions from available earthquake records and ambient vibration records were also calculated for comparison. The results obtained were presented in the form of response spectra and amplification functions for all the sites. The results suggest an amplification of the order of 2–3 with respect to actual motion at sites coming under site class D of NEHRP. No amplification was found at IMD site which suggest that it can be used as a reference site for Delhi. Other two sites with NEHRP site class C found to have amplification of the order of 1–2. Similarly, output response spectra, at surface, for all the sites except IMD and Kalkaji were found to be amplified with respect to the corresponding response spectra of IS 1893:2002.

Keywords Site amplification · Local site effects · Response spectra
Reference site · HVSR · Strong ground motion

B. Pandey · R. S. Jakka (✉) · A. Kumar
Indian Institute of Technology Roorkee, Rorkee 247667, India
e-mail: rsjakka@gmail.com

B. Pandey
e-mail: bhawesh.pandey@gmail.com

A. Kumar
e-mail: ashokeq@gmail.com

B. Adimoolam and S. Banerjee (eds.), *Soil Dynamics and Earthquake Geotechnical Engineering*, Lecture Notes in Civil Engineering 15,
https://doi.org/10.1007/978-981-13-0562-7_15

1 Introduction

It is a well-known fact that earthquakes pose a real threat to northern and north-eastern parts of the country. Most of the areas of northeast region, northern Bihar, northern Uttarakhand, Himachal Pradesh, Jammu and Kashmir, and some parts of Kutch are mapped in seismic Zone V as per IS 1893-2002. Similarly most of the regions adjacent to Zone V, such as entire Gangetic plains, southern region of Uttarakhand, and some parts of Rajasthan, are in seismic Zone IV (IS 1893-2002).

Himalayan region being one of the most seismically active regions of the world has experienced several earthquake disasters in past. Indian peninsula has experienced several earthquakes more than magnitude 8.0, such as, Kutch Gujarat, 1819; Assam earthquake, 1897; Kangra earthquake, 1905; Bihar–Nepal earthquake, 1934; Assam earthquake, 1950 in last 200 years. Also, very recent Nepal earthquake of 7.9 M in Nepal on April 25, 2015 cannot be overlooked. This earthquake was followed by more than 100 aftershocks with the largest being the one with a magnitude of 7.3 (as per Indian Meteorological Department website www.imd.gov.in) on May 12, 2015. These earthquakes continuously remind us to improve our preparedness for future earthquakes so that a large number of valuable lives can be saved.

Learning from past earthquakes, Indian agencies are continuously working on the methods and technologies to improve our preparedness for any large earthquake in future.

Delhi, the national capital of India, confronts a high seismic threat, as it is very close to the seismically active regions of Himalayas. Along with that there are other tectonic features close to Delhi which may be a cause of medium-intensity earthquakes in Delhi region. Delhi sits in the Seismic Zone IV as per IS-1893:2002. Delhi's high population density and proximity to various tectonic features warrant us to be more proactive in our preparedness for any future seismic event.

Thickness of the alluvium is variable in Delhi, and seeing such varieties is critical in evaluating site amplification. Various studies had been carried out in Delhi region related to site amplification. These studies are based on (a) micro-tremors (e.g. Mukhopadhyay et al. 2002), (b) earthquakes (Singh et al. 2002; Nath et al. 2003; Mittal et al. 2013a, b), (c) standard penetration test (Iyengar and Ghosh 2004), and (d) numerical modeling of wave propagation (e.g., Parvez et al. 2006).

Mukhopadhyay et al. (2002) worked on estimation of fundamental frequency and corresponding amplification factors at various locations of Delhi using micro-tremor data. Singh et al. (2002) and Mittal et al. (2013a, b) used standard spectral ratio (SSR) method for estimation of amplification. They reported high amplifications at different sites in Delhi which were of the order of 10–20. Iyengar and Ghosh (2004) evaluated amplification at 17 sites in Delhi by using standard penetration test (SPT) to obtain shear wave velocity profiles which were further used to obtain local site effects. They reported amplification of the order of 2–3 at the natural frequency of the sites. Singh et al. (2010) studied implications of Delhi

Table 1 V_{s30} and NEHRP site class

S. No.	Site name	V_{s30} (m/s)	NEHRP site class
1.	GEC Jaffarpur	338	D
2.	Jamia	346	D
3.	Kalkaji	564	C
4.	Shivaji College Raja Garden	346	D
5.	Vikaspuri	360	D
6.	IMD Ridge	543	C
7.	NPTI, Badarpur	322	D
8.	JNU	565	C

earthquake of November 25, 2007, on seismic hazard and concluded that the recorded PGA values at different locations were poorly explained by the estimates provided by previous studies.

It can be seen clearly that none of the study comes to a consensus; hence, an attempt is made in this study to estimate site amplification factors for eight locations of Delhi. Five locations, viz. Shivaji College, NPTI, Vikaspuri, Jamia Millia Islamia University, and GEC Jaffarpur, are situated on over deep alluvium and are in site class D of NEHRP (Pandey et al. 2016). Another set of sites were IMD, Kalkaji, and JNU which are shown as site class D of NEHRP as per the site characterization study conducted by Pandey et al. (2016). Site class and respective V_{s30} values, calculated as per site characterization study of Pandey et al. (2016), for these sites were shown in Table 1.

2 Methodology

There are various methods which can be used for estimation of site amplification. There are methods which use earthquake records or ambient vibration records from the sites of interest such as standard spectral ratio technique and horizontal-to-vertical spectral ratio technique. Other than this, there are theoretical methods such theoretical ground response analysis. For this study, theoretical ground response analysis (GRA) using equivalent linear approach has been used. For conducting GRA, a software STRATA (Kottke et al. 2013) has been used in this study.

For GRA, two kinds of input are required; one is information regarding soil strata and another is input motion. All the sites chosen for this study are having strong motion recording stations (Mittal et al. 2012), and the site characterization report for these sites is available (Pandey et al. 2016). This gives us the benefit of having details of underlying soil strata of the location and availability of earthquake records at the same location. The available records from these sites were used for comparison of transfer functions from GRA and HVSR (Nakamura 1989). Along

Fig. 1 Locations of eight strong motion recording stations considered in this study

with this, response spectra obtained from GRA and response spectra from recorded motion were also compared with response spectra in IS 1893:2002. For input ground motion, response spectra compatible motion was generated compatible with response spectra for rock sites. As IS 1893:2002 provides response spectra for outcrop/surface motion, the response spectra compatible motion was corrected for free surface effect and then was applied at top of the layer having shear wave velocity more than 750 m/s. The location of sites is shown in Fig. 1.

3 Results

Results of this study were compiled in two forms; one is transfer function comparison and another is response spectra comparison. For ease of comparison, the results were divided into two groups as per their NEHRP site class.

For transfer function comparison (Figs. 2 and 3), four kinds of transfer functions were generated—(a) transfer function between surface and half space (top of the layer having velocity greater than 750 m/s) from GRA, (b) transfer function between surface and bedrock outcrop (top of the layer having velocity greater than 750 m/s), (c) transfer function obtained from HVSR of earthquake records available from these stations, and (d) transfer function from HVSR using ambient vibration records.

For response spectra comparison (Figs. 4 and 5), normalized response spectrum obtained at surface from GRA and normalized response spectrum obtained from earthquake records were compared with response spectrum as per IS 1893:2002.

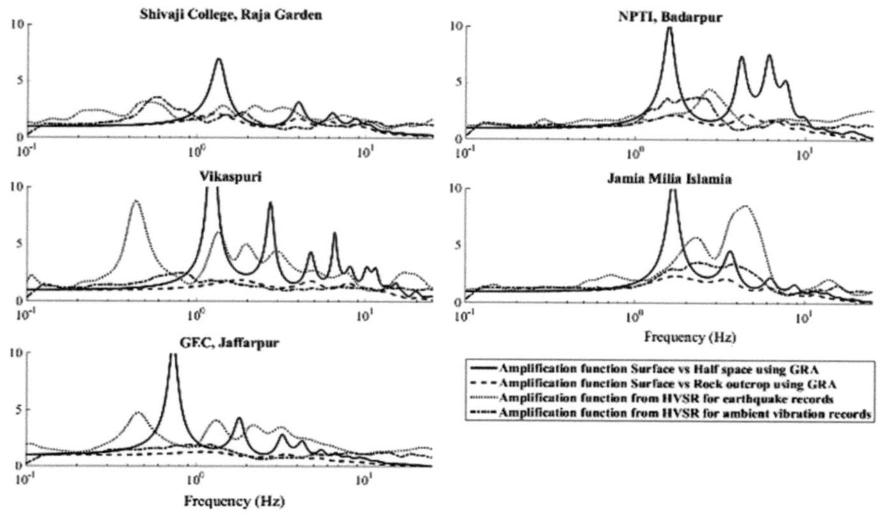

Fig. 2 Comparison of transfer functions for sites in class D of NEHRP

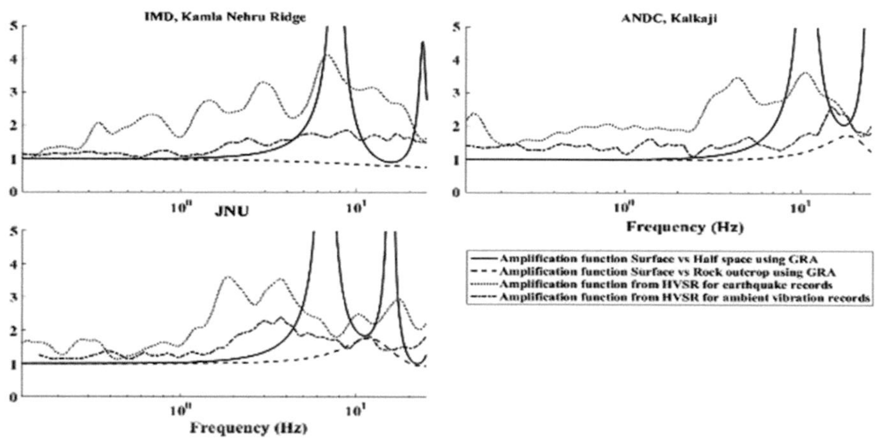

Fig. 3 Comparison of transfer functions of sites belonging to NEHRP site class C

For sites of site class D (Fig. 1), it can be clearly seen that all the four kinds of transfer functions are very close to each other in respect to frequency. Also amplification function between surface/half space and surface/outcrop is similar in terms of frequency although amplitude is very low for surface/outcrop amplification function which is of the order of 2–3.

Similarly, for site class C sites (Fig. 2), amplification functions match very well with each other in terms of frequency. Here also remarkable difference can be seen

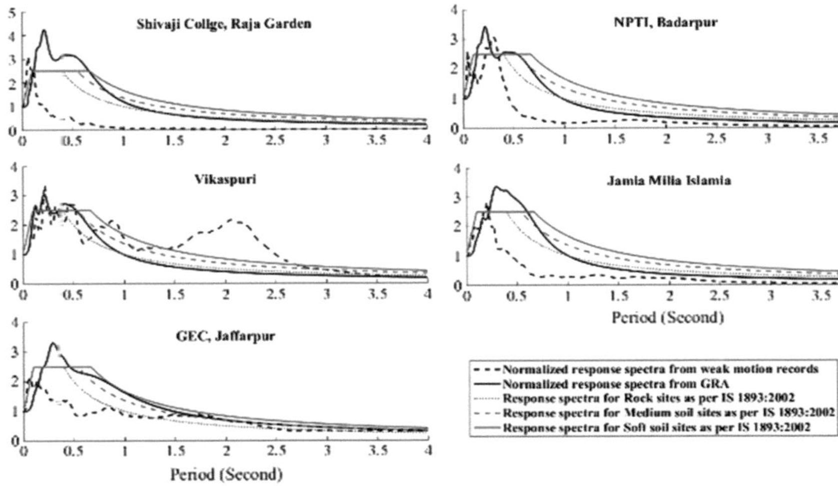

Fig. 4 Response spectra comparison for sites of NEHRP site class D

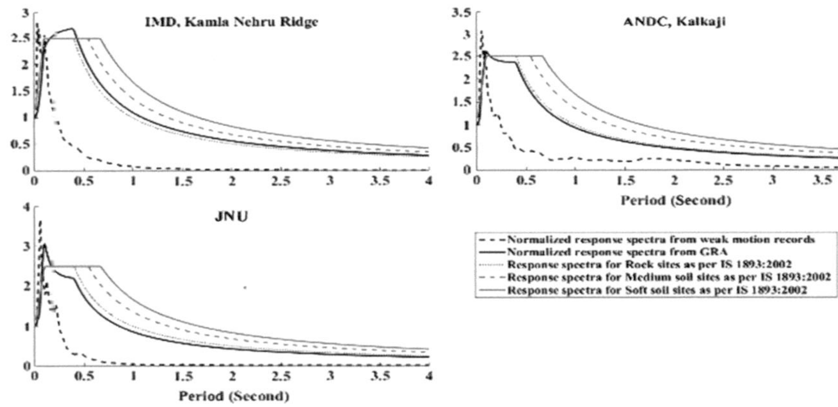

Fig. 5 Response spectra comparison for sites belonging to NEHRP site class C

between surface/half space function and surface/outcrop function. Another interesting observation which can be easily made from Fig. 2 is surface/outcrop transfer function for IMD is almost flat with value 1, which means there is no amplification between surface motion obtained after GRA and original response spectra compatible motion. The same thing is also evident from response spectra comparison in Fig. 5 where response spectra obtained for surface motion after GRA are almost matching with response spectra from IS 1893:2002 for rock sites. It will be interesting to note that almost similar results were obtained for Kalkaji site (Figs. 3 and 5) except at higher frequency (or lower period) where some amplification of 1.7 was found. For third site in class C group, i.e., JNU, amplification of 1.8 is found in surface/outcrop transfer function at 10 Hz frequency. The similar kind of amplification is also evident in response spectra plotted for JNU. These results clearly shows that IMD can comfortably be used as a reference site for earthquake engineering studies but Kalkaji can only be used for such studies at lower frequencies. Another inference which can be made from this analysis is even though a site looks like a rock outcrop, as is the case with JNU, and have high V_{s30} value, still care should be taken in considering any site as reference site.

Comparing the response spectra of all the sites shows that all the response spectra are amplified. Hence, by observing amplification of response spectra it can be clearly said that expected acceleration would be much higher than response spectra proposed in IS 1893:2002 for the maximum considered earthquake. Also the mismatch between response spectra obtained from earthquake records and GRA is due to the huge difference between intensity of recorded motion and input motion for GRA as the recorded motion has very low PGA values.

4 Conclusion and Discussion

Lots of variations were found in amplification ratios of different transfer function for each site. Hence, it can be concluded from this study that amplification ratios must be chosen with care whenever any analysis for seismic hazard assessment is to be. However, not much difference is observed in fundamental frequency estimated by different transfer functions. Also the amplification function obtained from surface/outcrop seems to be more realistic for engineering purposes as normally we tend to compare to amplification of a site with reference to other site location. Surface/half space transfer function can be very useful for studies related to attenuation studies as bedrock motion is very important for such studies.

Another inference from this study is that sites which look like rock outcrop may not be an actual rock outcrop as it was shown from the result of amplification function and response spectra of JNU in Figs. 3 and 5. Also it can be concluded that IMD can be used as a reference rock outcrop site for motion with PGA of order of 0.24 g. Kalkaji can also be used as reference outcrop if study deals with low frequency range. Using IMD as a reference site, it can be suggested that an amplification of the order of 2–3 can be expected at sites having site class D with respect to ground motion at IMD for a ground motion of intensity as used in this study.

Comparison of response spectra also conforms the amplifications observed through transfer functions. This study also suggests higher acceleration values for soil sites at almost all the periods. Hence, provision must be made in response spectra provided in IS 1893:2002 for amplification as well. Currently, the difference in response spectra provided in IS 1893:2002 for rock sites and soil sites is just in frequency bandwidth, and it does not provide any provision for amplification.

References

IS-1893:2002 (Part-1). *Indian Standard criteria for earthquake resistant design of structures, Part 1—General provisions and buildings (Fifth Edition).* Bureau of Indian Standards, New Delhi, 42.

Iyengar, R. N., & Ghosh, S. (2004). Microzonation of earthquake hazard in greater Delhi area. *Current Science, 87,* 1193–1202.

Kottke, A. R., Wang, X., & Rathje, E. M. (2013). *Technical manual of strata*. Geotechnical Engineering Center, Department of Civil, Architectural and Environmental Engineering, University of Texas.

Mittal, H., Kumar, A., & Rebecca, R. (2012). Indian strong motion instrumentation network and its site characterization. *International Journal of Geosciences, 3*(6), 1151–1167.

Mittal, H., Kumar, A., & Kumar, A. (2013a). Site effects estimation in Delhi from the Indian strong motion instrumentation network. *Seismological Research Letters, 84*(1), 33–41.

Mittal, H., Kumar, A., & Singh, S. K. (2013b). Estimation of site effects in Delhi using standard spectral ratio. *Soil Dynamics and Earthquake Engineering, 50,* 53–61.

Mukhopadhyay, S., Pandey, Y., Dharmaraju, R., Chauhan, P. K. S., Singh, P., & Dev, A. (2002). Seismic microzonation of Delhi for ground-shaking site effects. *Current Science, 87,* 877–881.

Nakamura, Y. (1989). Method for dynamic characteristics estimation of subsurface using microtremor on the ground surface. *Quarterly Report of Railway Technical Research Institute (Japan), 30*(1), 25–33.

Nath, S. K., Sengupta, P., Srivastav, S. K., Bhattacharya, S. N., Dattatrayam, R. S., Prakash, R., et al. (2003). Estimation of S-wave site response in and around Delhi region from weak motion data. *Proceedings of the Indian Academy of Sciences (Journal of Earth System Science), 112,* 441–463.

Pandey, B., Jakka, R. S., Kumar, A., & Mittal, H. (2016). Site characterization of strong-motion recording stations of Delhi using joint inversion of phase velocity dispersion and H/V curve. *Bulletin of the Seismological Society of America, 106*(3), 1254–1266. https://doi.org/10.1785/0120150135.

Parvez, I., Vaccari, F., & Panza, G. F. (2006). Influence of source distance on site-effect in Delhi city. *Current Science, 91,* 827–835.

Singh, S. K., Mohanty, W. K., Bansal, B. K., & Roonwal, G. (2002). Ground motion in Delhi from future large/great earthquakes in the central seismic gap of the Himalayan arc. *Bulletin of the Seismological Society of America, 92,* 555–569.

Singh, S. K., Kumar, A., Suresh, G., Ordaz, M., Pacheco, J. F., Sharma, M. L., et al. (2010). Delhi earthquake of 25 November 2007 (Mw 4.1): Implications for seismic hazard. *Current Science, 99*(7), 939–947.

Mitigation of Dynamic Loading Effects on Retaining Walls Using Recycled Tire Chips

S. Bali Reddy and A. Murali Krishna

Abstract This paper presents the results of shaking table tests on retaining wall models using recycled tire chips as compressible inclusions. Scrap tire derived tire chips of 10 × 10 mm size and about 20 mm length have been used in the study. The 600-mm-height model wall is constructed in a Perspex container and instrumented with pressure sensors and LVDTs at different locations. Sinusoidal dynamic excitations were applied on the model walls. The dynamic response of the retaining walls with the variation in the acceleration and frequency of base shaking has been monitored and discussed. It is observed from these tests results that the horizontal displacement and incremental lateral earth pressures are significantly reduced by the inclusion of tire chips as cushion in between the wall and backfill. Reduction of the lateral earth pressure and displacement implies a lower design requirement which implies lesser dimensions of the retaining wall with reduced material cost.

Keywords Retaining wall · Recycled tire chips · Compressible inclusion
Reduction earth pressures · Wall displacements

1 Introduction

Earth retaining structures such as retaining walls, bridge abutments, bulk heads, braced excavations, and mechanically stabilized walls play a very important role in public life (Reddy and Krishna 2015). These structures are frequently representing the main elements of transportation system, port, harbors, and other infrastructure systems. Sometimes these civil engineering structures are often being affected by dynamic loading, like earthquakes. With frequent occurrences of natural disasters in

S. B. Reddy (✉) · A. M. Krishna
Department of Civil Engineering, Indian Institute of Technology Guwahati, Guwahati 781039, India
e-mail: sodom@iitg.ernet.in

A. M. Krishna
e-mail: amurali@iitg.ernet.in

B. Adimoolam and S. Banerjee (eds.), *Soil Dynamics and Earthquake Geotechnical Engineering*, Lecture Notes in Civil Engineering 15,
https://doi.org/10.1007/978-981-13-0562-7_16

India (Bhuj 2001, Sikkim 2011, and other earthquakes) and the associated sustainable implications, earthquake mitigation techniques are becoming matters of interest. Performances of retaining wall under static and seismic loading conditions depend upon the type of backfill soil, generally clean granular cohesion less backfill material is preferred. However, new lightweight fills' materials like shredded tire chips, EPS geofoam, fly ash are being explored as alternative backfill materials nowadays (Bathurst et al. 2007; Lal and Mandal 2012; Reddy and Krishna 2013, 2015). These materials are beneficial in reducing earth pressures and lateral displacements of the retaining walls.

Use of scrap tire derived geomaterials (tire shreds, tire chips, and tire crumbs) has been found increasing in various geotechnical engineering applications (Humphrey 1993; Cecich et al. 1996; Tweedie et al. 1998; Lee and Roh 2006; Hazarika and Yasuhara 2007; Anbazhagan et al. 2011; Sheikh et al. 2013; Dammala et al. 2015; Boominathan et al. 2015; Reddy and Krishna 2015). Different researchers (Inglis et al. 1996; Ertugrul and Trandafir 2011; Reddy and Krishna 2013) were studied on reduction of lateral earth pressures on retaining wall using EPS geofoam as compressible inclusion. In this paper, recycled tire chips are used as compressible inclusions just behind the retaining wall to mitigate the dynamic loading effects. Shaking table tests on retaining wall models using recycled tire chips as compressible inclusions were conducted in this study.

2 Equipment and Materials Used

A computer-controlled servo-hydraulic shaking table facility with a single degree of freedom (horizontal) was used to simulate the horizontal shaking action, associated with seismic and other vibration conditions. The shaking table can be operated for frequency up to 10 Hz with the amplitude of ± 250 mm. Models of retaining walls have been built in a Perspex container fitted with compressible foam as back boundary to reduce the boundary effect. Retaining wall model of 600 mm height was constructed in a model container of 1200 mm $\times$ 600 mm in plan and 1000 mm height (Fig. 1). The model container was made of Perspex sheets of 10 mm thickness and braced by a steel frame made of steel angle sections that also facilitates for easy lifting and handling.

Locally available dry sand having specific gravity of 2.62 has been used as the backfill material. The sand has maximum unit weight of 16.1 kN/m^3 and minimum unit weight of 13.26 kN/m^3. The values of $e_{\max}$ and $e_{\min}$ are 0.94 and 0.64, respectively. The sand is classified as poorly graded sand with letter symbols SP as per the Unified Soil Classification System. Scrap tire derived tire chips of 10 $\times$ 10 mm size and about 20 mm length have been used as compressible inclusion for the study. Specific gravity and unit weight of the tire chips were determined as 1.08 and 6.45 kN/m^3, respectively.

Fig. 1 Retaining wall model container mounted on shake table

3 Model Preparation and Testing Procedure

The model wall of 600 mm high and 580 mm length was made with eight hollow rectangular steel sections, each of 25 mm wide and 75 mm height and 580 mm length, which were joined together using steel rods of 12 mm diameter. These steel rods were further connected to a bottom plywood base forming a rigid connection.

Two types of test cases were considered as shown in Fig. 2. In Case A, a retaining wall model with a conventional sandy backfill was used (control case); in Case B, behind the model wall, two different thickness of tire chips were placed vertically as compressible inclusions. The thickness chosen here was ($t/H = 0.15$ and 0.30) 0.15, 0.3 times of the wall height. Here, t = thickness of tire chips and H = height of the wall. The compressible inclusion was made by recycled tire chips. Backfilling was done by stagewise by free-falling technique (pacing the calculated amount of sand based on layer thickness and target density) and compacting manually to achieve the target density. For Case B, the compressible inclusion layer was prepared by filling the recycled tire chips inside a selected size

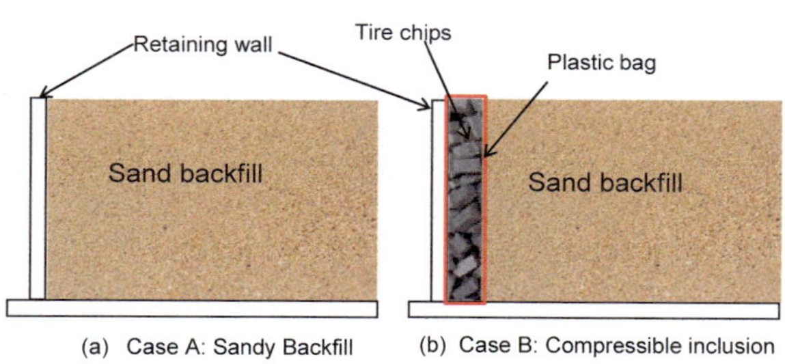

Fig. 2 Test case

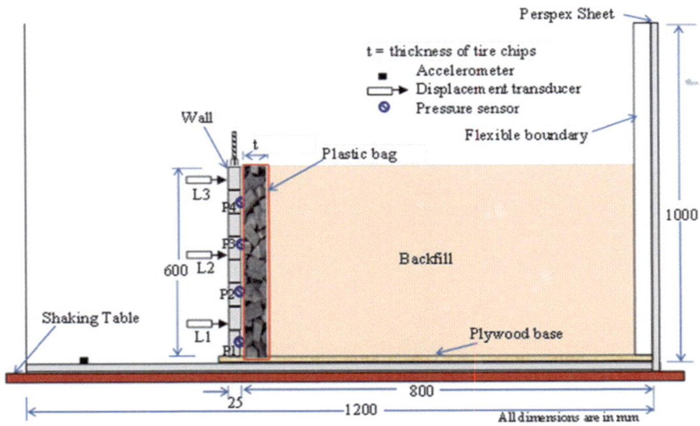

Fig. 3 Schematic diagram of retaining wall model

bag made of plastic sheet. The bag is required to wrap the tire chips so that they do not mix with the surrounding soils. Such confinement also makes the execution of backfilling easier. The average dry unit weight of tire chips achieved after filling and tamping was 6.345 kN/m^3.

For all tests, four pressure sensors (each of 50 kPa capacity), P1, P2, P3, and P4, were placed inside the wall, in contact with the facing at different elevations 487, 337, 187, and 37 mm, respectively. Three LVDTs L1, L2, and L3 were positioned at elevations 125, 380, and 580 mm, respectively, along the facing. Figure 3 shows the schematic diagram of retaining wall model. After completing the model preparation with external support to the facing, model was placed on shaking table. Then external support of facing was removed, and shaking table has been operated. Each model wall was subjected to 20 cycles of sinusoidal shaking at specified frequency and acceleration. The parameters varied in different model tests are given in Table 1. The dynamic response of each model wall, in terms of incremental lateral earth pressure and the displacement responses at different elevations at facing, is monitored using a data-acquisition system.

4 Results

Variation of horizontal displacements at different elevations of the wall with increasing number of dynamic loading cycles is shown in Fig. 4, for the test T1. The figure shows the nonlinearly increasing trend of displacements with increase in number of cycles. Further, higher displacements at higher elevations are noticed. Wall displacements of 0.61, 0.38, and 0.12 mm are seen from the figure, at 580, 380, and 125 mm elevation, respectively.

Table 1 Test parameters for different test

Test No.	*t*/*H*	Base acceleration, g	Frequency, Hz
T1	0.00	0.1	3
T2	0.15	0.1	3
T3	0.30	0.1	3
T4	0.0	0.3	5
T5	0.15	0.3	5
T6	0.30	0.3	5

t thickness of tire chips, *H* height of wall

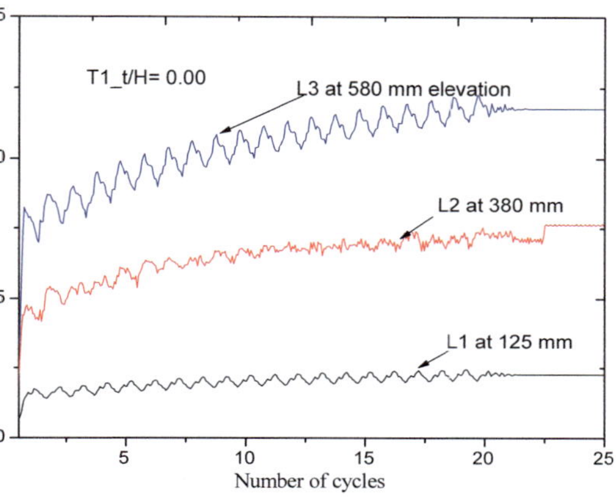

Fig. 4 Typical variation of displacement with number of cycles for the test T1

Figure 5a shows the displacement profiles, at the end of 20 cycles of dynamic motion (0.1 g acceleration and 3 Hz frequency), revealing the influence of compressible inclusion on the horizontal displacement response. Here the elevation (z) is represented in non-dimensional form after normalizing by the full wall height (*H*). It is observed, from the figure, that the displacements are significantly reduced with the inclusion of the compressible inclusion just behind the retaining wall. The maximum top displacement of 0.61 mm in the case of control model wall (T1) is reduced to 0.20 mm in the case of compressible compression model wall test (T3).

Incremental lateral earth pressures are observed during the dynamic excitation along the height of the wall in different tests. Figure 5b shows the measured incremental lateral earth pressures along the height of the wall for T1, T2, and T3 tests. It is observed, from the figure, that incremental lateral earth pressures are depicting the diminishing trend with increase in inclusion of compressible inclusion thickness. Maximum earth pressures are observed at bottom of the wall.

The measured top displacements and calculated percentage reduction are reported in Table 2, which shows around 68% reduction in top displacement. Maximum earth pressure values and calculated percentage reduction from earth pressures are reported in Table 2. Incremental earth pressures are reduced around 68% compared to test T1. This type of behavior can be observed when vertical compressible inclusion placed behind the retaining wall.

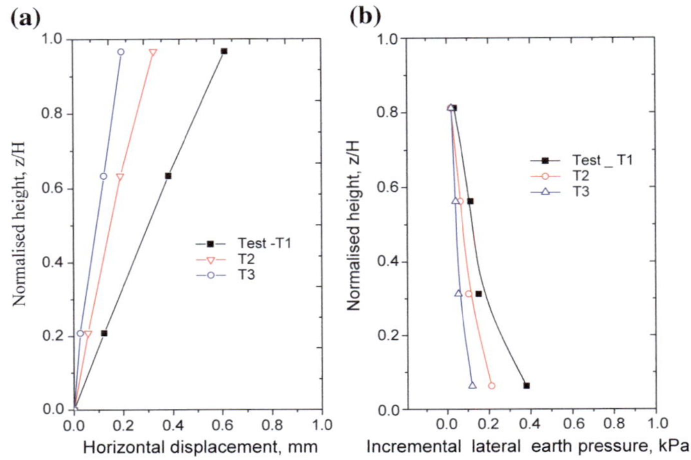

Fig. 5 Comparison of mode response for T1, T2, and T3 tests; **a** displacement profile, **b** earth pressure profile

Table 2 Displacement, earth pressure, and percent reduction for tests T1, T2, and T3

Parameter	Compressible inclusion, *t*/*H*		
	0.00	0.15	0.30
Top displacements, mm	0.613	0.324	0.195
% reduction of displacement	–	47.14	68.19
Maximum incremental earth pressure, kPa	0.379	0.211	0.119
% reduction of earth pressure	–	44.33	68.60

Further, the base acceleration of 0.3 g and 5 Hz frequency has been applied on different model walls to observe the effect of tire chips as compressible inclusion material. The response of the model wall is measured in terms of horizontal displacement and lateral earth pressures on model wall with different configurations and shown in Fig. 6. Wall top displacements of 3.05, 1.64, and 0.73 mm are seen from the figure, for T4, T5, and T6, respectively. From the laboratory model tests, large reductions in lateral earth pressures were achieved when $t/H = 0.30$ thickness tire chips were used as the vertical compressible inclusion.

Reduction of the lateral earth pressure and displacement implies a lower design requirement which implies lesser dimensions of the retaining wall with reduced material cost. However, the cushion thickness may vary depending upon many parameters such as wall rigidity, height of wall, stiffness of tire chips, and ground motion effect. However, model test results cannot be directly extrapolate to the response of field walls because of gravity and scaling effects, and the results obtained from this study will help in understanding the relative performance of retaining walls subjected to base shaking with compressible inclusions.

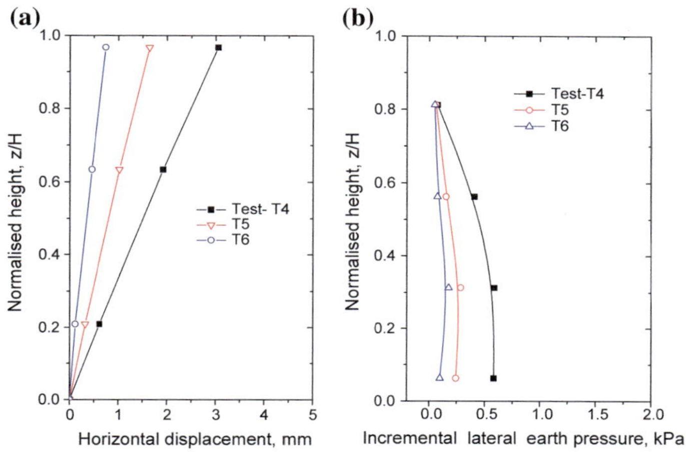

Fig. 6 Comparison of mode response for T4, T5, and T6 tests; **a** displacement profile, **b** earth pressure profile

5 Conclusions

A series of model tests were conducted using shaking table to examine the performance of retaining walls with recycled tire chips as compressible inclusion for mitigation of dynamic loading effects. Inclusion of recycled tire chips as vertical compressible inclusion just behind the retaining wall resulted in drastic reduction of the wall displacements, lateral earth pressures. The maximum horizontal displacement of model wall with compressible inclusions (recycled tire chips) is reduced by around 65% compared to the control test wall. Incremental earth pressure is reduced by around 75% compared to the control tests. This implies the dynamic performance of retaining wall improves with the use of recycled tire chips as cushion material.

References

Ahn Il-S., Cheng, L. (2014) Tire derived aggregate for retaining wall backfill under earthquake loading. *Construction and Building Materials* 57:105–116

Anbazhagan, P., Mamatha, M., Soumyashree, P., Sushyam, N., Bharatha, T. P., & Vivekan, R. W. (2011). Laboratory characterization of tire crumbs soil mixture for developing low cost damping materials. *International Journal of Earth Sciences and Engineering, 4*(6), 63–66.

Bathurst, R. J., Zarnani, S., & Gaskin, A. (2007). Shaking table testing of Geofoam seismic buffers. *Soil Dynamics and Earthquake Engineering, 27,* 324–332.

Boominathan, A., Banerjee, S., & Dhanya, J. S. (2015). Performance of soil-rubber tyre scrap mixture as seismic base isolators for foundations. In *6th International Conference on Earthquake Geotechnical Engineering,* Christchurch, New Zealand.

Cecich V., Gonzales, L., Hoisaeter, A., Williams, J., & Reddy, K. (1996). Use of shredded tires as lightweight backfill material for retaining structures. *Waste Management and Research, 14,* 433–451.

Dammala, P., Reddy, S. B., & Krishna, A. M. (2015). Experimental investigation of applicability of sand tire chip mixtures as retaining wall backfill. In *IFCEE, ASCE Geotechnical special publication* (pp. 1420–1429). https://doi.org/10.1061/9780784479087.128.

Ertugrul, O., & Trandafir, A. (2011). Reduction of lateral earth forces acting on rigid nonyielding retaining walls by EPS Geofoam inclusions. *Journal of Materials in Civil Engineering, 23*(12), 1711–1718.

Hazarika, H., & Yasuhara, K. (2007), *Scrap tire derived geo materials—Opportunities challenges.* Taylor and Francis, UK.

Humphrey, D. N., Sandford, T. C., Cribbs, M. M., & Manion, W. P. (1993). Shear strength and compressibility of the tyre chips for use as retaining wall backfill. *Transportation Research Record, 1422,* 29–35.

Inglis, D., Macleod, G., Naesgaard, E., & Zergoun, M. (1996). Basement wall with seismic earth pressures and novel expanded polystyrene foam buffer layer. In *Proceedings of the 10th Annual Symposium of the Vancouver Geotechnical Society*, Vancouver and B.C.

Lal, R. R. B., & Mandal, J. N. (2012). Feasibility study on fly ash as backfill material in cellular reinforced walls. *EJGE, 17,* 1437–1458.

Lee, H. J., & Roh, H. S. (2006). The use of recycled tire chips to minimize dynamic earth pressure during compaction of backfill. *Construction Building Materials, 21*(5), 1016–1026.

Reddy, S. B., & Krishna, A. M. (2013). Numerical simulations of earth-retaining structures using EPS geofoam inclusions. In *Indian Geotechnical Conference,* IITRoorkee.

Reddy, S. B., & Krishna, A. M. (2015). Recycled tire chips mixed with sand as lightweight backfill material in retaining wall applications: An experimental investigation. *International Journal of Geosynthetics and Ground Engineering*. https://doi.org/10.1007/s40891-015-0036-0.

Sheikh, M. N., Mashiri, M. S., Vinod, J. S., & Tsang, H. H. (2013). Shear and compressibility behavior of sand-tire crumb mixtures. *Journal of Materials in Civil Engineering, ASCE, 25*(10), 1366–1374.

Tweedie, J. J., Humphrey, D. N., & Sandford, T. C. (1998). Tire shreds as retaining wall backfill, active conditions. *Journal of Geotechnical and Geoenvironmental Engineering, ASCE, 124* (11), 1061–1070.

[TH-05] An Experimental Study on Seismic Soil-Pile Foundation-Structure Interaction in Soft Clay

Rajib Saha, Animesh Pandey and Richi Prasad Sharma

Abstract Dynamic soil structure interaction (DSSI) is an important seismic design consideration for sustainable design of structures. Further, such interaction has been reported to be detrimental consequences superseding seismic codes in structure situated mainly in soft clay and liquefiable deposit as observed during past earthquakes, such as, 1964 Niigata earthquake, 1985 Mexico city earthquake, and 1995 Kobe earthquake. In practice, design is made based on assumption of fixed base condition with a perception of beneficial effect on the response of system based on few seismic codal guidelines and to avoid complexity in SSI modeling. In this context, present study is a humble effort to investigate the influence of DSSI in pile foundation-supported structure through scaled model study in very soft clay using shake table tests. This limited study indicates that lengthening of period is insignificant for stiff period of structure in soft clay which is in-general counter-intuitive. The reason may be foundation stiffness is considerably higher than superstructure. Further, it is observed from the acceleration signals that elastic forces may be significantly high in column as compared to pile head in particularly stiff period structures supported by pile foundation in soft clay, while an opposite trend is observed in moderately longer period of structure supported by similar foundation.

Keywords Fundamental period · Kaolin · Shake table

R. Saha (✉) · A. Pandey · R. P. Sharma
Civil Engineering Department, National Institute of Technolgy Agartala, Agartala 799046, Tripura, India
e-mail: rajib.iitbbsr@gmail.com

A. Pandey
e-mail: animeshxperia@gmail.com

R. P. Sharma
e-mail: richisharma.sharma@gmail.com

B. Adimoolam and S. Banerjee (eds.), *Soil Dynamics and Earthquake Geotechnical Engineering*, Lecture Notes in Civil Engineering 15,
https://doi.org/10.1007/978-981-13-0562-7_17

1 Introduction

Complexity involved with seismic soil-pile-structure interaction (SSPSI) problem and unavailability of standard and validated analysis techniques routinely results in ignoring or simplifying the presence of pile foundations for the structural design. In fact, non-incorporation or simplification of SSPSI may not warrant actual prediction of behavior of the whole structural system under seismic loading as highlighted in many studies (e.g., Jeremic et al. 2004; Boulanger et al. 1999; Tokimatsu et al. 2005). Further, these studies also highlighted the importance of SSPSI in estimation of design forces in pile foundation-supported structural system. However, post-analysis of failure of pile foundation-supported structure during past earthquakes such as 1965 Niigata earthquake, 1985 Mexico city earthquake, 1989 Loma Prieta earthquake, 1995 Kobe earthquake indicated the detrimental effect of dynamic soil structure interaction (DSSI) (Meymand 1998). Interestingly, it is observed in these case studies that DSSI has led to detrimental response in pile-supported structure during an earthquake event, mainly in liquefied and soft clayey-type soil deposit (Bhattacharya et al. 2014; Meymand 1998). This acted as prime motivation behind present study. In fact, few investigators have already mentioned the significance of experimental study since different phenomenon related to DSSI could not be predicted while studied through prototype structure attributing numerical analysis (e.g., Meymand 1998). Limited studies based on shake table tests pointed out design implications resulting from such interaction in non-liquefied (Tokimatsu et al. 2005) as well as liquefied soil (Lombardi and Bhattacharya 2013). Further, a number of experimental studies were also found in this direction which verifies the theoretically developed modeling issues and the effect of complex SSPSI (e.g., Meymand 1998; Boulanger et al. 1999; Tokimatsu et al. 2005; Boominathan and Ayothiraman 2005; Chau et al. 2009; Lombardi and Bhattacharya 2013). However, a recent shake table study by the first author along with other co-authors (Saha et al. 2015) reported increase in shear force at pile head considering dynamic soil-piled raft-structure interaction.

From the above view point, present study is an attempt to investigate the effect of SSPSI on dynamic characteristics and seismic force distribution in a model pile foundation-supported structural system embedded in soft clay by carrying out shake table experiment considering steady-state harmonic motion. Percentage lengthening of period and an approximate lateral shear force distribution are calculated based on experimental observation for a 3×3 flexible pile group-supported structural system. Hence, the outcome of present study may give insight into the problem in order to modify the existing design guidelines for a pile foundation-supported structural system.

2 Experimental Program

2.1 Structural Model

Simplistically, the superstructure is modeled as a SDOF structure in the present study which is supported by a model 3 × 3 flexible pile group in soft clay. Two different periods of vibration in fixed base condition, i.e., $T_{\text{fixed}} = 0.3$ and 0.7 s representative of shorter to moderately longer range period of structure is considered herein. The model column is prepared by two steel scales welded together and fixed with a base plate which carries lumped weights made of circular MS plates bolted at the top of the scale. The theoretically calculated weights are accordingly adjusted in order to simulate 0.30 s ($m_1 = 230$) and 0.70 s ($m_2 = 1000$) period of structure which is further verified by free and forced vibration tests. Hollow piles made of steel alloy rigidly connected with pile cap made of 30-mm-thick MS plate are used to support the superstructure load. A 3 × 3 frictional long pile group is designed to support the maximum load of superstructure (i.e., $m_2 = 1000$ g) and pile cap with a factor of safety of 2.0 considering very soft clay bed. Length to diameter of pile is kept as 33.33 which may be considered as long pile as per IS2911-Part 1 (2010). However, the detailed specification of superstructure, cap, and piles are presented in Table 1. Figure 1 presents the whole structural model.

The objective of the scale modeling procedure for DSSI experiments is to achieve "dynamic similarity," where model and prototype experience homologous forces. Meymand (1998) presented a detailed review on similitude rule and

Table 1 Detailed specification of superstructure, pile cap, and pile

Specification/parameters	Values
Length of steel column	300 mm
Width of column	20 mm
Thickness of column	2.0 mm
Buckling load of column	2.08 kg
Lateral stiffness of column	83 N/m
Young's modulus of steel column	5.64×10^4 N/mm^2
Mass of single pile	120 gm
Length of pile	400 mm
Outer diameter of pile	12 mm
Inner diameter of pile	10 mm
Thickness of pile	2 mm
Young's modulus of pile	2×10^5 N/mm^2
Moment of inertia of hollow pile	527×10^{-12} m^4
Buckling load of pile	10.34 kN
Size of pile cap	100 mm × 100 mm
Thickness of pile cap	30 mm
Mass of pile cap	1350 gm

Fig. 1 Model structural system

suggested a rational scaling law which is followed in the present study. Accordingly, a 3 × 3 frictional pile group is designed as per IS 2911 (2010) considering maximum superstructure load exhibited by 0.7 s SDOF along with pile cap. The thickness of pile cap is provided as per IS 2911 (2010).

2.2 Model Container and Soil Preparation

A lightweight cylindrical container made of alloy of tin attached with commercially available foam at periphery as a boundary-absorbing material (Lombardi and Bhattacharya 2013) is used to prepare the consolidated soft clay bed for carrying out the shake table experiment. The container is strengthened by mild steel jacket in order to withstand maximum vertical consolidation pressure of 1 kN/m^2. The dimension of the container is considered as 500 mm (diameter) × 700 mm (height) based on the maximum payload capacity of shake table as 100 kg and to ensure minimum distance of one and half times of length of rigid pile cap from the edge of boundary. The consolidation unit is fabricated by different shapes of mild steel sections with an attachment of inverted hydraulic jack of capacity of 10 ton which may generate desired level of consolidating pressure. One-dimensional consolidation is carried out in the present study by keeping drainage arrangements at the bottom of the chamber. A 30-mm-thick wooden plunk grooved into four perpendicular channels having 20 mm depth is placed at the base of the container which is further attached with four drainage valves at the outside of the chamber. At the top of the plank, a cotton cloth is placed to separate the 30-mm-thick filter material placed at the top of plank. Graded coarse sand pertaining to following filter characteristics, i.e., D_{60} filter/D_{10} filter = 5, D_{15} filter/D_{85} soil = 3, is used in this study.

Further, a filter paper is provided at the boundary of very soft kaolin layer and graded coarse sand. Commercially available kaolin having moderate plastic property is used for preparation of soil bed in the present study.

Total 90 kg of kaolin was mixed at 1.2 times L.L, i.e., 63 kg of water to form a kaolin slurry using a rotary cement mixer. A dispersing agent solution made of 45 g sodium hexametaphosphate mixed with 1000 ml of distilled water, known as "Calgon solution," is added with each batch of kaolin-water mix in a proportion of 10% of the total weight of water. Calgon solution is used to avoid flocculation and to produce a uniform consistency of kaolin paste. The whole mixing is performed in three batches to fill up the container. A model clay sample of total height of 550 mm is prepared. An initial consolidation pressure of 10 kPa was given for 7 days, and further it was increased to 40 kPa for next 7 days. Total settlement of 100 mm is observed after 15 days which comprises of 1D consolidation settlement and expulsion of air voids. After the preparation of kaolin sample, it was covered with a layer of 30 mm of wax so that the moisture content of soil is retained and it can be stored for further experiments. Further, the average water content, average vane shear strength, and shear wave velocity measured on the top of the prepared sample are obtained as 68.56% (before consolidation) and 61.54 (after consolidation), 1.1 kPa (before) and 2.37 kPa (after) and 27 m/sec (before) and 30 m/sec (after), respectively.

2.3 Shake Table, Instrumentation, and Input Motion

A simplified motor operated crank-shaft mechanism based shake table with payload capacity of 9800 N (size: 1.0 m × 1.0 m) developed in the laboratory of NIT Agartala is used to carry out the seismic experiments. This shake table is capable to generate only sine and swept sine signals with maximum displacement of ±25 mm. A tachometer is used to generate the different frequencies of swept sinusoidal motion with a range of 1–10 Hz and amplitude of 0.02 g in the present study. Accelerometer sensors are used to measure the dynamic response of structure and soil. Figure 2 presents the schematic view of whole experimental setup in shake table.

3 Results and Discussion

3.1 Lengthening of Fundamental Period Due to SSI

Figure 3i, ii presents the acceleration response in frequency domain for T_{fixed} of 0.3 and 0.7 s supported by a 3 × 3 pile group foundation embedded in soft clay. Real-time acceleration response recorded at superstructure mass under both SSI and fixed base condition, pile cap, and pile head are presented herein. It is observed

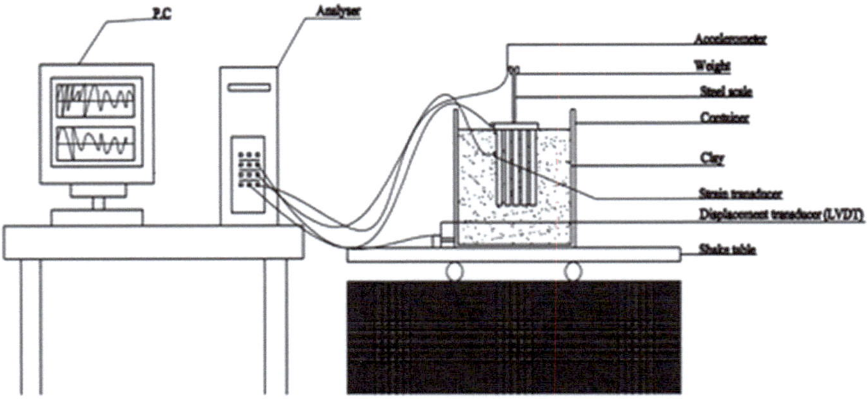

Fig. 2 Schematic view of the experimental setup

Fig. 3 Acceleration signals in frequency domain for **i** $T_{\text{fixed}} = 0.3$ s and **ii** 0.7 s with 3 × 3 pile group

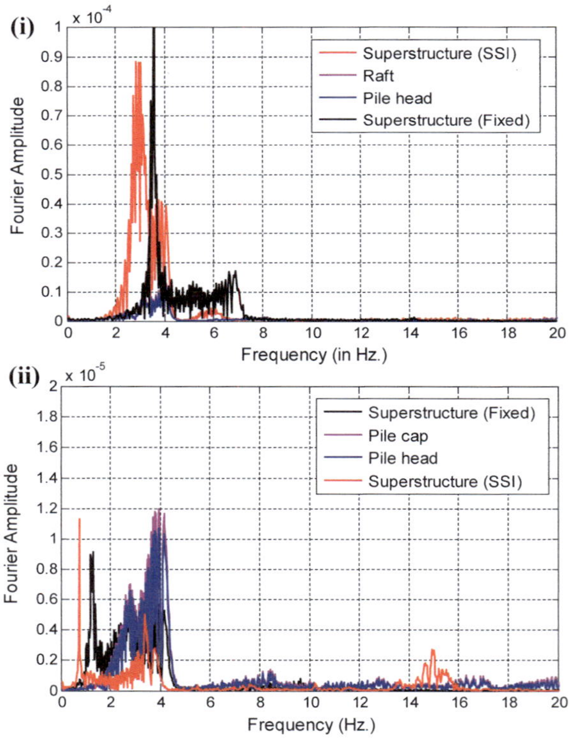

from the figures that the percentage lengthening of superstructure having $T_{\text{fixed}} = 0.3$ and 0.7 s is calculated in order of 10 and 25% (approx.), respectively, which is marginal. The reason for marginal lengthening may be due to higher lateral stiffness of in order of 35,480 N/m attributed by the model 3 × 3 pile group used in present study as compared to column stiffness of 83 N/m. In fact, a significant increase in period for stiffer period of structure in general was reported in many studies by the author along with other co-researchers. However, the present study indicates an opposite trend in case of $T_{\text{fixed}} = 0.3$ s. The relative stiffness of column to pile foundation in this case is found to be lesser which seems to be an important parameter may govern the SSI effect on dynamic response of structure supported by pile foundation even in very soft to soft clayey layer. This issue may be further investigated in detail which may help in resolving design issues related to SSI problem.

3.2 Dynamic Response in Model Soil-Pile Foundation-Structure System

Acceleration response recorded at different level of structure having $T_{\text{fixed}} = 0.3$ and 0.7 s for both SSI and fixed base condition is presented in Fig. 4i, ii, respectively.

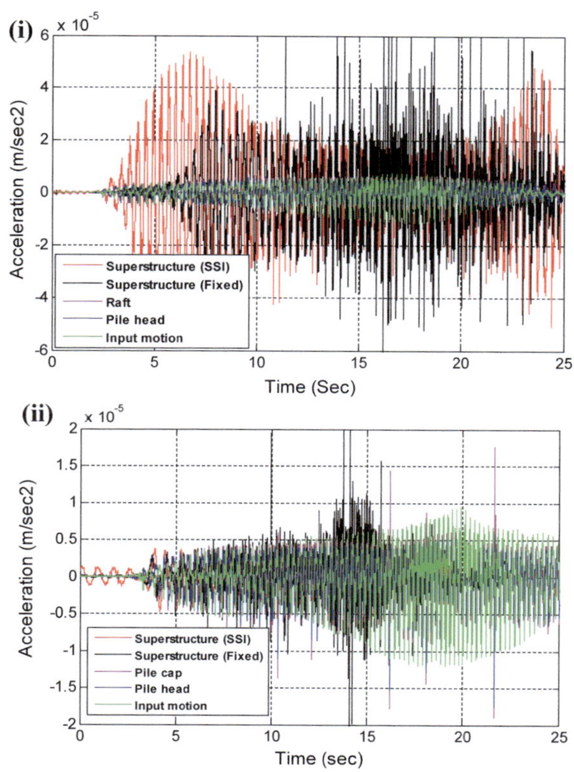

Fig. 4 Real-time acceleration for **i** $T_{\text{fixed}} = 0.3$ s and **ii** 0.7 s with 3 × 3 pile group

Acceleration response gives an approximate idea of elastic force transmitted to different elements of structure. It is observed in Fig. 4i that acceleration response is amplified significantly in superstructure having $T_{\text{fixed}} = 0.3$ s for both SSI and fixed base condition, while the response in pile cap and head are found to be subdued. On the other hand, a subdued response is observed in superstructure having relatively longer period, i.e., $T_{\text{fixed}} = 0.7$ s, which may be due to the beneficial effect of SSI. However, in case of $T_{\text{fixed}} = 0.7$ s structure, the response in pile cap and head is found to be considerably higher than superstructure response as observed in Fig. 4i, ii. Hence, this limited study indicates that the longer period structure supported by pile foundation may attract higher shear force at pile head due to such complex soil-pile foundation-structure interaction which needed to be confirmed by a detailed study. This may be due to higher mode effect.

4 Conclusions

Summarily, the present study is a limited effort to assess the effect of dynamic interaction of soil-pile foundation-structure on seismic response of structure. It may be concluded from this limited study that rigid pile group exhibiting higher foundation stiffness or lower ratio of lateral stiffness of superstructure and pile foundation may significantly reduce period lengthening of the stiffer structure embedded in very soft clayey layer which may be equivalent to fixed base condition. This implies that structure supported by conservatively designed pile group in soft soil may be designed as fixed base condition under seismic loading rather neglecting SPSI. Further, this limited study indicates that transfer of shear force at pile head may be a function of ratio of period of superstructure and pile group which may be studied in detail for the different soil conditions and frequency of motion.

References

Bhattacharya, S., Tokimatsu, K., Goda, K., Sarkar, R., Shadlou, M., & Rouholamin, M. (2014). Collapse of Showa Bridge during 1964 Niigata earthquake: A quantitative reappraisal on the failure mechanisms. *Soil Dynamics and Earthquake Engineering, 65,* 55–71.

Boominathan, A., & Ayothiraman, R., (2005). Dynamic behaviour of laterally loaded model piles in clay. *ICE*, GE-4, 207–215.

Boulanger, R. W., Curras, C. J., Kutter, B. L., Wilson, D. W., & Abghari, A. (1999). Seismic soil-pile-structure interaction experiments and analyses. *Journal of Geotechnical and Geoenvironmental Engineering, 125*(9), 750–759.

Chau, K. T., Shen, C. Y., & Guo, X. (2009). Nonlinear seismic soil–pile–structure interactions: shaking table tests and FEM analyses. *Soil Dynamics and Earthquake Engineering, 29*(2), 300–310.

IS: 2911 (Part 1/Sec 4). (2010). *Indian standard code of practice for design and construction of pile foundation: Part 1: Concrete piles, section 4 bored precast concrete piles*. Bureau of Indian Standards, New Delhi, India.

Jeremic, B., Kunnath, S., & Xiong, F. (2004). Influence of soil-foundation-structure interaction on seismic response of the I-880 viaduct. *Engineering Structure, 26*(3), 391–402.

Lombardi, D., & Bhattacharya, S. (2013). Modal analysis of pile-supported structures during seismic liquefaction. *Earthquake Engineering and Structural Dynamics*. https://doi.org/10.1002/eqe.2336.

Meymand, P. J. (1998). *Shaking table scale model tests of nonlinear soil-pile-superstructure interaction in soft clay* (Doctoral dissertation, University of California, Berkeley).

Saha, R., Dutta, S. C., & Haldar, S. (2015). Influence of dynamic soil-pile raft-structure interaction: An experimental approach. *Earthquake Engineering and Engineering Vibration, 14*(4), 625–645.

Tokimatsu, K., Suzuki, K., & Sato, M. (2005). Effects of inertial and kinematic interaction on seismic behaviour of pile with embedded foundation. *Soil Dynamics and Earthquake Engineering, 25,* 753–762.

Part II
Analytical and Numerical Modeling in Geomechanics

Numerical Modelling of Mechanised Tunnelling in Clay

K. Vineetha, A. Boominathan and Subhadeep Banerjee

Abstract Mechanised tunnelling method using Tunnel Boring Machine (TBM) is widely used in urban settings due to the lesser construction duration and minimal surface disruption. However, field-monitoring studies report ground movements and damages to the buildings under which the TBM ploughed through. Hence, it is important to attain a detailed understanding of the TBM–ground interaction before the initiation of any new tunnel construction activity so as to predict the variation in the in situ ground conditions, avoid any damage to the structures adjacent to the tunnel alignment and optimise the TBM performance. This paper comprises a brief review of various studies on the simulation of tunnelling in soft soils followed by the numerical back-analysis of a typical field-monitoring study involving underground tunnelling in clay using the general purpose finite element suite, ABAQUS. A circular tunnel with 8.16 m diameter, running straight and embedded in clay with 19.0 m overburden is considered. The deformation and pore pressure response is monitored along the ground surface and near the tunnel crown, respectively, for each excavation stage of 1.5 m. The numerical model clearly demonstrated the progressive changes in these parameters as the excavation advances.

Keywords Mechanised tunnelling · TBM · Finite element · ABAQUS
Ground movement

K. Vineetha (✉) · A. Boominathan · S. Banerjee
Department of Civil Engineering, Indian Institute
of Technology Madras, Chennai 600036, India
e-mail: vineetha.kombiyil22@gmail.com

A. Boominathan
e-mail: boomi@iitm.ac.in

S. Banerjee
e-mail: subhadeep@iitm.ac.in

B. Adimoolam and S. Banerjee (eds.), *Soil Dynamics and Earthquake Geotechnical Engineering*, Lecture Notes in Civil Engineering 15,
https://doi.org/10.1007/978-981-13-0562-7_18

1 Introduction and Background

The increasing demand for transportation facilities in the urban set-ups has urged the need for extensive underground networks. Mechanised tunnelling techniques involving Tunnel Boring Machine (TBM) are widely used for building up these networks due to the lesser construction duration and minimal surface disruption. Conversely, several field-monitoring studies (Chen et al. 2011; Standing and Selemetas 2013; Houhou et al. 2016) report ground disturbance and structural damage induced by the mechanised tunnelling methods. Hence, it is important to understand the ground–TBM interaction in order to predict the variation in the in situ ground conditions, avoid structural damage and control the TBM operational parameters to minimise the ground disturbances.

Majority of the researches on underground tunnelling in soft grounds are more focused on the conventional open-face tunnelling concept (Mair et al. 1993; Addenbrooke and Potts 2001; Ng and Lee 2005; Marshall et al. 2012; Pinto and Whittle 2014). Kasper and Meschke (2004), Lambrughi et al. (2012), Comodromos et al. (2014), Do et al. (2014) and Zheng et al. (2015) are a few studies discussing the ground deformation and stress transfer mechanism during closed-face tunnelling, which is the most commonly used tunnelling concept nowadays. It was noted that these studies involve essentially numerical techniques like finite difference and finite element methods due to their flexibility in accommodating the TBM as well as ground variables compared to the experimental and analytical approaches. However, attempts of rigorous modelling by incorporating the different TBM components like cutter-head torque, face pressure and grout pressure are still evolving. This research paper describes the numerical modelling procedure for mechanised tunnelling through stiff clay by appropriately accounting for all the TBM components. The results are then compared with an actual field-monitoring data in order to demonstrate the efficiency of the numerical model to predict the ground responses as the tunnel advances.

2 Development of the Computational Domain

In order to investigate the tunnelling-induced ground responses, a full three-dimensional model with circular twin-tunnel configuration was developed using the general purpose finite element suite, ABAQUS. Details of tunnelling and site conditions were referred from Standing and Selemetas (2013). The tunnels are of 8.16 m diameter, running straight with 16.0 m centre-to-centre distance and embedded in clay at a depth of 19.0 m. Figure 1 shows the cross section of the twin-tunnel alignment and the site stratigraphy up to a depth of 35.0 m below the ground surface.

The dimensions of the computational domain were chosen to be $H + 12\ D$ along the X-axis (across the direction of tunnelling), $H + 15\ D$ along the Y-axis (along the

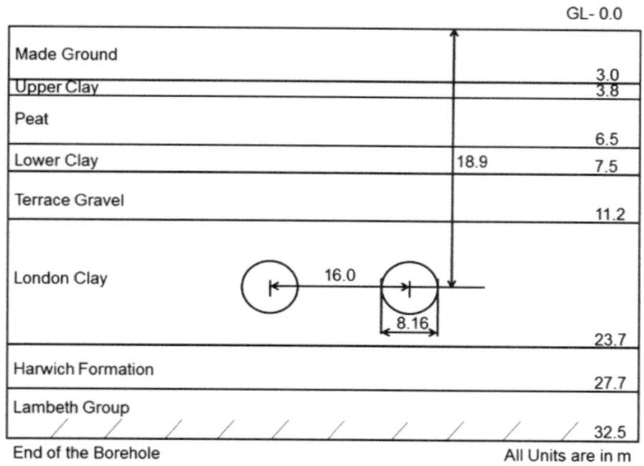

Fig. 1 Cross section of the twin-tunnel alignment

direction of tunnelling) and $H + 6\ D$ along the Z-axis, where H is the height of the overburden with respect to the tunnel axis and D is the tunnel diameter.

2.1 Finite Element Mesh

Figure 2 shows the 3D finite element mesh for the underground tunnel model considered in this study.

It consists of a heterogeneous soil section with eight distinct layers, TBM shield, lining and grout components. C3D8RP (eight-noded linear brick element with trilinear displacement, pore pressure and reduced integration scheme) was used to discretise soil and grout sections, whereas S4R (four-noded conventional stress/

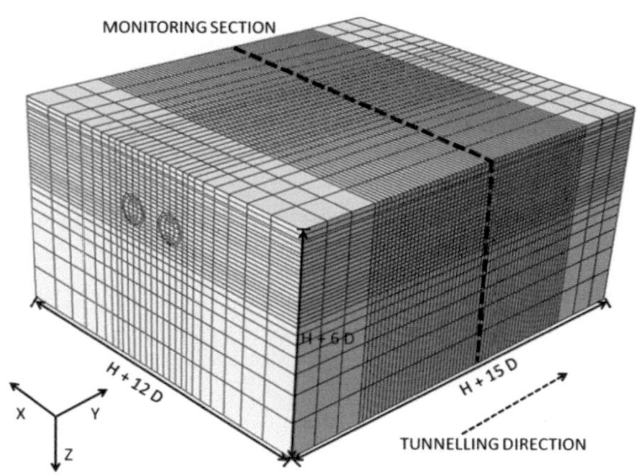

Fig. 2 Three-dimensional finite element mesh for the underground tunnel model considered in this study

displacement shell element with reduced integration scheme) was used for the TBM shield and tunnel lining.

Translations in *X*- and *Y*-directions were constrained along the vertical boundaries of the soil domain in *YZ*-plane and *XZ*-plane, respectively. Translations in all directions were constrained along the bottom boundary of the model. All the three rotational degrees of freedom were constrained for the TBM shield and lining.

2.2 Materials and Constitutive Relations

Soil properties used in the analysis are shown in Table 1. Groundwater table is assumed at the ground level. TBM shield, grout and tunnel lining material parameters used in the analysis are presented in Table 2.

Drucker–Prager constitutive relation with a cap is used to model the undrained behaviour of clay deposits. The addition of a cap yield surface helps to control volume dilatancy when the material yields in shear. Drucker–Prager constitutive relation without a cap is used to model the terrace gravel. The friction angle is 39° for the terrace gravel layer. Shield, lining and grout are modelled as simple linear elastic materials with an assumption that elastic strains developed in the material are less than 5%.

Table 1 Soil parameters

Material	Depth (m)	γ (kN/m^3)	E_u (MPa)	ν	e_0	c_u (kPa)
Made ground	0–3.0	18	3	0.49	1.08	20
Upper clay	3.0–3.8	15	0.9	0.49	8.1	20
Peat	3.8–6.5	15	0.45	0.49	9	20
Lower clay	6.5–7.5	15	0.9	0.49	1.62	20
Terrace gravel	7.5–11.2	20	24	0.2	0.79	–
London clay	11.2–23.7	19	54	0.49	0.68	60 + 3z[a]
Harwich formation	23.7–27.7	20	60	0.49	0.68	97.5 + 3z[a]
Lambeth group	27.7–32.5	21	70	0.49	0.68	109.5 + 3z[a]

[a]z = depth below top of London clay

Table 2 TBM shield, grout and tunnel lining material parameters

Material	Outer diameter (m)	Thickness (m)	γ (kN/m^3)	E (GPa)	ν
Shield	8.12	0.1	80	1500	0.13
Grout	8.12	0.155	20	0.5	0.4
Lining	7.81	0.35	25	26.3	0.2

Table 3 Loads considered in the mechanised tunnelling simulation procedure

Load type	Magnitude[a]	Region of application
Cutter-head torque	1.12 MN m	Excavation face at every 1.5 m interval
Face pressure	2.0 bar	Excavation face at every 1.5 m interval
Traction from shield skin	5.54 MN	Excavated soil surface in contact with the shield
Tail grout pressure	1.6 bar	Excavated soil surface after shield removal

[a]Magnitudes are taken from Standing and Selemetas (2013)

3 Modelling Procedure for Mechanised Tunnelling

The in situ stress conditions were initially established with the introduction of gravity load in the soil domain. To eliminate boundary effects, initially tunnel is excavated 25.0 m away from the front boundary in the *X*–*Z* plane. Loads/pressures are applied as summarised in Table 3. Tunnel is excavated for a length of 1.5 m, equivalent to one lining ring width. The step time period is 3.9 h which is equivalent to the rate of advance of the TBM (0.375 m/h). A coupled pore fluid diffusion and stress analysis offered by ABAQUS/Standard (ABAQUS 2012) was carried out. Accordingly, the porous medium is modelled as a multiphase material considering the presence of two fluids, i.e. wetting liquid and gas. The finite element mesh is attached to the solid phase allowing the fluid flow through this mesh. An effective stress principle is used to describe the porous medium behaviour.

4 Results and Discussion

Ground response discussed in this section includes vertical displacement and pore pressure variation at selected locations, monitored for each excavation stage of 1.5 m during the excavation of the left tunnel. Results from the finite element study are compared with the field-monitoring data reported in Standing and Selemetas (2013). Figure 3 shows the vertical displacement profile on the ground surface along the monitoring section against the distance of the TBM face from the monitoring section as the tunnel advances. The coordinates $(x, y, z) = (0, 0, 0)$ correspond to the location on the ground surface above the centreline of the left tunnel lying in the monitoring section. Both the field data and finite element analysis results show an initial heave (<1 mm) as the TBM approaches the monitoring section with the distance of face at about $y = -20.0$ m. This is followed by a progressive settlement of about 3 mm in the field study and about 6 mm in the numerical study.

Figure 4 shows the peak pore pressure responses obtained from the finite element study, during the withdrawal of tail grout pressure and installation of the tunnel lining. Excess pore water pressures close to 100 kPa was recorded in both the field as well as finite element study as the TBM face approaches the monitoring

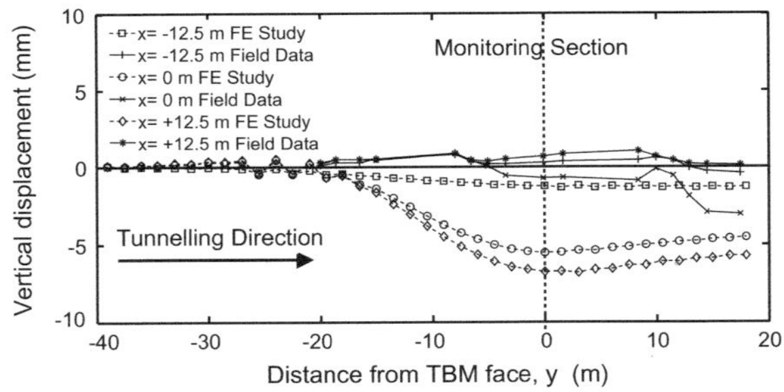

Fig. 3 Vertical displacement at selected locations during the tunnel advancement plotted against the distance of the TBM face from the monitoring section

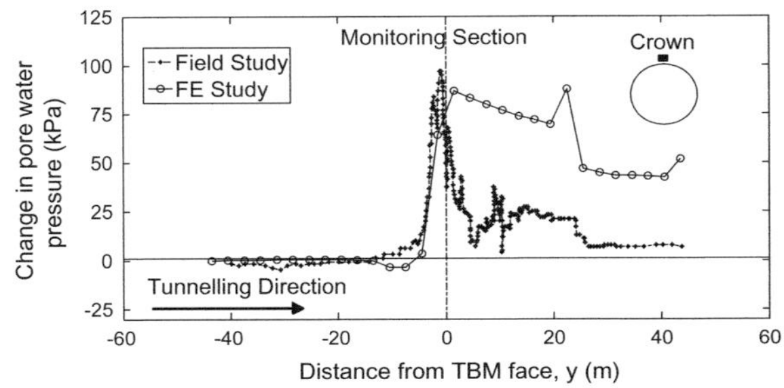

Fig. 4 Pore water pressure variation near the tunnel crown during the tunnel advancement plotted against the distance of the TBM face from the monitoring section

section. This can be attributed to the sudden settlement of soil above the tunnel crown due to the withdrawal of shield and installation of lining together with the fresh grout. As the tunnel advances beyond the monitoring section, it was observed that the pore pressure reduce gradually which can be attributed to the stress release above the tunnel due to grout hardening and transfer of overburden effectively to the tunnel lining.

The finite element study overestimates the pore water pressure variation beyond the monitoring section which might be due to the undrained analysis type. The adoption of tail grout pressure values gauged directly from the pump would have also influenced the pore pressure response. The values gauged from the pump will not be actually exerted in the field due to the different losses which include the grouting pipe and outlet friction resistance. However, the dissipation pattern of pore water pressure beyond the monitoring section in the finite element study is similar to the field study.

5 Conclusion

A three-dimensional finite element model is developed for simulating the mechanised shield tunnelling procedure in soft ground incorporating all the relevant components like torque, face pressure, shield skin friction, tail grout pressure, lining installation, backup trailer weight, grout hardening and excavation. Profiles were plotted demonstrating the ground responses in terms of deformation and pore pressure variation as the tunnelling progresses. The results were found to be in good agreement with the actual field data. The pore pressure variation from the developed model highlights the significance of properly accounting for the relevant TBM components which are usually ignored in the numerical simulation of mechanised tunnel construction problems. Modelling methodology proposed in this study is relevant in the estimation of ground behaviour for situations with similar ground conditions and excavation procedure.

References

ABAQUS. (2012). Inc. ABAQUS user's manual, Version 6.12. Pawtucket, R.I.

Addenbrooke, T. I., & Potts, D. M. (2001). Twin tunnel interaction: Surface and subsurface effects. *The International Journal of Geomechanics*, *1*(2), 249–271.

Chen, R. P., Zhu, J., Liu, W., & Tang, X. W. (2011). Ground movement induced by parallel EPB tunnels in silty soils. *Tunnelling and Underground Space Technology, 26*(1), 163–171.

Comodromos, E. M., Papadopoulou, M. C., & Konstantinidis, G. K. (2014). Numerical assessment of subsidence and adjacent building movements induced by TBM-EPB tunneling. *Journal of Geotechnical and Geoenvironmental Engineering (ASCE), 140*(11), 0401406-1-12.

Do, N., Dias, D., Oreste, P., & Djeran-Maigre, I. (2014). Three-dimensional numerical simulation of a mechanized twin tunnels in soft ground. *Tunnelling and Underground Space Technology, 42,* 40–51.

Houhou, M. N., Emeriault, F., & Vanoudheusden, E. (2016). Three-dimensional back-analysis of an instrumented shallow tunnel excavated by a conventional method. *Geotechnical and Geological Engineering, 34*, 1101–1117.

Kasper, T., & Meschke, G. (2004). A 3D finite element simulation model for TBM tunnelling in soft ground. *International Journal for Numerical and Analytical Methods in Geomechanics, 28*, 1441–1460.

Lambrughi, A., Rodriguez, L. M., & Castellanza, R. (2012). Development and validation of a 3D numerical model for TBM–EPB mechanised excavations. *Computers and Geotechnics, 40*, 97–113.

Mair, R. J., Taylor, R. N., & Bracegirdle, A. (1993). Subsurface settlement profiles above tunnels in clays. *Geotechnique, 43*(2), 315–320.

Marshall, A. M., Farrell, R., Klar, A., & Mair, R. (2012). Tunnels in sands: the effect of size, depth and volume loss on greenfield displacements. *Geotechnique, 62*(5), 385–399.

Ng, C. W. W., & Lee, G. (2005). Three-dimensional ground settlements and stress-transfer mechanisms due to open-face tunnelling. *Canadian Geotechnical Journal, 42*, 1015–1029.

Pinto, F., & Whittle, A. J. (2014). Ground movements due to shallow tunnels in soft ground. I: Analytical solutions. *Journal of Geotechnical and Geoenvironmental Engineering (ASCE), 140*(4), 04013040-1-17.

Standing, J. R., & Selemetas, D. (2013). Greenfield ground response to EPBM tunnelling in London clay. *Geotechnique, 63*(12), 989–1007.

Zheng, G., Lu, P., & Diao, Y. (2015). Advance speed-based parametric study of greenfield deformation induced by EPBM tunnelling in soft ground. *Computers and Geotechnics, 65*, 220–232.

Numerical Analysis of Stone Columns in Soft Clay with Geotextile Encasement and Lime Stabilization

K. S. Amith and V. Ramana Murthy

Abstract Stone columns have been widely accepted as a method of foundation to support large area loads over soft clay beds. Laboratory studies were carried out in test tanks to study the behavior of stone columns in soft clay with different initial consistencies. The study was conducted without and with geotextile encasement, and also the influence of pretreatment of soft clay with lime on the behavior of stone columns was investigated. This paper presents the numerical analysis of the above stone column systems for load-settlement behavior and compares with laboratory test results. This analysis revealed that the numerical analysis using PLAXIS 3D software is useful to predict the behavior of stone columns under different field conditions.

Keywords Stone columns · Soft clay · Geotextile encasement
Lime stabilization · PLAXIS 3D

1 Introduction

Soft clays are present in several world nations, and these deposits are characterized by high natural water content, low shear strength and high compressibility (Murugesan and Rajagopal 2010). Construction of infrastructure facilities in such deposits is a challenge to civil engineers. Deep foundations have been widely used in such deposits despite heavy investment (Priebe 1995; Wood et al. 2000). Several ground improvement techniques such as preloading with vertical drains, electro-osmotic drainage, and stone columns were promulgated to improve these soil deposits. Several project sites are immensely benefitted by these techniques; especially to support area loads such as embankments and even to support building

K. S. Amith (✉) · V. Ramana Murthy
NIT-Warangal, Warangal, India
e-mail: amith.kadaba@gmail.com

V. Ramana Murthy
e-mail: vrm_nitw@yahoo.com

B. Adimoolam and S. Banerjee (eds.), *Soil Dynamics and Earthquake Geotechnical Engineering*, Lecture Notes in Civil Engineering 15,
https://doi.org/10.1007/978-981-13-0562-7_19

foundations (Ambily and Gandhi 2007). Stone columns not only improve the load carrying capacity of ground but also promote consolidation. However, in case of very low strength (C_u < 25 kPa) soils, difficulty in installation of stone columns was reported (Chummar 2000). The load-settlement behavior of stone columns was studied previously in the test tanks without and with geotextile encasement. Also, the influence of pretreatment of soft clay with lime on stone column behavior was studied. The present study is an effort to model the above systems in PLAXIS 3D software to predict the load-settlement behavior in relation to the laboratory test results.

2 Materials and Methodology

Analysis was carried out using PLAXIS 3D software to compare the load-settlement behavior of stone columns with different test conditions considering Mohr–Coulomb model. The dimensions of tanks used were 300 mm diameter and 300 mm height for installing single stone column and 500 mm × 500 mm × 500 mm for group of stone columns. The initial consistencies of clay bed adopted were 25, 40, and 55%. A stone column of 50 mm diameter and 280 mm height was made at the center of the clay bed with the help of a PVC pipe at a density of 1.6 g/cc corresponding to 70% relative density, and at the top of it, a leveling course of 20-mm-thick sand is placed and loaded with a plate having diameter twice the diameter of stone column. The tensile strength of geotextile (PD381) is 18 kN/m, which is used in the study. The clay properties are as follows: gravel = 4%, sand = 26%, fines = 70%, liquid limit = 56%, and plastic limit = 18%. Pressure-settlement plots were drawn, and the ultimate capacity is obtained by drawing the tangents to the initial and final straight line portions. For the formation of group three-stone columns in the test tank, three-stone columns were placed at the spacing of two times the diameter of the stone column and the remaining process is similar to that for single stone column. The input parameters for modeling the stone columns in soft clay are given in Tables 1 and 2. The bulging patterns of stone columns are shown in Figs. 1 and 2 without and with encasement.

Table 1 Input parameters for modeling stone columns in soft clay

Materials	Test consistencies (%)	*W* (%)	*E* (kPa)	Poissons ratio μ	C_u (kPa)	Φ (°)	Ψ (°)	Υ_{dry} (kN/m^3)	Υ_{sat} (kN/m^3)
Clay	25	46.5	1400	0.47	2.8			11.54	16.87
	40	38	2750	0.45	5.5			12.37	17.42
	55	35.1	4000	0.43	8			13.33	18.01
Stone aggregate			55,000	0.3		38	10	16.62	
Sand			20,000	0.3		30	4	15.6	

Table 2 Input parameters for modeling stone columns in lime stabilized soft clay

Materials	Test consistencies (%)	*W* (%)	*E* (kPa)	Poissons ratio μ	C_u (kPa)	Φ (°)	Ψ (deg)	Υ_{dry} (kN/m^3)	Υ_{sat} (kN/m^3)
Clay	25	28	90,000	0.3	180			14.7	18.85
	40	26	100,000	0.3	200			15.18	19.15
	55	23.8	140,000	0.3	280			15.7	20.68
Stone aggregate			55,000	0.3		38	10	16.62	
Sand			20,000	0.3		30	4	15.6	

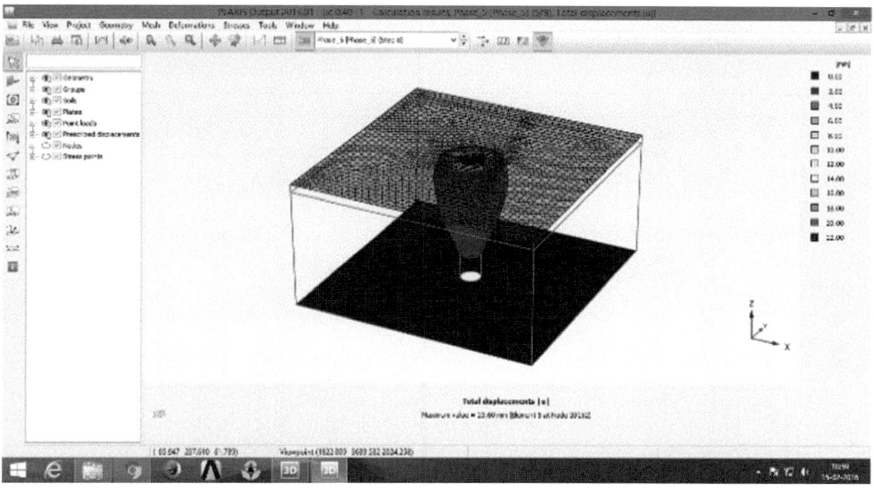

Fig. 1 Stone column in soft clay without geotextile encasement

3 Results and Discussions

3.1 *Pressure-Settlement Behavior of Individual Stone Columns at Different Consistencies of Clay*

Figure 3 shows the load-settlement plot for single stone column without encasement. It can be inferred from this figure that at $I_c = 0.25$, the ultimate load capacity is 20 kPa with the corresponding settlement of 7 mm. Similarly at $I_c = 0.40$, the ultimate load capacity is 28 kPa with corresponding settlement of 6 mm. At $I_c = 0.55$, the load carrying capacity is 55 kPa with corresponding settlement of 4 mm. These patterns indicate that the load capacity of stone columns increases with increasing consistency of clay bed. The relative variation of the

Fig. 2 Stone column in soft clay with geotextile encasement

Fig. 3 Pressure-settlement plots of individual stone column at different consistencies of clay

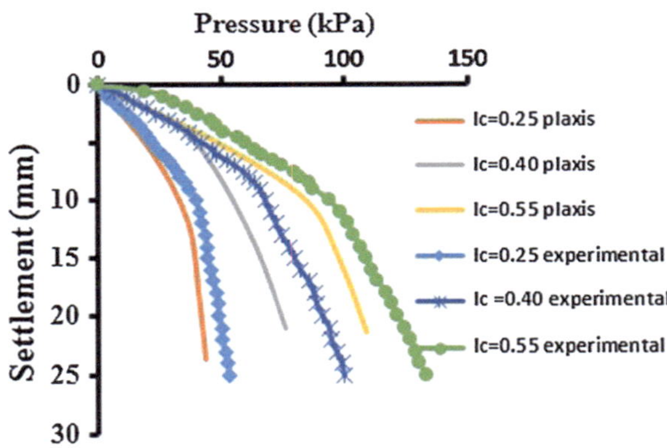

load-settlement plots at different initial consistencies is also shown in Fig. 3 for ready comparison of numerical analysis and laboratory test results.

The load-settlement plots for individual encased stone column obtained through numerical analysis are presented in Fig. 4. It can be observed from this figure that the ultimate load carrying capacity of encased stone column is about 60 kPa at $I_c = 0.25$ with the corresponding settlement of 6 mm. At $I_c = 0.40$, the ultimate load carrying capacity is 75 kPa with corresponding settlement of 4 mm. However, at $I_c = 0.55$ the ultimate load capacity is 102 kPa with corresponding settlement of 3 mm. This can be attributed to the fact that as confinement increases, the load carrying capacity also increases. The relative variation of the load-settlement plots

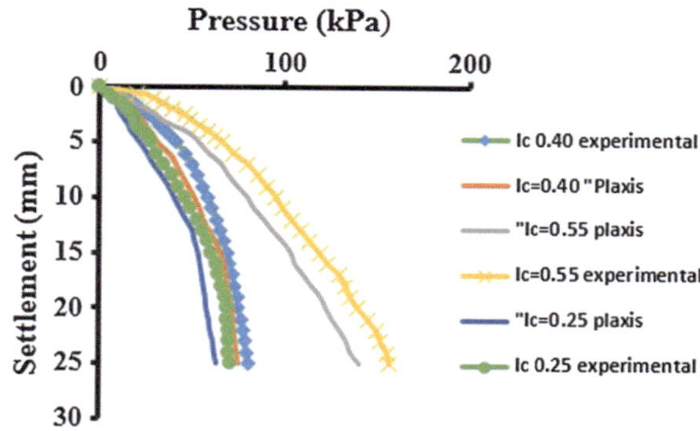

Fig. 4 Pressure-settlement plots of individual encased stone column at different consistencies of clay

at different consistencies is also shown in Fig. 4 for ready comparison of numerical analysis and laboratory test results.

3.2 *Pressure-Settlement Behavior of Three Group of Stone Columns at Different Consistencies of Clay*

Figures 5 and 6 show the pressure-settlement behavior of group of stone columns without and with encasement, respectively. From Fig. 5, it can be observed that at $I_c = 0.25$, the ultimate load is 36 kPa; at $I_c = 0.40$, the ultimate load carrying capacity is 55 kPa, and at $I_c = 0.55$, the ultimate load capacity is increased to 92 kPa. In all these cases, the settlement varied from 4 to 7 mm. Similarly, for encased stone column groups, the ultimate capacity is in the range of 85–175 kPa with the corresponding settlements of 6–8 mm. From these observations, it is revealed that the group capacity is altogether different from the mere multiplication of single column capacity with the number of columns in a group. This behavior could be attributed to the mutual confining effect in groups compared to single stone column.

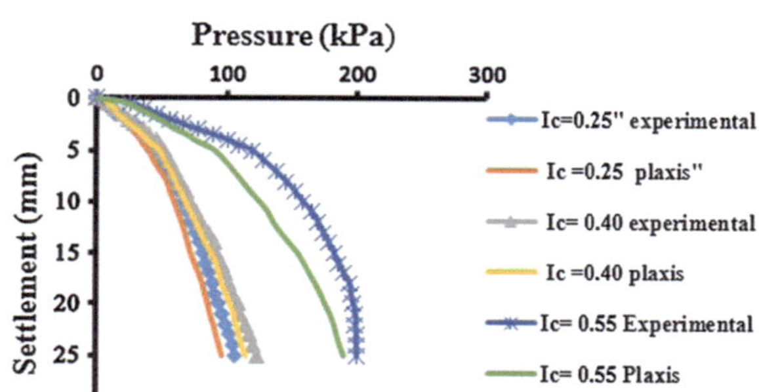

Fig. 5 Pressure-settlement plots of three-stone column groups at different consistencies of clay

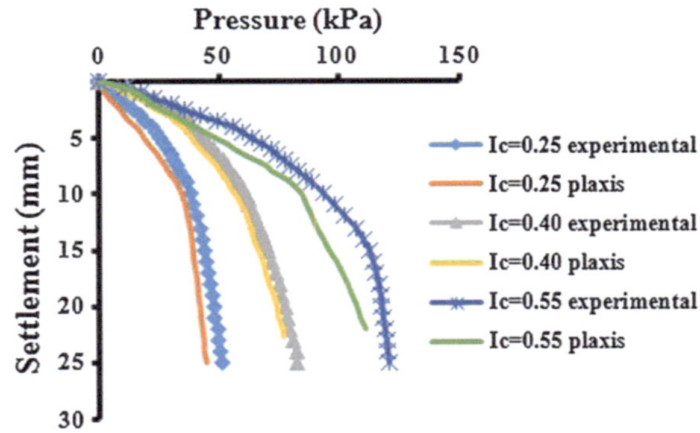

Fig. 6 Pressure-settlement plots for group of encased stone column at different consistencies of clay

3.3 Pressure-Settlement Behavior of Stone Column Groups in Lime Stabilized Soft Clay at Different Initial Consistencies

Figures 7, 8, and 9 show the load-settlement behavior of group of three-stone columns installed in 3% lime stabilized soft clay. It can be observed that at $I_c = 0.25$, the ultimate load carrying capacity obtained from numerical analysis is about 1400 kPa, whereas the laboratory value is 1500 kPa. At $I_c = 0.40$, numerical analysis showed that the ultimate load carrying capacity is 1000 kPa and the laboratory value is about 1200 kPa. Also at $I_c = 0.55$, the ultimate load carrying capacity obtained from numerical analysis is around 700 kPa, whereas the value is 750 kPa from laboratory testing. For all these cases, the settlement corresponding to ultimate loads varied from 5 to 9 mm. In lime stabilized clay, similar trends are observed for individual stone column also. The decrease in degree of improvement in lime stabilized clay with increasing consistency could be attributed to transition of failure from bulging to vertical displacement of stone column in lime stabilized

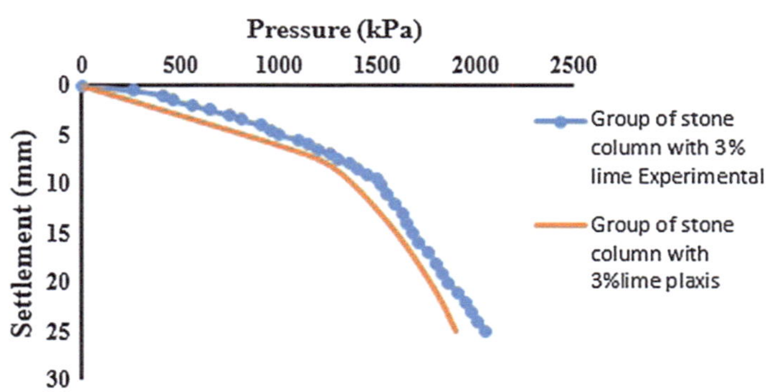

Fig. 7 Pressure-settlement plot for three-stone column group in soft clay stabilized with 3% lime at $I_c = 0.25$

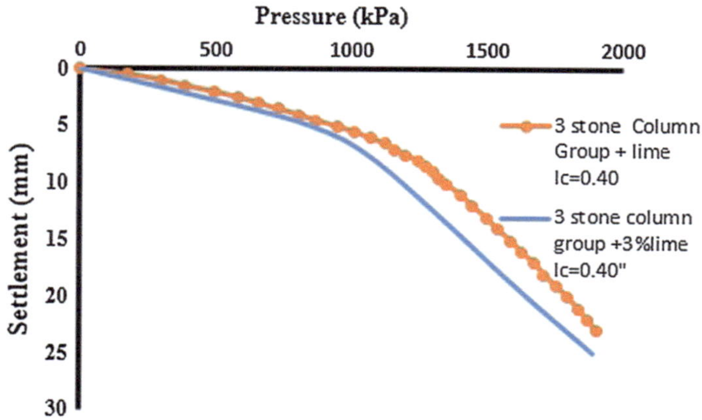

Fig. 8 Pressure-settlement plot for three-stone column group in soft clay stabilized with 3% lime at $I_c = 0.40$

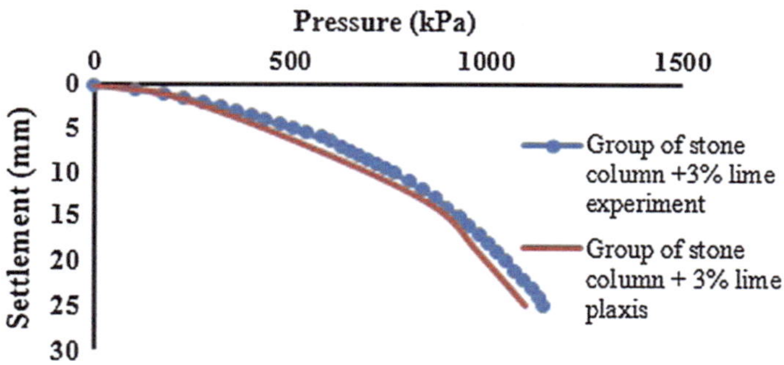

Fig. 9 Pressure-settlement plot for three-stone column group in soft clay stabilized with 3% lime at $I_c = 0.55$

clay bed. From the above observations, it is understood that the behavior of stone columns under different field conditions can be predicted using PLAXIS 3D software.

4 Conclusions

The following conclusions are drawn based on numerical investigations carried out on stone columns installed in soft clay without and with lime stabilization and geotextile encasement.

The ultimate load capacity of stone columns is increased with increasing undrained strength Cu of soft clay.

The load carrying capacity of geotextile-encased stone column is enhanced by almost 3 times that without encasement. The group capacity is not the multiplicative value of a single stone column capacity at any consistency with or without geotextile encasement.

The bulging of stone columns is reduced both with geotextile encasement and lime stabilization.

The degree of improvement in load carrying capacity of stone columns with lime stabilization is lesser in case of clay with high initial stiffness.

From PLAXIS 3D analysis, it is found that the load carrying capacity of stone columns is significantly increasing with increasing Young's modulus. This is true with increasing consistency of clay.

The load-settlement plots obtained from laboratory testing compared well with numerical method of finite element analysis for ordinary stone columns, geotextile-encased stone columns, and stone column in lime stabilized clay, indicating the potential use of PLAXIS 3D software to predict the behavior of stone columns in the field.

References

Ambily, A. P., & Gandhi, S. R. (2007). Behavior of stone columns based on experimental and FEM analysis. *Journal of Geotechnical and Geo Environmental Engineering, ASCE, 133*(4), 405–415.

Chummar, A. V. (2000). Ground improvement using stone columns: Problems encountered. In *Proceedings of GeoEng 2000* (pp. 19–24), Australia.

Murugesan, S., & Rajagopal, K. (2010). *Journal of Geotechnical and Geoenvironmental Engineering, 136*(1), 129–139.

Brinkgreve, R. B. J., Engin, E., & Swolfs, W. M. (2013). PLAXIS 3D 2013 user manual. Plaxis bv, Delft.

Priebe, H. J. (1995). The design of vibro replacement. *Ground Engineering, 28*(10), 31–37.

Wood, D. M., Hu, W., & Nash, F. T. (2000). Group effects in stone column foundations: Model tests. *Geotechnique, 50*(6), 689–698.

Stability Analysis of a Typical Sunderban Embankment During Diurnal Tidal Cycle

Kaustuv Bhattacharya, Gupinath Bhandari and Sibapriya Mukerjee

Abstract Earth embankments are utilized as protective structures along river, coastal banks and maybe thus exposed to steady or transient seepage induced by tides. In the Indian part of Sunderban region, the tidal head variation is 4–6 m in cycle time of 12 h. A model has been developed in FEM-based software, GeoStudio, to study the effect of transient seepage in such an embankment. The response to tidal head was found to be limited to a certain zone within the embankment. The cyclic pore water pressure distribution was seen to differ in various regions in this zone. Also, even though the cyclic FOS against stability was adequate, serviceability check indicates chances of failure near toe region.

Keywords Transient seepage · Unsaturated permeability · Sunderban embankment · Finite element method · Tidal cycle

1 Introduction

Earth embankments are used as efficient and economic forms of protective structures along rivers and coasts. These are exposed to hydrostatic and hydrodynamic forces as per location, topography, weather conditions, etc. The common hydrostatic element affecting the design life of such structure is seepage, which may be steady and/or transient in nature. Water retaining structures along rivers and coasts are exposed to tidal heads which may induce transient seepage, characterized by river water head which is a function of tidal cycle time. A single tidal cycle usually

K. Bhattacharya (✉) · G. Bhandari · S. Mukerjee
Department of Civil Engineering, Jadavpur University, Kolkata 700032, West Bengal, India
e-mail: kaustuv_poirot@yahoo.com

G. Bhandari
e-mail: g.bhandari.ju@gmail.com

S. Mukerjee
e-mail: sibapriya.mukherjee@gmail.com

B. Adimoolam and S. Banerjee (eds.), *Soil Dynamics and Earthquake Geotechnical Engineering*, Lecture Notes in Civil Engineering 15,
https://doi.org/10.1007/978-981-13-0562-7_20

consists of the phenomenon of rise-up, or ascent of river water from Low Tide Level (LTL) to High Tide Level (HTL), and further drawdown, or descent of the same from High Tide Level back to the Low Tide Level. The time period of rise-up and drawdown half cycles are termed as rise-up half cycle time and drawdown half cycle time, respectively.

2 Study Area

The Sunderban delta is formed at the confluence of the rivers Ganga, Brahmaputra and their distributaries. This region is shared by South 24 Pgs, West Bengal (India) and Khulna division (Bangladesh) with the latter having a share of about 60%. Network of creeks and channels divides the entire area into several small and large islands. The ground level of most of the islands is above the Low Tide Level (LTL) but below the High Tide Level (HTL); hence, there is requirement of embankments in order to protect the islands from tidal water inundation. The construction of these embankments (shown in Fig. 1) dates back to the eighteenth century, when human settlement started to flourish here, and is said to have been done mainly using locally available topsoil, which is predominantly silty clay/clayey silt. This study is based on the Indian part of Sunderban where the total length of existing earth embankments is approximately 3520 km. The water head difference due to diurnal tidal cycle is about 4–6 m, across various sites of the Sunderban. The time period for a single cycle is about 12 h, comprising 6 h of rise-up and drawdown half cycle time, respectively. Table 1 shows the soil properties from a bore-log in Patharpratima site.

Fig. 1 Typical Sunderban embankment at Sonakhali

Table 1 Soil properties from Patharpratima, West Bengal

Soil description and type	Grey clayey silt, CI
Volume compressibility, m_V (per kPa)	0.048
Bulk density (kN/m^3)	18.2
Saturated porosity (m^3/m^3)	0.465
D_{60} (mm)	0.034
Percentage of sand (%)	10
Percentage of silt (%)	62
Percentage of clay (%)	28
Liquid limit (%)	47
Saturated conductivity (K_{sat}) (m/s)	10^{-8}
Residual water content	0.18
Effective cohesion, c' (kN/m^2)	5
Effective angle of internal friction, Φ' (°)	25
Modulus of elasticity (with $v = 0.3$) (kPa)	2083

3 Numerical Analysis

A model was developed in the popular finite element-based software, GeoStudio. The soil properties were taken as per Table 1. The volumetric water content (VWC) function, shown in Fig. 2, has been developed as per Aubertin et al. (2003) using grain size distribution, liquid limit, porosity, coefficient of volume compressibility. The hydraulic conductivity function, shown in Fig. 3, has been estimated from the volumetric water content function using the Van Genuchten methodology (1980), with saturated permeability and residual water content as the input. The air entrant value and residual water content have been obtained from the volumetric water content curve. The landside groundwater table was taken at the existing ground level. The steady-state condition was established between the groundwater table and the Low Tide Level on the riverside. The tidal effect was simulated on the basis of hydraulic boundary conditions as a function of tidal cycle time. The difference between Low Tide Level and High Tide Level was taken as 4 m, and the full cycle time was taken as 12 h, consisting of 6 h each of rise-up and drawdown, respectively.

The finite element model with 489 nodes and 473 elements has been shown in Fig. 4. The shape of mesh was triangles and quadrilaterals with unstructured origin and individual element size of 0.5 m in the embankment and 1 m in other parts. A linear time stepping was applied for the transient stage analysis.

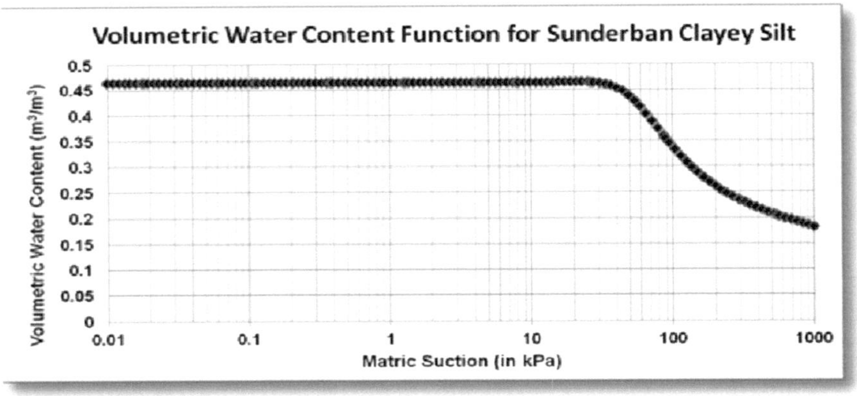

Fig. 2 VWC function for Sunderban clayey silt

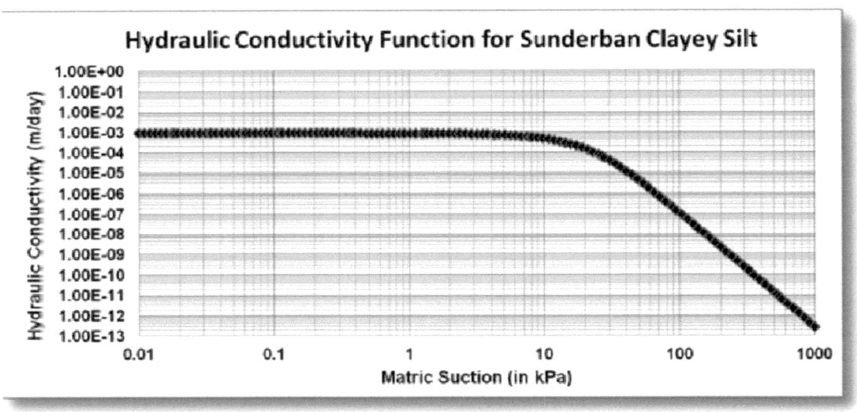

Fig. 3 Conductivity function for Sunderban clayey silt

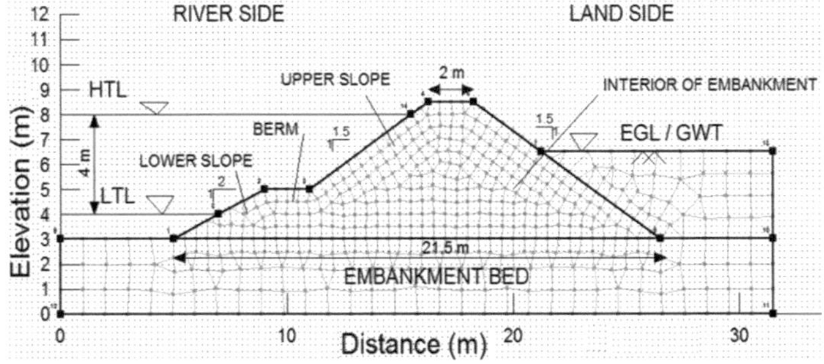

Fig. 4 FEM model of Embankment in GeoStudio

4 Results and Discussions

4.1 Dynamics of the Phreatic Surface

The dynamics of the phreatic surface during rise-up and drawdown of single tidal cycle has been shown in Figs. 5 and 6, respectively. It was observed that during rise-up (t = 1–6 h), the phreatic surface slowly developed along the lower slope, berm and upper slope, till High Tide Level at t = 6 h. The developed phreatic surface started to recede during drawdown phase (t = 6–12 h). However, at the end of a single tidal cycle, a cumulative effect in phreatic surface was seen as can be seen from Fig. 6, where the phreatic surface has moved up slightly encompassing more area above the berm region, compared to initial condition at t = 0 h.

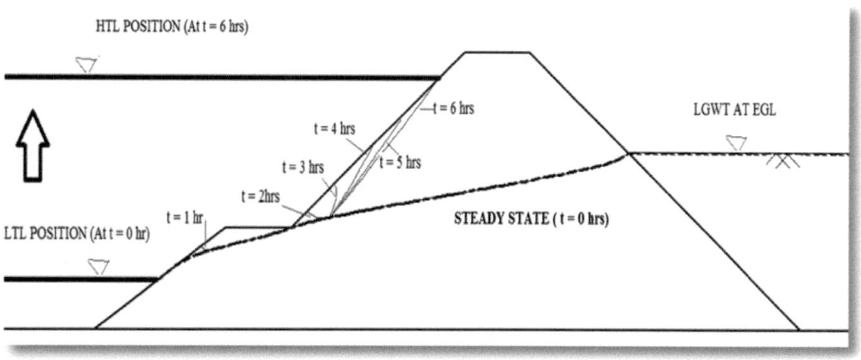

Fig. 5 Phreatic surface dynamics during rise-up

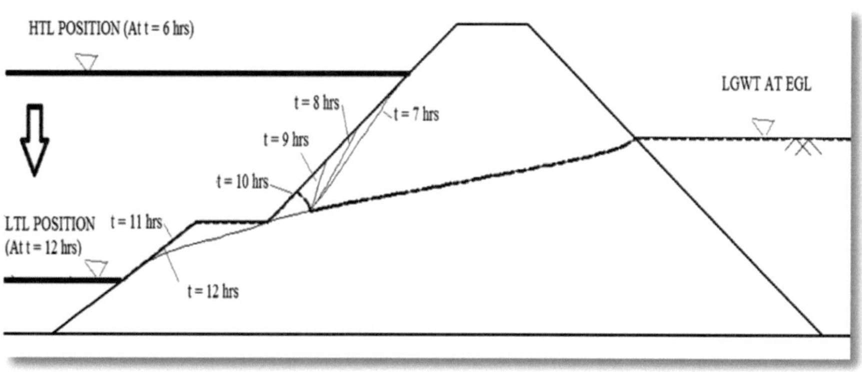

Fig. 6 Phreatic surface dynamics during drawdown

4.2 Distribution of Equipotential Contour

The equipotential total head contour due to transient seepage during a single tidal cycle has been shown in Figs. 7, 8, 9, 10 and 11. A zone is seen to develop along the river–embankment interface, through the rise-up stage. The equipotential total head is observed to slowly increase in this zone with rise-up half cycle time. This zone is termed as "Zone of Influence of Rise-up" (ZIR). During drawdown, this zone dissipates slowly till the end of a single tide cycle. However, cumulative effect is observed at the end of a single cycle (t = 12 h). It is seen that the zone has dissipated along lower slope; berm portion yet still persists in parts of the upper slope as the river water prepares for the next tidal cycle. This indicates that during a single tidal cycle, the excess head induced due to transient seepage during rise-up does not fully dissipate at the end of drawdown stage, for certain regions within this zone. No change in total head induced during tidal cycle is observed in the interior part of the embankment which are regions outside the Zone of Influence of Rise-up.

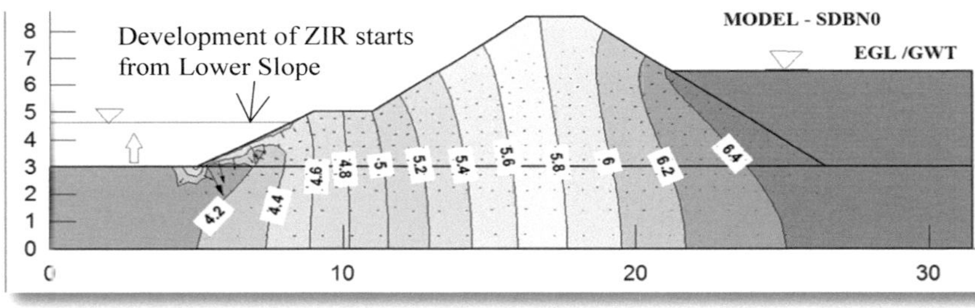

Fig. 7 Total head contour during rise-up, t = 1 h

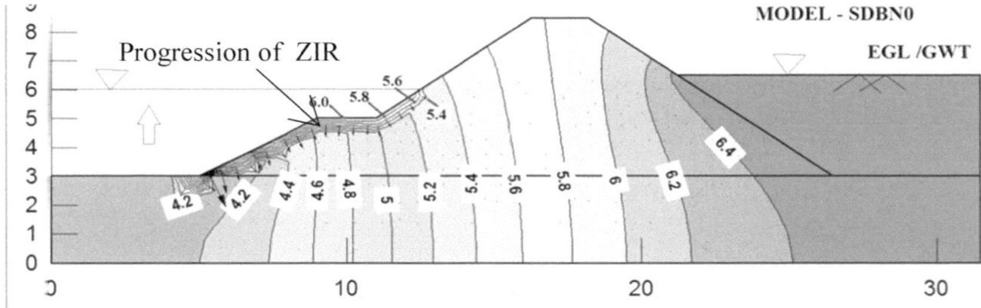

Fig. 8 Total head contour during rise-up, t = 3 h

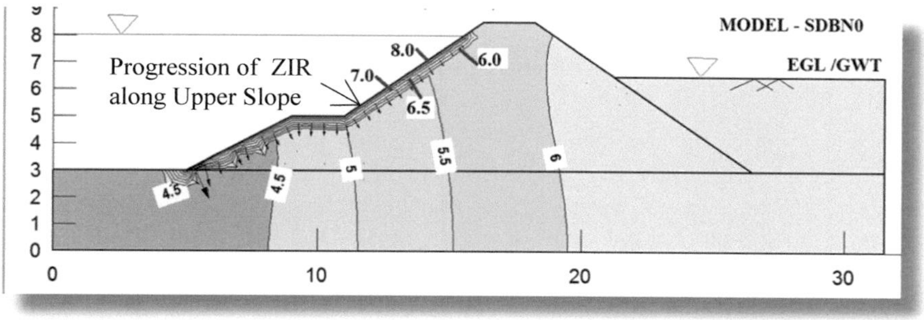

Fig. 9 Total head contour at HTL, $t = 6$ h

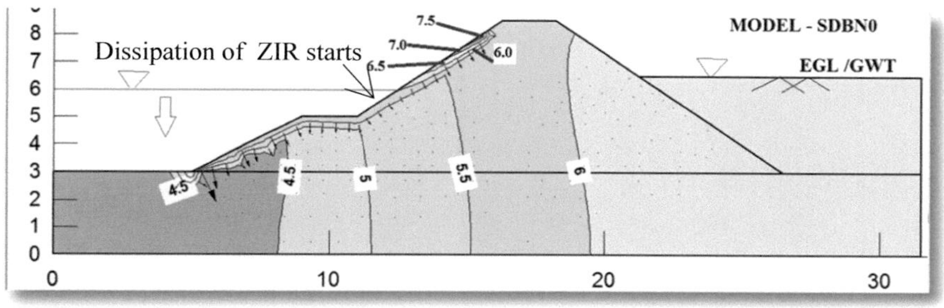

Fig. 10 Total head contour during drawdown, $t = 9$ h

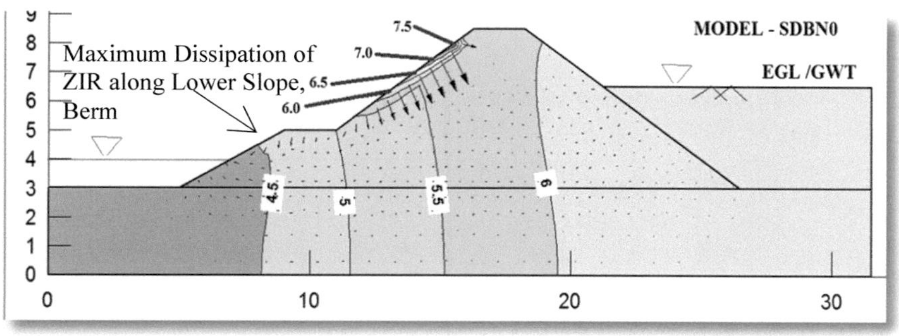

Fig. 11 Total head contour at end of one cycle, $t = 12$ h

4.3 Pore Water Pressure Distribution for Various Zones Within the Embankment

The excess pore water pressure (PWP) distribution in lower slope, berm, upper slope (shown in Fig. 4) for five tidal cycles has been shown in Figs. 12, 13 and 14, respectively. The parts of these regions that lie within the Zone of Influence of Rise-up have shown changes in PWP due to cyclic tidal behaviour. It is seen that there is gradual build-up of excess PWP during rise-up half cycle, followed by release of the same during drawdown half cycle. Similar pattern is observed for

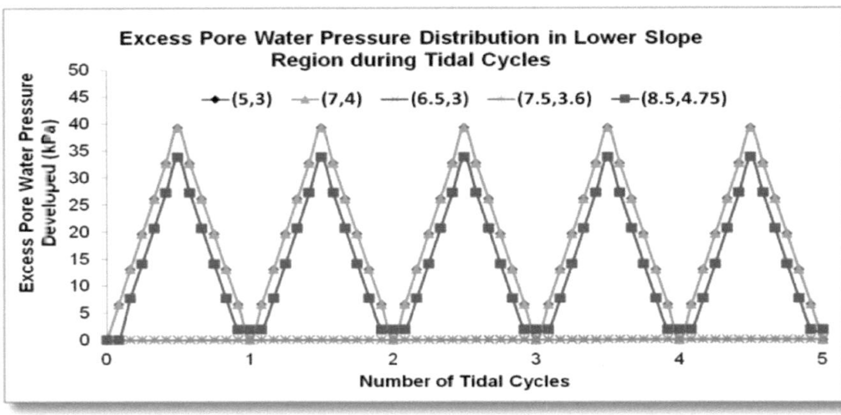

Fig. 12 Excess PWP distribution in lower slope region

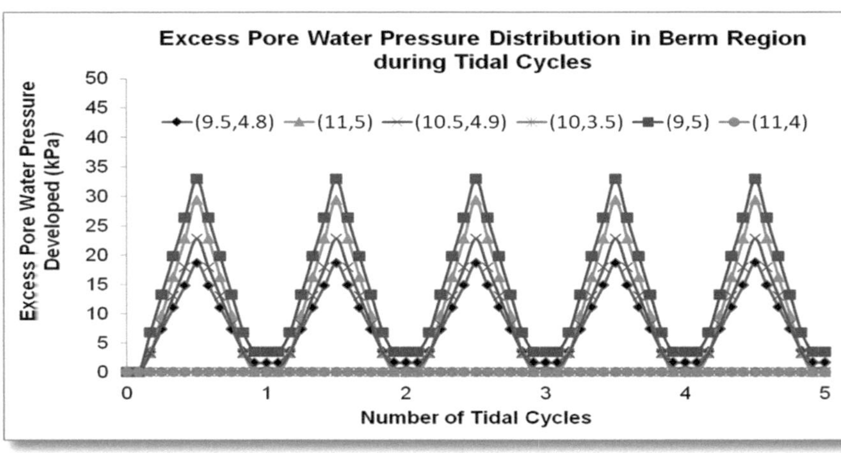

Fig. 13 Excess PWP distribution in berm region

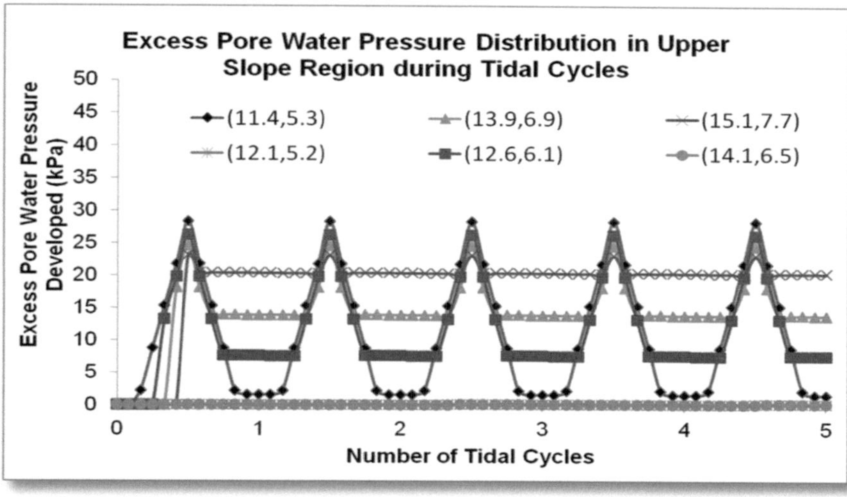

Fig. 14 Excess PWP distribution in upper slope region

multiple cycles, however, with a cumulative effect. It is seen that the maximum variation in excess PWP is observed for the lower slope followed by berm and upper slope region. The pattern varies as per region and is dependant on area of unsaturation, difference in level between HTL and LTL and tidal cycle time. Negligible variation in PWP was observed for points lying in the interior of the embankment as they lie outside the Zone of Influence of Rise-up.

4.4 *Stability Factor*

A slope stability analysis using Spencer methodology was performed using SLOPE/W. The minimum factor of safety plotted against number of tidal cycles is shown in Fig. 15. The Cyclic Factor of Safety Ratio (CFOS) is the ratio of the transient FOS (during tidal cycle) to the initial FOS, i.e. during steady state. The same was seen to increase during rise-up half cycle time and decrease during drawdown half cycle time. It was observed that the FOS remained constant more or less at the end of each cycle compared to initial conditions.

4.5 *Serviceability*

Additionally, a deformation analysis was performed using SIGMA/W. It was observed that even though the stability factor of safety is adequate for the typical

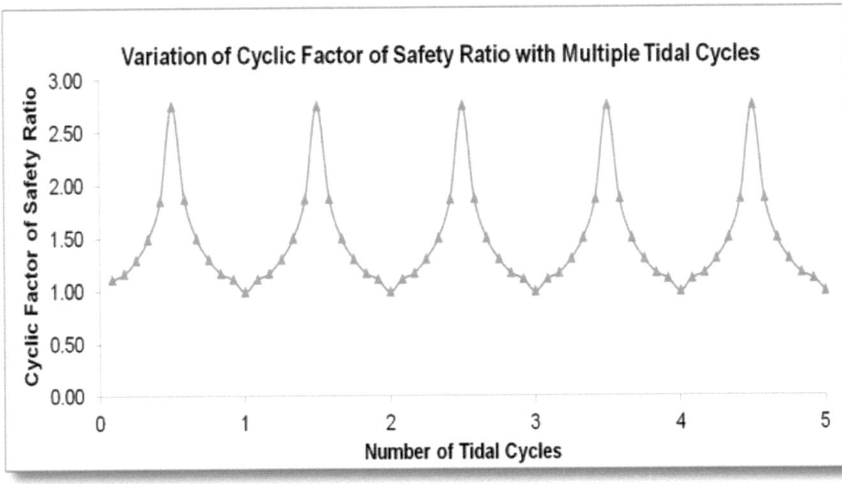

Fig. 15 Variation of CFOS with multiple tidal cycles

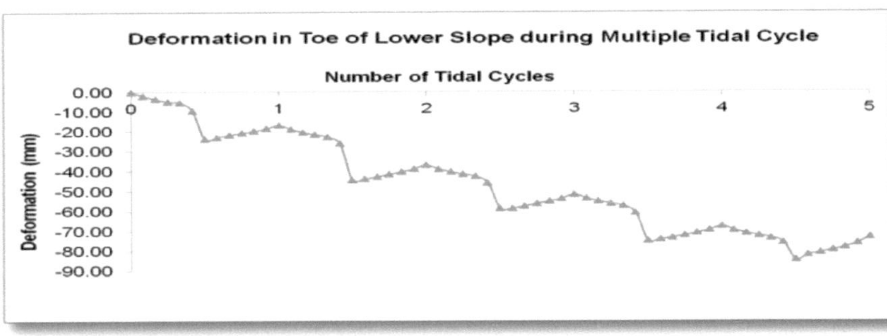

Fig. 16 Deformation pattern in toe of lower slope

Sunderban embankment during tidal cycles, the deformation of the toe region of the lower slope showed significant vertical settlement during rise-up and elastic rebound during drawdown, with a cumulative effect over multiple cycles. This has been shown in Fig. 16. This may be attributed to the highly plastic nature of the soil used in construction of the embankment. This may lead to deformation-based failure, instances of which have been reported from sites like Patharpratima, Gosaba, Sonakhali and other sites of Sunderban region. A typical lower slope failure in Sunderban embankment is shown in Fig. 1.

4.6 Comparison with Other Studies

The result patterns observed were in good agreement with similar type of study carried out by Desrosiers et al. (2013) for a tidal front embankment in Boston, UK.

5 Conclusions

(i) Changes in location of phreatic surface and variation in pore water pressure due to transient seepage induced by tidal behaviour are observed within the Zone of Influence of Rise-up only. The variations are influenced by the initial conditions, tidal head difference between LTL and HTL, tidal cycle time. Regions lying outside the ZIR show negligible response to the tide-induced transient seepage.

(ii) The excess pore water pressure variation pattern is different across regions within the ZIR, as observed for a typical Sunderban embankment with maximum variation in lower slope region during full tidal cycle.

(iii) Even though the stability factor of safety may be adequate for an embankment exposed to tidal behaviour, it is important to check for deformation at critical areas. For the Sunderban embankment, the toe region appears to be most susceptible to serviceability failure.

References

Aubertin, M., Mbonimpa, M., Bussière, B., & Chapuis, R. P. (2003). A model to predict the water retention curve from basic geotechnical properties. *Canadian Geotechnical Journal, 40*(6), 1104–1122.

Desrosiers, T., Bennett, V., Abdoun, T., & Simm, J. (2013) Comprehensive real-time field monitoring at active embankment subjected to tidal loading. *ASCE Geo-Congress*, pp. 1556–1565.

Van Genuchten, M. T. (1980). A closed-form equation for predicting the hydraulic conductivity of unsaturated soils (Masters' thesis). *Soil Science Society of America Journal, 44*, 892–898.

Rock Mass Slope Stability Analysis Under Static and Dynamic Conditions in Mumbai, India

Ahmed Shaz, Endalu Tadele Chala and K. Seshagiri Rao

Abstract The rock slope stability has gained much importance in the recent years owing to the development of infrastructure in/on rock slopes areas in the form of high-rise building, roads, railways, dams, and other rock engineering projects. Rock slope stability is generally affected by the property of rock material along with the discontinuities in the rock mass. It is also dependent on slope angle, slope height, surcharge, groundwater conditions, rainfall, and dynamic forces like earthquake. If these factors, coupled with deep weathering of the rock mass, complicate the entire behavior of the rock slope and makes the analysis of rock slopes challenging task. For the present work, fieldwork was carried out for joint mapping on differentially weathered volcanic rock of selected slope sections at Mumbai in Deccan trap regions. Block and core samples were collected from each category of the rocks and their respective geotechnical properties measured experimentally and presented her in. The rock slope stability analysis was done using numerical modeling techniques, finite element method (FEM)-based two-dimensional analyses (RS2), or Phase2 9.0 software from Rocscience. The rock slope having four different quality of rock mass was analyzed under static and pseudo-static conditions. Finally, the optimum support measures were provided to the slope for stabilizing it and the corresponding analysis was also done.

Keywords Rock slope stability · Rock mass properties · Numerical modeling RS2 · Stabilization

A. Shaz (✉)
Larsen & Toubro Ltd., Chennai 600089, India
e-mail: syedahmadshaz@gmail.com

E. T. Chala · K. Seshagiri Rao
Department of Civil Engineering, Indian Institute of Technology Delhi, New Delhi, India
e-mail: endu2020@gmail.com

K. Seshagiri Rao
e-mail: raoks@civil.iitd.ac.in

B. Adimoolam and S. Banerjee (eds.), *Soil Dynamics and Earthquake Geotechnical Engineering*, Lecture Notes in Civil Engineering 15,
https://doi.org/10.1007/978-981-13-0562-7_21

1 Introduction

Rock slope stability among many factors generally depends on the strength of the rock materials and characteristics of discontinuities (e.g., roughness, wall strength, aperture, fill material, and persistence). In tropical climate, weathering of rock material and discontinuity walls also affects the rock slope stability and significantly influences the response of rock masses to loadings and unloading of excavations.

In recent years, numerical methods such as finite element method (FEM), finite difference method (FDM), and discrete element method (DEM) are being more commonly applied to slope stability problems. Hudson and Feng (2007) provided a flowchart for numerical modeling of rock engineering. It is a common practice to model the behavior of discontinuities by a linear Coulomb relation using the parameters c and ϕ. However, the researchers in the field of rock mechanics and rock engineering well recognized that the shear strength parameters are not truly constant, and it depends upon the normal stress and scale (Bhasin and Kaynia 2004).

Hoek et al. (2002) has given c and ϕ in terms of Hoek parameters particularly to solve rock slope stability problems. For the present analyses and design of rock slopes, different degree of weathered blocks and core samples were collected from the rock slope of Deccan trap region, India. Specimens were prepared as per ISRM standards and tested for tensile, compression, and triaxial strengths. The average UCS and material constant of each weathered rock material was determined from experimental results. The GSI value of each of weathered rock mass was used for the estimation of rock mass properties (i.e., equivalent rock mass properties) for the analysis and design of slopes.

2 Geological Description and Specimen Prepared

The rock slope is located at the eastern scarp of Cumballa Hill in Tardeo, Mumbai. Regionally, the rocks in Maharashtra and around Mumbai are of volcanic origin known as Deccan trap. The Deccan trap is composed of very hard, tough, and compact rocks. The rocks are susceptible to weathering which begins on the exposed surface and along joints, cracks, and fissures. With time, it penetrates deep into the rock masses and resulting in complete loss of original skeleton structure of the rock mass. Chala et al. (2016) studied the strength properties of different degree of weathered basaltic rocks of some locations in Deccan trap and observed that weathering has drastically reduced the strength of rocks. They found the strength of slightly weathered basalt to be 50% less than the fresh basalt strength (Figs. 1 and 2).

For the present rock slope stability analysis, efforts were made to incorporate the effect of chemical weathering and discontinuity characteristics on rock mass through detail fieldwork such as joint mapping and laboratory investigations. Summary of the geotechnical properties of rocks is presented in Tables 1 and 2.

Fig. 1 Present rock slope appearance on out crops from eastern to western corner of the slope **a** dominant joint sets, discoloration, and staining of rock material due to weathering clearly visible **b** already stabilized portion of the rock slope

Fig. 2 Specimen prepared for testing under tensile, uniaxial, and triaxial loading

Table 1 Basalt intact rock properties and equivalent rock mass properties of fine breccia, coarse breccia, and tuff

Properties	Fine breccia	Coarse breccia	Tuff	Basalt
GSI	20	40	40	Intact
γ dry (kN/m^3)	21.79	21.50	20.56	27.91
γ sat (kN/m^3)	22.39	22.740	22.40	28.21
σt (MPa)	0.005	0.007	0.01	19.670
E (MPa)	517.14	514.0	892.1	34,937
c peak (MPa)	0.812	1.152	1.62	5.365
c residual (MPa)	0.0812	0.1152	0.016	0.5365
$\varphi^{(\circ)}$ peak	23.92°	41.51°	38.3°	65.7°
$\varphi^{(\circ)}$ residual	12°	20°	19°	32.53°

Table 2 Joint property of basalt

Joint property	Value
c (saw cut) (MPa)	0.0397
$\varphi^{(\circ)}$ (saw cut)	36.479°
Normal stiffness (MPa/m)	29,700
Shear stiffness (MPa/m)	12,270

3 Making Sections from Fence Diagram

Five sections of the hill slope were plotted for chainage 15, 35, 65, 90, and 110 m. After studying, the core samples were obtained from already drilled borehole (22.5–50 m deep) below existing ground level (EGL). From the borehole data, the rock present at various depths of the slope in a borehole of a particular section is marked and joined with the same rock in another borehole of the same section, to get the inclination of various rock strata beneath the EGL of the hill slope.

From the borehole study, it was found that basalt forms the top strata of the hill slope, as was observed on the site. Below the basalt layer is the layer of fine volcanic breccia (FVB) which forms the lower portion of the hill slope, it is followed by a layer of coarse volcanic breccia (CVB) and then welded tuff (Fig. 3).

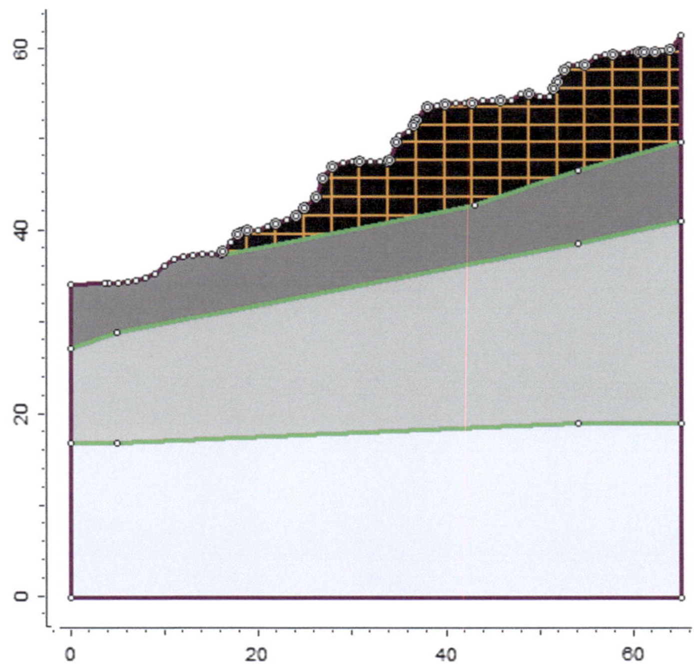

Fig. 3 Geometry and formation of the four rock masses at 65 m chainage

The angle of the slope varies throughout the slope section, and the average angle of the slope was found to be 30°.

4 Determining Rock Mass Properties

Rock material properties were obtained from laboratory testing including index properties tests, uniaxial, triaxial, Brazilian, and point load test. The joint properties were calculated using direct shear test.

For the top basalt layer, rock material property (σ_c = 68.85 MPa and E = 34.937 GPa) and joint property had been calculated in the laboratory and used in the analysis given (Tables 1 and 2). Normal stiffness and shear stiffness were taken from Kainthola et al. (2014).

Since none of the underlaying strata was exposed for the joint mapping, the equivalent rock mass properties were calculated based upon GSI Index chart (Marinos and Hoek 2000).

The structure of the rock mass (intact, blocky, very blocky, disturbed, disintegrated, and laminated) was studied along with the surface condition of the rock mass. Finally, GSI was quantified accordingly for fine volcanic breccia (FVB), coarse volcanic breccia (CVB), and welded tuff (Tuff) as 20, 40, and 40, respectively.

Rock data 5.0 software from Rocscience was used to determine the equivalent rock mass properties. The triaxial test data was given as an input to the software along with GSI value and uniaxial compressive test results (σ_c = 26.1, 33.3, 32.4 MPa and E = 11.32, 3.22 and 5.58 GPa) for FVB, CVB, and Tuff, respectively.

Also, the residual values of cohesion and friction angle were taken as 10 and 50% of the peak values, respectively, for the slope stability analysis. The intact and equivalent rock mass properties are provided in Tables 1 and 2.

5 Slope Stability Analysis

The geometry was made for different slope sections at respective chainage and each rock strata in every slope section was assigned with corresponding rock mass and joint properties accordingly in the Phase2 model. Then the boundary conditions were assigned to the model (Fig. 4).

The ground surface of the slope was made free in both X- and Y-direction. The vertical boundaries of the section were free to move in Y-direction only and restrained in X-direction. The base of the section of slope was fixed and was restricted to move in both X- and Y-direction. A uniform mesh was used in the FEM model for the analysis of all the five slope sections.

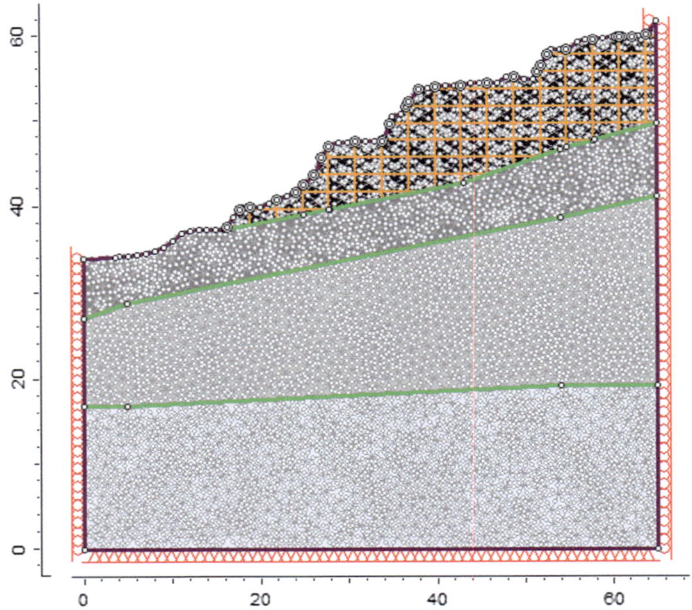

Fig. 4 Discretized Phase2 model of slope section at 65 m chainage

5.1 *Static Analysis*

Model of the slope sections was loaded with field stresses by taking gravitational load from actual ground surface. All the analyses were done using Mohr–Coulomb failure criterion (Table 3).

The critical SRF for the slope section at 65 m chainage was found to be 1.75 from the normal static gravitational analysis without any seismic conditions (Fig. 5).

Table 3 Results of the static analysis

Chainage of the slope section (m)	Factor of safety	Max. total displacement (m)
15	2.65	0.0073
35	1.89	0.0027
65	1.75	0.0019
90	2.85	0.0079
110	2.34	0.0024

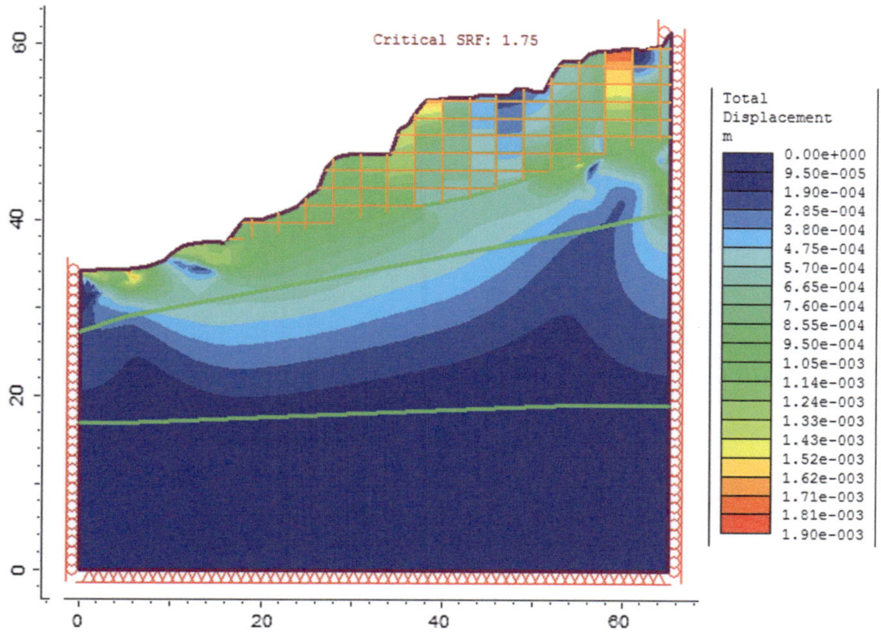

Fig. 5 Total displacement contour plot at critical SRF of section at 65 m chainage

5.2 *Dynamic Analysis*

The present rock slope is located at Mumbai which falls under seismic zone 3 for which the horizontal seismic coefficient value (K_h) is prescribed as 0.16 for moderate seismic intensity (I.S.1893.1.2002). Numerical analysis was done under pseudo-static loading using the horizontal seismic coefficient to model their stability under any seismic event. From static analysis, it has been concluded that the slope section at chainage 65 m has the minimum factor of safety. In this section, this slope section at 65 m chainage has been analyzed under different pseudo-static loading conditions along with the gravitational forces using FEM-based Phase2 software (Fig. 6).

The pseudo-static analysis was carried out for four cases with different seismic loading conditions along with the gravitational load, for the stability analysis of the hill slope at 65 m chainage. The result of the analysis is provided in Table 4.

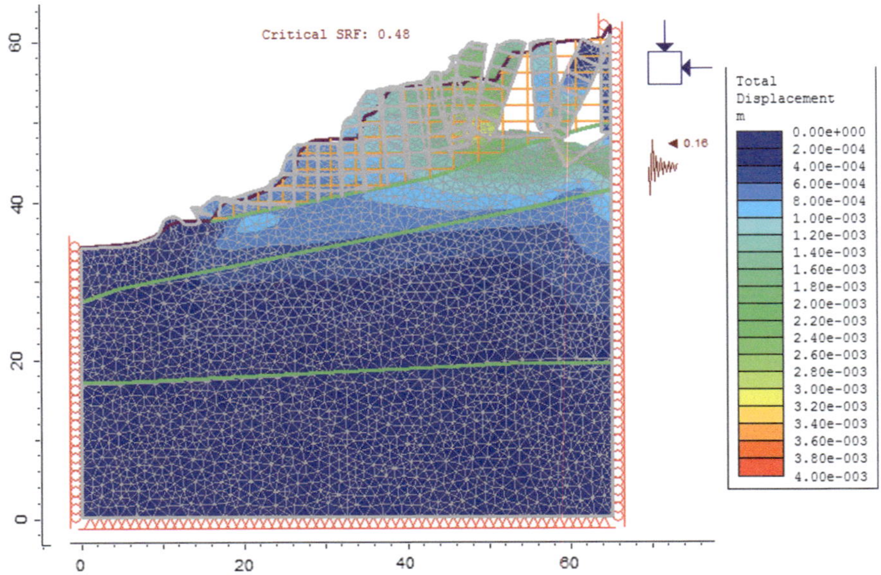

Fig. 6 Maximum deformed shear strain plot of 65 m chainage section with $K_h = 0.16$

Table 4 Factor of safety of the slope section with corresponding K_h value

Horizontal seismic coefficient (K_h)	Factor of safety
0 (static condition)	1.75
0.03	1.54
0.08	0.99
0.1	0.94
0.16	0.48

6 Stabilization of the Slopes

From the above analysis, it is clear that the slope is failing at $K_h = 0.16$ as prescribed by I.S.1893.1.2002 for seismic zone 3 areas. Hence, there was a need to provide optimum support system in order to stabilize the slope. The hill slope at 65 m chainage was therefore stabilized with shotcrete and fully bonded bolts so that the factor of safety can be reached greater than one in dynamic condition. Various trials had been done with different reinforcement to stabilize the slope at the corresponding loading condition.

The optimum support measures were provided using shotcrete ($f_{ck} = 35$ MPa and 100 mm thickness) and fully bonded bolts (32 mm Diameter, 10 m length with 3 × 3 m spacing) due to which the factor of safety of the most critical section(at 65 m chainage) was raised from 0.48 to 1.2 at pseudo-static and gravitational loading conditions. The same support was provided to all the sections and the corresponding FOS was 1.82, 1.61, 1.2, 2.5, and 2.44 for slope sections at chainage 15, 35, 65, 90, and 110 m, respectively.

7 Conclusion

The fieldwork was carried out at the hilly site of Cumballa Hill in Tardeo, Mumbai, to get the joint mapping and other site parameters. Four different types of borehole rock samples were collected from various already drilled boreholes drilled 22.5–50 m below the existing ground level (EGL).

The laboratory testing was carried out to get the physical and strength parameters of the rock material. The summary of geotechnical properties is given in Tables 1 and 2. Then the borehole data was studied, and the slope sections were prepared at chainage 15, 35, 65, 90, and 110 m. Since the joint pattern in the bottom three rock layers was not exposed, the rock mass property was estimated from the structure and surface condition of the rock mass using GSI Index chart.

From the numerical modeling done using RS2 software on all the prepared sections for gravitational loading in dry condition, it was found that the slopes are in stable condition. The slope section at chainage 65 m was found to have the lowest factor of safety = 1.75. This slope section was further analyzed under pseudo-static loading along with gravitational load. It was found that the slope comes in critical condition at horizontal seismic coefficient (K_h) = 0.08. Since the site is located at Mumbai which comes under seismic zone 3 area, hence the slope was analyzed for K_h = 0.16 as recommended by I.S.1893.1.2002. The slope was found to be failing with factor of safety = 0.48. Hence, there was a need of providing support measures for slope stabilization. The optimum support measures were provided using shotcrete (f_{ck} = 35 MPa and 100 mm thickness) and fully bonded bolts (32 mm diameter, 10 m length with 3 × 3 m spacing) due to which the factor of safety for the most critical section (at 65 m chainage) was raised from 0.48 to 1.2 at the dynamic loading conditions. The same support was provided to all the slope sections and all the corresponding factor of safety was found to be 1.82, 1.61, 1.2, 2.5, and 2.44 at chainage 15, 35, 65, 90, and 110 m, respectively.

References

Bhasin, R., & Kaynia, A. M. (2004). Static and dynamic simulation of a 700-m high rock slope in western Norway. *Engineering Geology, 71*, 213–226.

Chala, E. T., Seshagiri Rao, K., & Malik, A. (2016). Estimating engineering properties of weathered rocks from Deccan trap. In *Indorock 2016 Proceedings, 6th Rock conference,* IIT Bombay, June 17–18.

Hoek, E., Carranza-Torres, C., & Corkum, B. (2002). Hoek-Brown failure criterion-2002 edition. In *Proceedings of NARMS-Tac* (Vol. 1, pp. 267–273).

Hudson, J. A., & Feng, X. T. (2007). Updated flowcharts for rock mechanics modelling and rock engineering design. *International Journal of Rock Mechanics and Mining Sciences, 44*, 174–195.

Kainthola, A., Singh, P. K., & Singh, T. N. (2014). Stability investigation of road cut slope in basaltic rock mass, Mahabaleshwar, India. *Geoscience Frontiers, 6*(6), 837–845.

Marinos, P., & Hoek, E. (2000, November). GSI: A geologically friendly tool for rock mass strength estimation. In *ISRM International Symposium*.

Numerical Analysis of MSE Wall Using Finite Element and Limit Equilibrium Methods

S. S. Konnur, A. M. Hulagabali, C. H. Solanki and G. R. Dodagoudar

Abstract Mechanically stabilized earth (MSE) retaining walls are the most suitable design alternatives to the conventional retaining walls due to their simple, rapid, and cost-effective construction, reduced right-of-way acquisition, etc.; hence, the MSE walls are used in many central, state, and private sector projects. But the design and analysis is a challenging task for geotechnical engineers. This paper deals with the study of stability and wall movement of a existing MSE wall constructed on a major state highway in central Texas, using a finite element (FE) analysis and limit equilibrium (LE) slope stability analysis program GEO5 2016. The detailed analyses for both internal and external stabilities were obtained from the finite element and limit equilibrium analysis, with a critical failure surfaces and the wall movement of a MSE wall. The factors of safety obtained from both analyses were compared. The study shows that the factors of safety obtained from finite element and the limit equilibrium analysis, for a given problem, match in an acceptable range with a different critical failure surfaces. Also, this paper deals with the effect of backfill soil and reinforcements on stability and excessive movements of MSE wall.

Keywords MSE retaining wall · Reinforcements · Finite element (FE) analysis
Limit equilibrium (LE) analysis · Wall movement

S. S. Konnur (✉) · A. M. Hulagabali
Department of Civil Engineering, Basaveshwar Engineering College, Bagalkot, Karnataka, India
e-mail: konnursangamesh1991@gmail.com

A. M. Hulagabali
e-mail: anandmhulagabali@gmail.com

C. H. Solanki
Applied Mechanics Department, SVNIT Surat, Surat, Gujarat, India
e-mail: chs@amd.svnit.ac.in

G. R. Dodagoudar
Department of Civil Engineering, IIT Madras, Chennai, Tamilnadu, India
e-mail: goudar@iitm.ac.in

B. Adimoolam and S. Banerjee (eds.), *Soil Dynamics and Earthquake Geotechnical Engineering*, Lecture Notes in Civil Engineering 15,
https://doi.org/10.1007/978-981-13-0562-7_22

1 Introduction

Soil is an most widely used construction material, which is strong in compression and very weak in tension, and similarly, the steel is widely used as reinforcing material in almost of all civil engineering infrastructure projects which is very strong in tensile strength. Hence, the combination of both these soil and reinforcement will give very good engineering properties than properties of individual materials. Basically, reinforced soil has two components which are soil and reinforcement with different properties but basic concept is that the embedded reinforcement in soil provides tensile strength to the soil it is because of higher stiffness of the reinforcement. The basic mechanism of MSE is that while reinforcement is in the soil system the friction is formed between them and due to this friction soil movement is hold on the surface of reinforcement and then the shear stress is developed which produces tension in reinforcement this leads to confinement to the soil and results in decrease in soil deformation and increase in shear strength of the soil.

The complete design analysis of a mechanically stabilized earth (MSE) retaining wall includes evaluation of internal, external, and global stabilities as well as horizontal and vertical wall displacements (Abdelouhab et al. 2011). In these three parameters, the internal stability of MSE wall by means of pullout resistance and tensile strength of the reinforcement mainly depends on the type of reinforcement, spacing and length of the reinforcement, and the external stabilities by means of failure modes such as sliding, overturning, wall bearing, and global stability failure. Hence, these both internal and external failure modes are considered as slope failure through sliding surfaces, and the overall stability of an MSE wall can be evaluated using slope stability analysis (Payeur et al. 2015).

The two common and general methods of stability analysis are limit equilibrium and numerical methods in which limit equilibrium method considers the soil as a perfectly plastic and rigid material. The safety factor of limit equilibrium technique is calculated by Spencer method, Fellenius method, Morgenstern method, Janbu method, Bishop Method, etc. As we know that numerical modeling involves the calculation of stresses/displacements by method of discretization of continua and then assembling properties of each element and nodes by imposing boundary conditions to those elements and finally stresses/displacements are calculated by solving governing or system equations. But it is not possible to get exact safety factor by numerical method, but it can be determined by shear strength reduction technique (SSR).

In this paper, the stability analysis of an existed MSE wall project using both limit equilibrium and numerical method is presented. The slope stability analysis and stress/deformation analysis programs, GEO5 MSE (for LE) and GEO5 FEM (for numerical) are used in the present study. The calculated results from limit equilibrium method and numerical method are compared with each other.

2 Limit Equilibrium and Numerical Analysis Programs

GEO-5 fine is a powerfull program for solving geotechnical problems based on traditional and FE method. It is mainly designed to solve most geotechnical tasks, from the basic one to highly specialized programs. It works basically on the technique that initially the structure is designed analytically and then it is transferred analytical model into FEM program where the structure is verified by the FE method. GEO-5 fine contains two options, one for analysis internal and external stabilities, slope stability using a variety of popular limit equilibrium methods including the Bishops method, Spencer method, the Morgenstern-Price method, Janbu method and another one for analysis of stresses, displacements/deformations of wall. GEO-5 fine also provides explicit options for reinforcing elements including piers/piles, tiebacks (anchors), soil nails. In this study, the Spencer method is used in the analysis due its wide acceptance and the fact that it satisfies both force and moment equilibrium.

3 Mechanically Stabilized Earth Wall Project

The MSE wall project selected in this study was constructed on a major state highway in central Texas. The wall was required to support a new bridge abutment for an overpass on a new roadway passing over an existing roadway. The project contained numerous walls of varying heights. The wall analyzed in this study was the tallest wall with the steepest back slope and was the most critical case. A minimum FS of 1.3 was required in the design. The retained fill was placed simultaneously with the reinforcement (Abdelouhab et al. 2011).

The proposed geometry of the MSE wall and the soil profile are presented in Fig. 1. Soil properties used in the limit equilibrium and numerical analyses are listed in Table 1. The reinforcement consisted of galvanized metal ribbed straps

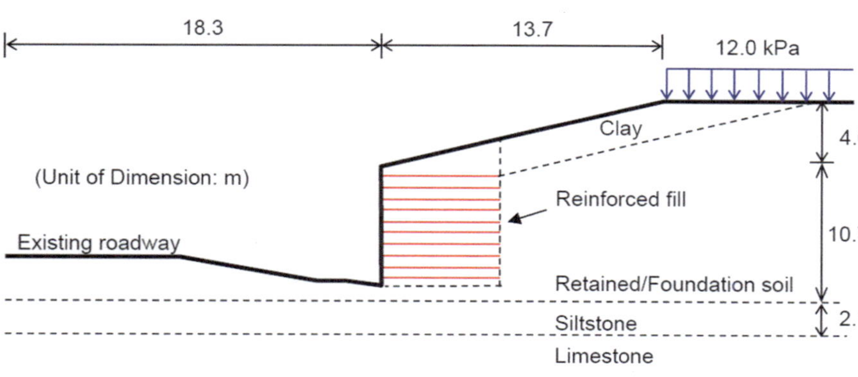

Fig. 1 Soil profile and slope geometry (Abdelouhab et al. 2011)

Table 1 Soil properties (Abdelouhab et al. 2011)

Soil description	Moist unit weight (kN/m^3)	Saturated unit weight (kN/m^3)	Cohesion (kPa)	Friction angle (°)
Siltstone	20.4	20.4	0	28
Limestone	21.7	21.7	56	30
Reinforced fill	19.6	20.1	0.05	34
Retained/ foundation soil	19.6	20.1	4.8	28
Clay	19.6	20.1	4.1	26

with a nominal width of 50 mm and a nominal thickness of 4 mm. The straps were 5.7 m in length with both horizontal and vertical spacing of 0.76 m. The allowable tendon strength of each metal strap was 89.6 kN. The allowable pullout strength of each metal strap was determined according to the FHWA design standard. A surcharge load of 12 kPa was considered at the back of the wall (as shown in Fig. 1) to account for the traffic load (Abdelouhab et al. 2011).

4 Stability Analysis

As mentioned earlier, the Spencer method was used in the limit equilibrium method in the GEO-5 MSE software. In the limit equilibrium analysis, trial failure surfaces were assumed to pass through the toe of the wall, since it is considered the most critical failure surface. A total of 232 circular failure surfaces were evaluated for the MSE wall, and the factor of safety for each failure surface was calculated. The most critical failure surface with a factor of safety of 1.5 is illustrated by the yellow line in Fig. 2 (Table 2).

The numerical analysis of the MSE wall was performed using the finite element program, GEO5 FEM. The numerical model was built with the same dimensions of that shown in Fig. 1. Horizontal displacement was fixed at the side boundaries, and the vertical displacement was fixed at the bottom boundary. In this study, the facing material of the wall was not physically modeled thus exclusion zone was set so that no failure surface passes through the facing of the wall.

Compared to the limit equilibrium method, the numerical analysis does not need a presumed failure surface. More information (stress, strain, displacement, etc.) can be extracted from a numerical model rather than a single factor of safety output. For example, the contour plot of shear strain increment is a good indicator of the critical failure surface of the MSE wall, as shown in Fig. 4. For comparison purposes, the critical failure surface determined by the GEO5 MSE software was also plotted. It is shown that in the numerical analysis the failure surface is similar to an active earth pressure failure case. The failure surface passed through the back toe of the reinforced zone, and the reinforced zone behaves as a coherent mass. The factor of

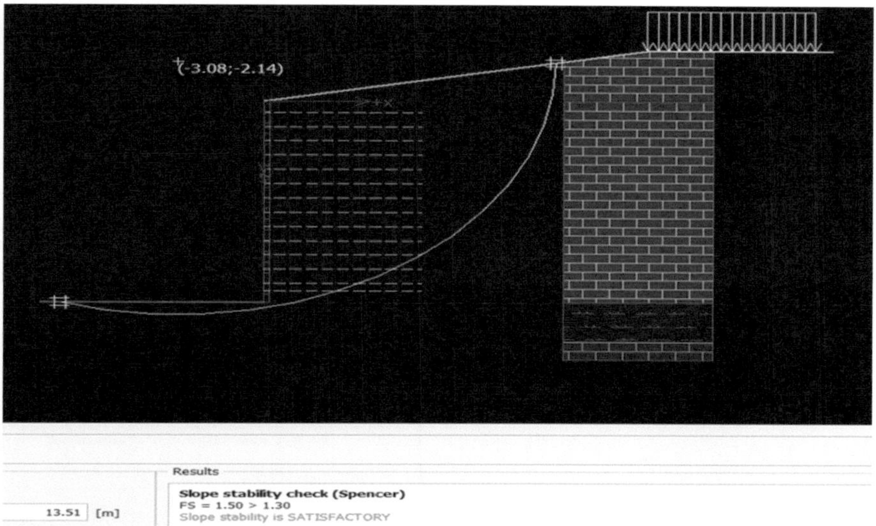

Fig. 2 Critical failure surfaces and the factor of safeties

Table 2 FOS for each trial critical failure surfaces

Critical failure surfaces	1	2	3	4	5	6	7	8	9	10
Factor of safety	1.5	1.5	9.6	1.5	7.2	1.5	1.5	1.5	1.5	1.5

safety determined by the shear strength reduction method is 1.56, which is close to that determined by the limit equilibrium method.

Figure 3 shows the discretized MSE wall with 3209 nodes and 1956 elements (1090 regions, 219 beams, 657 interfaces) in the numerical model and the Fig. 4 (displacement plot) indicated that entire reinforced zone is rotating around the front toe of wall, which confirmed that failure model of the wall is an active earth pressure failure. Figure 5 shows the comparison between the numerical solution and the limit equilibrium solutions, in which the factor of safety from numerical solution is 1.56 and factor of safety from limit equilibrium method is 1.5. The different predicted critical failure surfaces predicted from the two programs are due to the different underlying assumptions associated with the two different types of analyses.

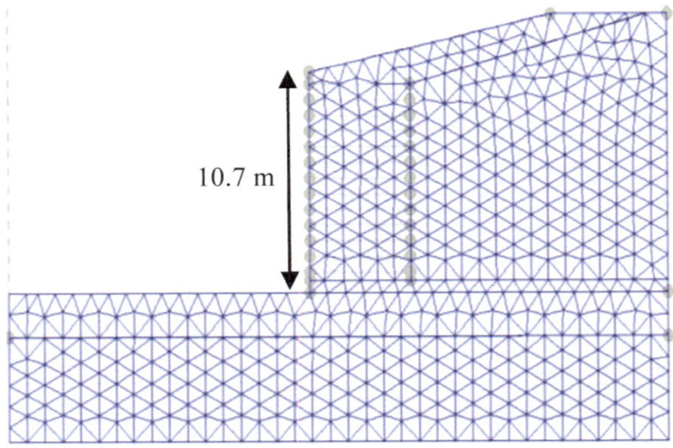

Fig. 3 Discretized MSE wall (Meshing)

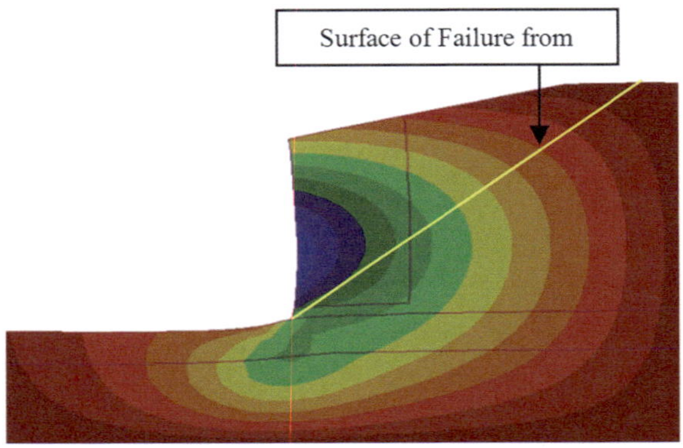

Fig. 4 Failure surface predicted by the numerical method

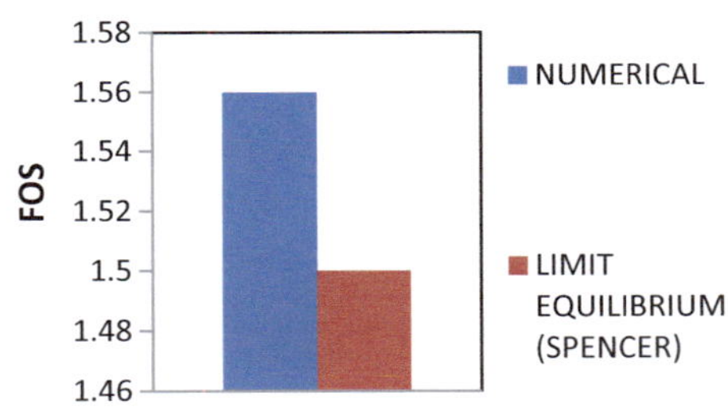

Fig. 5 FS of numerical and limit equilibrium (Spencer)

5 Conclusion

In this study, the stability of an MSE wall is analyzed using both limit equilibrium and numerical methods. It was found that the two analysis programs give different factor of safety for the retaining wall. This type of difference in factor of safety is due to the different underlying assumptions made with the two different analysis methods. From FEM and limit equilibrium methods, FEM gives little higher factor of safety than LE method. This is because of discretization of single structure into number of nodes, element, and regions. So it gives more convenient results than other methods. Analyzed values from both methods are in well acceptable ranges and are in good agreement. For the difficulties of many real-world projects, it is suggested that two or more different analysis methods should be considered in order to get a better analysis of the MSE retaining wall projects.

References

Abdelouhab, A., Dias, D., & Freitag, N. (2011). Numerical analysis of the behaviour of mechanically stabilized earth walls reinforced with different types of strips. *Geotextiles and Geomembranes, 29*, 116–129.

Payeur, J. B., Corfdir, A., & Bourgeois, E. (2015). Dynamic behaviour of a MSEW under harmonic loading: Experimental characterization and 3D FE model. *Computers and Geotechnics, 65*, 199–211.

Field and Numerical Investigation on Time-Dependent Behavior of Jute Geotextile (JGT) Reinforced Rural Road

Souvik Patra and Ashis Kumar Bera

Abstract A rural road section reinforced with Jute Geotextiles (JGT) sandwiched between thin sand layers has been constructed to study the strength improvement of subgrade soil. In this present investigation, an attempt also has been made to develop a three-dimensional (3D) finite element (FE) model of the JGT reinforced and unreinforced rural road sections using general purpose FE software ABAQUS. Numerical simulation of time-dependent behavior for JGT reinforced rural road has been carried out using the data obtained from field investigation at regular time interval. In this present investigation, subgrade soil has been modeled using Mohr–Coulomb plasticity model, whereas JGT has been assumed as membrane type of material. From FE analysis, pavement responses under traffic loading such as rut depth, maximum tensile strain developed at JGT, and vertical compressive strain developed at the top of subgrade soil has been evaluated. From the field investigation, it has been found that the subgrade California bearing ratio (CBR) value is increased with increase in time under study. It has been found that the reinforced rural road sections are performing well in terms of rut depth than the unreinforced section.

Keywords Rural road · Jute geotextile · 3D-FE analysis · ABAQUS
Subgrade

S. Patra (✉) · A. K. Bera
Department of Civil Engineering, Indian Institute of Engineering Science and Technology, Shibpur, Howrah 711103, India
e-mail: souvik4rmjgec@gmail.com

A. K. Bera
e-mail: ashis@civil.iiests.ac.in

B. Adimoolam and S. Banerjee (eds.), *Soil Dynamics and Earthquake Geotechnical Engineering*, Lecture Notes in Civil Engineering 15,
https://doi.org/10.1007/978-981-13-0562-7_23

1 Introduction

1.1 Background of Present Study

Construction of rural road subgrade with poor local soil is causing an excessive deformation to the subgrade soil and results in the failure of rural roads. In recent days, strengthening of the road subgrade with the help of natural fibers such as jute and coir is in practice. The problem of use of fabric made of natural fibers in civil engineering field is that loss of its tensile strength in a short time. From the previous literature (Ramaswamy and Aziz 1989; Basu et al. 2009), it has been found that the JGT may be efficiently used to strengthen the rural road subgrade. Field study on JGT reinforced road section by Rao (2003) indicated that the subgrade soil gain sufficient strength with time. In this present investigation, an attempt has been made to study the behavior of the JGT reinforced rural road section under traffic loading by using FE-based software "ABAQUS" based on field measured data.

1.2 Description of the Field Rural Road Section

In March 2013, a rural road section reinforced with JGT has been constructed from Kanksa to Bati (KB) at Murshidabad District in West Bengal, India. In this JGT reinforced rural road section, JGT has been placed in between granular subbase (GSB) and subgrade soil with the presence of sand layers on each side of it. Detail dimensions of the JGT reinforced rural road section have been shown in Fig. 1. A monitoring schedule has been followed at regular interval of six month for two years. Soil parameters of the subgrade such as CBR value, dry density, cohesion, and angle of internal friction have been determined for each monitoring. These field-measured parameters are used in FE analysis to define the conditions of the subgrade soil at initial condition (after laying of JGT), 12-month, 18-month, and at 24-month.

1.3 Plan for Numerical Analysis

In this present investigation, two series (series A and series B) of model have been considered. In the series A, unreinforced rural road sections have been considered with varying time. JGT reinforced rural road sections have been taken under series B. For both the series of models (series A and series B), the rural road subgrade CBR values have been taken from the field investigation. Based on the study

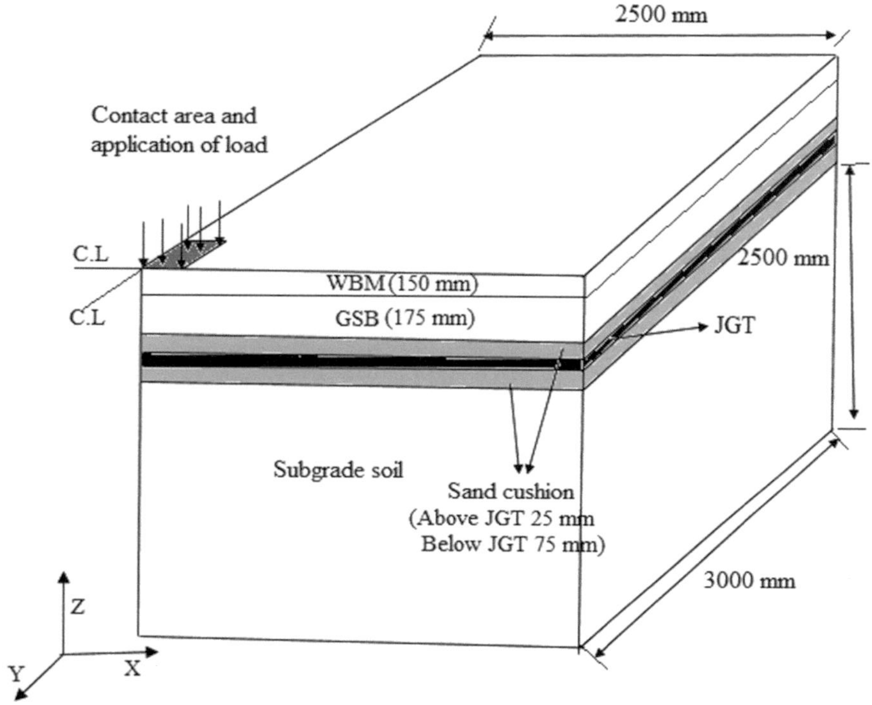

Fig. 1 Schematic diagram of reinforced section at initial condition

Table 1 Plan for numerical study

FE model series	Model name	Time from installation of JGT	Subgrade CBR value
Series A (unreinforced)	UR0	0	2.8
	UR12	12	3.8
	UR18	18	4.35
	UR24	24	5.57
Series B (reinforced)	R0	0	2.8
	R12	12	3.8
	R18	18	4.35
	R24	24	5.57

conducted by Bera and Roy (2012) and Saha et al. (2012), half-life of JGT has been considered as 12 month and full degradation time of JGT has been taken as 18 month. Details' plan and corresponding models' name for the FE analysis have been presented in Table 1.

2 FE Modeling Methodology

2.1 *Idealization of Geometry*

In this numerical study, geometry of the model has been idealized as a 3D model. Here only one-fourth of the model has been taken for computational purpose (Saad et al. 2006). On the basis of the previous literature (Gupta et al. 2014), the depth of the subgrade soil has been taken as 2.5 m in the present investigation.

2.2 *Material Constitutive Behavior*

In the present paper, subgrade soil has been considered as an elasto-plastic material. Plastic behavior of the subgrade soil has been defined by Mohr–Coulomb plasticity model, and parameters have been obtained from field investigation data. Resilient modulus of the subgrade soil and top granular layers (water-bound macadam, WBM and granular subbase, GSB) has been determined as per IRC 37:2012. In the present work, JGT has been considered as a membrane type material. In this investigation, JGT having a tensile strength of 20 kN has been used and elastic modulus of JGT has been determined at 2% strain from laboratory-wide width tensile strength test. It has been assumed that the elastic modulus of JGT at half-life will be half of the initial elastic modulus of 61.22 MPa. Poisson's ratio of JGT has been assumed as 0.25. It has been also assumed that the sand used in the present study is medium dense sand.

2.3 *Load and Boundary Condition*

In this work, traffic load has been simulated by triangular wave with a duration of 0.1 s based on the previous literature (Saad et al. 2006; Al-Khateeb et al. 2011). Peak wheel load has been assumed as 40 kN, and a tire contact pressure of 0.55 MPa has been applied on rectangular contact area. In the present paper, the length and breadth of the rectangular contact area have been calculated and the corresponding values are 406.5 and 177.6 mm, respectively. In this present, FE analysis bottom of the model has been considered as fixed support. Vertical walls adjacent to the loading area have been assumed as plane of symmetry along the direction of axis. Similar boundary condition has been used successfully by Perkins and Edens (2002). Horizontal displacements have been restricted, and only vertical displacements have been allowed on vertical faces away from loading area.

2.4 Meshing Criteria and Interaction Properties

Finest mesh has been created near the loaded area to capture the step stress and strain gradient more precisely in these areas. Discretization method used in this study has been shown in Fig. 2. In the present FE analysis, implicit direct integration method has been adopted. Here interaction between pavement layers has been defined by tie constraints. Interaction between JGT and sand layer has been simulated by coefficient of friction. In this study, JGT has been discretized by using M3D4R (a four-node quadrilateral membrane element with reduced integration). An eight-node linear brick element (C3D8R) has been used to mesh the other pavement layers.

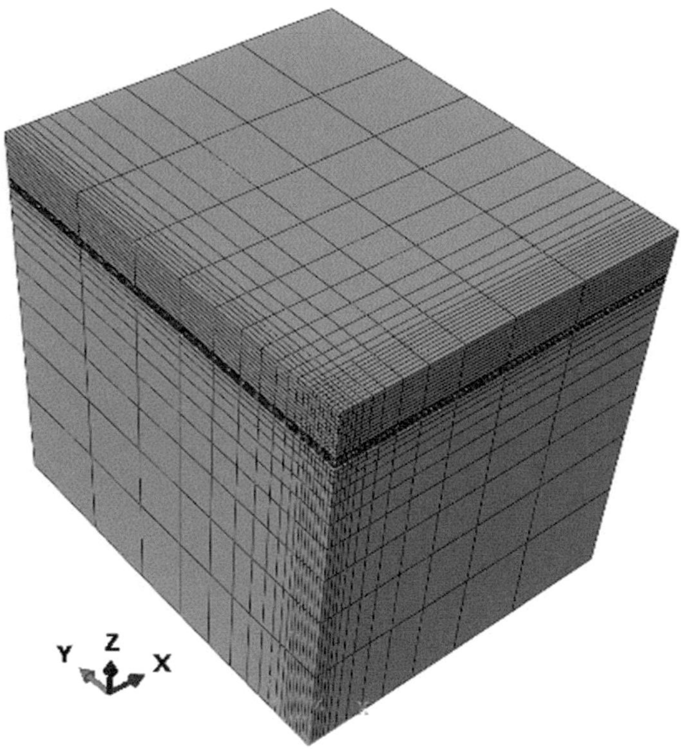

Fig. 2 Meshing technique adopted in the present investigation

3 Results and Discussion

3.1 Verification of the Model

The 3D finite element model has been qualitatively verified in two aspects: (i) vertical stress distribution and (ii) vertical compressive strain distribution along depth under the center line of the loading area. For this purpose, unreinforced rural road (UR0) has been idealized as a one-layer pavement system. Results obtained from ABAQUS analysis and vertical stress and strain computed from theoretical solution available in Ahlvin and Ulery (1962) have been presented in Fig. 3.

3.2 Study on Strength Increment of Subgrade Soil

Results obtained from regular field investigation indicate that the subgrade CBR value has been increased with time (Table 1). From Table 1, it has been seen that in 24 months the CBR value of subgrade soil increased about 1.98 times than the initial CBR value. Dry unit weight obtained from field monitoring with time is presented in Fig. 4. From Fig. 4, it has been also found that with time the values of

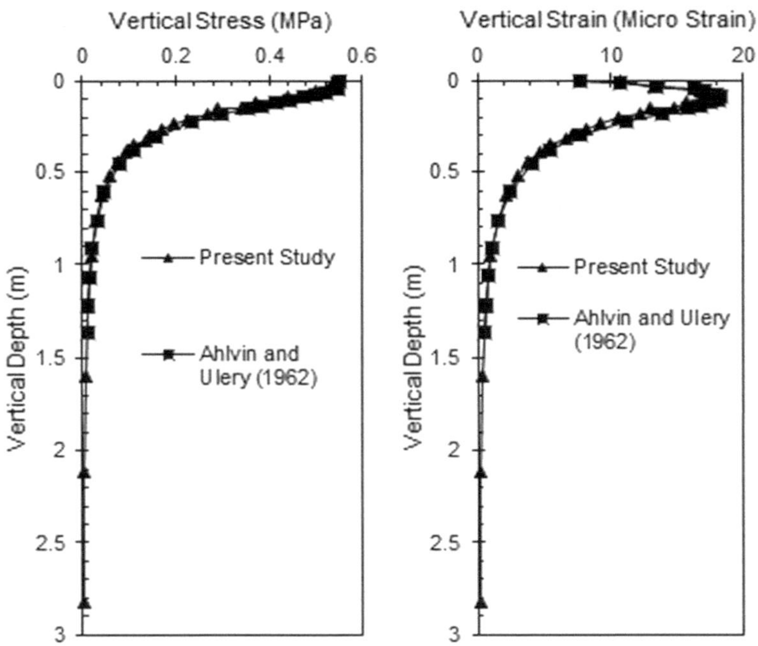

Fig. 3 Comparison between results obtain from FE analysis and theoretical solution

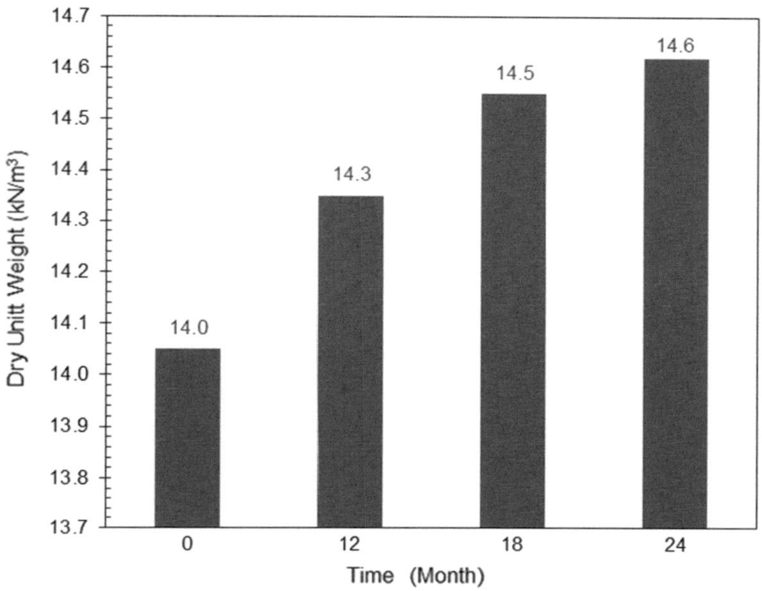

Fig. 4 Dry unit weight versus time curve for field JGT reinforced rural road section

dry unit weight of the subgrade soil is increases. Taskiran (2010) identified dry unit weight as the prime parameter influencing the CBR value of subgrade soil. Figure 4 also indicates that the dry unit weight has been increased about 1.04 time of its initial dry unit weight. Authors suggested that the increase in dry unit weight and subgrade CBR value may be happen due to consolidation process and compaction of the subgrade soil under traffic loading.

3.3 *Utilization of JGT Strength Under Traffic Loading*

In this present paper, reinforced section R0 and reinforced section R12 have been taken to study the utilization of JGT tensile strength under traffic loading. Maximum tensile stress and strain developed in JGT beneath the loading area has been taken from ABAQUS analysis. Utilization of JGT strength under traffic loading has been presented in Fig. 5. From Fig. 5, it has been seen that only a small fraction of tensile strength of JGT has been mobilized. In case of JGT, reinforced section at initial condition only 0.6% of tensile strength of JGT has been mobilized. It has been also observed that only 0.58% of tensile strength has been mobilized after 12 month of laying of JGT. Tensile strain has been developed around 0.1% in both the reinforced section R0 and reinforced section R12. Hufenus et al. (2006) reported that permanent strain in reinforcement under trafficking was usually below

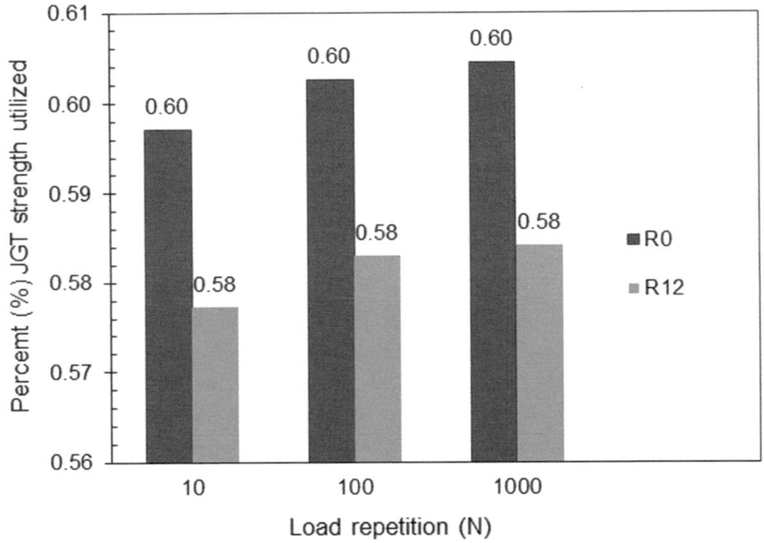

Fig. 5 Utilization of JGT strength under traffic loading

0.5% and exceed 1% only in extreme cases. Numerical study performed by Patra et al. (2016) also reported that the tensile mobilized in reinforcement is in the range 0.57–1.2% for reinforced rural road section. It may be suggested that as the subgrade soil gain strength with time which allows the JGT to serve its desired function at lower tensile strength.

3.4 Study on Usefulness of JGT on Subgrade Deformation Behavior of Rural Road Sections

Figure 6 represents the subgrade deflection taken at the top of the subgrade soil under the loading area for unreinforced and JGT reinforced rural road sections considered in the present study. It has been seen that the subgrade rut has been reduced in the range of 22–26% for the reinforced section R0 and reinforced section R12 with respect to the unreinforced section UR0 and unreinforced section UR12. In case of reinforced section R18 and reinforced section R24, after degradation of JGT subgrade rut depth has been reduced about 10–11% with respect to the unreinforced section UR18 and unreinforced section UR24. Based on the result of the present FE analysis, it has been clearly understood that the presence of JGT reduces subgrade deformation significantly under traffic loading and allows the rural road section to sustain on subgrade soil with low-CBR value due to membrane mechanism of the JGT. After degradation of JGT, subgrade soil already gains

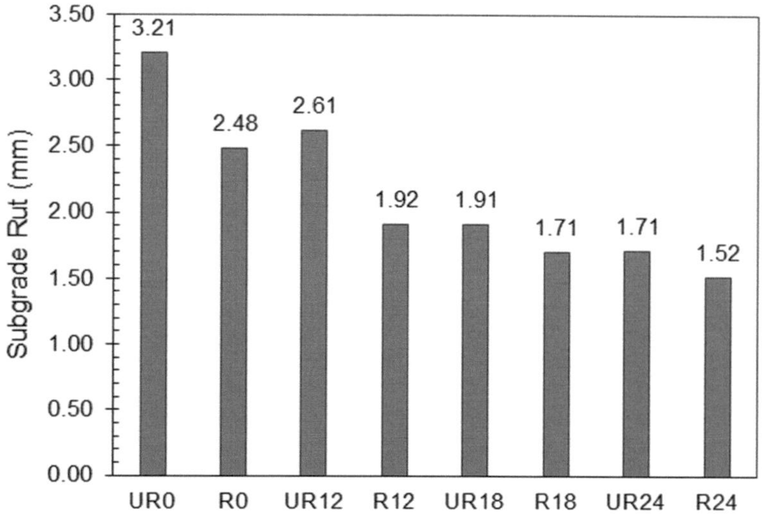

Fig. 6 Subgrade rut depth measured at top of subgrade soil beneath the loading area

sufficient strength to become independent of JGT. In this study, it has been also seen that after degradation of the fabric, although the unreinforced and reinforced rural road sections have same subgrade soil property, the subgrade deformation is reduced in reinforced section than the unreinforced section. Authors suggested that it is due to the presence of sand layers which allow the subgrade to face lower stress magnitude during traffic loading. In this present investigation, it has been observed that development of vertical compressive subgrade strain is reduced in the range of 16.96–35.73% in the JGT reinforced rural road section with respect to the unreinforced section.

4 Conclusions

Based on the present field investigation and FE analysis, following conclusions has been drawn.

- Reduction of subgrade rut depth of JGT reinforced rural road section than the unreinforced section is within the range of 22–26% and 10–11% before degradation of JGT and after degradation of JGT, respectively.
- In case of JGT reinforced rural road section, the vertical compressive subgrade strain has been reduced in the range of 16.96–35.73% with respect to the unreinforced section.

- Under traffic loading, only a small fraction of JGT strength around 0.58–0.6% has been mobilized.
- From the field investigation, the subgrade CBR value increases with increase in time under study.

Acknowledgements Authors express their sincere gratitude to National Jute Board (NJB), Kolkata, India, for their active support in the field investigation work. Authors also express their heartfelt thankfulness to Common Fund for Commodities (CFC), the Netherlands, for providing the facility of numerical analysis.

References

Ahlvin, R. G., & Ulery, H. H. (1962). Tabulated values for determining the complete pattern of stresses, strains and deflections beneath a uniform circular load on a homogeneous half space. *Highway Research Board Bulletin*, 342.

Al-Khateeb, L. A., Saoud, A., & Al-Msouti, M. F. (2011). Rutting prediction of flexible pavements using finite element modelling. *Jordan Journal of Civil Engineering, 5*, 173–190.

Basu, G., Roy, A. N., Bhattacharyya, K. S., & Ghosh, S. K. (2009). Construction of unpaved rural road using jute-synthetic blended woven geotextile—A case study. *Geotextiles and Geomembranes, 27,* 506–512. Technical Note, Elsevier.

Bera, A. K., & Roy, A. (2012). Engineering properties of jute geotextile and its efficacy to the consolidation of soil. *EJGE, 17,* 2631–2645.

Gupta, A., Kumar, P., & Rastogi, R. (2014, December). Performance evaluation and design criteria for low volume road. *Indian Highways*, 35–42.

Hufenus, R., Rueegger, R., Banjac, R., Mayor, P., Springman, S. M., & Bronnimann, R. (2006). Full scale tests on geosynthetic reinforced unpaved roads on soft subgrade. *Geotextiles and Geomembranes, 24,* 21–37.

Indian Road Congress 37 (2012). Giudelines for the design of flexible pavements (pp. 370–380). New Delhi, India.

Patra, S., Roy, S. K., Bera, A. K., Chakraborty, S., & Ghosh, A. (2016). Study on effectiveness of jute geotextiles reinforced rural road using ABAQUS. In *National Workshop on Jute Geotextiles* (pp. 17–18), Kolkata, India.

Perkins, S. W., & Edens, M. Q. (2002). Finite element and distress models for geosynthetic-reinforced pavements. *International Journal of Pavement Engineering, 3*(4), 239–250. Taylor & Francis.

Ramaswamy, S. D. & Aziz, M. A. (1989). Jute geotextiles for roads. In *International Workshops on Geotextile* (pp. 137–143), India.

Rao, A. S. (2003). Jute geotextile application in kakinada port area. In *Proceedings, National Seminar on Jute Geotextile & Innovative Jute Products.*

Saad, B., Mitri, H., & Poorooshasb, H. (2006). 3D FE analysis of flexible pavement with geosynthetic reinforcement. *Journal of Transportation Engineering* 402–415. ASCE.

Saha, P., Roy, D., Manna, S., Adhilkari, B., Sen, R., & Roy, S. (2012). Durability of transesterified jute geotextiles. *Geotextiles and Geomembranes, 35*, 69–75. Elsevier.

Taskiran, T. (2010). Prediction of California bearing ratio of fine grained soil AI methods. *Advances in Engineering Software, 41*, 886–892. Elsevier.

Parametric Study on the Behavior of Combined Pile Raft Foundation Founded on Multi-layered Soil Using PLAXIS 3D

Riya T. Johnson, Renjitha Mary Varghese and Jerin Joseph

Abstract The uses of pile raft foundations are common in recent years. The combined effect of pile and raft as foundation can lead to the reduction in total and differential settlements in problematic soil conditions. Many researchers have been done studies to predict the effect of combined pile raft foundations (CPRF) on the response of soil under heavy loads. But most of the numerical studies were done by considering soil as a single layer. The present study aims to find out the effect of pile raft in multi-layered soil using a three-dimensional finite element software PLAXIS 3D. The current parametric study includes the effect of diameter and spacing of piles with varying raft thickness on settlement. It is observed that raft foundation undergoes more settlement than pile raft foundation. The effect of raft thickness has a very little effect on reducing total and differential settlements. The effect of pile diameter and pile spacing was also given useful insight into the behavior of multi-layered soil.

Keywords Combined pile raft foundation · PLAXIS 3D analysis
Multi-layered soil

1 Introduction

The scarcity of land for the future construction has made the man to build high-rise buildings. These buildings will impart huge axial and lateral loads to the soils underneath their foundations. The load transfer mechanism from superstructure to

R. T. Johnson · R. M. Varghese · J. Joseph (✉)
Department of Civil Engineering, Albertian Institute of Science and Technology,
Kalamassery, Kochi, Kerala, India
e-mail: jerinjoseph@aisat.ac.in

R. T. Johnson
e-mail: riya.rosemary93@gmail.com

R. M. Varghese
e-mail: renjithamv@aisat.ac.in

B. Adimoolam and S. Banerjee (eds.), *Soil Dynamics and Earthquake Geotechnical Engineering*, Lecture Notes in Civil Engineering 15,
https://doi.org/10.1007/978-981-13-0562-7_24

substructure is usually done by providing suitable foundations. The foundation is the first element of the building where construction starts, but when it fails, it can cause many defects in the building including failure or collapse of the building. The repair of defect in the foundations is most difficult and very costly, so it is most important to design foundation to avoid such failure of building. There are different types of foundations to enhance load transfer mechanism. Raft foundations are one among them to carry the loads from columns and distribute the load in the whole base area of building. When allowable soil pressure is low, or the building loads are heavy, the use of spread footings would cover more than one-half of area and it may prove more economical to use raft foundation. The raft foundation can satisfy the bearing capacity requirement but fails to keep differential as well as total settlement below the maximum allowable limit. Another alternative to the huge load transfer is giving deep foundations. Pile foundation may be adopted instead of a raft foundation where no firm bearing strata exist in any reasonable depth and the loading is uneven. The coastal region of India is found mainly by soft soil, and pile foundations are adopted as the best solution to overcome the low bearing capacity of soft soils. The pile foundation is a deep foundation in which the loads are taken to deeper level. Sometimes the length of the pile foundations will be very large to achieve the load-bearing capacity. The challenges from raft or pile foundation alone can be avoided using the combination of both which is generally known as the combined pile raft foundation (CPRF) system. In piled rafts, the raft directly interacts with soil and it is supported by group of piles of various configurations. The recent projects proved that the pile raft foundation will give most economical length of piles with minimum differential settlement.

2 Review of Literature

Burland et al. (1977) studied the CPRF system and found that pile is the only load-bearing part. But later studies showed that the raft system also takes load. The load sharing by the pile and the raft was studied by many researchers (Shukla et al. 2013; Davids 2008; Randolph 1994) and made designs according to that. Recent studies carried out by Nirmal et al (2014) reported that CPRF is best to provide larger diameter piles to reduce the maximum and differential settlement. Few researches have done studies on the behavior of combined pile raft foundations in layered soils which is very common in the field. Gopinath et al (2010) presented that the increase in raft thickness decreases the overall settlement of pile raft foundation when it is found on layered soils. While it is mentioned that even though the raft thickness does not have any significant effect in layered soils, the raft thickness has a major contribution in reducing the differential settlement (Oh et al. 2008). So this particular paper tries to give more insight into the response of CPRF on layered soil using finite element method.

3 Soil Model

The general stratum of subsoil consists of two soil layers mainly a loose sand layer of 15 m deep overlaid by the next layer of stiff clay of thickness 15 m. The properties of soil layers taken for the model studies are given in Table 1. The groundwater table was at 3 m below ground surface. Extensive parametric studies were carried out by varying surface loads of intensities 200, 400, 600, and 800 kN/m^2 to simulate the loads coming from high-rise buildings.

4 Finite Element Modeling in PLAXIS 3D

The combined response of sand and clay layer was analyzed using finite element method. Three-dimensional analysis was conducted using the software PLAXIS 3D AE. The soil behavior of soil was captured by Mohr–Coulomb model which is an elastic perfectly plastic model. The piles were modeled as linear elastic models. The length and width of the soil model were taken as 50 m, and the thickness was given as 30 m. The boundary conditions were fixed in such way that the boundary effects to the model should be minimal. The sides of the model were allowed to move freely in vertical direction but restrained in both *X* and *Y* directions. The bottom was fixed in all direction, and top of the soil layer was free to displace according to the external loads. The finite element model was analyzed with total stress approach.

5 Validation of the Model

The model was validated for various spacing between the piles under 200 kPa surface load. The model includes five layers of soil. The spacings of pile were taken as 3*d*, 4*d*, 5*d,* and 6*d* with pile diameter as 1 and 0.8 m thick raft. The results obtained from the present studiy matches well with results reported by Oh et al. (2008). Figure 1 shows the validation of the model by plotting the variation of pile spacing with settlement under surface load of 200 kN/m^2.

Table 1 Properties of subsoil used for model studies

Properties	Loose sand	Stiff clay
Thickness (m)	15	15
Unit weight (kN/m^3)	15	16
Saturated unit weight (kN/m^3)	18	19
Undrained cohesion (kN/m^2)	0	80
Friction angle (°)	28°	0
Young's modulus (kN/m^3)	6000	20000
Poison's ratio	0.3	0.35

Fig. 1 Average settlement versus pile spacing

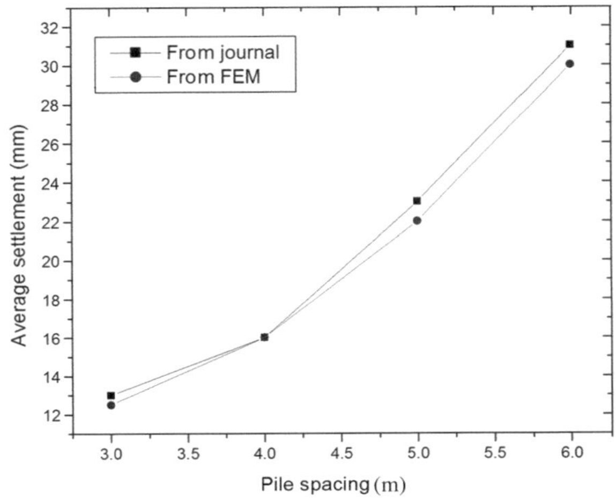

6 Parametric Studies

Parametric studies were planned to investigate the settlements of piled raft under the changes in parameters such as diameter of pile, spacing of piles, and raft thickness. The raft was modeled as concrete element, while piles are modeled as series of concrete embedded piles. Various case studies were taken into consideration to analyze the exact behavior of pile raft foundation system. The area of pile raft foundations and length of piles were fixed as 20 m × 20 m and 18 m, respectively. All studies were carried out considering the two soil strata. The details are listed below:

- Case 1: The effect of raft thickness was analyzed in Case 1 with a constant pile spacing of 4*d*, where diameter (*d*) of the pile was taken as 0.6 m. In this case, piled rafts with varying raft thickness as 0.3, 0.4, 0.6, 0.8, 1 m were modeled and found out the maximum and differential settlements under the surface loads ranging from 200 to 800 kN/m^2 with an increment of 200 kN/m^2.
- Case 2: Case 2 is mainly focused to find out the effect of pile spacing under different loading intensities. In Case 2, pile raft with raft thickness of 0.8 m was modeled by varying the pile spacing as 3*d*, 4*d*, 5*d*, 6*d*, 7*d*. The same surface load increments, and pile diameter was taken as that of Case 1.
- Case 3: Case 3 was included to study the influence of pile diameter on the response of CPRF systems. The other parameters were adopted as the same as that of Case 1.

7 Results and Discussions

The influence of various parameters such as the raft thickness, pile spacing, and pile diameter in a CPRF system is described in detail in the following sections.

7.1 *Effect of Raft Thickness*

The results obtained from Case 1 analysis are shown in Fig. 2 which shows the variation of maximum settlement with respect to raft thickness. There was a significant increase in both the settlements with increase in surface loads. The maximum settlement was found to be 20 cm under all surface loads except at the maximum surface load of 800 kN/m^2. It can be concluded that the maximum settlement decreases as the raft thickness increases in all loading cases (Fig. 2). On the other hand, the comparison of raft thickness on maximum settlement shows that there was little effect on the maximum settlement when raft thickness was varied from 0.3 to 1 m. In addition to that, the construction becomes uneconomic if raft thickness increases beyond 1 m. In view of these reasons, it can be summarized that the increasing the raft thickness does not have much significant influence in combined pile raft foundation and it is uneconomical after certain limits. The maximum and differential settlements were increased with increase in surface loads under multiple layers of soils (Figs. 2 and 3). Figure 3 also defines the variation differential settlement with raft thickness.

Fig. 2 Maximum settlement with varying raft thickness

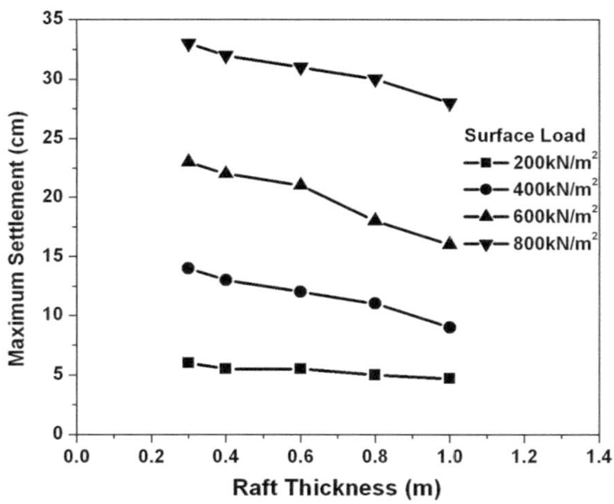

Fig. 3 Differential settlement with varying raft thickness

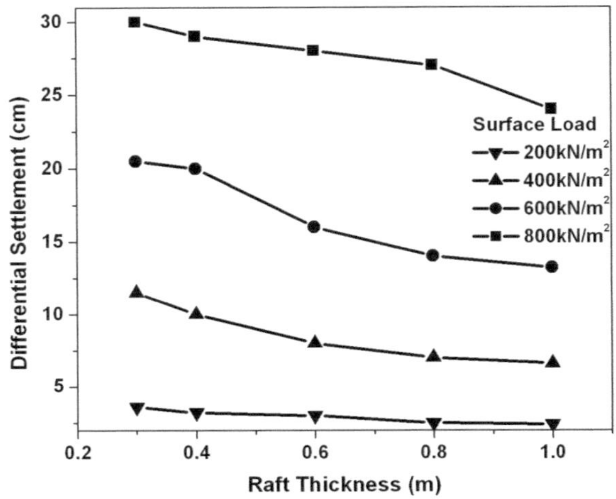

7.2 *Effect of Pile Spacing*

The variations in settlement due to different pile spacing are shown in Figs. 4 and 5. When the intensity of loading greater is than 200 kN/m^2, the increase in pile spacing has an adverse effect on the maximum and differential settlement under various loads.

The results showed that for 200 and 400 kN/m^2, both settlements were less than 10 cm, while the spacings between the piles were taken as 3*d* and 4*d*.

The differential settlements were found to be less than the maximum settlement in all cases. When the surface loads were increased from 200 to 800 kN/m^2, the

Fig. 4 Maximum settlement with varying pile spacing

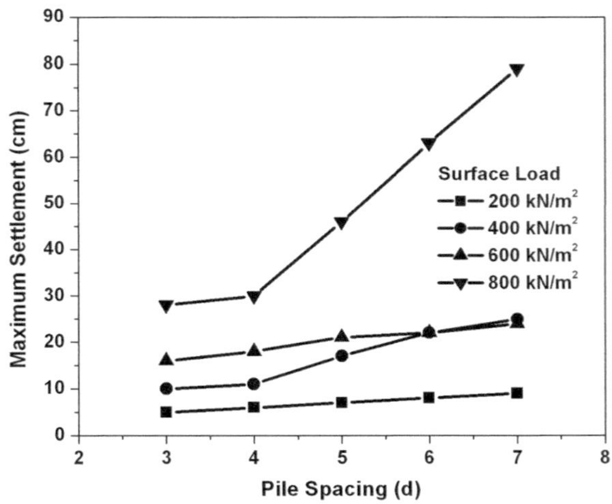

Fig. 5 Differential settlement with varying pile spacing

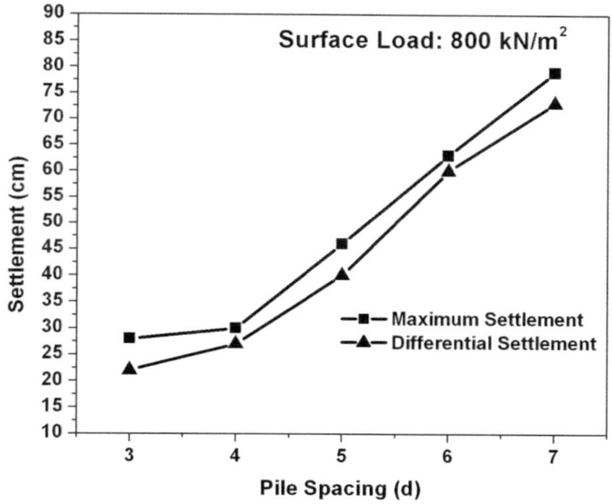

analysis with multi-layer soil strata has proved that the difference between the total settlement and differential settlement was decreased significantly (Fig. 5).

7.3 *Effect of Pile Diameter*

The diameter has a major effect on reducing the settlements. Figure 6 depicts the variations of settlement under various pile diameters and surface loads. It was found that both differential and total settlements of structure increase drastically at high values of vertical load intensity. The variation is found to be nonlinear (Fig. 7).

Fig. 6 Maximum settlement with varying pile diameter

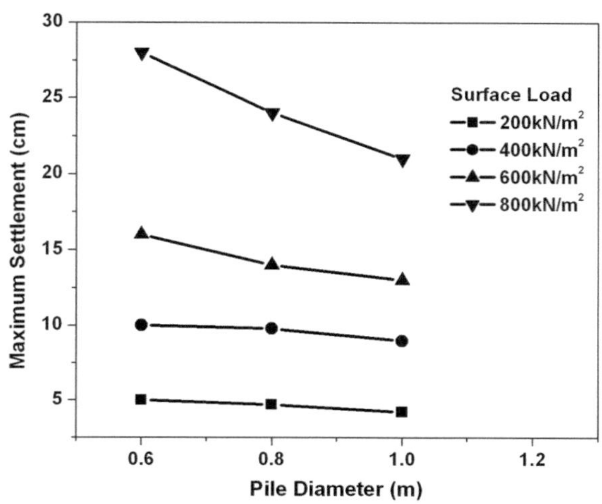

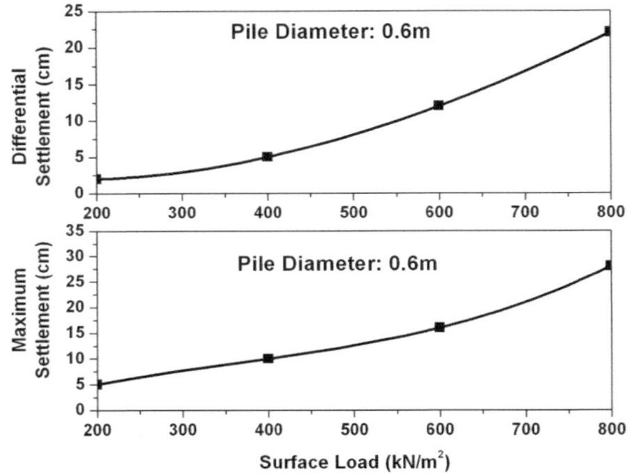

Fig. 7 Nonlinear variation of settlement with load

When pile diameter increased from 0.6 to 1 m, the reduction of both settlements was less than 20% for all vertical load intensities except 800 kN/m^2. The surface load of 800 kN/m^2 gave very high value of both maximum and differential settlements.

8 Conclusions

A set of studies were conducted on pile raft foundation in multi-layered soil to measure the soil response. The following conclusions can be drawn from the present studies.

- The raft thickness does not have significant effect on maximum and differential settlements.
- Pile spacing has important role in the performance of piled raft foundation. Both maximum settlement and differential settlement reduce with decrease in pile spacing.
- The pile diameter also plays an important role in reducing the settlements. Increase in pile diameter shows a decrease in maximum and differential settlements under all loading intensities.

References

Burland, J. B., Broms, B. B, & Demello, V. (1977). Behavior of foundation and structures. In *Proceedings of 9th ICSMFE*, Tokyo (pp. 495–546).

Davids, A. (2008). A postcard from Dubai design and construction of some of the tallest buildings in the world. In *Proceedings of the CTBUH 8 the World Congress*, 3–5 March, Dubai.

Gopinath, B., Junija, A., & Agarwal, A. (2010). Numerical modeling of piled raft foundation in soft clays. In *Indian Geotechnical Conference*, December 16–18, 2010, IIT Bombay.

Nirmal, J. J., & Hashifa, H. (2014). 'Study on settlement characteristics of combined pile raft foundation founded on sands with various arrangements of piles using Plaxis 3D. *International Journal of Emerging Technology and Advanced Engineering, 4*.

Oh, E. Y. N., Buli, Q. M., Surarak, R., & Balasubramaniam, B. (2008). Finite element modeling for piled raft foundation in sand. In *Eleventh East Asia-Pacific Conference on Structural Engineering & Construction*, November 19–21, Taiwan.

Randolph, M. F. (1994). *Design methods for pile groups and piled raft* (S.O.A. Report, 13 ICSMFE, New Delhi) Vol. 5, pp. 61–82.

Shukla, S. J., Desai, A. K., & Solanki, C. H. (2013). A dynamic behavioral study of 25 storey building with piled raft foundation with variable subsoil. *International Journal of Advance Research in Science and Engineering, 2*(1), 119–130.

Numerical Analysis of Seismic Response of a Piled Raft Foundation System

Ramon Varghese, A. Boominathan and Subhadeep Banerjee

Abstract Piled raft foundation is gaining popularity as a composite foundation system economizing foundation design for tall, heavy structures. The seismic response of a piled raft foundation would involve complex raft–soil–pile interaction as well as pile–soil–pile interaction, which cannot be estimated efficiently by existing methods available for pile groups. This paper discusses the development of a numerical model to study the seismic response of piled raft foundation system by substructure method-based seismic soil-structure interaction (SSSI) analysis using SASSI 2010 program. The flexible volume substructure method is used to analyze SSSI problem. Published data from a centrifuge test to study earthquake response of a piled raft system in soft clay was selected for the simulation. Results from the analysis are compared with experimental data.

Keywords Soil-structure interaction · Piled raft · Substructure method
Flexible volume method · SASSI

1 Introduction

The concept of a piled raft foundation that shares load between raft and pile is not new. Several papers published during the 1970s threw light onto the behavior and analysis of piled raft (PR) systems. Over the years, the use of piled raft foundation systems for buildings has gained popularity owing to an increase in the availability of case histories and guidelines. In regions where seismic activity is high, the design

R. Varghese (✉) · A. Boominathan · S. Banerjee
Department of Civil Engineering, Indian Institute of Technology Madras, Chennai 600 036, India
e-mail: ramonvarghese@gmail.com

A. Boominathan
e-mail: boomi@iitm.ac.in

S. Banerjee
e-mail: subhadeep@iitm.ac.in

B. Adimoolam and S. Banerjee (eds.), *Soil Dynamics and Earthquake Geotechnical Engineering*, Lecture Notes in Civil Engineering 15,
https://doi.org/10.1007/978-981-13-0562-7_25

of any foundation system has to be done keeping in mind the seismic response characteristics. Ignoring soil-structure interaction effects could potentially lead to unsafe design or even may cause failure. For composite foundation systems like the piled raft particularly founded in multi-layered soil system, dynamic analysis cannot be done by simple analytical techniques. However, a seismic response analysis requires the use of numerical tools like the finite element or finite difference methods for reliable results. This study aims to develop a 3D model to estimate the seismic response of a piled raft system considering both inertial and kinematic interactions.

2 Literature Review

Several methods of static analysis of PR foundation systems have been published in the literature. Soil-structure interaction becomes significant for heavy structures, which may significantly interact with the ground. Hence, the practice of using ground surface acceleration for seismic analysis cannot be deemed safe. Research on horizontal/seismic loads on piled rafts is limited. In comparison to pile groups, piled rafts usually have lesser number piles. This can result in higher loads on piles, under seismic loading compared to a pile group. To better understand the performance of piled rafts under seismic loading, several researchers including Yuksekol et al. (2015) have performed 1-g shake table tests. They observed an increase in horizontal load resistance and the influence of vertical displacement due to horizontal loads on the vertical resistance of piles.

Centrifuge studies to evaluate seismic response of piled rafts were carried out by Sawada and Takemura (2014) and Horikoshi et al. (2002). They observed settlement of the piled raft during horizontal loading and shaking table tests, as well as reduction in raft resistance to horizontal load compared to a raft alone. Banerjee (2009) investigated the response of piled raft in soft clay subjected to seismic excitation. The piles used were of short nature, and the finding suggests that surrounding soil imposed an inertia load to the piled raft structure, thereby lengthening its resonance period. The 3D numerical simulation performed by the author could replicate the centrifuge model reasonably well.

Seismic response of a single degree of freedom superstructure on an idealized piled raft foundation was studied by Saha et al. (2015) using a simplified numerical model involving Winkler foundation concept. Mayoral et al. (2009) compared actual seismic response of an urban bridge support system on soft clay with computed results using SASSI 2000 program by flexible volume method.

A review of the literature suggests that there is limited understanding of the seismic behavior of piled raft foundation systems. Given the capabilities of modern computational facilities, it is feasible to study the seismic response of piled raft systems using numerical simulations. The present study was aimed at developing a numerical model for a substructure-based seismic analysis of a piled raft foundation system using SASSI 2010.

3 Details of Piled Raft System

Experimental data from the centrifuge study on seismic response of a piled raft foundation in system in soft clay by Banerjee (2009) was chosen for the numerical simulation. The PR foundation considered for modeling in the present study has a raft dimension of 12.5 m × 7.5m × 0.5m, and four piles with diameter 0.9 m and length of 13 m. These prototype dimensions were modeled in the centrifuge test using a rigid steel plate of 25 cm × 15 cm × 1 cm and the four solid steel piles of 1.8 cm diameter and 26 cm long with the test being performed at 50 g. The vertical load applied on the raft in prototype scale would be 605 tonnes.

4 Flexible Volume Substructure Method

The flexible volume substructuring method is based on the concept of partitioning the total soil structure system into three substructures (SASSI 2010) as presented in Fig. 1. The first substructure consists of free-field site, second substructure consists of excavated soil volume, and third substructure consists of structure, of which the foundation represents excavated soil volume.

The three substructures together form the soil-structure interaction system. The flexible volume method considers interaction between free-field site and excavated soil volume at the boundary as well as within the excavated soil volume.

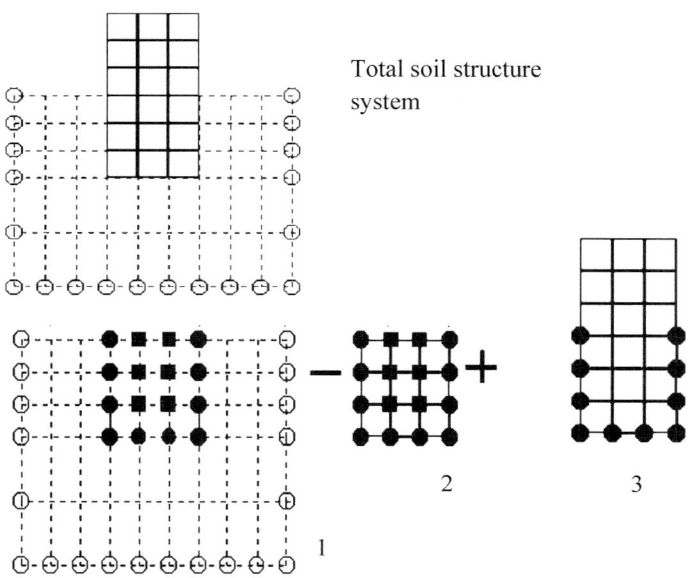

Fig. 1 Substructuring in the flexible volume method (Ostadan and Deng 2010)

The substructuring in the SASSI code is formulated using the complex response method and finite element technique. Free-field soil is modeled using horizontal layers of viscoelastic material, resting on a viscoelastic half-space. Both horizontal layering and viscoelastic material property are inherent limitations of the SASSI code. It should also be noted that analysis in frequency domain is several times faster than an analysis using the direct method.

5 Description of Numerical Model

5.1 *Foundation*

The raft is modeled using four-noded brick elements with stiffness property similar to that of steel used in the experiment. Material properties of structural elements used in the analysis for the PR system are presented in Table 1. The load of 605 tonnes was applied uniformly on the top surface of the raft. There are several techniques of modeling piles including the use of beam elements, volume elements, and inter-pile elements available in the SASSI finite element library. Beams could be used for modeling piles owing to their flexural characteristics; however, due to their inability to simulate actual soil resistance, the model is predicted to be less accurate. Nakaia et al. (2004) suggest that beam element modeling underestimates impedance functions and overestimates foundation input motions.

In the present study, piles in the piled raft system were modeled using brick elements. Flexural rigidity of the prototype pile dimension is 10,308,351 kN m^2 (Banerjee 2009). The modulus of elasticity of pile was then back-calculated using the flexural rigidity and moment of inertia of the pile section. Figure 2 shows the finite element mesh of the piled raft system with both raft piles modeled using four-noded brick elements. The total number of interacting nodes in the model was 1550.

5.2 *Soil Properties*

The modeling of soil layers using viscoelastic layers makes it a linear model, which cannot predict the cyclic changes in dynamic soil properties. In order to account for modulus reduction and damping increase, due to shear strain developed as the shear wave propagates, an equivalent linear ground response analysis is first performed

Table 1 Material properties used for structural elements

Component	Modulus of elasticity (GPa)	Poisson's ratio	Unit weight (kN/m^3)
Raft	200.0	0.25	78.5
Pile	320.1	0.25	78.5

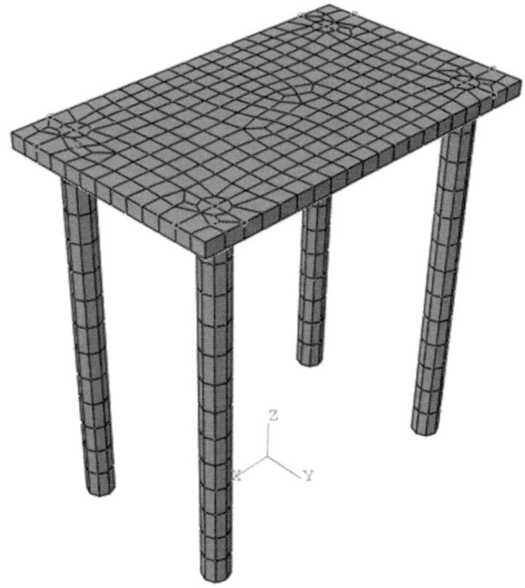

Fig. 2 Finite element mesh of the piled raft foundation

using the program SHAKE 2010. The soft kaolin clay used in the laminar shear box for the centrifuge test had a bulk unit weight of 16 kN/m^3, water content of 66%, liquid limit of 80%, and plastic limit of 35%. Initial modulus and damping values are taken from the results of dynamic tests on the kaolin clay as reported in Banerjee (2009). The variation of G_{max} with depth (or mean effective stress p') was calculated based on Eq. (1) suggested by Banerjee 2009. The resulting s-wave velocity, V_s, p-wave velocity, V_p, and damping ratio, ξ, for each layer are then used in the SASSI analysis.

$$G_{\text{max}} = 2060 p'^{0.653} \tag{1}$$

The dynamic soil properties obtained from ground response analysis for the earthquake motion are presented in Table 2. In short, non-linearity of soil is accounted for partially by using soil properties that that will result after a seismic event. A total of 15 soil layers were defined in addition to the half-space below the 15th layer. Viscous boundary developed by Lysmer and Kuhlemeyer (1969) is used in the SASSI program.

5.3 *Input Motion*

The earthquake motion used in the centrifuge test as reported in Banerjee (2009) was used in the present study. The time history of acceleration was synthetically developed using typical response spectra of Sumatran earthquakes measured at rock sites in Singapore. The input motion time history with peak ground acceleration

Table 2 Dynamic soil properties used in the analysis

Layer no.	Thickness (m)	V_s(m/s)	V_p(m/s)	ξ
1	0.5	45.71	76.56	0.001
2	1	64.53	108.08	0.006
3	1	75.30	126.12	0.009
4	1	83.53	139.90	0.011
5	1	90.19	151.06	0.012
6	1	95.83	160.50	0.013
7	1	100.84	168.89	0.013
8	1	105.40	176.53	0.013
9	1	109.60	183.56	0.014
10	1	113.51	190.11	0.014
11	1	117.19	196.28	0.014
12	1	120.68	202.12	0.014
13	1	124.00	207.68	0.014
14	1	127.17	212.99	0.014
15	0.5	124.94	201.86	0.056

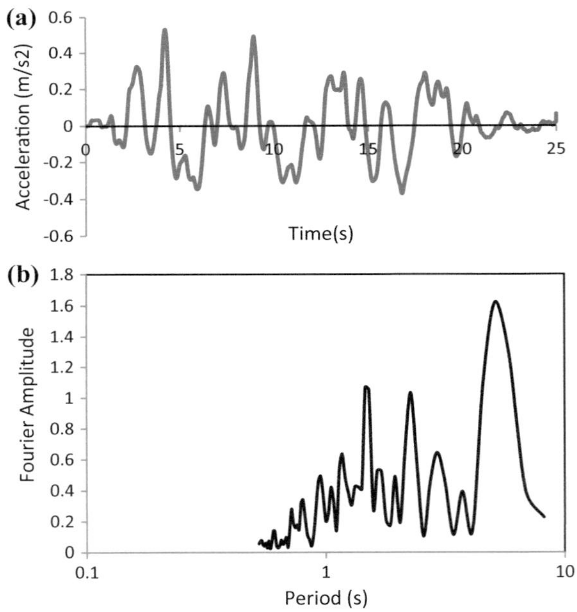

Fig. 3 **a** Time history of acceleration and **b** Fourier amplitude spectrum of input motion

(PGA) of 0.053 g is presented in Fig. 3a along with its Fourier spectrum in Fig. 3b. The earthquake has typical characteristics of far-field events including long periods and long durations.

6 Results and Discussion

The results extracted from the program after analysis presented in this paper include time history acceleration and response spectrum at the top of raft. The time history of acceleration is obtained at the top of raft along with time history obtained from the centrifuge test and presented in Fig. 4. The peak acceleration value recorded at the top of raft in the centrifuge test was 0.69 m/s^2, while the simulation produced a peak acceleration of 0.57 m/s^2.

Fourier spectra of acceleration recorded at the top of raft from both experiment and the simulation are presented in Fig. 5. The simulation has been able to reproduce most of the predominant frequencies, however, with some variation in Fourier amplitude. It is observed from the Fourier spectra that in the simulation the amplitude is de-amplified by about 10–20% at the range of period from 2 to 3 s, whereas at a period of around 5 s the amplification is about 23%. Response spectra obtained at the top of raft from both experiment and simulation are presented in Fig. 6. For the simulation, the highest peak spectral acceleration is observed near a period of 1.46 s, which is close to the peak from centrifuge test.

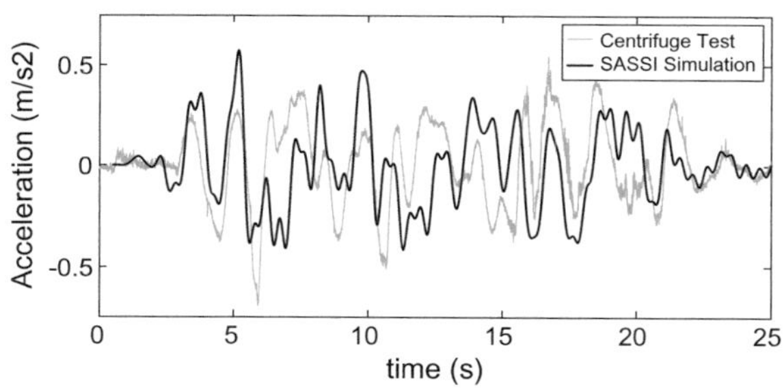

Fig. 4 Comparison of time histories of acceleration at top of raft for second earthquake

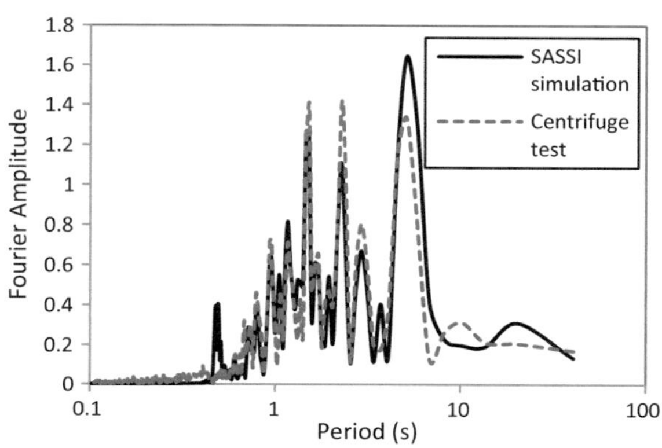

Fig. 5 Fourier spectra of acceleration at the top of raft

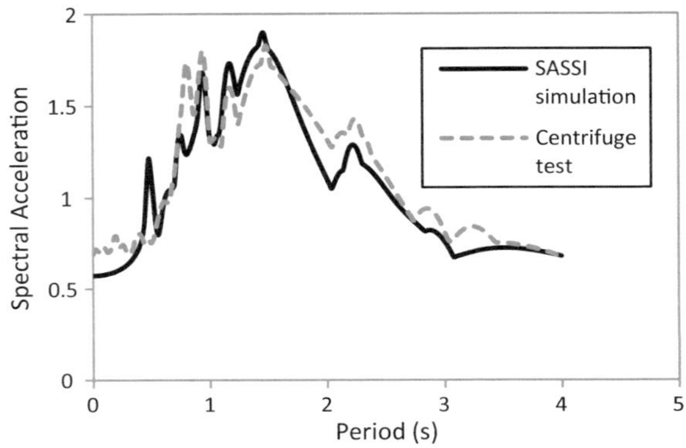

Fig. 6 Response spectra at the top of raft

7 Conclusion

A model to study seismic response of piled raft foundation is developed in the substructure-based soil-structure interaction program SASSI 2010. Published data from a dynamic centrifuge test on a piled raft system in soft clay subjected to an earthquake excitation was selected for the simulation. A pile raft foundation system with four piles in homogeneous soft clay was subjected to earthquake excitation, and its behavior was studied. Degraded soil properties from a one-dimensional ground response analysis were used in the SSSI analysis. The results in terms of time history of acceleration, Fourier spectrum, and response spectrum are compared with data from the centrifuge test. The simulation was found to be in reasonable agreement with measured data. The applicability of the model needs to be further verified by analysis using different input motions.

References

Banerjee, S. (2009). *Centrifuge and numerical modelling of soft clay-pile-raft foundations subjected to seismic shaking*. PhD dissertation, National University of Singapore.

Horikoshi, K., Matsumoto, T., Watanabe, T., & Fukuyama, H. (2002). Performance of piled raft foundations subjected to seismic loads. In *Foundation design codes and soil investigation in view of international harmonization and performance* (pp. 381–389).

Lysmer, J., & Kuhlemeyer, R. L. (1969). Finite element model for infinite media. *Journal of Engineering Mechanics Division, 95,* 859–877.

Mayoral, J. M., Alberto, Y., Mendoza, M. J., & Romo, M. P. (2009). Seismic response of an urban bridge-support system in soft clay. *Soil Dynamics and Earthquake Engineering, 29*(5), 925–938.

Nakaia, S., Katoa, H., Ishidaa, R., Manob, H., &Nagatac, M. (2004). Load bearing mechanism of piled raft foundation during earthquake. In *Proceedings of 3rd UJNR Workshop on Soil-Structure Interaction*, March 29–30.

Ostadan, F., & Deng, N. (2010). *SASSI 2010, User's Manual*. Berkeley: Geotechnical Engineering Division, Civil Engineering Department, University of California.

Saha, R., Dutta, S. C., & Haldar, S. (2015). Seismic response of soil-pile raft-structure system. *Journal of civil engineering and management, 21*(2), 144–164.

Sawada, K., & Takemura, J. (2014). Centrifuge model tests on piled raft foundation in sand subjected to lateral and moment loads. *Soils and Foundations, 54*(2), 126–140.

Yuksekol, Y. U., Matsumoto, T., & Shimono, S. (2015). Shaking table tests of piled raft and pile group foundations in dry sand. In *Proceedings of 6th International Conference on Earthquake Geotechnical Engineering*, 1–4 November, Christchurch, New Zealand. ICEGE.

Ultimate Pullout Capacity of Isolated Helical Anchor Using Finite Element Analysis

P. Ghosh and S. Samal

Abstract In this paper, the ultimate pullout capacity of single isolated helical anchor resting in homogeneous soil deposit with different helix configurations is determined using finite element analysis. The anchor is pulled to its ultimate failure controlling the displacement. Eight different types of anchor configuration are considered in the analysis, where mainly the number of helical plates, the depth of upper- and lowermost helical plates, and the ratio of spacing between the helical plates to the diameter of the plate are varied. The variation of load–displacement curve for each anchor configuration is obtained, and subsequently, the ultimate uplift capacity of each anchor is determined by using double-tangent method. The soil is assumed to follow Mohr–Coulomb failure criteria. The present theoretical observations are generally found in good agreement with those theoretical and experimental results available in the literature.

Keywords Finite element analysis · Helical anchor · Homogeneous soil
Plasticity · Pullout capacity

1 Introduction

The anchor is a foundation system generally designed and constructed to transmit any uplift force and overturning moments coming from a structure to the underlying soil. Anchors are important for many engineering applications such as transmission towers, suspension bridges, tall chimneys, high-rise structures which experience lateral load like wind load, buried pipelines under water, offshore structures as well as tunnel construction. As compared to the installation of conventional plate anchor, helical anchor is quite easy to install and is quite cost-effective, which may be used both in tensile and compressive loading condition. Basically, helical anchors are

P. Ghosh (✉) · S. Samal
Department of Civil Engineering, Indian Institute of Technology Kanpur, Kanpur 208016, India
e-mail: priyog@iitk.ac.in

B. Adimoolam and S. Banerjee (eds.), *Soil Dynamics and Earthquake Geotechnical Engineering*, Lecture Notes in Civil Engineering 15,
https://doi.org/10.1007/978-981-13-0562-7_26

geotechnical foundations consisting of central steel shaft and number of helical plates welded along the shaft. These helical plates are generally made up of steel and are formed with a definite pitch. The anchor shaft is used to transmit torque during installation and to transfer loads to the helical plates. A number of investigations have been performed by several researchers to predict the uplift resistance of the plate anchor with the help of different numerical as well as experimental investigations. However, the work on the helical anchor is still scarce. From the literature, it has been seen that few research works (Ghaly et al. 1991; Rao and Prasad 1993; Ghaly and Clemence 1998; Hanna et al. 2007; Lutenegger 2011; Mittal and Mukherjee 2013; Nazir et al. 2013; Wang et al. 2013; Demir and Ok 2015) have been carried out to determine the ultimate uplift capacity of helical anchor theoretically as well as experimentally. From the reported investigations, it can be clearly understood that the ultimate uplift capacity of helical anchor depends upon the number of helical plates, the depth of upper- and lowermost helical plates, the ratio of spacing between the helical plates to the diameter of the plate, and embedment depth.

In the present investigation, a numerical study on the pullout capacity of single isolated helical anchor embedded in homogeneous soil layer has been carried out using finite element method developed in the framework of three-dimensional failure domain in ABAQUS 6.13.

2 Definition of the Problem

A single isolated helical anchor with multiple helical plates of diameter, D, is embedded in a homogeneous soil layer with an embedment ratio, $\lambda = H/D$, where H is the embedment depth of the uppermost helical plate (Fig. 1). The helical anchor is made up of steel, and the surface is assumed to be perfectly rough. The soil is assumed to obey Mohr–Coulomb failure criteria. The objective is to determine the magnitude of uplift capacity of the helical anchor, where the anchor is pulled with an incremental velocity in the upward direction for calculating the uplift loading capacity. Full failure domain is considered for the analysis as the helical anchor is not truly an axi-symmetric member.

3 Material and Geometry

3.1 Material

To determine the uplift capacity of the helical anchor, a single-layer homogeneous soil deposit is considered. The properties of the soil deposit have been taken from

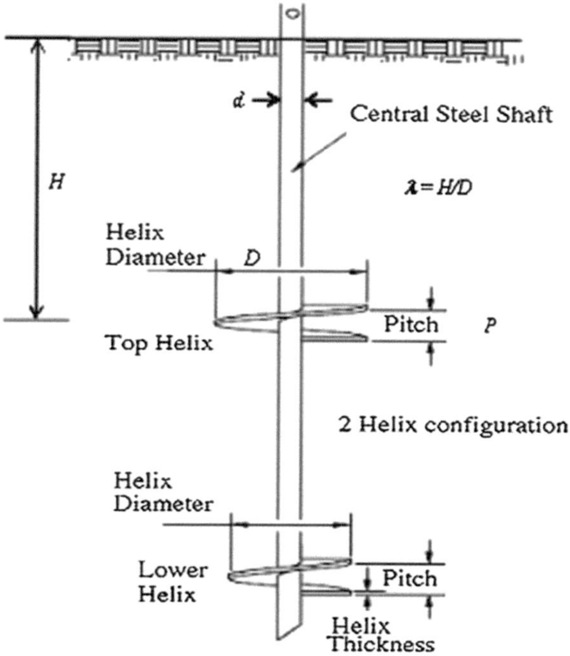

Fig. 1 Schematic diagram of helical anchor

Table 1 Properties of soil deposit (Wang et al. 2013)

Young's modulus (E) (kN/m^2)	Undrained cohesion (c_u) (kN/m^2)	Unit weight (γ) (kN/m^3)	Poisson's ratio (υ)	Angle of internal friction (ϕ°)
12.75×10^3	12.75	16	0.4	0

Table 2 Properties of steel anchor

Young's modulus (E) (kN/m^2)	Density (ρ) (kg/m^3)	Poisson's ratio (υ)
2.1×10^8	7800	0.25

Wang et al. (2013) and are mentioned in Table 1, whereas the properties of steel anchor used in the analysis are given in Table 2.

3.2 Anchor Configuration

Eight different anchor configurations have been considered for the analysis (Fig. 2) whose geometric specifications (Table 3) are taken from Wang et al. (2013). It can

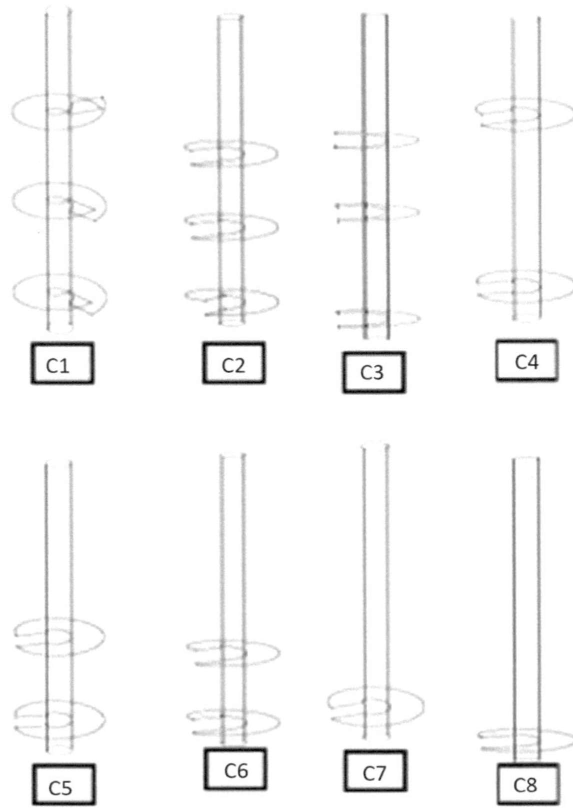

Fig. 2 Configuration of helical anchors

be seen from Fig. 2 that C1–C3 consist of three helical plates, C4–C6 consist of two helical plates, and C7–C8 comprise of single helical plate. For any particular anchor, the spacing between two plates (S_p) is kept constant except C3, where the spacing is unequal among the plates. The diameter of the helical plate and the central shaft is considered as 2.4 and 0.4 m, respectively, whereas the pitch of the helix (P) is taken as 0.46 m. The thickness of the helical plate is assumed as 0.1 m, which is considered to be negligible as compared to the diameter of the helical plate.

4 Analysis

The present finite element analysis has been performed using commercially available software ABAQUS 6.13. The numerical analysis has been carried out in 3D model. The whole soil domain is assumed to be at undrained condition. The soil

Table 3 Geometric details of different anchor configuration (Wang et al. 2013)

Configuration	Number of helical plates	Depth of uppermost plate, H (m)	Depth of lowermost plate (m)	Spacing between plates, S_p (m)	S_p/D
C1	3	2.65	9.85	3.6	1.5
C2	3	2.65	7.45	2.4	1.0
C3	3	2.65	8.65	2.4, 3.6	1, 1.5
C4	2	2.65	9.85	7.2	3
C5	2	6.25	9.85	3.6	1.5
C6	2	5.05	7.45	2.4	1
C7	1	6.25	6.25	–	–
C8	1	9.85	9.85	–	–

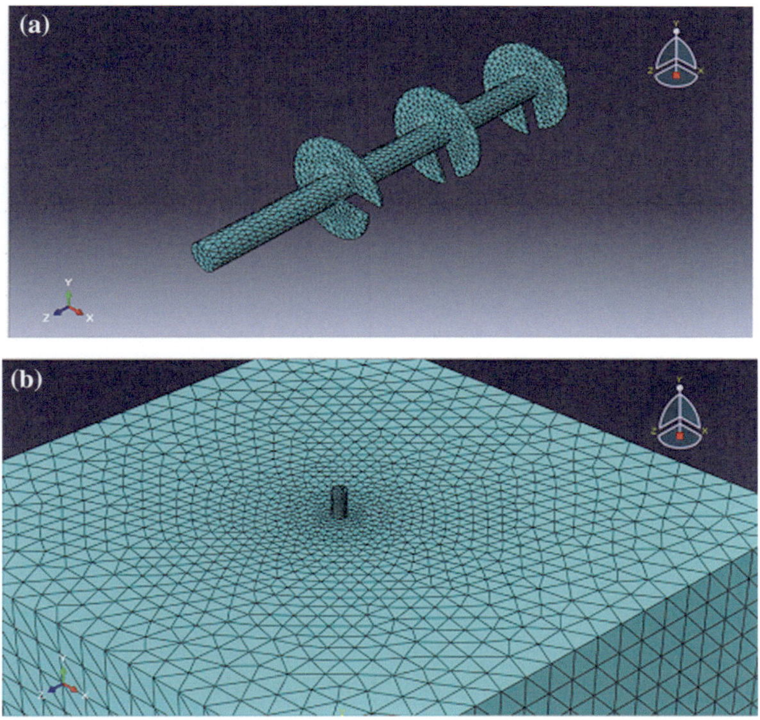

Fig. 3 Overview of **a** AutoCAD drawing and finite element meshing of C1 anchor, **b** finite element meshing of soil domain

deposit is assumed to follow Mohr–Coulomb failure criteria available in ABAQUS. Due to complicated geometry, the helical anchor is first designed in AutoCAD and then imported to ABAQUS. The finite element meshing has been done using four

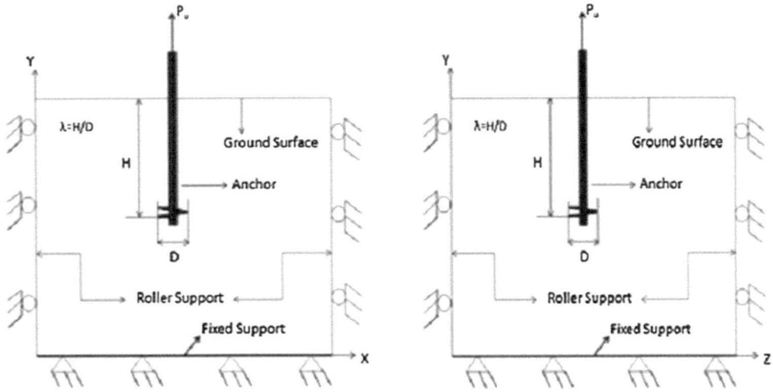

Fig. 4 Boundary conditions along XY and YZ plane

nodded tetrahedral elements. The AutoCAD drawing of helical anchor and the details of finite element meshing are shown in Fig. 3. In the analysis, general contact properties are used to simulate the interaction among all the surfaces. All the interacting surfaces are considered as penalty for tangential behavior and hard for normal behavior. The penalty behavior denotes that there is a finite coefficient of friction value between the surfaces and no slip occurs. In this analysis, the penalty value is taken as 0.1 as the soil is cohesive one (Table 1). The following boundary conditions are considered in the present analysis (Fig. 4).

- Displacements along *X*-axis, u_X are set to zero on the vertical boundaries parallel to YZ plane.
- Displacements along *Z*-axis, u_Z are set to zero on the vertical boundaries parallel to XY plane.
- All displacements are set to zero on the bottom boundary, i.e., $u_X = u_Y = u_Z = 0$.

Sensitivity analysis has been performed in order to determine the optimum domain size. For finding out the optimum domain size, anchor C1 and anchor C8 are chosen. The main reason behind choosing these two configurations is that C1 anchor configuration is having three helical plates with uniform spacing between the plates, and the lower plate is embedded at a depth of 9.85 m, which is the maximum depth considered in the present analysis, whereas C8 configuration is having single helical plate with embedment depth of 9.85 m. In this study, the depth of the soil deposit is fixed at 12.5*D*. It is found that beyond 7*D* in the horizontal direction, no significant change in the ultimate uplift capacity gets observed, and hence, the optimum domain size in the horizontal direction is selected as 7*D* from the center of the anchor.

5 Results and Discussion

In Figs. 5 and 6, the normalized load–displacement curves for isolated helical anchor placed in single-layer homogeneous soil deposit are shown for different configurations from C1 to C8, where P is the uplift capacity and is the vertical displacement of the anchor. The ultimate uplift capacity (P_u) of the anchor has been obtained by considering double-tangent method. It can be seen from Fig. 5a that for the same embedment depth of the lowermost helical plate, the anchor having three helical plates provides higher uplift capacity as compared to that with double and single helical plates. It can be observed from Figs. 5b and 6a that with the same number of helical plates, the uplift capacity increases with increase in S_p/D ratio. It can be also noticed from Fig. 6b that for the anchor with single helical plate the uplift capacity increases with increase in embedment depth ($\lambda = H/D$). The variation of ultimate uplift capacity for different anchor configurations is shown in Fig. 7. It can be seen that the magnitude of P_u increases with increase in S_p/D and the depth of lowermost helical plate.

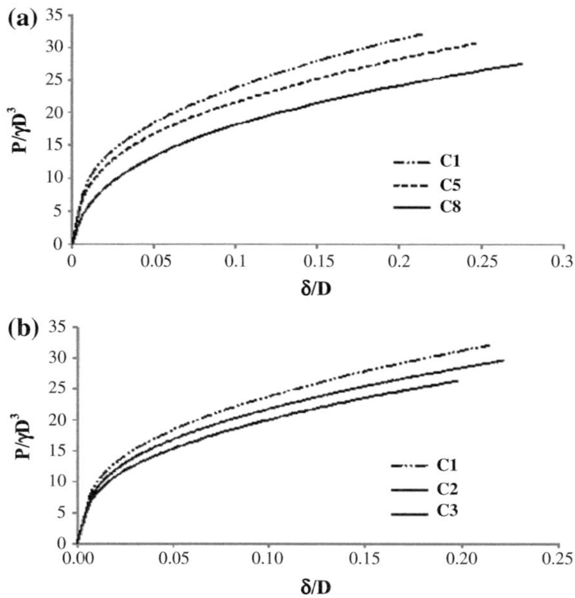

Fig. 5 Normalized load–displacement curve for isolated helical anchor with **a** same depth of lowermost helical plate, **b** three helical plates

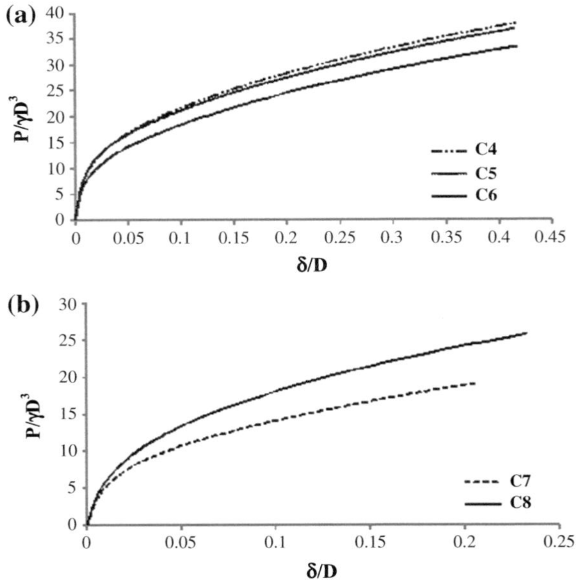

Fig. 6 Normalized load–displacement curve for isolated helical anchor with **a** two helical plates, **b** single helical plate

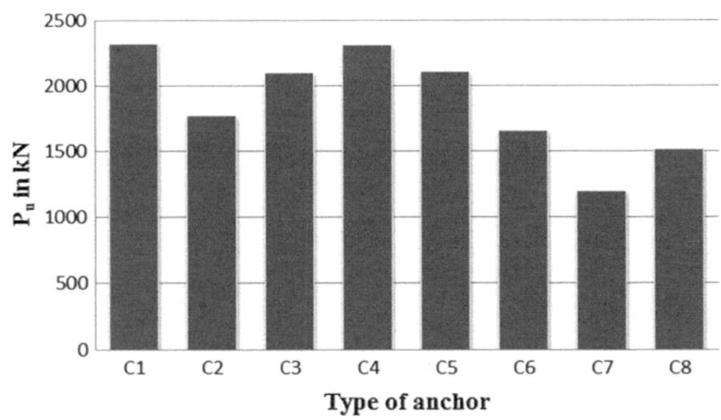

Fig. 7 Variation of P_u for different anchor configurations

6 Comparison

In Table 4, the magnitude of P_u for different anchor configurations obtained from the present finite element analysis is compared with the values reported by Wang et al. (2013). It can be seen that the present values compare reasonably well with the experimental and numerical results proposed by Wang et al. (2013).

Table 4 Comparison of P_u for different anchor configurations

Configuration	P_u (kN)		
	Present analysis	Wang et al. (2013)	
		Experimental analysis	Numerical analysis
C1	2320	1987	2381
C2	1775	1780	1755
C3	2100	2003	2068
C4	2310	1971	2343
C5	2110	1930	2166
C6	1660	1513	1666
C7	1200	1351	1171
C8	1520	1715	1504

7 Conclusions

In the present study, the ultimate uplift capacity of single isolated helical anchor is determined using finite element analysis. Eight different configurations of helical anchor are considered in the analysis. The uplift capacity of helical anchor is found to increase with increase in S_p/D ratio and λ. The magnitude of P_u is found to increase with increase in S_p/D and the depth of lowermost helical plate. The present results compare well with the results available in the literature.

References

Demir, A., & Ok, B. (2015). Uplift capacity of multi-plate helical anchors in cohesive soil. *Geomechanics and Engineering, 8*(4), 615–630.

Ghaly, A., Hanna, A., Ranjan, G., & Hanna, M. (1991). Helical anchors in dry and submerged sand subjected to surcharge. *Journal of Geotechnical Engineering, ASCE, 117*(10), 1463–1470.

Ghaly, A. M., & Clemence, S. P. (1998). Pullout performance of inclined helical screw anchors in sand. *Journal of Geotechnical and Geoenvironmental Engineering, ASCE, 124*(7), 617–627.

Hanna, A., Ayadat, T., & Sabry, M. (2007). Pullout resistance of single vertical shallow helical and plate anchors in sand. *Geotechnical and Geological Engineering, 25*(5), 559–573.

Lutenegger, A. J. (2011). Behavior of multi helix screw anchor in sand. In *Pan-Am CGS Geotechnical Conference*.

Mittal, S., & Mukherjee, S. (2013). Vertical uplift capacity of a group of helical screw anchors in sand. *Indian Geotechnical Journal, 43*(3), 238–250.

Nazir, R., Chuan, H. S., Niroumand, H., & Kassim, K. A. (2013). Performance of single vertical helical anchor embedded in dry sand. *Measurement, 49,* 42–51.

Rao, S. N., & Prasad, Y. V. S. N. (1993). Estimation of uplift capacity of helical anchors in clays. *Journal of Geotechnical Engineering, ASCE, 119*(2), 352–357.

Wang, D., Merifield, R. S., & Gaudin, C. (2013). Uplift behaviour of helical anchors in clay. *Canadian Geotechnical Journal, 50,* 575–584.

Part III
Geohazards: Analysis, Mitigation and Management

Landslide Risk Assessment and Mitigation—A Case Study

A. R. Amashi, A. M. Hulagabali, C. H. Solanki and G. R. Dodagoudar

Abstract Landslides are defined as down slope movement of rock, debris and earth under the influence of gravity. Slide is one of the most common types of hazards on slopes, which might lead to considerable casualties and economic loss. The study and analysis of slope is essential in understanding their performance and, in particular, their stability, reliability and deformations. The aim of the present study is to analyse the typical slope by varying different soil parameters like cohesion, angle of internal friction, water table and also slope geometry such as slope angle and slope height. The stability of the slope is analysed based on numerical simulation using three geotechnical software namely Slope/W, SV-Slope and Plaxis 2D. The study further is directed towards the assessment of stability of landslide along State Highway-72, near Mahabaleshwar, Maharashtra. The analysis of slope is carried out using Slope/W and Plaxis 2D. It was found through the analysis that the slope was unstable under the saturated condition. In an attempt of suggesting suitable remedial measures for the same, soil nailing is provided for the slope and simulation is carried out using Geo-5/Nailed slope software. As a result, the slope was found stable after the application of soil nailing.

Keywords Landslide · Numerical simulation · Factor of safety
Remediation

A. R. Amashi (✉) · A. M. Hulagabali
Department of Civil Engineering, Basaveshwar Engineering College,
Bagalkot, Karnataka, India
e-mail: ashwini.amashi@gmail.com

A. M. Hulagabali
e-mail: anandmhulagabali@gmail.com

C. H. Solanki
Applied Mechanics Department, SVNIT Surat, Surat, Gujarat, India
e-mail: chs@amd.svnit.ac.in

G. R. Dodagoudar
Department of Civil Engineering, IIT Madras, Chennai, Tamil Nadu, India
e-mail: goudar@iitm.ac.in

B. Adimoolam and S. Banerjee (eds.), *Soil Dynamics and Earthquake Geotechnical Engineering*, Lecture Notes in Civil Engineering 15,
https://doi.org/10.1007/978-981-13-0562-7_27

1 Landslide

1.1 Introduction

Landslides are a serious geologic hazard that involves all varieties of mass movements of hill slopes resulting from the failure of the materials such as rock, soil, debris, artificial fills driven by the force of gravity. These landslides not just cause an extensive damage to the property but also cause loss of lives. The triggering factors responsible for landslides are rainfall, rapid snowmelt, earthquakes and anthropogenic activities. A comprehensive evaluation of slope stability has an essential theoretical and practical significance. Analysis of slope stability has always been an important research content in the field of geotechnical engineering. It develops in an endless stream in the analysis method currently.

1.2 Methodology

Present study deals with analysis of slopes using Limit Equilibrium Method (LEM)-based software namely Geostudio (Slope/W), Geo-5 as well as Finite Element (FE)-based numerical tool Plaxis 2D. The study has been carried out in two parts. First part deals with analysis of a general homogeneous slope by varying different soil parameters and the second part includes the analysis of case study with slope failure. Suitable remedial measures are also recommended and demonstrated for the slope. The input data required to analyse the case study with slope failure is obtained from previously published paper that is mentioned in the later part. Analysis gives a single value of factor of safety (FOS) which is a ratio of available shear resistance to that required for equilibrium.

2 Parametric Study

2.1 Introduction

Based on the principles of soil mechanics, there are three general groups of parameters which determine the strength or instability of slope such as strength of a soil, geometry of a soil and pore water pressure which includes seepage forces. In this study, our approach is to keep some of the parameters constant and to investigate effects of remaining parameters such as slope geometry, cohesion, friction and variation in groundwater table on slope stability. In this manner, we can understand the effects of various parameters on the stability of slopes.

2.2 Simulation

Series of investigations are performed considering slope having simple homogeneous geometry as shown in Fig. 1. The parametric analysis has been carried out using Slope/W, SV-Slope and Plaxis 2D with phi-c reduction approach. To validate and obtain more accurate results, three software was used. The variations of FOS values with different soil parameters are shown in Figs. 2, 3, 4, 5 and 6.

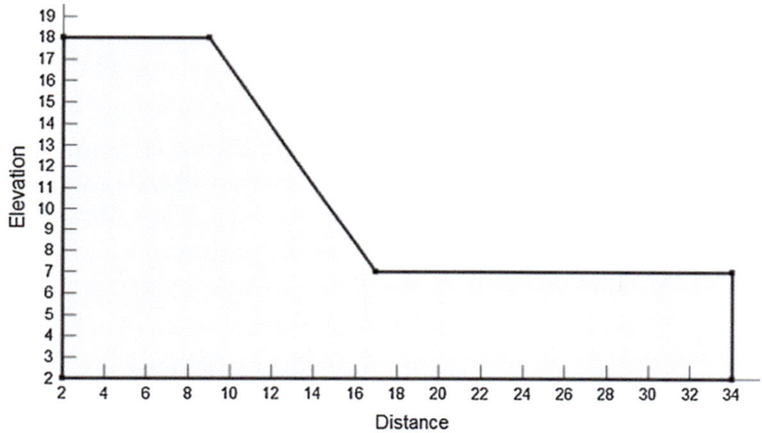

Fig. 1 Slope adopted for parametric analysis

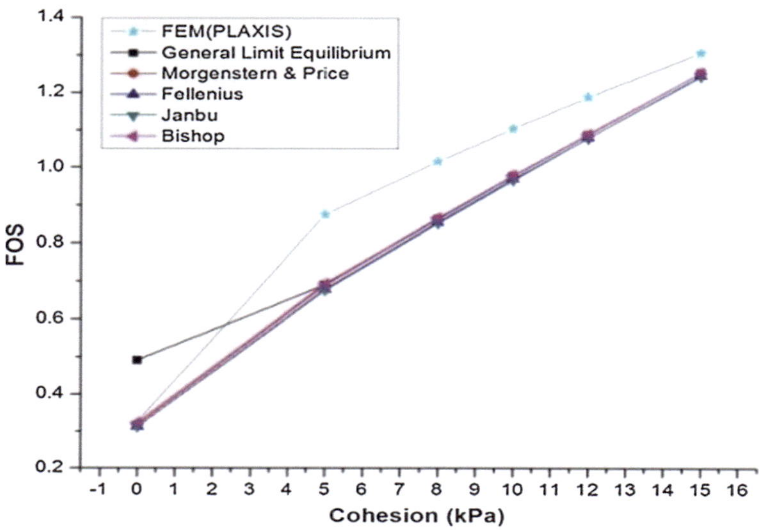

Fig. 2 Variation of cohesion with constant friction angle $\varphi = 25°$ using Slope/W and Plaxis 2D

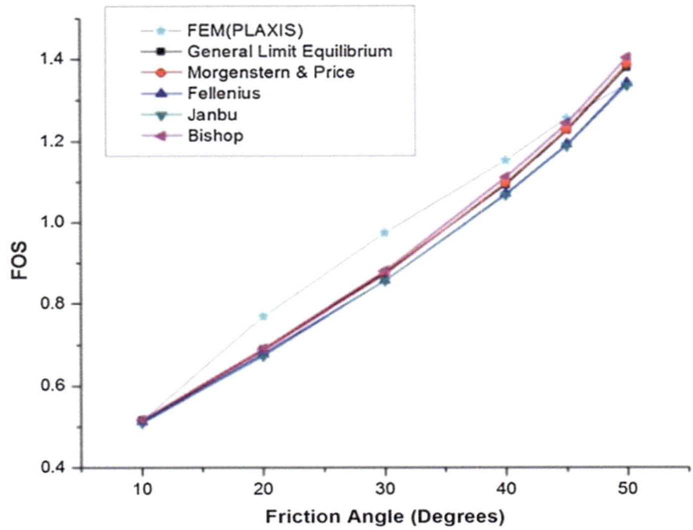

Fig. 3 Variation of friction angle with constant cohesion $c = 5$ kPa using Slope/W and Plaxis 2D

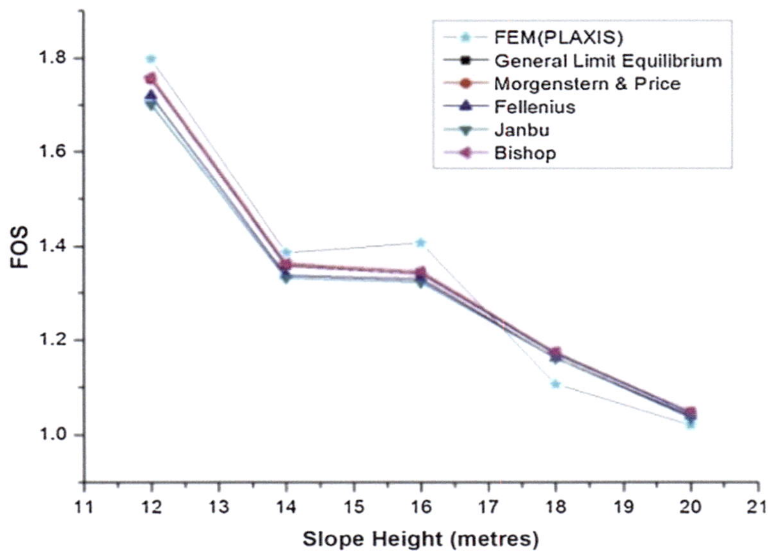

Fig. 4 Variation of slope height with $c = 10$ kPa, $\varphi = 30°$ using Slope/W and Plaxis 2D

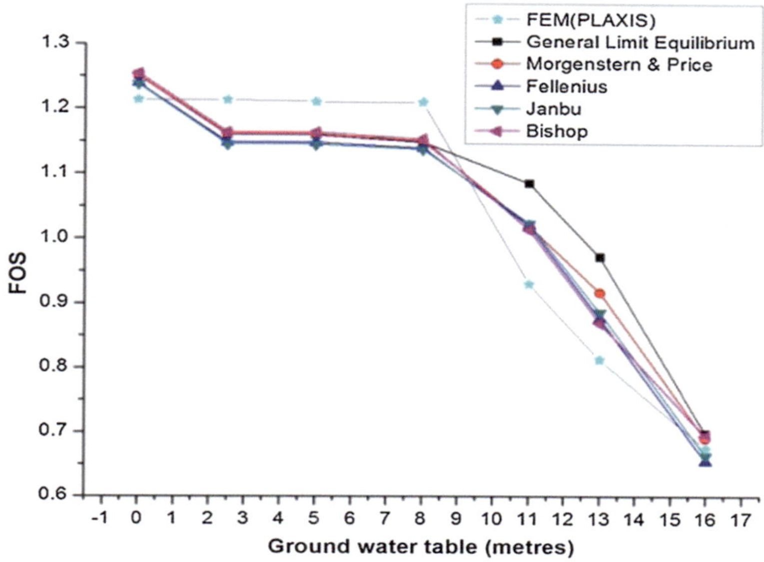

Fig. 5 Variation of groundwater table using Slope/W and Plaxis 2D

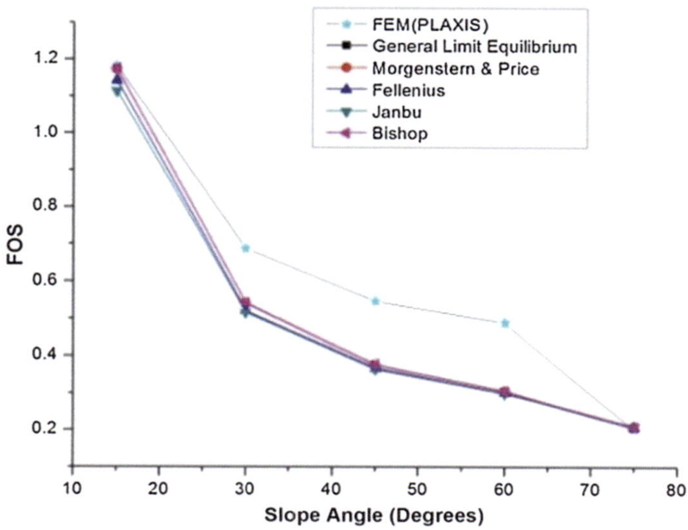

Fig. 6 Variation of slope angle with $c = 2$ kPa, $\varphi = 15°$ using Slope/W and Plaxis 2D

The parametric study shows that as the cohesion and internal friction increase, FOS increases but as water table, slope height and slope angle increase the FOS decreases. The FOS varies linearly as the cohesion increases but it is not the same with other parameters like slope height, slope angle, water table and friction angle. The values of FOS obtained from different LEMs such as Bishop, Fellinius, Janbu and Morgesteron price (M-P) methods showed almost same results with less than 10% variations, whereas General Limit Equilibrium (GLE) method tends to overestimate the FOS. The analysis was also carried out using SV-Slope, and the results were within acceptable limits.

3 Case Study SH72 Mahabaleshwar

3.1 Study Area

The study area is chosen from the previous research work by M. Ahmad et al. 2013, which was focused on State Highway (SH-72) between Poladpur and Mahabaleshwar which is a part of the Deccan traps. The SH-72 connects the famous hill station in Mahabaleshwar, located in Satara district of Maharashtra and lies at a distance of about 120 and 285 km from Pune and Mumbai, respectively. The study area falls between latitude 17° 52′ 30″–18° 00′ 00″ and longitude 73° 26′ 15″–73° 41′ 15″ (Ahmad et al. 2015). The area is prone to weathering action, formation of soils and presence of bole beds in between two consecutive basaltic flows; in addition to this, joints, sub-vertical-to-vertical slopes, heavy rainfalls, etc., lead to slope instability. The area falls under Seismic Zone III. The basalts found in this area are fine-grained, massive and jointed in nature (Kainthola et al. 2012).

3.2 Numerical Simulation

The data required to carry out the analysis of this area is obtained from the previous paper published by Ahmad et al. (2013). They carried out the slope stability analysis using Slide-6 numerical tool. The details are as follows; five different types of soil have been identified on the basis of physical nature and colouration. Through field survey, they have considered a general slope having maximum height as 15 m above the road and the average slope angle of 70°. To assess the impact of five different soils, 4.5 m soil bed on the top of the slope and about 3 m soil layer has been taken above the slope in their study (Ahmad et al. 2015). The five different geo-engineering soil properties are independently applied on the slope as shown in Fig. 7 and analysed using Geostudio and Plaxis 2D. The aim of this study is to analyse the stability of the slope having basaltic soil formations and the effect of these soil on slope stability (Tables 1 and 2).

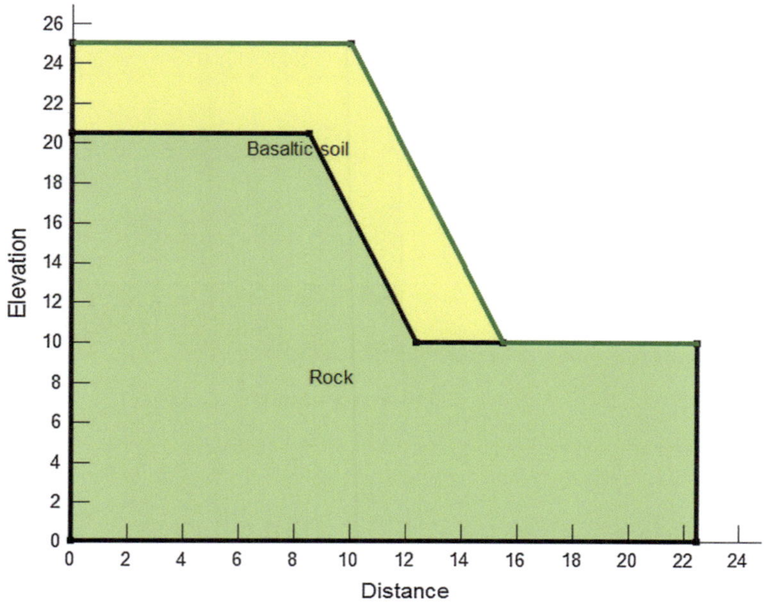

Fig. 7 Geometry of slope

Table 1 Properties of soil

Soil sample	Cohesion (kPa)		Friction angle (°)		Density (kN/m^3)	
	Dry	Wet	Dry	Wet	Dry	Wet
S1	70	35	23	18	18	19.5
S2	44	27	17	13	17.5	18.5
S3	62	83	18	17	19	20
S4	88	52	27	22	22	24
S5	60	31	19	16	19.5	22.5

Table 2 Rock soil sample details

Parameters	Dry	wet
Average density (kN/m^3)	27	28.5
Average cohesion (MPa)	11.6	8.4
Average friction angle (°)	37	30
Average Young's modulus (GPa)	40.1	36.3
Average Poison's ratio	0.24	0.22

3.2.1 Simulation Using Slope/W and Plaxis 2D

The results of slope stability analysis along SH-72 Mahabaleshwar by using Slope/W and Plaxis 2D under dry and saturated condition are given in Figs. 8 and 9.

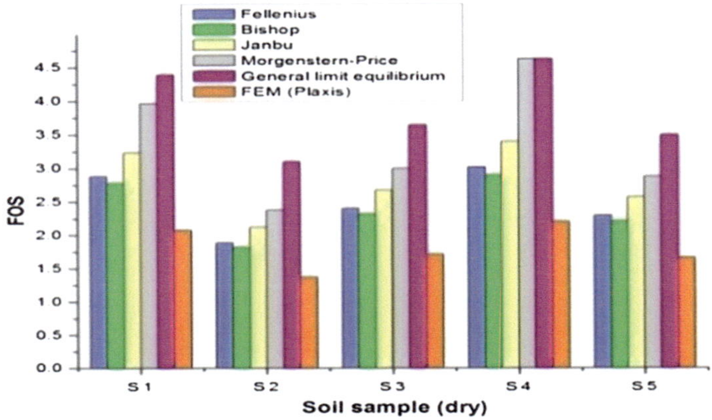

Fig. 8 Simulation using Slope/W and Plaxis 2D under dry condition

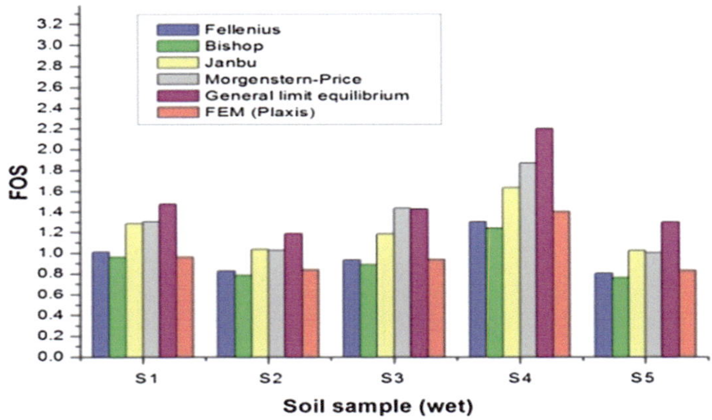

Fig. 9 Simulation using Slope/W and Plaxis 2D under wet condition

The results obtained through the analysis say that the LE methods such as Fellenius, Bishop and FE method showed FOS values with less variation whereas Janbu, GLE and M-P overestimated the FOS values. It is observed that the resultant FOS obtained from the stability analysis is more than 1 for all 5 soil samples under dry condition which implies the slope is stable. In wet condition, the FOS obtained is less than 1 for soil samples S2, S3, S5 which indicates that they are unstable and FOS values are nearly equal to 1 for sample S1 which says it is critically stable. The FOS obtained for soil sample S4 is greater than 1 and is found to be stable.

3.3 Remediation

The highway that is part of study area remains closed due to frequent landslides that occur in small and large slopes during monsoon caused due to heavy rainfall and vibrations produced by the traffic. In the present study, an attempt is made to demonstrate the application of remedial measure (soil nailing) for the mitigation of landslide using LE-based software tool Geo-5 (Nailing). Soil nailing is provided to stabilize the slope with different types of basaltic soils formed due to weathering on the surface of hills. The analysis is carried out for three field conditions, i.e., under dry condition, wet condition and simulation of earthquake (pseudo-static) in dry and wet condition. The horizontal and vertical acceleration coefficients considered for pseudo-static analysis are $k_h = 0.15$ and $k_v = 0.075$. Properties of soil nails are nail diameter 25 mm, spacing 1.2 m, yield strength 0.28 MPa, ultimate bond 120 kPa, length 1.5 m. The variation of results for different LE methods for soil samples 1–5 is shown in Fig. 10.

The resultant FOS obtained from slope stability analysis using Geo-5 (Nailing) tool with Bishop Method is around 2–3 after providing soil nailing. The provision of soil nailing is shown in Fig. 11.

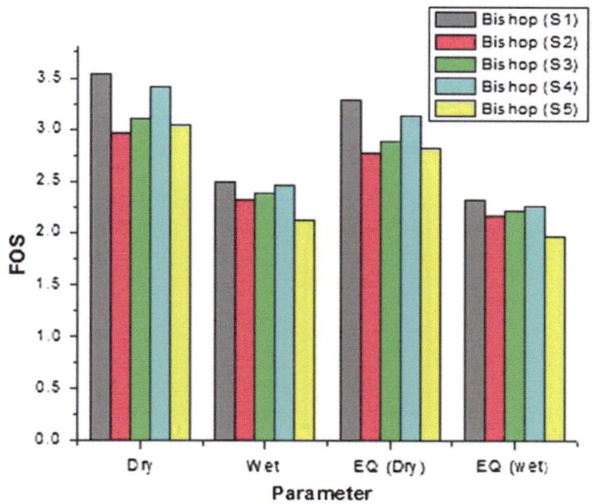

Fig. 10 FOS values after remediation

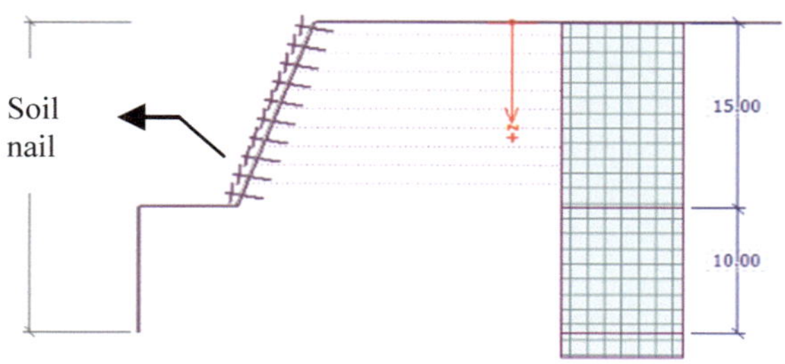

Fig. 11 Provision of soil nailing

4 Conclusions

- The parametric study shows that as the cohesion and internal friction increase, FOS increases, but as the water table, slope height and slope angle increase the FOS decreases. The FOS varies linearly by increasing the value of cohesion, whereas for the other parameters, the FOS does not vary linearly.
- When we compare FOS values using Limit Equilibrium Method and Finite Element Method, later method gives higher FOS values.
- The FOS obtained from Bishop Method and Ordinary Fellinius Method is almost similar, whereas GLE, Janbu's and M-P methods overestimate the FOS values.
- From the analysis of case study, it was found that the slope was unstable or on the verge of failure. Hence to increase the stability of the slope, suitable remedial measure was demonstrated.
- The FOS values obtained after providing remediation were around 2–3 and slopes were found stable.

References

Ahmad, M., Ansari, M. K., & Singh, T. N. (2015). Instability investigations of basaltic soil slopes along SH-72, Maharashtra, India. *Geomatics, Natural Hazards and Risk, 6*(2), 115–130.

Chakraborty, D., & Choudhury, D. (2009). Investigation of the behavior of tailings earthen dam under seismic conditions. *American Journal of Engineering and Applied Sciences*, pp. 559–564.

Fawaz, A., Farah, E., & Hagechehade, F. (2014). Slope stability analysis using numerical modeling. *American Journal of Civil Engineering, 2,* 60–67.

Hamdhan, N., & Schweiger, H. F. (2013). Finite element method-based analysis of an unsaturated soil slope subjected to rainfall infiltration. *International Journal of Geomechanics, ASCE, 13* (5), 653–658.

Kainthola, A., Singh, P. K.,Wasnik, A. B., & Singh, T. N. (2012). Distinct element modelling of Mahabaleshwar road cut hill slope. *Geomaterials, 2,* 105–113.

Rabie, M. (2014). Comparison study between traditional and finite element methods for slopes under heavy rainfall. *Housing and Building National Research Center Journal*, pp. 160–168.

Xue, D., Li, T., & Wei, Y. (2013). Numerical modeling of Giant Badu landslide reactivated by excavation in China. *Journal of Earth and Science Engineering, 2,* 109–115.

A Catastrophe Theory for Planar Sliding Slope

S. Bala Padmaja, G. V. Narasimha Reddy and E. Saibaba Reddy

Abstract For safe design of man-made or natural slopes, stability analysis of slopes is carried out using various conventional methods. In these methods, factor of safety is considered to be greater than or equal to one for stable slopes. Natural disasters like landslides are caused due to sudden and discontinuous phenomenon of slopes failures. But in practice, the failures may occur even when the factor of safety is greater than 1, as in the case of landslides. To overcome these situations, catastrophe theory can be used where conventional stability analysis methods have certain limitations to analyze these types of slope failures. In the present paper, the stability analysis of landslides using catastrophe theory for a mathematical model with planar sliding slope is carried out. Instability mechanism of slope is based on the material stiffness along the sliding slope. Critical factor of safety of the slope depends on the stiffness of the soil medium.

Keywords Catastrophe theory · Stiffness · Landslides

1 Introduction

Landslide is a natural disaster that threatens the human life, property, and restricts socioeconomic development of a local region. Landslide evolution mechanism has been studied to satisfy the increasing demand for disaster mitigation and control. In addition to geologic and geomorphologic conditions, it is the rainfall that was the meteorological cause for landslides or debris flow to occur in many cases.

S. Bala Padmaja (✉) · G. V. Narasimha Reddy
Department of Civil Engineering, JNTUH College of Engineering, Hyderabad 500085, India
e-mail: s.balapadmaja@gmail.com

G. V. Narasimha Reddy
e-mail: gvnreddy@jntuh.ac.in

E. Saibaba Reddy
VSS University of Technology, Burla 768018, Odisha, India
e-mail: esreddy1101@gmail.com

B. Adimoolam and S. Banerjee (eds.), *Soil Dynamics and Earthquake Geotechnical Engineering*, Lecture Notes in Civil Engineering 15,
https://doi.org/10.1007/978-981-13-0562-7_28

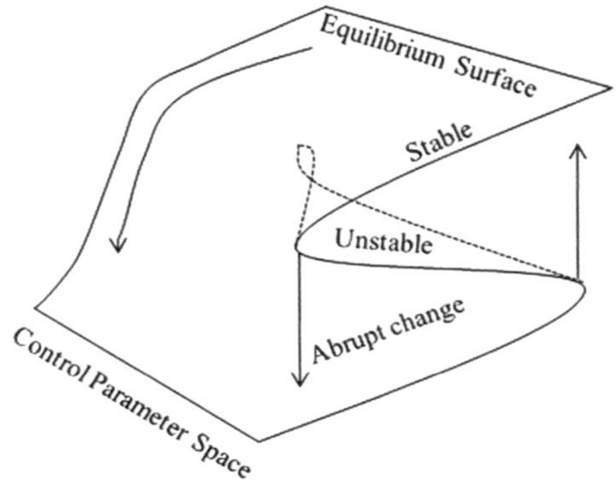

Fig. 1 Cusp catastrophe model

Many methods have been developed for slope stability analysis, but are unable to predict mechanism of catastrophic failures like landslides. Two strange cases of landslides are reported by Qin et al. (2001) where landslide occurs when the factor of safety is greater than one and landslide does not occur when factor of safety is less than one. Catastrophe theory (Rene Thom, 1960) is one of its kinds which describe the sudden and dramatic failure of slopes as in the case of landslides. This mathematical theory studies and classifies the small sudden changes which occur at unpredictable timing and also the magnitude of a landslide. Catastrophe theory can reflect the real situation (Yang et al. 2008) when compared to the limit equilibrium methods. In rock mass stability analysis, the application of catastrophe theory is vastly carried out (Li and Ping 2005; Zhang et al. 2014). The effect of local precipitation and human activity on intensity of landslides is studied using cusp catastrophe theory (Tao et al. 2013). The cusp catastrophe model which is one fundamental type of catastrophe theory was characterized by sudden and smooth changes, divergent and bimodal behaviors, hysteresis, and stability of structure. Among various applications, the cusp catastrophe theory plays a vital role in landslide analysis. In Fig. 1, the cusp model is presented where gradual potential decrease takes place along the smooth equilibrium surface and the control space consists of two control parameters. It is understood from the model that the failure does not occur instantly as soon as the displacement occurs along the failure slope. But, it fails at certain displacement which is defined in the potential function equation of the mechanical model.

2 Landslide Model

A mechanical model of planar sliding slope is considered, which is an inhomogeneous intercalation composed of two media with different strengths, where one media is elastic medium and the other is elasto-plastic medium. Assuming that the

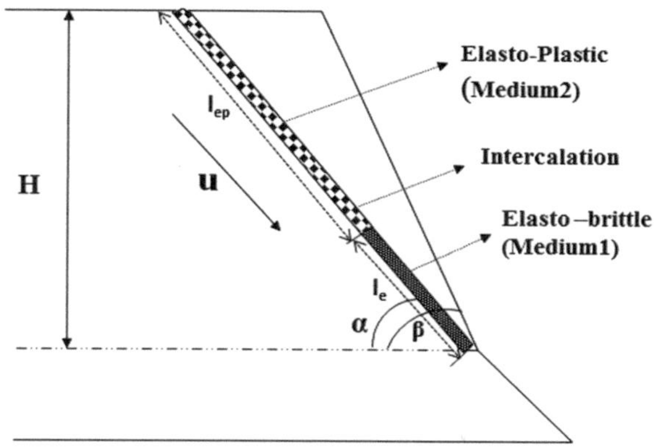

Fig. 2 Landslide mechanical model of planar sliding slope

sliding slope surface is inclined at β with horizontal and the soil mass above a sliding surface is a rigid mass (Fig. 2).

The potential function equation for the mechanical model is developed based on governing equations given by Qin et al. (2001), in which the inhomogeneous intercalation is composed of strain softening medium. In the present paper, instead of strain softening it is considered as elasto-plastic to stimulate a different field condition. The overall potential energy (V) of the system is given by,

$$V = l_{ep} \int_{0}^{u} G_{ep} \frac{u}{h} e^{-\left(\alpha \frac{u}{u_o}\right)} du + \frac{1}{2} \frac{G_e l_e}{h} u^2 - mgu \sin \beta \quad (1)$$

where G_{ep} is the initial shear modulus; G_e is the shear modulus; u is displacement; u_o is the displacement value at peak stress (Fig. 3); α is the displacement factor; l_{ep} and l_e are lengths of the sliding surface for the elasto-plastic medium and

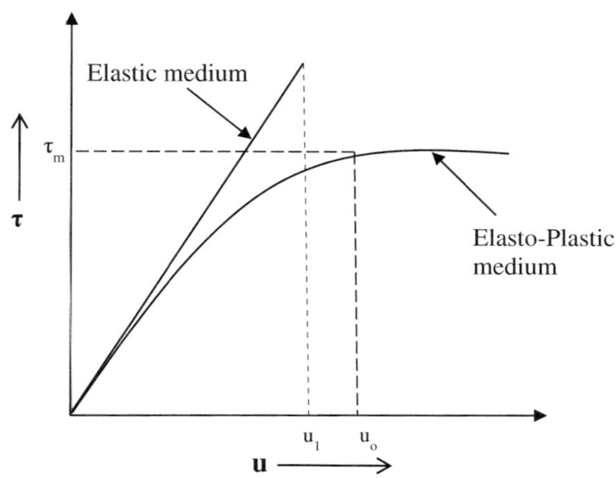

Fig. 3 Shear stress versus displacement curve for two media

elasto-brittle medium, respectively; H is the height of the rock mass, h is the thickness of the intercalation; β is the failure slope angle, and mg is the weight of the rock mass.

From Eq. (1), the necessary cusp condition is derived for displacement ratio (u/u_o) in terms of displacement factor (α) to define the stiffness ratio (k). The stiffness ratio is defined as the stiffness of elastic media to the stiffness of the elasto-plastic media. Stiffness of the soil medium is based on the shear modulus of soil and the lengths of the medium.

$$k = \frac{G_e l_e}{G_{ep} l_{ep} \mathrm{e}^{-\left(\alpha \frac{u}{u_o}\right)}} \tag{2}$$

Stiffness ratio ($k \leq 1$) is the required condition for catastrophic landslide occurrence according to cusp theory. At cusp, the displacement ratio (u/u_o) is assumed to be 2. The critical factor of safety (FS_c) defined by the ratio of the resisting force to the driving force at displacement u is:

$$FS_c = \frac{\left(1 - \sqrt{2}(1-k)^{1/2}/2\right)\left(\mathrm{e}^{\sqrt{2(1-k)}} + k\right)}{1 + k - \frac{\sqrt{2}}{3}(1-k)^{\frac{3}{2}}} \tag{3}$$

The factor of safety is based on the stiffness property of soil at failure. The failure mechanism of landslides with critical factor of safety greater than one is discussed in the proceeding section.

3 Results and Discussions

The critical factor of safety (FS_c) depends on the stiffness ratio (k) and the displacement ratio (u/u_o). The effect of various parameters such as lengths' ratio (l_e/l_{ep}) and displacement factor (α) on k is analyzed. Hence, parametric study has been carried out to know the effect of various parameters on critical factor of safety (FS_c).

The displacement factor (α) is used in evaluation of the shear stress of medium 2, to predict soil behavior under various loading conditions, and describes the behavior of the curve after peak stresses. This stress distribution will be more relevant in landslide analysis for field stimulation. In Fig. 4, the variation of shear stresses (t) versus displacement ratio (u/u_o) for different displacement factors (α) is plotted. From Fig. 4, it is observed that as α is increasing the shear stress is decreasing gradually and also decreases with increase in u/u_o. As u/u_o is increasing, it implies that the strain softening of the soil increases, thus increasing the percolation of rainwater affecting the shear stresses.

Figure 5 shows the variation of stiffness ratio with respect to length ratio for various displacement factors (α). The stiffness ratio increases linearly with length ratio and displacement factor. The lengths' ratio (l_e/l_{ep}) depends on the planar

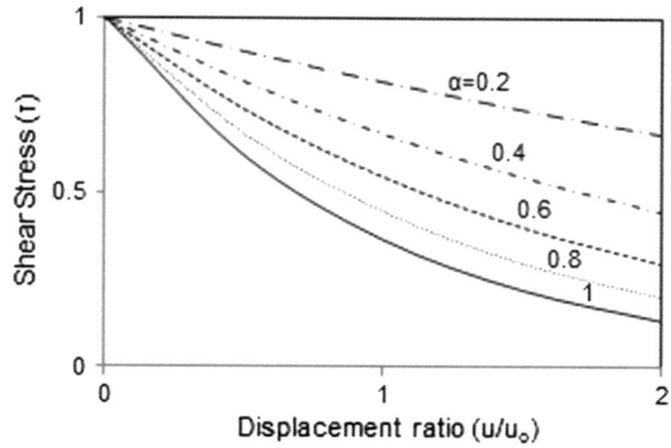

Fig. 4 Shear Stress variation with respect to displacement ratio (u/u_o) with different displacement factors

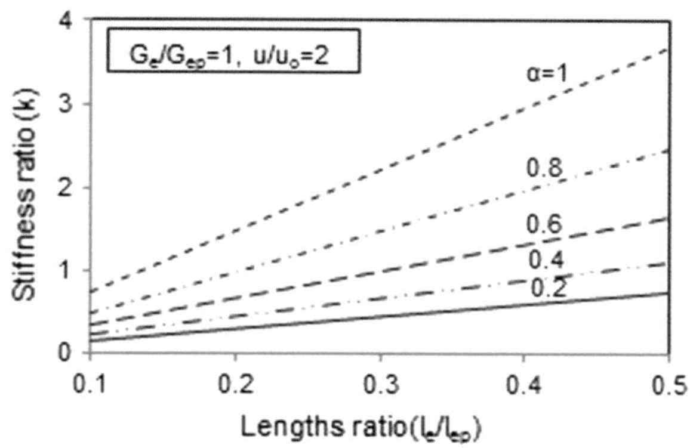

Fig. 5 Variation of stiffness ratio with lengths' ratio for various displacement factors

sliding slope of inclination (b). The increase in l_e/l_{ep} ratio results in increase of stiffness ratio along the sliding plane. According to cusp theory, if $k \geq 1$, the slope is stable. In this graph, $k \geq 1$ at certain l_e/l_{ep} ratio, still slope fails due to seepage of rainwater. In such rainfall-induced landslides, the shear modulus of soil media varies based on the pore water pressure resulting decrease in $k \leq 1$.

Figure 6 shows the variation of stiffness ratio (k) with displacement factor for various length ratios (l_e/l_{ep}) at shear modulus ratio ($G_e/G_{ep} = 1$) and displacement ratio ($u/u_o = 2$). It is understood that as α increases the stiffness ratio decreases nonlinearly and also with decrease in length ratio. The l_e/l_{ep} ratio decreases with increase in length of medium 2 along the intercalation plane making the slope prone to a landslide. The stiffness ratio is directly dependent on l_e/l_{ep} ratio, that is why landslide occurs only when $l_e < l_{ep}$.

The critical factor of safety (FS_c) variation with displacement ratio (u/u_o) for varying α value is shown in Fig. 7. The critical factor of safety decreases with increase in displacement ratio (u/u_o) and α. The increase in displacement ratio is possible when the shear stress along the sliding plane decreases due to various loading conditions. The displacement factor (α) is high for dense soil which

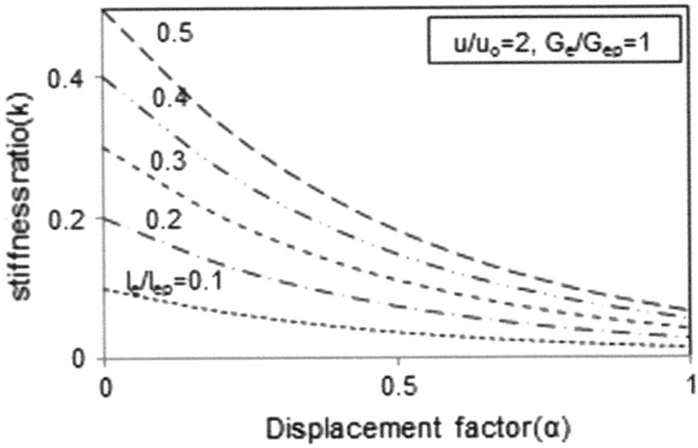

Fig. 6 Variation of stiffness ratio with displacement factor for varying lengths' ratio values

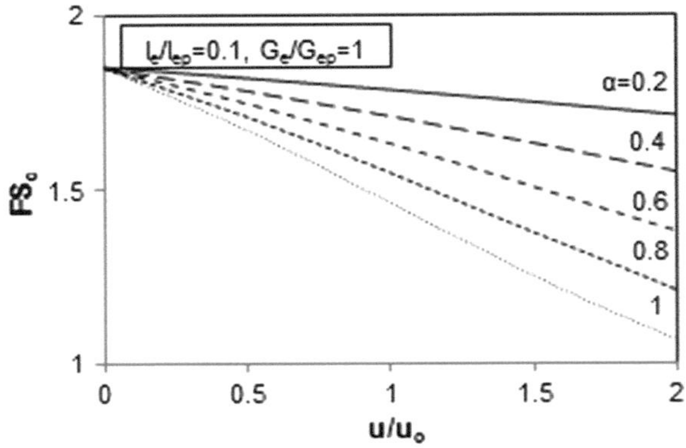

Fig. 7 Critical factor of safety versus displacement ratio for different displacement factor (α) values

indicates that the displacement increases along the failure slope and decreasing the critical factor of safety. Even though the $FS_c = 1.37$, for $\alpha = 0.6$ and $u/u_o = 2$, there is a probability of landslide occurrence according to cusp theory.

In Fig. 8, the effect of length ratio on critical factor of safety for varying displacement factor is studied. It is analyzed that the critical factor of safety (FS_c) is decreasing with increase in length ratio (l_e/l_{ep}) and displacement factor (α). The slope is assumed to be unstable up to an $FS_c = 1.71$ at $l_e/l_{ep} = 0.1$ and $\alpha = 0.2$. But when $l_e/l_{ep} = 0.5$ and $\alpha = 0.2$, the $FS_c = 1.06$, which says that the slope is stable, if $FS_c \geq 1.06$. As the l_e length increases, the maximum shear stresses also increase resulting decrease in the critical factor of safety. The length of the elastic medium (l_e) on the failure plane will increase when the percolation of rainwater is less due to less porosity.

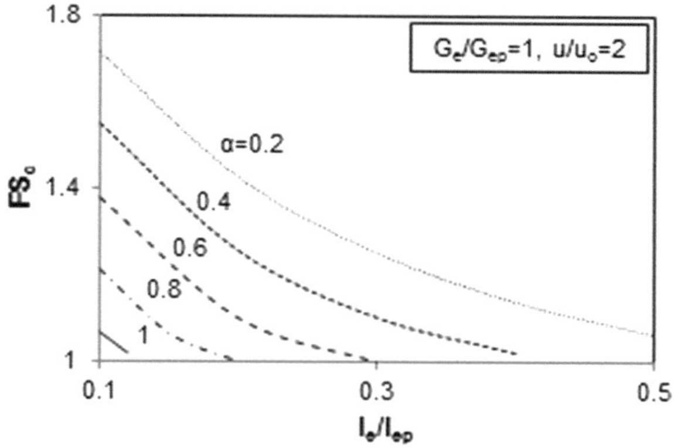

Fig. 8 Critical factor of safety versus length ratio for different displacement factors (α)

4 Conclusions

- The instability mechanism of a landslide using cusp theory gives insight of critical factor of safety, which is based on stiffness ratio and displacement of the sliding plane.
- Using the cusp mathematical model, the displacement criterion is more relevant to field conditions to determine the instability mechanism of the planar sliding slope.
- This method is used to predict the instability of landslides with critical factor of safety greater than one.
- The stiffness ratio increases with increase in l_e/l_{ep} ratio and displacement factor (α). As stiffness ratio is increasing, the critical factor of safety decreases gradually with increasing the stability of the slope structure.

Acknowledgements We are thankful to Professor M. R. Madhav who moderated this technique and guided in carrying out research work.

References

Gilmore, R. (1993). *Catastrophe theory for scientists and engineers*. New York: Dover Publications Inc.

Li, J-T., & Ping, C. A. O. (2005, August). Cusp catastrophe model of instability of pillar in asymmetric mining. *Applied Mathematics and Mechanics, 26*(28). ISSN 0253-4827.

Qin, S. Q., Jiao, J. J., Wang, S. J., & Long, H. (2001). A nonlinear catastrophe model of instability of planar-slip slope and chaotic dynamical mechanisms of its evolutionary process. *International Journal of Solids and Structures, 38*, 8093–8109.

Tao, Y., Cao, J., Hu, J., & Dai, Z. (2013). A cusp catastrophe model of mid-long-term landslide evolution over low latitude highlands of China. *Geomorphology, 187*, 80-85.

Yang, K., Shi, C., & Wang, J. F. (2008). Applying catastrophe theory to slope reliability analysis. In *Boundaries of rock mechanics*. London: Taylor & Francis. ISBN 978-0-415-46934-0.

Zhang, J., Shu, J., Zhang, H., Ren, X., & Qi, J. (2014). Study on rock mass stability criterion based on catastrophe theory (Vol. 2015, Article ID 850604, 7 pages). Hindawi. http://dx.doi.org/10.1155/2015/850604.

Properties of Sand–Rubber Tyre Shreds Mixtures for Seismic Isolation Applications

B. R. Madhusudhan, A. Boominathan and Subhadeep Banerjee

Abstract Disposal of scrap rubber tyres is a big challenge and a serious environmental problem. In India, it is estimated that one waste tyre per person is produced annually. The percentage of recycling is very less when compared to the stockpiling every year. It is widely known that the rubber is a good damping material. Consequently, there is a big scope for the usage of this scrap rubber for earthquake engineering purposes. The present study involves the characterization of sand–rubber tyre shreds mixtures for the application in seismic base isolation of low-rise buildings. The sand–rubber tyre shreds mix is characterized with respect to dynamic properties for various percentages of rubber tyre shreds such as 0, 10, 30, 50 and 100% by weight. In this regard, a series of cyclic triaxial tests were carried out to evaluate strain-dependent moduli and damping of the mixtures. It was found that 10% sand–rubber tyre shreds mixture yielded satisfactory dynamic properties required for seismic isolation of low-rise buildings.

Keywords Sand–rubber tyre mix · Seismic isolation · Shear modulus
Damping ratio

1 Introduction

Since the existing methods of seismic isolation of buildings are very expensive for their use in low-rise buildings, it becomes very important to find alternative ways. Study on sand–rubber mixtures was carried out by various researchers for various

B. R. Madhusudhan (✉) · A. Boominathan · S. Banerjee
Department of Civil Engineering, Indian Institute
of Technology Madras, Chennai 600036, India
e-mail: mssimha1990@gmail.com

A. Boominathan
e-mail: boomi@iitm.ac.in

S. Banerjee
e-mail: subhadeep@iitm.ac.in

B. Adimoolam and S. Banerjee (eds.), *Soil Dynamics and Earthquake Geotechnical Engineering*, Lecture Notes in Civil Engineering 15,
https://doi.org/10.1007/978-981-13-0562-7_29

geotechnical applications like usage as backfill materials in retaining structures (Humphrey et al. 1997; Lee et al. 1999), as construction or fill materials (Edil and Bosscher 1994; Foose et al. 1996; Bosscher et al. 1997; Zornberg et al. 2004). Few studies on this material were carried out keeping in view the protection of buildings from earthquake-induced vibrations (Tsang 2007; Anastasiadis et al. 2011; Senetakis et al. 2011, 2012). As far as dynamic soil properties such as shear modulus and damping ratio are concerned, previous researches are not conclusive on optimum percentage of sand–rubber tyre shreds mixtures yielding satisfactory strain-dependent shear moduli and damping ratios. Hence, it is imperative to find the behaviour of these mixtures under dynamic loading. This paper presents the results of strain-controlled cyclic triaxial tests. The strain-dependent shear moduli and damping ratios for various percentages of rubber shreds mixed with sand are presented. The rubber content which yields better results at large strain levels is also presented.

2 Materials Tested and Testing Equipment Used

2.1 Materials

River sand which is widely used for construction purposes was utilized for experimental investigation. The rubber shreds were derived from scrap tyres of vehicles. The size of shreds adopted was 2 mm and down. The percentages of rubber content adopted were 0, 10, 30, 50 and 100%. These percentages are gravimetric. Particle size analyses were carried in accordance with IS 2720 (Part 4)—1985. The materials are classified as poorly graded in accordance with Unified Soil Classification System and can be denoted by SP. Figure 1 shows the particle size distribution of the materials. Figure 2 shows the rubber shreds.

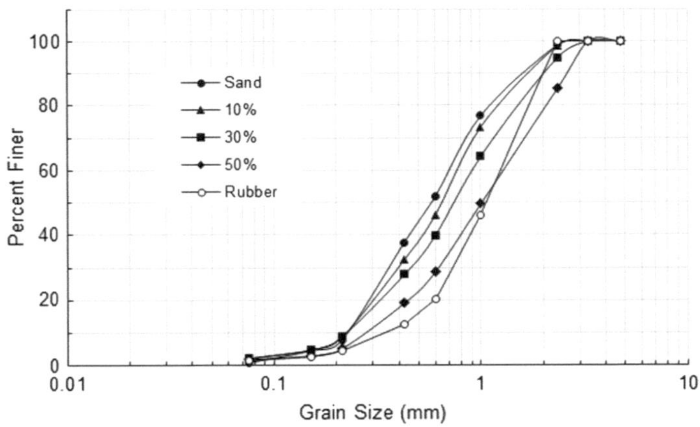

Fig. 1 Particle size distribution curves for sand–rubber tyre shreds mixtures

Fig. 2 Rubber tyre shreds

Table 1 Properties of sand–rubber tyre shreds mixtures

Sand–rubber tyre shreds mixture percentage (by weight)	Property		
	Specific gravity	γ_{dmax} (kN/m^3)	γ_{dmin} (kN/m^3)
0	2.68	17	15
10	2.30	14.1	12
30	1.90	9	7.2
50	1.56	7.5	5.1
100	1.14	5.1	2.7

The specific gravity, the maximum and minimum unit weights of the materials were found in accordance with IS 2720 (Part 3)—1980 and IS 2720 (Part 14)—1983, respectively, and the results are summarized in Table 1.

2.2 *Testing Equipment*

The cyclic triaxial tests were carried out on sand–rubber tyre shreds mixture specimens using Wykeham Farrance, UK, make servo-controlled advanced cyclic triaxial apparatus available at the Soil Dynamics and Earthquake Engineering Laboratory, IIT Madras (Fig. 3).

Fig. 3 Cyclic triaxial apparatus

3 Sample Preparation and Testing Methodology

3.1 Preparation of Sample

The samples were prepared by dry deposition and tamping method. The relative density adopted was in the range of 75–85%. The samples were prepared in five layers with each layer given equal number of blows. Since the materials are cohesionless, a vacuum pressure of up to 10 kPa was used as aid for preparing samples. The diameter and height adopted were 50 and 100 mm, respectively, by using vacuum split mould. Figure 4 shows the specimen prepared using the above-mentioned procedure.

3.2 Testing Procedure

The samples prepared were saturated till Skempton's pore pressure coefficient (B-Value) greater than 0.95 was achieved. Then, consolidation was done under an effective confining pressure of 100 kPa. A series of strain-controlled cyclic triaxial tests were carried out under undrained conditions in accordance with ASTM D3999/D3999M-11 (2011). All the tests were carried out at a frequency of 1 Hz with 25 number of loading cycles.

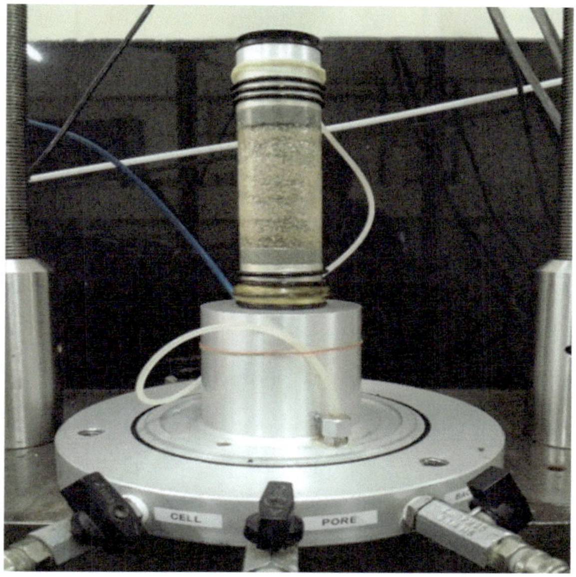

Fig. 4 Specimen prepared from sand–rubber tyre shreds mixture

4 Results and Discussion

Typical hysteresis loops obtained from the cyclic triaxial tests carried out on 50% sand–rubber tyre shreds mixture are shown in Fig. 5. The shear strain (γ) is calculated from the axial strain (ε) using the formula:

$$\gamma = \varepsilon(1 + \mu) \tag{1}$$

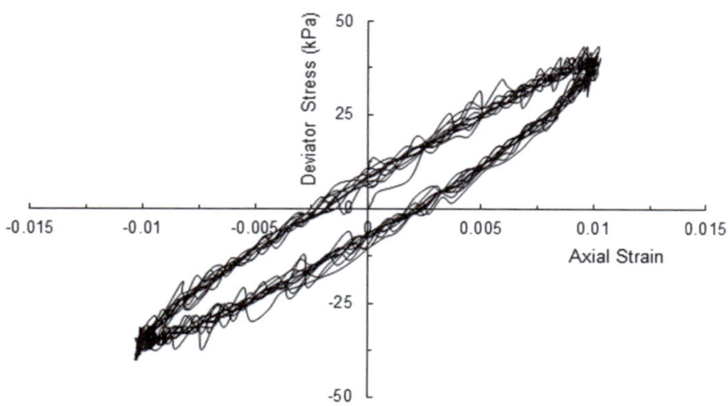

Fig. 5 Typical hysteresis loops for 50% rubber content showing 25 number of cycles at 1% strain

where μ is Poisson's ratio and its value is 0.5 for undrained conditions. The secant modulus (E_{sec}) is calculated by taking the slope of the line joining the ends of the loops. The secant shear modulus (G_{sec}) is calculated by using the formula:

$$G_{sec} = \frac{E_{sec}}{2(1+\mu)} \tag{2}$$

The damping ratio is calculated by using the formula:

$$D = \frac{1}{4\pi}\frac{A_{loop}}{A_{triangle}} \tag{3}$$

where A_{loop} is the area of the hysteresis loop, and $A_{triangle}$ is the area of the triangle (area of the shaded portion in Fig. 6). The shear moduli and damping ratios for all the mixtures were calculated for the fifth cycle.

Figure 7 shows the plot of shear modulus versus shear strain for all the mixtures. As shear strain increases, the shear modulus decreases for all the rubber contents. The modulus reduction is highest for the pure sand and least for the pure rubber. The sand–rubber shreds mixture with 10% rubber content shows higher value of shear modulus in comparison with other mixtures but less than that of pure sand. Figure 8 plots the damping ratio versus shear strain for all rubber contents. It can be observed that the damping ratio increases with shear strain. At any shear strain level, with increase in the percentage of rubber, the damping ratio is decreasing. It is only for the 10% rubber content, the maximum damping ratio is observed at all strain levels. It is intriguing to note that the pure rubber is showing the least damping ratio at all the strain levels. Authors believe that at 10% rubber content, the interaction between the soil and rubber particles is highest and this may be due to the size of the sand and rubber particles adopted for experimental investigation which is very small.

Fig. 6 Schematic diagram of hysteresis loop

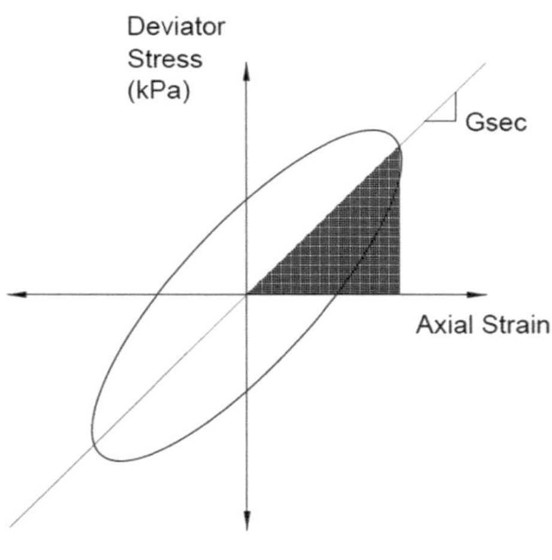

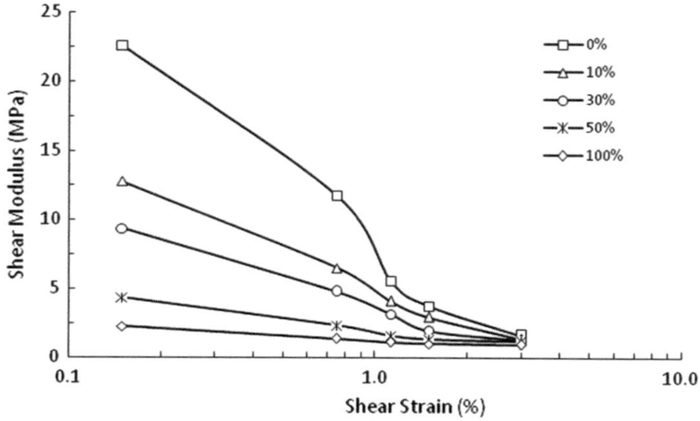

Fig. 7 Plot of shear modulus versus shear strain for sand–rubber tyre shreds mixtures

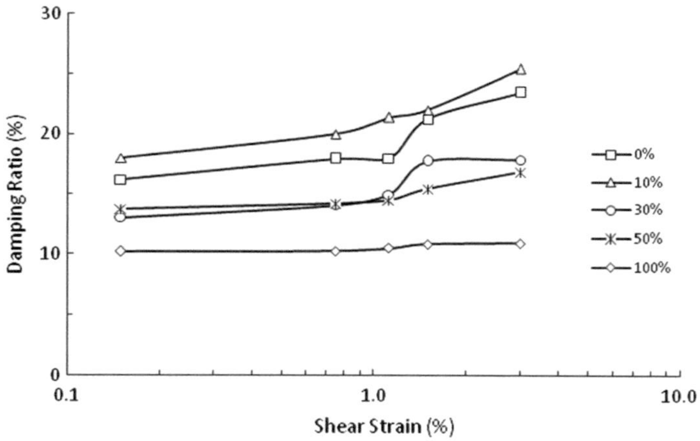

Fig. 8 Plot of damping ratio versus shear strain for sand–rubber tyre shreds mixtures

5 Conclusions

From the experimental study, the following conclusions can be made:

With increase in shear strain, the shear modulus reduces for all the percentages of rubber in sand–rubber tyre shreds mixtures.

The modulus reduction increases with a decrease in percentage of rubber. The modulus reduction is highest for the sand and lowest for the rubber.

The damping ratio increases with increase in strain levels. Lesser damping ratios are observed for higher percentages of rubber. The pure rubber shows least damping ratio. This behaviour is mainly due to the size of the sand and rubber particles adopted for experimental investigation.

The 10% rubber content by weight shows higher shear modulus and damping ratio values compared to all other percentages. Hence, 10% rubber content possesses satisfactory shear moduli and damping ratios at large strain levels and may be effectively used as a material for seismic base isolation applications.

References

Anastasiadis, A., Senetakis, K., Pitilakis, K., Gargala, C., & Karakasi, I. (2011). Dynamic behavior of sand/rubber mixtures. Part I: Effect of rubber content and duration of confinement on small-strain shear modulus and damping ratio. *Journal of ASTM International, 9*(2).

ASTM D3999/D3999M-11. (2011). Standard test methods for the determination of the modulus and damping properties of soils using the cyclic triaxial apparatus. In *Annual Book of ASTM Standards*. West Conshohocken, PA: ASTM International.

Bosscher, P. J., Edil, T. B., & Kuraoka, S. (1997). Design of highway embankments using tire chips. *Journal of Geotechnical Engineering, 123*(4), 295–304.

Edil, T. B., & Bosscher, P. J. (1994). Engineering properties of tire chips and soil mixtures. *Geotechnical Testing Journal, 17*(4), 453–464.

Foose, G. J., Benson, C. H., & Bosscher, P. J. (1996). Sand reinforced with shredded waste tires. *Journal of Geotechnical Engineering, 122*(9), 760–767.

Humphrey, D., Cosgrove, T., Whetten, N. L., & Herbert, R. (1997). Tire chips reduce lateral earth pressure against the walls of a rigid frame bridge. In *ASCE Technical Seminar on Renewal, Rehabilitation and Upgrades in Civil and Environmental Engineering*.

IS 2720 (Part 3). (1980). *Methods of test for soils: Determination of specific gravity*. New Delhi, India: Bureau of Indian Standards.

IS 2720 (Part 4). (1985). *Methods of test for soils: Grain size analysis*. New Delhi, India: Bureau of Indian Standards.

IS 2720 (Part 14). (1983). *Methods of test for soils: Determination of density index (relative density) of cohesionless soils*. New Delhi, India: Bureau of Indian Standards.

Lee, J. H., Salgado, R., Bernall, A., & Lovell, C. W. (1999). Shredded tires and rubber-sand as lightweight backfill. *Journal of Geotechnical and Geoenvironmental Engineering, 125*(2), 132–141.

Senetakis, K., Anastasiadis, A., Pitilakis, K., & Souli, A. (2011). Dynamic behavior of sand/rubber mixtures. Part II: Effect of rubber content on G/Go-γ-DT curves and volumetric threshold strain. *Journal of ASTM International, 9*(2).

Senetakis, K., Anastasiadis, A., & Pitilakis, K. (2012). Dynamic properties of dry sand/rubber (SRM) and gravel/rubber (GRM) mixtures in a wide range of shearing strain amplitudes. *Soil Dynamics and Earthquake Engineering, 33*, 38–53.

Tsang, Hing-Ho. (2007). Seismic isolation by rubber-soil mixtures for developing countries. *Earthquake Engineering and Structural Dynamics, 37*, 283–303.

Zornberg, J. G., Cabral, A. R., & Viratjandr, C. (2004). Behaviour of tire shred-sand mixtures. *Canadian Geotechnical Journal, 41*, 227–241.

MIX
Papier aus verantwortungsvollen Quellen
Paper from responsible sources
FSC® C105338

If you have any concerns about our products,
you can contact us on
ProductSafety@springernature.com

In case Publisher is established outside the EU,
the EU authorized representative is:
Springer Nature Customer Service Center GmbH
Europaplatz 3, 69115 Heidelberg, Germany

Printed by Libri Plureos GmbH
in Hamburg, Germany